D0897765

Bryophyte Ecology

Bryophyte Ecology

Edited by A.J.E. Smith

LONDON NEW YORK
CHAPMAN AND HALL

First published 1982 by Chapman and Hall Ltd
11 New Fetter Lane, London, EC4P 4EE

Published in the USA by Chapman and Hall
in association with Methuen, Inc.
733 Third Avenue, New York, NY 10017

Photoset by Enset Ltd
Midsomer Norton, Bath, Avon

Printed in Great Britain
at the University Press, Cambridge

ISBN 0 412 22340 6

British Library Cataloguing in Publication Data

Bryophyte ecology.
1. Bryophytes
I. Smith, A.J.E.
588 QK 533.7

ISBN 0–412–22340–6

Contents

Contributors

J.W. Bates, Department of Pure and Applied Biology, Imperial College Field Station, Silwood Park, Ascot, Berks., SL5 7PY, Great Britain.

H.J.B. Birks, Botany School, Downing Street, Cambridge, CB2 3EA, Great Britain.

D.H. Brown, Department of Botany, University of Bristol, Woodland Road, Bristol, BS8 1UG, Great Britain.

R.S. Clymo, Department of Botany and Biochemistry, Westfield College, Kidderpore Avenue, London, NW3 7ST, Great Britain.

Patricia Geissler, Conservatoire et Jardin Botaniques, Case Postale 60, CH–1292 Chambesy/Genéve, Switzerland.

Uri Gerson, Faculty of Agriculture, Hebrew University of Jerusalem, Rehovot, Israel.

P.M. Hayward, Department of Botany and Biochemistry, Westfield College, Kidderpore Avenue, London NW3 7ST, Great Britain.

R.E. Longton, Department of Botany, Plant Science Laboratories, The University of Reading, Whiteknights, Reading, RG6 2AS, Great Britain.

K. Mägdefrau, 8024 Deisenhoffen bei Munchen, Waldstrasse 11, Germany.

T. Pócs, Research Institute for Botany of the Hungarian Academy of Sciences, 2163 Vacratot, Hungary.

M.C.F. Proctor, Department of Biological Sciences, University of Exeter, Hatherly Laboratories, Prince of Wales Road, Exeter, EX4 4PS, Great Britain.

D.N. Rao, Department of Botany, Banaras Hindu University, Varanasi-221005, India.

G.A.M. Scott, Botany Department, Monash University, Clayton, Victoria, Australia.

A.J.E. Smith, School of Plant Biology, University College of North Wales, Memorial Buildings, Bangor, LL57 2UW, Great Britain.

Preface

There has been an increasing interest in bryophyte ecology over the past 100 or so years, initially of a phytosociological nature but, additionally, in recent years, of an experimental nature as well. Early studies of bryophyte communities have led to detailed investigations into the relationships between the plants and their environment. Ecological papers, the large number of which is evidenced by the length of the bibliographies in the subsequent chapters, have appeared in numerous journals. Yet, apart from review chapters, by H. Gams and P.W. Richards in *Manual of Bryology*, edited by H. Verdoorn in 1932 and chapters in E.V. Watson's *Structure and Life of Bryophytes*, Prem Puri's *Bryophytes – A Broad Perspective* and D.H.S. Richardson's *The Biology of Mosses*, published in 1972, 1973 and 1981 respectively, no general accounts of bryophyte ecology have been published.

Although the Bryophyta is a relatively small division of plants, with between 14000 and 21000 species the interest that they have aroused is out of all proportion to the size either of the plants or of the division. It is evident, however, that despite their relative insignificance they play an important ecological role, especially in extreme environments and, in the case of bryophytes in tropical cloud forests and of *Sphagnum*, may even be a dominant factor in the ecology of the area concerned. Analysis of bryophyte communities presents particular problems; aspects of their physiological ecology, especially drought resistance and nutrition, are topics of special interest, and the recent upsurge of interest in atmospheric pollution has revealed a new aspect of the relevance of the group.

In attempting to edit a book on bryophyte ecology one is limited by space and the availability of appropriate and willing authors. Hence it has not proved possible to cover all aspects of the subject but it is hoped that a sufficient diversity of topics has been dealt with to provide an adequate overview. Some of the chapters are general reviews, others are more detailed accounts of the authors' own research interests. In either case it is evident that there remains a great deal of research to be done on all aspects of bryophyte ecology and it is hoped that this volume, as well as providing an account of what is known, will stimulate further study.

Chapter 1

Quantitative Approaches in Bryophyte Ecology

J.W. BATES

1.1 INTRODUCTION

Studies of bryophyte vegetation are normally undertaken to answer one or more of the following questions:

(1) Which species occur and in what quantities and spatial arrangement?
(2) How does community structure vary with variation in the intensities of environmental factors?
(3) How do different species manage to co-exist in the community?
(4) How does the community maintain itself through time given the available resources?

These four topics, community structure, community variation, species strategies and community function, form a series which poses increasing technical difficulties to the would-be investigator. The difficulties are increased by certain peculiarities of bryophytes themselves; their small size, the often fragmentary nature of their communities, the tendency of many species to grow on highly irregular surfaces, their frequent presence as relatively low biomass components in communities dominated by other types of vegetation and their lack of economic importance leading to restricted financial support for research. It is, therefore, perhaps not surprising that, although simple descriptive accounts of particular bryophyte communities are reasonably common in the literature, relatively few quantitative studies of correlation between vegetation and environmental factors have been undertaken. More sophisticated studies of functional and strategic aspects of bryophyte communities are scarce.

The history of bryophyte ecology, in common with other branches of ecology, shows an increasing tendency towards the use of an objective, quantitative approach. This must be regarded as a healthy sign and results principally from a more general awareness of the power of statistical methods as aids in interpretation and decision-making, and the much wider availability of computational facilities than formerly (Jeffers, 1972a). In the present review reference will be made to areas of bryophyte ecology, where a quantitative approach has permitted, or might permit, a greater understanding of species distribution patterns and community organization. Special emphasis will be given to the various techniques of numerical analysis and some suggestions about their application to less conventional situations in bryophyte ecology. Discussion will be under the following headings; sampling methods, community recognition, species diversity, correlation with environment, and experimental applications of multivariate methods.

1.2 SAMPLING METHODS

The methods used to sample bryophyte vegetation will depend on the objectives of each study. In this section it is assumed that either presence–absence records (qualitative data) or a measure of abundance is required and perhaps also measurements of environmental factors. Bryophytes are susceptible to the same basic sampling approaches as higher plants and no quantitative exercise should be undertaken without full consideration of the meanings and limitations of the various abundance measures and methods of sample location (Greig-Smith, 1964; Chapters 1 and 2).

An important decision which the investigator must make at the onset is whether to include other types of plant in the study because bryophytes rarely form pure communities. This is particularly true in the case of saxicolous and epiphytic vegetation where lichens may be equally abundant or dominant, and may actively compete with bryophytes for resources. To attempt to delimit communities of just bryophytes, or lichens, in these situations would be extremely artificial.

The bryophyte ecologist often has to name diminutive or scrappy specimens, growing in suboptimal conditions, which the casual collector would normally pass by. Frequently this necessitates a system of collection of voucher specimens and the use of descriptive names until accurate determinations can be made later with a microscope. When quantitative data are being recorded there is no easy way of overcoming the difficulty produced when a voucher 'species' later proves to consist of more than one taxon (Scott, 1966, 1972). However, most of the special problems which arise in sampling are due more to features of the habitats of bryophytes rather than the plants themselves. Rock surfaces, tree boughs and aquatic habitats yield practical sampling problems not often encountered in higher plant ecology and a number of ingenious sampling techniques have been devised.

Most work has involved the use of some kind of 'quadrat' sample within which cover or frequency has been determined. However, it may be extremely difficult to position and record this on an irregular or vertical rock face, as in a limestone pavement grike (Yarranton and Beasleigh, 1968), or on the cylindrical surface of a tree trunk. The selection of an appropriate quadrat size may also present problems which are not solved by the minimal area type of approach (Greig-Smith, 1964). Bryophytes are well known to be sensitive to subtle differences in micro-environment, and in habitats such as the trunk of a tree, where steep environmental gradients exist in both the vertical and circumferential dimensions, the researcher may feel that his quadrat is rather an embarrassment. A quadrat of, say, 20×20 cm may be too small to average out random fluctuations in epiphyte abundance but too large to allow sufficiently fine resolution of the operative environmental factors. The alternative, in such cases, would appear to be the use of point sampling for both vegetation and environment (cf. Yarranton, 1967a, c and d).

An appraisal of sampling procedures applicable in particular habitat types follows. Measurement of environmental factors is outside the scope of this article and the reader is referred to other chapters for further details.

1.2.1 Ground vegetation

Bryophyte communities occurring on the soil surface, as in grasslands, the woodland ground-layer, heathlands and bogs, usually do not require special sampling methods. In grasslands, and possibly other habitats, allowance should be made for the possibility that occurrence and abundance of bryophytes may be related to the density of higher plant foliage (cf. Watson, 1960a; Morton, 1977). Where representative samples are required for making statistical comparisons some form of random sample location is essential but for mapping, and other studies where statistical inferences regarding the whole community will not be made, a form of systematic or regular sampling may suffice. Scott (1966) has described a systematic sampling procedure for ground bryophyte vegetation when insufficient time is available for true random sampling. Quadrats are staggered to left and right along a series of short transect lines through the middle of the stand. A similar regular approach was adopted by Watson (1960a) in his pioneer quantitative studies of chalk grassland bryophytes. The relative merits of systematic and random sample location are discussed fully by Greig-Smith (1964) and Lambert (1972).

1.2.2 Saxicolous vegetation

Rock surfaces often pose difficult sampling problems because of their topographic irregularity. Besides making quadrat positioning troublesome, topographic irregularities lead to a highly fragmented micro-environment in which different niches may closely neighbour one another. Another problem, particularly in xeric situations, is the sparsity of bryophyte vegetation requiring very large numbers of samples to obtain trustworthy abundance estimates.

Difficulties due to rough surfaces may be partly overcome by the use of a small quadrat. Foote (1966), for instance, employed 5×5 cm quadrats located at random within 1 m^2 stands to determine local frequency. Also, to enable accurate recording of the smaller and more critical species, the entire rock surface was removed from within each 25 cm^2 subunit and examined in the laboratory. Working in an extremely difficult habitat, the grikes in limestone pavement, Yarranton and Beasleigh (1968) used local frequency as an abundance measure. Presence of species was recorded in sets of nine 15 × 15 cm quadrats arranged horizontally in the grikes, and specimens in the narrower fissures were recovered by means of a metal scraper.

The plotless sampling technique devised by Yarranton (1966) for use in rocky habitats deserves special mention. This method was designed specifically to record joint occurrences of species. The sampling unit as originally described (Yarranton, 1966, 1967a, d), consisted of a one-inch (2.54 cm) herring net which was draped over the rock outcrop or boulder. Every fourth mesh intersection was regarded as a sample 'point' and the species touched by the point was recorded together with the species touching it nearest to the point. If no species was touching the contacted individual 'no contact' was recorded. The method has two advantages over other procedures; (1) the flexible sampling grid is easily positioned on irregular rock surfaces; (2) association of species is detected at the scale of the plants themselves and not at an arbitrary quadrat size. The data so obtained can be arranged into a contingency table and interspecific association indices calculated (Yarranton, 1966, 1967a, b, d) which, in turn, may be utilized for multivariate analyses of community variation. Further aspects of plotless sampling will be considered in later sections (pp. 12–13 and 29–31).

Cover data are generally preferable to frequency estimates because of their more certain biological meaning (Greig-Smith, 1964), but accurate determination of low cover values is difficult and time-consuming. In a study of the bryophyte vegetation of siliceous rocky shores, Bates (1975a) employed local frequency in preference to cover because of the low coverage of the vegetation. Local frequency, measured in grids of 100 4×4 cm squares, appeared to be reasonably well correlated with the density of individuals, at least in the case of cushion-forming species. However, there is a poor correlation between local frequency and density or cover in the case of straggly pleurocarpous mosses.

Fletcher (1973a, b) described a technique for determining cover of rocky seashore lichen communities which might also be applied to saxicolous bryophytes. The abundance of the rocky shore flora was established along transects employing a basic unit consisting of a square (50×50 cm) metal frame. The four corners and centre of each frame were sampled separately using a transparent Perspex quadrat (10×10 cm) ruled into 100 1×1 cm squares. Cover was obtained by counting the number of 1 cm squares which were more than half-covered by each species. Data from the five Perspex quadrats were combined to give an overall estimate of species cover at each position. In fact this method measures the frequency of individuals with over 50% cover in small squares. For most species this can be expected to approximate to percentage cover but for any unable to achieve a size greater than half the area of a small square, for instance some straggly, pleurocarpous mosses, this would not apply and cover would be underestimated.

Photographic and tracing methods have frequently been used in studies of growth in saxicolous lichens (references in Hooker and Brown, 1977) but only rarely for studies of rock-inhabiting bryophytes. Both Watson (1960b) and Stotler (1976) have used tracing methods in studies of saxicolous bryophyte communities. Photographic methods could prove valuable for demographic studies of saxicolous species.

1.2.3 Epiphytic vegetation

Epiphytic communities have received considerably more attention than ground or saxicolous bryophyte vegetation and a summary of the major European epiphyte communities, as defined by phytosociological principles, is provided by Barkman (1958). The latter work also summarizes much of the information on sampling bryophyte vegetation and environment which was then available. Part of the attraction of epiphyte vegetation lies in the solid, cylindrical nature of the tree trunk which provides a ready-made experiment for examining effects of aspect-correlated environmental factors.

Sampling methods have varied considerably. Where the aim has been to compare different aspects, the usual practice has been to sample different faces of the trunk separately (Hoffman, 1971; Bates and Brown, 1981). Other workers have removed the influence of aspect in their studies by the use of 'girdle' quadrats (Hale, 1955; Culberson, 1955), by restricting sampling to a single aspect (Slack, 1976), or by recombining data originally collected from the four cardinal points on each trunk (Rasmussen, 1975; Rasmussen and Hertig, 1977).

Many species show vertical zonation on the tree trunk, particularly in relation to potential evaporation gradients, and so height of samples needs to be carefully considered. Kershaw (1964) and Harris (1971) showed that zonation of epiphytic lichens on birch and oak trees, in North Wales and

Devon respectively, becomes compressed on shorter phorophytes. The same effect may apply in the case of bryophytes and so a sliding quadrat-positioning system should be contemplated if the intention is to compare particular communities at different sites where overall tree dimensions differ.

Frequency and subjective cover estimates have been applied widely in studies of epiphytic vegetation and Hoffman and Kazmierski (1969) and Pike *et al.* (1975) measured biomass of epiphytic bryophytes. Objective determination of cover of epiphytic bryophytes has been undertaken rather rarely. Two reasonably effective methods for estimating cover which have been used in a number of teaching exercises and research projects are described here. Both procedures assume that the community consists of a single layer of individuals.

The first method involves the use of a dressmaker's tapemeasure, marked in inches, which is placed around the trunk at a given height and held in position with a single drawing pin. The intersection of each inch mark with one edge of the tape is regarded as a point quadrat and the species immediately underneath is recorded. Percentage cover of each species is given by:

$$\text{total number of 'hits'/girth of tree (inches)} \times 100.$$

The second method, which is more appropriate if different aspects of the trunk are to be recorded separately, uses a quadrat (usually 15 × 15 cm) within which 100 regularly spaced points are recorded (Bates and Brown, 1981). The quadrat consists of a transparent cellulose acetate sheet marked with a grid of 100 equally spaced 3-mm diameter circles. The quadrat is attached to the bark with drawing pins and cover is estimated by recording the frequency with which each species occupies the visually estimated centres of the circles. An arbitrary score can be given to species present within the quadrat but which fail to achieve 1% cover.

Twigs and branches pose rather different sampling problems for which satisfactory answers have yet to be obtained. In studies of epiphyte colonization and succession on twigs, the annual increments, defined by bud scale scars provide a sampling unit (Degelius, 1964).

1.2.4 Aquatic vegetation

Bryophytes are often conspicuous elements of the macrophyte vegetation of flowing water courses and lakes but relatively few quantitative studies have been undertaken. Researchers have usually restricted themselves to the compilation of species lists due to the inaccessibility and irregularity of the habitat. An example of this approach is provided by Malme (1978) who obtained species lists for 23 Norwegian lakes. Malme used a combination of techniques to obtain his samples. For deep-water sites he obtained samples for biomass determination by means of an Eckman bottom grab operated along transects positioned at right angles to the shore. Empain and Lambinon (1974) obtained frequency scores for aquatic and semi-aquatic bryophytes at 11 stations on the River Sambre (Belgium) by combining information from marginally located transects with random grab samples from deep water. Effective methods for quantitative sampling of bryophytes in fast-flowing, rocky streams still remain to be devised.

1.3 COMMUNITY RECOGNITION

The bulk of ecological work to date has concerned the recognition and attempted classification of groups of species in relation to major habitat differences. Various approaches have been used including the methods of the continental schools of phytosociology, statistical studies of interspecific association, numerical classification and other mathematical approaches. Ordination methods have been used rather rarely as tools for community delimitation alone (e.g. Del Fabbro *et al.*, 1975) and, since these methods are most frequently used in studies of community variation in relation to environment, their application to bryophyte ecology will be discussed in later sections (pp. 22–35 and 37–38).

1.3.1 Phytosociological approaches

Much of the earlier, and a proportion of more recent, quantitative work on bryophyte communities has employed continental phytosociological methods. Reviews of some of these techniques have been provided by Poore (1955a–c, 1956), Becking (1957), Moore (1962) and Whittaker (1962); and Barkman (1958) gives a detailed account of these methods applied to bryophyte and lichen communities in his monograph on European epiphytic vegetation. Some examples of variations of the phytosociological approach applied to bryophyte communities include the studies of Richards (1938) on oceanic bryophyte communities in Ireland; a study of corticolous bryophytes in Michigan (Phillips, 1951); Yarranton's (1962) analysis of saxicolous species on Breidden Hill; an account of Norwegian oceanic bryophytes (Lye, 1966); Proctor's (1962) description of Dartmoor oakwood epiphytes; a study of Japanese saxicolous bryophytes (Nagano, 1969); work on Antarctic moss communities (Gimingham and Smith, 1970); studies of urban bryophyte communities in northern England (Gilbert, 1971); descriptions of a range of community types on Skye (Birks, 1973); and a study of the epiphytic cryptogams of *Populus deltoides* in North America (Hoffman and Boe, 1977). There is also a vast body of phytosociological literature dealing with higher plant-dominated communities in which bryophytes feature to a lesser or greater extent.

A major drawback of the phytosociological approach is the large element of subjectivity featuring during several stages of sampling and data processing (Greig-Smith, 1964). Generally, the phytosociologist only samples areas of vegetation which, by visual inspection, are considered 'homogeneous', and intermediate or heterogeneous stands are excluded from consideration. Many workers, in common with James *et al.* (1977) who have used these methods in a survey of major lichen community types in the British Isles, follow Poore (1955b, c) in their justification of the phytosociological approach, that is the individual 'associations' are claimed to represent noda in a multidimensional continuum of plant communities. At best, phytosociological methods may yield a rapid and efficient summary of the major assemblages of species in an area or habitat but, at worst, a biased, personal impression of communities may result. Perhaps the biggest danger of the phytosociological approach in practise is that the need to classify may become an end in itself and direct attention away from more fundamental and important aspects of the ecology of the species under consideration (McIntosh, 1967).

In some phytosociological accounts, including studies of bryophyte communities, attempts have been made to summarize objectively the ecological affinities of the subjectively defined 'associations' (Barkman, 1958; Yarranton, 1962; Lye, 1966). Poore (1955c), for instance, described the use of the Sörenson coefficient (based on presence or absence of species within each community) to examine affinities of a number of Scottish montane communities but remarked that the method was statistically unsatisfactory because of the nature of the data. Barkman (1958) used more sophisticated indices taking into account the total cover value (T.C.V.) of species within each community and produced constellation diagrams showing the relative affinities of bryophyte associations on trees in the Netherlands. During a phytosociological study of bryophyte communities on the rock exposures of Breidden Hill. Yarranton (1962) employed one of Barkman's coefficients of community affinity (D_2) to

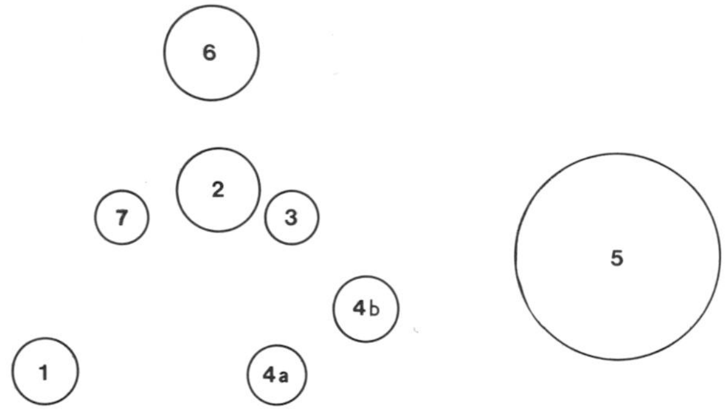

Fig. 1.1 Comparium showing inter-relationships of seven phytosociological groupings of bryophytes from communities on Breidden Hill. (Redrawn from Yarranton, 1962.)

construct a two-dimensional comparium (Fig. 1.1). Each circle represents a separate community, the distances between circles are given by log D_2 and the size of the circles approximately represents the degree of variability within each sociological grouping. Groups 2 and 3 occur close together within the comparium and share a number of species of high coverage values. Group 5, which has a low affinity to all other groups, was from an exposed situation and contained species (*Polytrichum piliferum, Racomitrium heterostichum, Grimmia trichophylla*) not found in abundance elsewhere. The greatest objection to this type of approach is that it is a hybrid between subjective and objective methods. In such cases one wonders whether objective methods of data collection and analysis should not have been applied from the onset.

1.3.2 Interspecific association

A more objective approach to the recognition of communities of bryophytes is through statistical studies of interspecific association. Species which are found co-occurring in quadrat samples more often than chance expectation are said to be positively associated, those co-occurring less often are negatively associated, and species occurring together as estimated by random expectation are termed unassociated (Greig-Smith, 1964). The different types of interspecific association pattern may reflect similarity or dissimilarity of ecological niche or competitive and allelopathic effects which are poorly understood at present. The statistical significance of an apparent association is normally tested by means of a χ^2 test of a 2×2 contingency table or the Fisher Exact test (see Greig-Smith, 1964 for details). Once the existence of statistically significant associations has been demonstrated, the strengths of association between different pairs of species may be compared by means of one of several coefficients of association (Southwood, 1966). This type of approach has been employed in several studies of bryophyte communities (Hale, 1955; Hopkins, 1957; Omura and Hosokawa, 1959; Redfearn, 1960; Yarranton, 1967a–d; Stringer and Stringer, 1974b; Bates, 1975a, b; Rasmussen, 1975; Siedel, 1976).

A simple illustration of the method was provided by Redfearn (1960), who investigated bryophyte communities on limestone rocks in a ravine at Pronto Springs, Florida. Frequency (presence–absence) was assessed in a sample of 78 20×50 cm quadrats and the Cole index of interspecific association (Cole, 1949) was used as an 'objective method of determining species with similar ecological amplitudes'. On the basis of significant positive associations two major species groupings were recognized; one occurring under conditions of infrequent inundation (*Fissidens taxifolius, Eurhynchium hians* and *Crossotolejeunea bermudiana*) and the other in semi-aquatic conditions (*Amblystegium varium, Dumortiera hirsuta* and

Fissidens minutulus). A further significant positive association between two wide-ranging species, *Barbula cruegeri* and *Dicranella varia* was believed to be due to the ability of both species to grow in small crevices which were not occupied by other bryophytes. A number of significant negative associations were discovered but little use was made of these in the analysis.

A more thorough analysis of associations was undertaken by Hale (1955) during a study of the lichen and bryophyte epiphytes of upland forests in southern Wisconsin. Association indices (Cole, 1949) were calculated between the 21 most important cryptogams, taken in all possible pairs, and entered into a half diagonal matrix. Both positive and negative values were included. The order of species in the rows and columns of the matrix was simultaneously rearranged to concentrate the highest positive values along the leading diagonal and to move negative indices furthest from it (Table 1.1). This has the effect of placing highly positively associated species close together and negatively associated ones far apart. Inspection of the reordered matrix shows that the species fall naturally into two groups, each with high positive interspecific associations and with many negative associations existing between them. Hale found that these groupings corresponded to different host tree species, Group I being species characteristic of *Quercus rubra* and *Q. velutina* and Group II were species with a general preference for *Q. alba* as host. Hale (1955) also used the Cole index to study the degree to which individual epiphyte species were associated with each type of host tree. The matrix reordering procedure described above has been used extensively by Yarranton (1966, 1967a, d) and Yarranton and Beasleigh (1968) in studies of saxicolous bryophyte vegetation.

An alternative method for summarizing the information contained in matrices of interspecific association coefficients is by means of plexus or constellation diagrams. A diagram or model is constructed in which species are represented as spheres and joined by lines whose lengths are inversely proportional to the intensity of association. The species constellation diagram gives an immediate visual impression of the major species groupings in a habitat and has been used in several studies of bryophyte communities (Hopkins, 1957; Omura and Hosokawa, 1959; Yarranton, 1966, 1967a, d; Bates, 1975a, b; Rasmussen, 1975; Siedel, 1976). However, the method has two important limitations which have restricted its wider application. Firstly, to give a completely accurate representation of associations, the plexus must theoretically be constructed in multi-dimensional space (in a maximum of $m-1$ dimensions, where m is the number of species). A plexus diagram drawn on paper in two dimensions will inevitably distort the representation of associations to some extent and the distortion will generally increase as the number of species is increased thus making interpretation more difficult. A second problem is the

Table 1.1 A re-ordered half-diagonal matrix of Cole's indices of association calculated between 21 upland epiphytic cryptogams from southern Wisconsin (modified from Hale, 1955).

		2	3	4	5	6	7	8	9	10	11	12	13	14	15	16	17	18	19	20	21
	1. *Lecanora subfuscata*	73	67	66	66	73	61	57	27	46	31	20	0	7	17	−52	2	−63	−65	−100	−58
	2. *Parmelia aurulenta*		60	57	52	61	61	39	35	30	22	46	7	39	29	−55	6	−7	−31	−18	−10
	3. *Arthonia caesia*			56	42	43	53	37	35	19	38	22	18	26	5	−72	−29	−40	−59	−67	−49
	4. *Platygyrium repens*				60	54	50	38	30	22	15	40	3	35	24	−43	0	−31	−32	−61	−25
	5. *Frullania eboracensis*					55	57	40	42	31	17	37	0	30	27	9	14	3	0	13	4
GROUP I	6. *Parmelia rudecta*						71	45	46	37	31	43	−12	29	15	−15	30	19	−1	9	11
	7. *Parmelia caperata*							67	56	53	40	14	−35	−11	2	−27	6	5	−48	−45	−28
	8. *Physcia millegrana*								68	36	49	9	−30	8	−1	−8	−26	−20	−40	−62	−33
	9. *Parmelia borreri*									55	54	−9	−44	−46	4	2	6	8	−21	−41	−4
	10. *Candelaria concolor*										35	17	−47	−10	29	16	34	30	11	14	19
	11. *Physcia alpolia*											−35	−40	−35	6	−2	−7	10	−42	−57	−21
	12. *Lepraria incana*												−16	24	45	13	26	15	17	40	22
	13. *Graphis scripta*													20	−11	−87	−64	−86	−91	−97	−81
	14. cf. *Lecanora symmicta*														54	−1	2	1	3	1	4
	15. *Phaeophyscia orbicularis*															67	46	49	49	40	54
	16. *Orthotrichum pumilum*																24	16	36	27	58
	17. *Physcia tribacoides*																	41	30	40	67
GROUP II	18. *Xanthoria fallax*																		46	26	68
	19. *Lindbergia austinii*																			59	56
	20. *Pylaisia selwynii*																				79
	21. *Physconia grisea*																				

difficulty of incorporating negative associations, which are theoretically as important as positive values, into the plexus in a satisfactory manner. A third problem which applies to all statistical studies of association, and which has often been overlooked, is that where large numbers of species are involved a certain number of significant associations will arise purely by chance (Greig-Smith, 1964). These drawbacks may often be overcome by using multivariate techniques for analysing community structure.

Before leaving the subject of interspecific association it is necessary to discuss the effect of quadrat size in detection of association. Fig. 1.2 shows a hypothetical relationship between index of association and quadrat size for two species which, for the sake of argument, may be regarded as positively associated in a stand of vegetation. At quadrat sizes close to, or smaller than, mean plant size the species in fact show negative association because, on average, only one individual may occupy the quadrat. A peak in the value of positive association occurs as quadrat size approaches the scale of the association but, at very large quadrat sizes, both species occupy a large proportion of the samples and so the index of association approaches zero. It follows, therefore, that the choice of quadrat size is of considerable importance in the detection of interspecific association. Moreover, the size of quadrat might have to be varied for different sizes of plant, even within one community. The majority of workers have ignored these problems and have chosen an arbitrary quadrat size for studying the whole community. However, cryptogams are known to occupy clearly defined niches within the mosaic of available micro-environments and such an approach may mask subtle vegetational differences and so a more objective method of quadrat size selection is desirable.

The most successful attempts at solving the quadrat-size problem involve the use of plotless sampling procedures and a range of methods is reviewed by Goodall (1965). The most relevant solution to date for studies

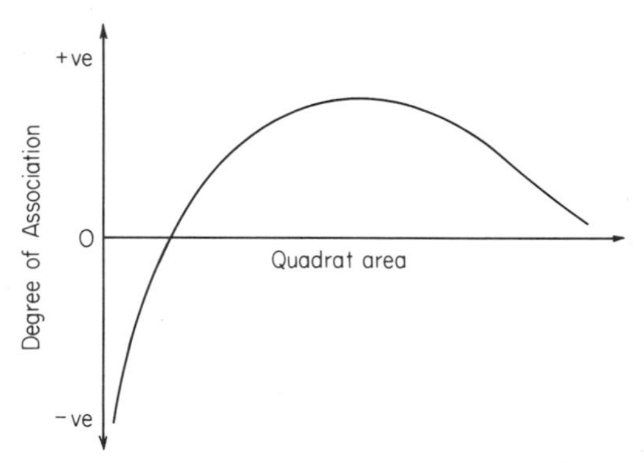

Fig. 1.2 Hypothetical relationship between index of association and quadrat area for two positively associated species.

of association is provided by the method of Yarranton (1966) which was outlined earlier. This technique evades the problem of quadrat size by detecting associations between species at the scale of the individuals themselves (p. 4). Plotless sampling has been employed in a number of studies of saxicolous bryophyte and macrolichen vegetation (Yarranton, 1966, 1967a–d; Bates, 1978) and also used to study the neighbour relationships of *Trifolium repens* in permanent pasture (Turkington and Harper, 1979). The latter authors have extended the original scope of the method to include intraspecific as well as interspecific associations. One possible criticism of the plotless sampling procedure is that species may be scored as 'in contact' when they, in fact, occupy distinct but neighbouring microhabitats (Yarranton, 1966; Fletcher, 1973a). However, this problem applies even more to situations where a bounded sample unit is used and, so far, has not been shown to be of great significance.

Association is defined on the basis of the presence and absence of species in samples but, in many instances, quantitative estimates of abundance may be available. In these cases, correlation and regression methods may be employed to test the existence of significant association between species (Greig-Smith, 1964).

1.3.3 Numerical classification

Techniques of numerical classification are an important aid to the taxonomist and, in recent years, have been widely used for vegetation classification (Greig-Smith, 1964; Pielou, 1969; Whittaker, 1973a; Orlóci, 1975). In common with ordination procedures (see pp. 22–35), many of these techniques represent developments of the simple matrix manipulation and plexus methods described in the preceding section. Numerical classification techniques frequently employ matrices of interspecific association or correlation coefficients, or various interstand similarity or distance measures (Williams, Lambert and Lance, 1966), as a starting point for the analysis. To the uninitiated, there appears to be a bewildering array of techniques and this arises partly from the various alternative theoretical standpoints which can be used to classify objects. In the following discussion, the objects to be classified are referred to as individuals and the criteria by which they are classified are called attributes. It is possible to classify the species on the basis of their scores in different quadrats, or to classify the quadrats on the species which they contain.

The simplest numerical classification procedures classify individuals on the basis of a single attribute at each step of the analysis (monothetic methods), whereas other techniques use more than one attribute simultaneously to classify individuals (polythetic methods). However, all methods are termed multivariate because they operate upon a large number of variables and attempt to produce a simple final structure – the classification. The classification may be made by lumping together individuals which are similar to form groups of differing composition (agglomerative classification) or, alternatively, by splitting the sample of individuals into smaller and smaller groups so as to maximize the variation between the groups and minimize variation within the groups (divisive classification). Generally, polythetic methods provide a more effective result than monothetic procedures because more of the available information is used in producing the classification. Divisive methods are also generally superior to agglomerative ones because the first and most important divisions of the classification are made on the basis of all available information and serious misclassifications of individuals are, therefore, less probable (Lambert *et al.*, 1973). The preferred methods are thus polythetic and divisive but these are also, unfortunately, the most demanding computationally (Lambert *et al.*, 1973) and so agglomerative polythetic and monothetic divisive methods are widely used. Recently, some computationally efficient polythetic divisive procedures have appeared (Noy-Meir, 1973; Lambert *et al.*, 1973; Hill, Bunce and Shaw, 1975), which make use of ordination to structure the data, and thus speed up the laborious business of finding the optimal positions for making divisions.

In bryophyte ecology, relatively little use has been made of numerical classification procedures to identify species groupings. In situations where these methods have been used, frequently this has been to give an additional perspective in multi-approach studies (e.g. Stringer and Stringer, 1974a; Bates, 1975b; Vitt and Slack, 1975; Pakarinen, 1976; Smith and Gimingham, 1976; Hoffman and Boe, 1977).

Smith and Gimingham (1976) used the monothetic, divisive technique, association analysis (Williams and Lambert, 1959, 1960, 1961), and an agglomerative, polythetic procedure, to assess the effectiveness of a classification of bryophyte- and lichen-dominated communities on Signy Island in the maritime Antarctic which had been produced by subjective phytosociological methods (Gimingham and Smith, 1970; Smith, 1972). The method employed involved intensive quadrat sampling of three sites (Thulla Point, Tern Cove and Mirounga Flats) within which the whole range of hypothetical communities occurred. The results of a normal association analysis of the combined quadrat data from all sites showed good general agreement with the subjective classification scheme. The first divisions of the analysis separated the quadrat data into four major groups representing the principal phytosociological 'sub-formations'. Further divisions yielded smaller groups corresponding reasonably well with the lower level 'sociations' of the subjective classification. As a result of these and other analyses, Smith and Gimingham (1976) concluded that the subjective classification scheme did not require amending, but they also noted that the use of numerical procedures had allowed a better understanding of inter-relations between the various community types than was afforded by a subjective approach.

Similarly, Hoffman and Boe (1977) employed two numerical classification procedures to test the validity of a simple phytosociological classification of epiphytic bryophyte and lichen communities on *Populus deltoides* in an area of North America. The communities recognized by subjective methods were *Phaeophyscia ciliata–P. stellaris* union on small trees and a *Phaeophyscia orbicularis–Candelaria concolor* union on large trees. The combined results from two different polythetic, agglomerative classifications broadly supported this arrangement but also highlighted a third category, a *Candelariella subdeflexa*-dominated community, not recognized by the phytosociological approach, and found to be characteristic of intermediate-sized trees.

Association analysis was one of several techniques employed by Bates (1975b) in studies of the ecology of maritime bryophytes on rocky shores. In a survey of maritime bryophyte communities on Cape Clear Island (Eire), quadrat samples were taken from along transects positioned around the coastline and from isolated rock outcrops in the central parts of the island. An association analysis of the combined quadrat data (107 samples) led to the recognition of eight major community types. While each community could be categorized on the basis of presence or absence of indicator species, the different types clearly formed a linear series in which each group merged into its two neighbours. The classification had in fact yielded a one-dimensional ordination of community types in relation to the over-riding environmental factor, salt spray deposition. This example serves to illustrate a problem encountered in the use of numerical classification procedures for community analysis. These methods may produce rigidly defined groupings in an objective fashion but do not always indicate whether classification is the correct course of action. In many cases, where continuous variation exists between communities a better view of inter-relationships between species, and with environment, may be obtained by applying ordination rather than classification techniques.

1.3.4 Other methods

Pattern analysis techniques (Greig-Smith, 1964; Hill, 1973a) allow the detection of the scale at which non-randomness occurs in species distributions. Non-randomness generally results from uneven propagule dissemination, vegetative spread from established plants, local variations in environmental conditions or interactions between species. Various modifications of the pattern analysis technique are available for detecting the scale at which positive and negative correlations between species may exist but these have not been applied to bryophyte vegetation (see p. 22).

An interesting method for studying species inter-relationships in a community has recently been described by Rasmussen and Hertig (1977). Following an association analysis classification of epiphytic bryophytes on *Fagus* and *Fraxinus* trees in Slotved Forest, Northern Jutland (Rasmussen, 1975), Rasmussen and Hertig undertook a re-examination of their data using a multiple regression approach. The regression analysis concerned only the five most abundant species, *Homalothecium sericeum, Hypnum cupressiforme, Isothecium myurum, Neckera complanata* and *Metzgeria furcata*. Little sociological information had been revealed about these species in the association analysis because of their high frequency in the quadrat sample. Cover-abundance of bryophytes was estimated at three heights (0.5, 1.5, 2.5 m) on both species of tree.

A series of multiple regression equations was calculated, for the different heights and phorophyte species, in which each species appeared separately as the dependent variable (y) and the other species were considered as independent variables (x_i). The analyses were performed in a stepwise manner. In the first step, the most abundant species, which was considered to have the largest competitive ability, was made the independent variable (x_1) in simple regression equations of the form,

$$y = a + b_1 x_1$$

where a is the intercept constant and b_1 the regression coefficient for x_1. In the second step, the next most abundant species was added as a second independent variable (x_2),

$$y = a + b_1 x_1 + b_2 x_2$$

Table 1.2 Regression coefficients from a stepwise regression analysis of interspecific relationships between epiphytic bryophytes on *Fagus* in northern Jutland (modified from Rasmussen and Hertig, 1977).

		Independent variable		
Height (m)	Dependent variable	*Hypnum* (step 1)	*Neckera* (step 2)	*Homalothecium* (step 3)
2.5	*Metzgeria*	−0.10	0.28	0.19
	Homalothecium	−0.48	0.20	—
	Neckera	−0.54	—	—
1.5	*Metzgeria*	−0.07	0.29	0.21
	Homalothecium	−0.45	0.41	—
	Neckera	−0.80	—	—
0.5	*Isothecium*	−0.17	−0.36	—
	Neckera	−0.63	—	—

and so on until further additions failed to account for significant additional variation in y. The partial regression coefficients (bs) for analyses of *Fagus* epiphytes are given in Table 1.2. Rasmussen and Hertig (1977) imply that strongly negative partial regression coefficients represent situations where there is intense interspecific competition. This would appear to be the case for the trio of species, *Hypnum, Homalothecium* and *Neckera* judging by the large negative partial regression coefficients obtained with *Hypnum* as the independent variable and the other two as dependent variables (Table 1.2). When *Metzgeria* was made the dependent variable the partial regression coefficient for *Hypnum,* on higher parts of the trunk of *Fagus,* was not significantly different from zero, indicating the absence of competition between these two epiphytes. Rasmussen and Hertig (1977) suggest that the special growth form of *Metzgeria,* its ability to grow strongly appressed to bark or to creep amongst other epiphytes, explains this behaviour. In a few cases, positive regression coefficients were obtained, for example

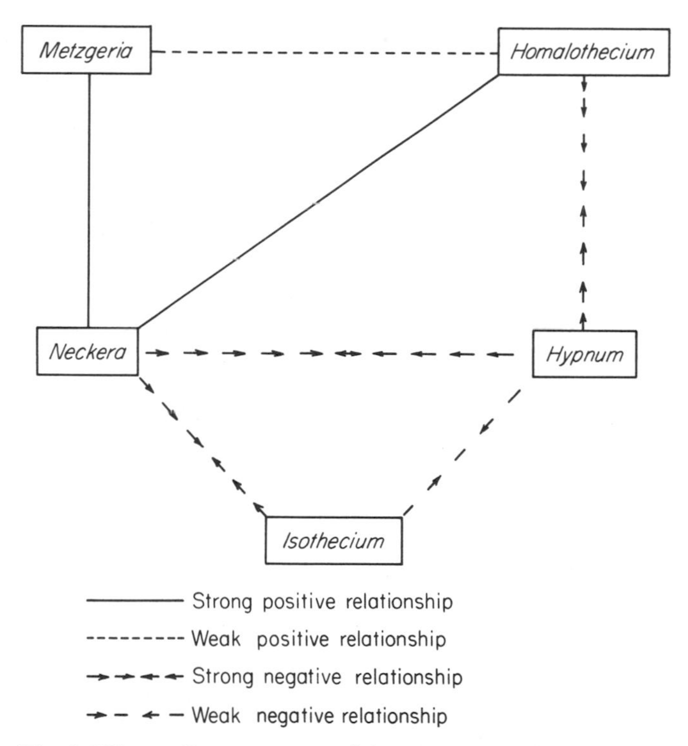

Fig. 1.3 Plexus diagram summarizing the results of a multiple regression study of interspecific relationships of some epiphytic bryophytes in north Jutland. (Redrawn from Rasmussen and Hertig, 1977.)

where *Homalothecium* was the dependent variable and *Neckera* the second independent variable. Here an increase in the abundance of one species was accompanied by an increase in the abundance of the other. Rasmussen and Hertig (1977) summarize these inter-relationships in the form of a species plexus diagram (Fig. 1.3). They consider that the analysis points to the existence of two communities, a tree base assemblage of *Hypnum, Isothecium* and *Neckera* and an upper trunk community consisting of *Homalothecium, Hypnum, Neckera* and *Metzgeria.* On the tree base, relationships are generally strongly negative indicating strong competitive effects. At higher levels, negative coefficients are only associated with *Hypnum* indicating that if *Hypnum,* presumably a very aggressive species, is absent a species-rich community may develop.

Clearly, this type of analysis needs careful interpretation to distinguish between environmental effects and competition. However, it is interesting because, unlike many currently popular approaches, it focusses attention on the interactions which may be occurring between the species in a community.

1.4 SPECIES DIVERSITY

The species diversity of a community is normally expressed in terms of two properties, the number of species present and their relative proportions. These two aspects, known respectively as species richness and evenness or equitability, are often combined in indices of diversity (Whittaker, 1965; Hill, 1973c). Much of the interest in species diversity derives from hypotheses about niche structure and from a supposed relationship of diversity to other community properties such as productivity and stability (e.g. Whittaker, 1965). The few published accounts of diversity in bryophyte vegetation refer mainly to exploratory studies, undertaken to see what differences in species richness and evenness exist under various environmental regimes (e.g. Hoffman, 1971; Slack, 1977; Winner and Bewley, 1978).

Slack (1977) made extensive investigations of bryophyte diversity in some upland communities in New York State. In general, the diversity of the bryophyte flora was not found to correlate well with individual environmental factors but increased, not surprisingly, in response to increasing habitat heterogeneity. Also, in most situations, no correlation existed between bryophyte and higher plant diversity, presumably because shading effects etc. caused by the latter are non-specific. Slack pointed out that it may be artificial to study the diversity of bryophytes independently when they occur in mixed communities with other types of plant.

Several authors have utilized diversity indices as measures of the suitability of particular environments for bryophyte vegetation. Hoffman

(1971) examined species richness and evenness, by means of several indices, in his study of epiphytes on *Pseudotsuga menziesii*. Overall epiphyte diversity was found to decrease towards the tree base due to dominance, in this mesic position, by one or a few bryophyte species. For similar reasons a fall in diversity occurred, at a standard position on the trunk, across a climatic gradient from xeric to mesic conditions. Winner and Bewley (1978) have used the Shannon–Weaver index of diversity to study the pattern of decline of mosses in *Picea glauca* woodland on approaching an SO_2 pollution source.

While a consideration of species diversity may occasionally provide a useful alternative picture of variation in plant communities in response to different environments (Slack, 1977) it seems unlikely that, alone, the approach will fundamentally improve our understanding of bryophyte community organization. Such an understanding is only likely to come from a combination of observational and experimental studies.

1.5 CORRELATION WITH ENVIRONMENT

The method of looking for correlations of species occurrences, abundances or performances in the field with levels of environmental factors is one of several approaches for investigating the causal factors for species distribution patterns. However, a statistically significant correlation is no proof of a cause–effect relationship. The evidence is circumstantial and the suspected relationship needs to be tested by careful experiment. Even then there remains the possibility that some factor has been overlooked during the experiment and so the latter approach also seldom produces irrefutable proof of a cause–effect relationship. The bryophyte ecologist must, therefore, steer a commonsense course and use a balanced combination of correlation studies and experimentation to understand species distribution patterns. In this section, the use of three methodological approaches for studying correlations will be examined, together with a consideration of techniques suitable for related studies in bryophyte geography. Some of the methods dealt with here are also reviewed by Austin (1972), and elementary statistical approaches, which will not be covered, are fully discussed by Greig-Smith (1964, Chapter 5).

1.5.1 Multiple regression

The use of a multiple regression approach by Rasmussen and Hertig (1977) to study interspecific relationships of corticolous bryophytes has already been discussed (pp. 16–17). A more familiar use of multiple regression is to examine the apparent response of individual species to groups of simultaneously operating environmental factors (Blackman and Rutter, 1946; Rutter, 1955; Greig-Smith, 1964). The end point of the analysis is the generation of an equation of the type below:

$$y = a + b_1x_1 + b_2x_2 \ldots + b_nx_n$$

where y is the abundance or performance of the species under consideration (the dependent variable), a is a constant, the x's are the different environmental factors (independent variables) and the b's are their partial regression coefficients (see Greig-Smith, 1964, for further explanation). The signs and magnitudes of the partial regression coefficients indicate the importance of each environmental factor in the hypothesized relationship. In situations where relationships are non-linear the terms b_nx_n may be replaced by a polynomial expansion, e.g. $b_1x_1 + b_{11}x_1^2 + b_{111}x_1^3$ (Yarranton, 1969) to increase the goodness of fit, although in such cases the biological meaning of the equation may then be exceedingly difficult to interpret (Greig-Smith, 1964; Mead, 1971).

The only major use of multiple regression in bryophyte ecology to date has been in a study of the cryptogamic vegetation of limestone pavement grikes undertaken by Yarranton and Beasleigh (1968, 1969) and Yarranton (1970) at Kemble in Ontario. This investigation was attempted primarily to examine the feasibility of simultaneously applying the multiple regression approach to all species in a community. Limestone pavement grikes were chosen because it was considered that marked variations would occur in relatively few environmental factors thus simplifying the numerical analysis. A range of potential combinations of environmental conditions was included in the study by selecting grikes running in different directions and of different widths. Also, both vegetation and environment were sampled on opposing faces of the fissures, and at various depths (Yarranton and Beasleigh, 1968).

Preliminary analyses of variance of the species frequencies were undertaken by partitioning thc scores on the basis of major microtopographic features (aspect, depth, width, etc.). The results indicated a significant effect of at least one microtopographic feature in 27 of the 31 species studied. A number of microclimatic variables (temperature, humidity, light intensity etc.), and the acidity of the rock surface, were measured periodically at each sample site (Yarranton and Beasleigh, 1969). Selection of suitable independent variables for the multiple regression analyses was made on the basis of a microtopographic analysis of variance approach similar to that used for the species. Seven independent variables were used for the construction of the equations, pH, mean temperature, mean relative humidity, mean light intensity and the variances of the latter three factors were also included as separate variables indicating the stability of each aspect of the environment (Yarranton, 1970).

Yarranton's initial regression equations for the 31 species were disappointing, from an ecological point of view, with the percentage of variation in species' frequencies accounted for by environmental factors ranging from 7% (*Lophocolea minor*) to 67% (*Verrucaria nigrescens*). In an attempt to improve this situation Yarranton (1970) tried the effects of various manipulations of the environmental data. A major problem was considered to be the existence of correlations between the environmental variables (cf. Mead, 1971). Where high intercorrelations existed, as in the case of the microclimatic variables, the correlated variables were replaced by single variables obtained by cross multiplication of the pairs of values. Various new regression solutions were obtained using these new independent variables and others obtained by the use of standard statistical transformations (logarithms, square roots, etc.). Generally, the increase in 'variance accounted for' by the environmental variables was small, only greater than 10% in a few species. In terms of the greater difficulty of interpreting combined environmental factors, it might be argued that the small increase in statistical precision was not worth the effort.

Yarranton (1970) tested the predictive power of his regression 'models' by making a second series of measurements of species frequency and environmental variables in some similar limestone grikes nearby. The environmental data were substituted into the equations and frequency scores of each species predicted for the new sites. The predicted frequencies were then compared with the observed scores using χ^2 tests. When the best regression equations (in terms of original 'variance accounted for') were employed, in only three species, *Anomodon attenuatus, Schistidium alpicola* var. *alpicola* and *Radula complanata,* was there a significant difference between the observed and predicted scores but, in general, the predictions of individual frequency values were poor.

A major difficulty encountered in this study was the presence of a large number of zero species frequencies in the data. The zeros, in fact, represent a truncation at the bottom end of the otherwise continuous abundance scale since they indicate absence of species but not their degree of absence. Under these conditions it is not surprising that the multiple regression predictions proved to be rather unreliable. Yarranton (1970) offers suggestions on how to overcome this problem; however, in most instances it would be wise to restrict the use of multiple regression to situations where the abundance scale is truly continuous and zero records infrequent.

The use of multiple regression, as described above, ignores the possibility that interactions between species may also be a contributory factor in controlling relative abundances of species in a community. Yarranton (1970) considered that the existence of significant competitive effects was improbable in limestone pavement grikes because of the sparsity of the vegetation. However, in some poorly vegetated rocky habitats it is quite possible that the number of suitable niches is very small and intense competition for occupancy may occur. Competitive influences might have to be included as separate independent variables in the multiple regression equation (cf. Rasmussen and Hertig, 1977).

An important point which should be considered in multiple regression studies, and other correlative investigations, is the possibility that communities may not be at static equilibrium with their physical and chemical environment. Cyclical changes have been observed in a number of bryophyte communities (Richards, 1938; Proctor, 1962) and these may confound the interpretation of blindly applied numerical analyses (Yarranton, 1970).

Multiple regression has also been employed by Callaghan *et al.* (1978) to relate photosynthesis by bryophytes to environmental factors, and by Slack (1977), who investigated correlations between bryophyte species diversity (see p. 18) and habitat factors along an altitudinal gradient.

1.5.2 Pattern analysis

Pattern analysis and the related techniques of pattern covariance analysis are well-established techniques in the study of higher plant ecology (Kershaw, 1957; Greig-Smith, 1964; Hill, 1973a). Pattern analysis is used to detect the scales at which non-randomness occurs in the distributions of individual species or environmental factors. Pattern covariance analysis allows the scale at which positive or negative co-relationships exist between different species, different environmental factors, or species and environmental factors, to be detected. As far as I am aware there are no published accounts describing the use of pattern analysis procedures to study bryophyte communities. Partly this may arise from the very fragmentary nature of many bryophyte habitats because sampling for pattern analysis requires uninterrupted runs of vegetation. However, there are a number of situations where these techniques might be straightforwardly and profitably applied. Many investigations, for instance, have been undertaken in tundra areas where bryophyte vegetation is often dominant and extensive. Indeed, Sheard (1968) has identified a very clear repeating vegetation pattern in moss and lichen communities which is correlated with topographic and climatic conditions in one such situation.

Yarranton and Green (1966), using pattern analysis, demonstrated a correlation between the distributional patterns of some crustose lichens and small variations in slope on limestone cliffs in Ontario. Similar studies could easily be performed on bryophytes and might well provide important information regarding microhabitat dimensions and preferences, competitive interactions and establishment of propagules.

1.5.3 Ordination and gradient analysis

The various methods of ordination or gradient analysis have become familiar tools in ecological research in recent years. Theoretically, ordination represents an alternative approach to numerical classification in which, instead of forcing individuals (normally quadrats) into a rigid classification scheme, the samples are allowed to reveal their floristic similarities and dissimilarities by means of a scatter diagram. Ordination methods were designed from the theoretical standpoint that vegetation is a continuum, that is, there is continuous variation between communities and no clear boundaries exist.

In one approach, known as direct gradient analysis (Whittaker, 1973b), the reference axes in the scatter diagram represent gradients of known environmental factors. In the alternative approach, termed indirect gradient analysis (cf. Whittaker, 1967) or, more commonly, ordination, the axes are positioned without knowledge of environmental conditions but in a manner which emphasizes the major trends of variation in the vegetation data. Generally, however, the major floristic trends are themselves reflections of trends in controlling environmental conditions and so the axes may be expected to represent ecological gradients in many cases.

(a) Direct gradient analysis

The method of direct gradient analysis has been largely pioneered by Whittaker (1967, 1973b) and involves a consideration of the response pattern of individual species in relation to selected environmental gradients. Where the over-riding importance of one or a few factors on vegetation structure is beyond doubt the approach is attractive because of its simplicity. However, in situations where the relative importance of different environmental factors is in doubt, as is frequently the case, direct gradient analysis may be of little use and more information would result from the use of an ordination approach. Direct gradient analysis should, therefore, be seen as a method for examining the response of vegetation to particular environmental factors rather than as an exploratory device for identifying environmental causes of vegetational variation.

Direct gradient analysis has been used in several bryological studies, mostly by American workers (Vitt and Slack, 1975; Siedel, 1976; Lee and La Roi, 1979). Some indications of the strengths and weaknesses of the technique will be given by considering a recently published study of bryophyte communities in an area of the Canadian Rocky Mountains (Lee and La Roi, 1979). One of the primary aims of this investigation was to examine the effect of increasing altitude on bryophyte communities. The main environmental variables were considered to be moisture regime and elevation, although it was recognized that the latter must represent a complex of factors and was not itself an operative factor. The quadrat samples consisted of cover-abundance records obtained from 30 stands distributed over the range of the moisture and elevation gradients.

Lee and La Roi (1979) plotted their stands with respect to elevation, measured by an altimeter, and a synthetic moisture index (SMI), obtained as follows: Firstly, each stand was tentatively assigned a moisture index in the range 1–5 (xeric–hygric) in the field. Next, an index was obtained for each vascular plant species present by averaging the values for all stands containing that species. Finally, the SMI value for each stand was obtained by averaging the indices for all vascular plant species present in the stand. Cover-abundance values for individual bryophyte species were overlaid on the resulting scatter diagram and areas of high abundance indicated by the use of contours (Fig. 1.4). The results show that different species occupy different habitats with respect to the two environmental gradients although there are considerable overlaps.

Lee and La Roi's study has several obvious shortcomings. The method of derivation of the SMI is at best only very approximate since vascular plants are likely to respond to moisture conditions rather differently to bryophytes. Also, the original stand moisture indices given in the field may

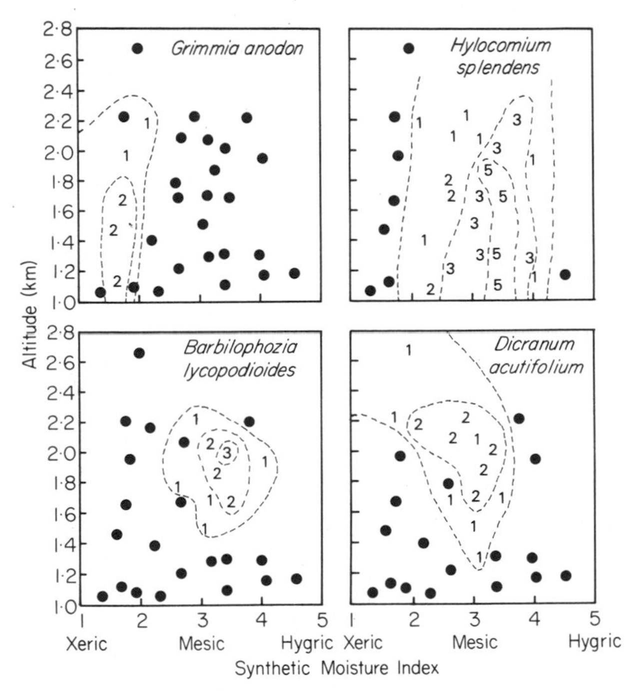

Fig. 1.4 Direct gradient analysis of selected bryophytes from Jasper National Park, Canadian Rocky Mountains. Numerals represent cover classes as follows: 1 = < 1%; 2 = 1–5%; 3 = 6–15%; 4 = 16–25%; 5 = 26–50%. Solid circles indicate that the species was absent. (Redrawn from Lee and La Roi, 1979.)

have been subconsciously affected by the bryophytes which were observed to be present, in which case the argument is circular. An important criticism is that the two environmental factors used to ordinate the samples are not translatable in terms of physical variables which may directly affect bryophytes. It is quite possible that moisture conditions may be an important component of the altitudinal gradient and a two-dimensional representation then becomes unnecessary. Whether the use of such crude environmental factors can lead to anything but a very superficial understanding of factors controlling the distributions of bryophytes is questionable since these plants are well known to possess precise micro-environmental requirements.

It is perhaps pertinent to mention here the hemispherical plot technique pioneered by Perring (1959, 1960) in his studies of chalk downland ecology and used extensively by Grime and Lloyd (1973). This is a type of direct gradient analysis method in which two factors, aspect and slope, are given special importance. Lye (1966) has used the method to show differences in habitat preferences of a number of oceanic bryophyte communities in Western Norway. Admittedly, neither aspect nor slope are themselves factors likely to affect the distribution of plants directly, but a number of important environmental variables correlate closely with them (see Grime and Lloyd, 1973). The method deserves wider use in bryological studies and could, for instance, be applied to the analysis of distributional patterns of species growing on rock outcrops, walls, roofs, gravestones, etc.

(b) Ordination

Indirect gradient analysis or ordination starts from a rather different point of view to direct gradient analysis since no assumption is made about the identities of environmental factors influencing the vegetation. The approach is exploratory and, as with all methods described here, aims to produce hypotheses about possible causal environmental factors which should be tested by appropriate experimentation. In applying ordination the assumption is made that some form of floristic gradient exists in the vegetation in response to environmental variation.

Fig. 1.5 shows examples of two possible types of floristic gradient in response to a single environmental gradient. In Fig. 1.5a the species exhibit linear relationships, both to the environmental gradient and to each other, and thus the existence of a floristic gradient will be signalled by the presence of a range of statistically significant, positive and negative inter-correlations between the species. The responses of species to the environmental gradient shown in Fig. 1.5b are in the form of bell-shaped curves and, as a result, their relationships to one another are extremely non-linear. The use of linear correlation methods to detect the floristic gradient would be inappropriate in the latter case (Van Groenewoud, 1976).

Ordination techniques aim to reconstruct vegetation response patterns of the types shown in Fig. 1.5 and this is done using just the species abundance information within quadrat samples. The dashed lines in Fig. 1.5a and b might, for example, represent particular quadrat samples and the problem is to rearrange and space these samples relative to one another in order to

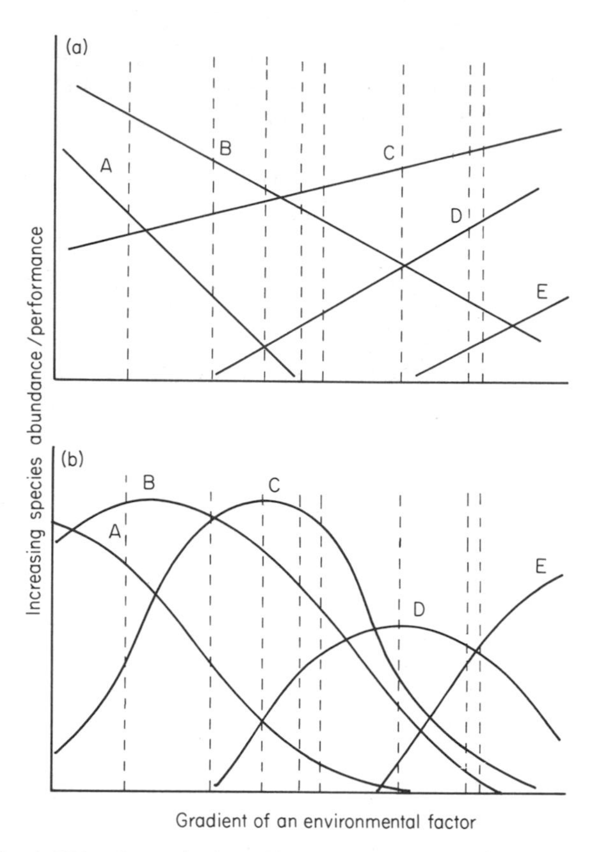

Fig. 1.5 Two hypothetical types of vegetational response to an environmental gradient. The solid lines (a) or curves (b) represent response patterns of individual species (A–E). Vertical pecked lines denote floristic composition within 'quadrats' recorded at random positions along the gradient.

express their floristic similarities and thus reconstruct the gradient. A major complication is that, in most situations, the individual species are simultaneously responding to several different environmental influences and the analysis of such situations requires sophisticated multivariate numerical procedures. For details of some of the more well-established ordination techniques the reader is referred to reviews by Greig-Smith (1964), Gittins (1969), Pielou (1969), Whittaker (1973a), and Orlóci (1975) which contain references to original papers. An important second stage in the analysis is to match the ordering of quadrats along the axes with different environmental variables in order to identify possible causes of the floristic gradients. This can often be achieved by correlation or regression techniques (Gittins, 1969).

Ordination methods have been used to study a range of bryophyte vegetation types including ground communities (Petit and Symonds, 1974; Lechowicz and Adams, 1974; Roux and Salanon, 1974; Smith and Gimingham, 1976; Morton, 1977), epiphytes (Hale, 1955; Beals, 1965; Bates and Brown, 1981), rock and cliff communities (Foote, 1966; Bunce, 1967; Yarranton, 1967a–d; Bates, 1975a, b, 1978; Pentecost, 1980), aquatics (Empain, 1973; Empain and Lambinon, 1974) and *Sphagnum*-bog dominated vegetation (e.g. Vitt and Slack, 1975).

One of the earliest applications of ordination to a cryptogam community was undertaken by Foote (1966) who studied the vegetation of limestone outcrops in Wisconsin. The quadrats were arranged into a linear sequence by applying the approximate technique of Curtis and McIntosh (1951) known as continuum analysis. The linear ordering of quadrats was obtained by placing stands sharing high frequencies of particular species close together. The final rank ordering of stands, obtained purely on a floristic basis, showed a close correspondence to a gradient of moisture conditions and this factor was, therefore, considered to be the major one influencing vegetation variation. Foote (1966) summarized the analysis by plotting smoothed curves of abundance of some of the species along the floristic gradient. Some species, such as the lichen *Lecanora dispersa*, showed peak abundance towards the 'dry' end of the floristic gradient, others, such as *Anomodon attenuatus*, at the 'moist' extreme and some, for example *Endocarpon pusillum*, were intermediate. *Schistidium apocarpum*, although most abundant in 'moist' quadrats, was present over a wide range of the floristic gradient probably due to its ability to grow in crevices and cracks in drier situations.

The continuum analysis procedure used in this study has several limitations of which the most serious is that no allowance is made for the existence of other, superimposed floristic gradients resulting from separate, unsuspected environmental influences. Foote (1966) also employed a simple multivariate method of ordination devised by Bray and Curtis (1957) and was able to expand his linear ordination into a three-dimensional model of vegetation structure. Information about environmental variables was limited in this study but the results suggest that, in addition to moisture conditions, other factors may have been influencing species distributions on the limestone outcrops.

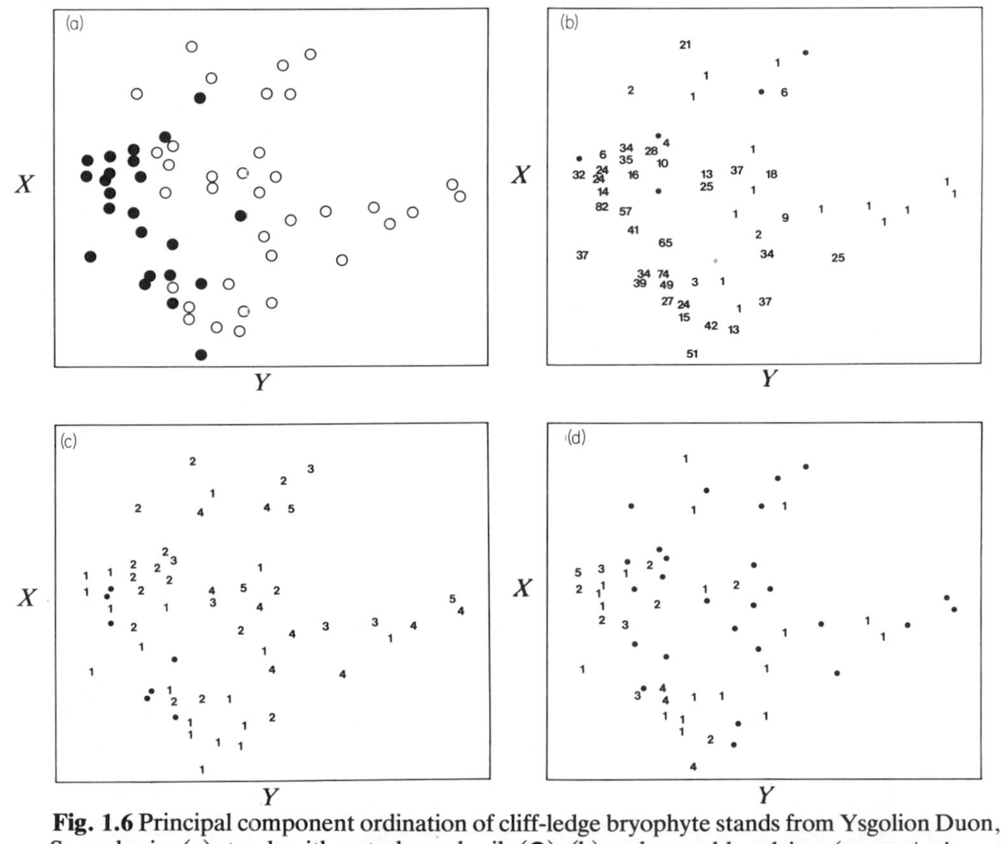

Fig. 1.6 Principal component ordination of cliff-ledge bryophyte stands from Ysgolion Duon, Snowdonia. (a) stands with waterlogged soils (●); (b) exchangeable calcium (p.p.m./unit volume of soil); and frequency classes (1 = 1–5; 2 = 6–10; 3 = 11–15; 4 = 16–20; 5 = 21–25) of (c) *Rhytidiadelphus loreus*; (d) *Rhizomnium punctatum*. (Redrawn from Bunce, 1967.)

Bunce (1967) performed a similar type of analysis on bryophyte vegetation data from Welsh mountain cliff ledges. Principal components analysis was used to obtain a two-dimensional ordination of the 58 rock-ledge stands (Fig. 1.6). Values for various environmental factors determined for each rock ledge were overlaid on the stand positions in the ordination in an attempt to identify the major environmental causes of floristic variation (Fig. 1.6a–b). By this means a principal axis (*Y*) of the ordination was interpreted as representing a gradient from dry, oligotrophic to wet, eutrophic conditions and found to relate topographically to the gulley and buttress system of the cliff. Tolerances of individual species with respect to this gradient were explored by overlaying species frequencies on the ordination plot (Fig. 1.6c–d). Some of the main points arising from this part of the analysis were that species composition varied continuously along the ordination axes, and different species exhibited peak abundances in different areas of the ordination.

In his studies of saxicolous bryophyte communities in south-west England, Yarranton (1967a–d) also employed principal components analysis as an ordination technique. Vegetation sampling was by the plotless sampling method described earlier (p. 4) and a primary objective was to obtain an ordination of species. In an investigation of bryophyte and macrolichen vegetation on granite outcrops on Dartmoor, Yarranton (1967d) calculated correlation coefficients, based on the χ^2 values from interspecific association contingency tests, and these were subjected to principal components analysis. The resultant ordination (Fig. 1.7) represents an alternative to the plexus diagram approach discussed on pp. 10–12. As in the plexus diagram, the proximities of the species to each other are directly related to their distributional similarities but, in addition, they are positioned according to their 'peak' abundances along the major floristic gradients.

Yarranton (1967d) also ordinated his sample grids. Their positions on each ordination axis were averages of the scores (after ranking) of the species recorded at each intersection of the grid. The sample grid ordination was used to provide a basis for examining correlations of environmental variables with the ordination axes in much the same manner as adopted by Bunce (1967). Several correlations were apparent and, in particular, the first ordination axis showed a trend with respect to soil depth, deeper accumulations being associated with grids having smaller axis loadings. The positions of species with respect to axis 1 may therefore represent their response to a gradient of soil depth and related factors (Fig. 1.7). In general, however, correlations between the ordination axes and environmental variables were poor in Yarranton's study. Partly, this may have arisen from distortions of the data due to the inclusion of 'no contact' as a pseudospecies in the principal components analysis, and perhaps also

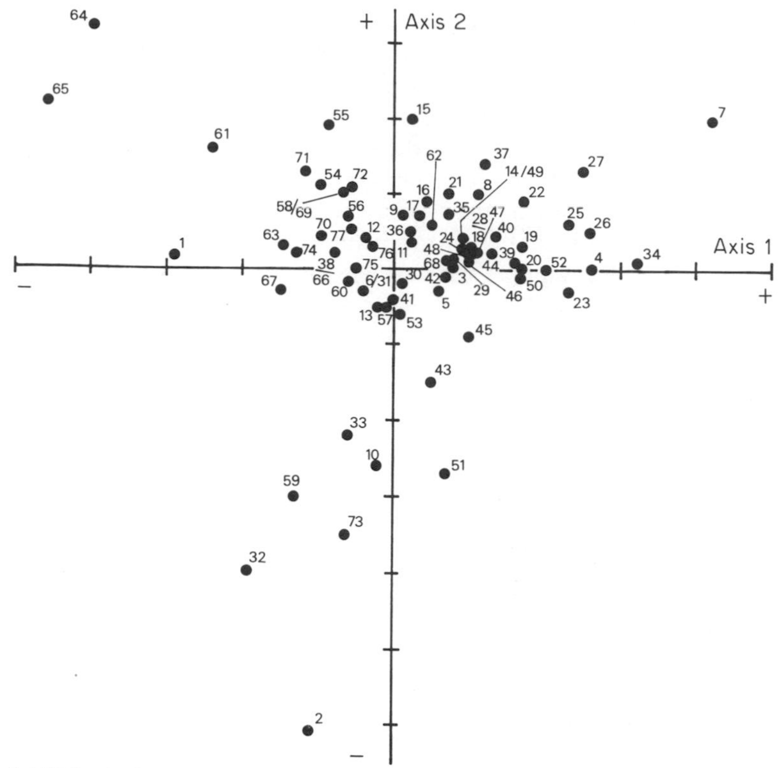

Fig. 1.7 Principal components analysis ordination of saxicolous bryophyte and macrolichen species of the Dartmoor granite. (Redrawn from Yarranton, 1967d.) Key to the species: 1, 'No contact'; 2, *Andreaea rothii*; 3, *Bryum alpinum*; 4, *Campylopus paradoxus*; 5, *Cephalozia* spp.; 6, *Dicranoweisia cirrata*; 7, *Dicranum scoparium*; 8, *Diplophyllum albicans*; 9, *Douinia ovata*; 10, *Dryptodon patens*; 11, *Frullania tamarisci*; 12, *Grimmia pulvinata*; 13, *Hedwigia ciliata*; 14, *Hylocomium splendens*; 15, *Hypnum cupressiforme*; 16, *Isopterygium elegans*; 17, *Isothecium myosuroides*; 18, *Leptodontium flexifolium*; 19, *Lophozia ventricosa*; 20, *Marsupella emarginata*; 21, *Barbilophozia attenuatus*; 22, *Pleurozium schreberi*; 23, *Pohlia nutans*; 24, *Polytrichum commune*; 25, *Polytrichum formosum*; 26, *Polytrichum juniperinum*; 27, *Polytrichum piliferum*; 28, *Pseudoscleropodium purum*; 29, *Ptilidium ciliare*; 30, *Racomitrium aquaticum*; 31, *Racomitrium fasciculare*; 32, *Racomitrium affine*; 33, *Racomitrium heterostichum*; 34, *Racomitrium lanuginosum*; 35, *Rhytidiadelphus loreus*; 36, *Scapania dentata*; 37, *Scapania gracilis*; 38, *Bryoria fuscescens*; 39, *Cladonia arbuscula*; 40, *Cladonia ciliata* var. *tenuis*; 41, *Cladonia coccifera*; 42, *Cladonia fimbriata*; 43, *Cladonia floerkeana*; 44, *Cladonia furcata*; 45, *Cladonia gracilis*; 46, *Cladonia macilenta*; 47, *Cladonia portentosa*; 48, *Cladonia pyxidata*; 49, *Cladonia rangiformis*; 50, *Cladonia squamosa*; 51, *Cladonia subcervicornis*; 52, *Cladonia uncialis*; 53, *Coelocaulon aculeatum*; 54, *Evernia prunastri*; 55, *Hypogymnia physodes*; 56, *Hypogymnia tubulosa*; 57, *Lasallia pustulata*; 58, *Parmelia caperata*; 59, *Parmelia conspersa*; 60, *Parmelia glabratula*; 61, *Parmelia glabratula* ssp. *fuliginosa*; 62, *Parmelia laevigata*; 63, *Parmelia mougeotii*; 64, *Parmelia omphalodes*; 65,

the crude nature of some of the environmental measurements.* A major factor, however, may have been the use of the sample grids as the basic units for examining the environmental correlations although the species data were obtained at the scale of the individual mesh intersections. This must mask subtle micro-environmental variations present at a smaller scale than the grids. In an almost identical study of culm outcrops at Steps Bridge, Devon, Yarranton (1967c) also made some environmental measurements at the individual sample points and these were found to be more informative than those obtained at the grid level. This approach still offers possibilities for the study of cryptogams which remain relatively unexplored.

One further reason for the rather inconclusive results obtained in Yarranton's studies may have been due to the use of principal components analysis for ordinating samples. This technique assumes linear relationships between species of the type shown in Fig. 1.5a but, in many instances, the true form of relationships may approximate to that shown in Fig. 1.5b. A problem closely related to this is that of vegetation heterogeneity. In situations where very wide ranges of one or more environmental factors are encountered it can be expected that vegetation types present at different environmental extremes will differ greatly. Quadrat samples collected from the extremes may have no species in common and this causes difficulties in the calculations of most similarity or distance coefficients used as the basis of ordination methods, including principal components analysis. In these instances the coefficients have, in fact, reached their upper or lower level of sensitivity and are unable to properly express the spatial positions of quadrats. This leads to distortion in the final ordination, for instance, a linear floristic sequence may be represented incorrectly as a two-dimensional, curvilinear arrangement of samples (e.g. Swan, 1970).

An example of this type of effect can be seen in a principal component ordination of seashore bryophyte and macrolichen vegetation (Bates,

*Jong, Aarssen and Turkington (*Oecologia (Berl.)* **45,** 322–34 (1980)) indicate that the use of the chi-square statistic as applied by Yarranton is invalid. The method leads to inflated expected values which will have a detrimental effect on the chi-squared calculation and subsequent multivariate analyses. Jong *et al.* present a correct method of calculation.

Parmelia saxatilis; 66, *Parmelia perlata*; 67, *Parmelia verruculifera*; 68, *Peltigera horizontalis*; 69, *Platismatia glauca*; 70, *Pseudevernia furfuracea*; 71, *Ramalina siliquosa*; 72, *Sphaerophorus globosus*; 73, *Stereocaulon vesuvianum*; 74, *Umbilicaria polyrrhiza*; 75, *Umbilicaria torrefacta*; 76, *Usnea fragilescens*; 77, *Usnea subfloridana*.

1975a) shown in Fig. 1.8a. The axes were interpreted as representing (1) increasing maritime influence due to salt spray deposition and (2) also increasing maritime influence but particularly the 'wetness' element. A re-analysis of the data (Bates, 1976b) by reciprocal averaging (Hill, 1973b), a method much less susceptible to these distortions (Gauch *et al.*, 1977), is shown in Fig. 1.8b. The reciprocal averaging ordination represents the variation, which was summarized by the first two axes of the principal component solution, in the form of a single axis (Axis 1) and the second axis is equivalent to the third axis of the principal components analysis. Re-examination of the principal component result clearly indicates that the linear floristic sequence has been folded over. In this example, the distortion produced by principal components analysis was recognized but, in other cases, erroneous interpretations might result. Heterogeneity in the data is signalled by the presence of many zeros in the data matrix (see also p. 21) and in such cases the investigator is advised to apply more robust methods (e.g. Kruskal, 1964) or those designed specifically for vegetation analysis (Hill, 1973b; Hill and Gauch, 1980; Ihm and Van Groenewoud, 1975; Williamson, 1978).

In the examples cited above ordination has been employed mainly as an exploratory technique to provide a framework for examining correlations between overall vegetational variation and environmental factors. Several investigators, however, have used the ordination approach to help answer more specific questions about vegetation.

Bates (1978) and Pentecost (1980) applied ordination in situations where a major environmental feature, rock type, potentially shows discontinuous

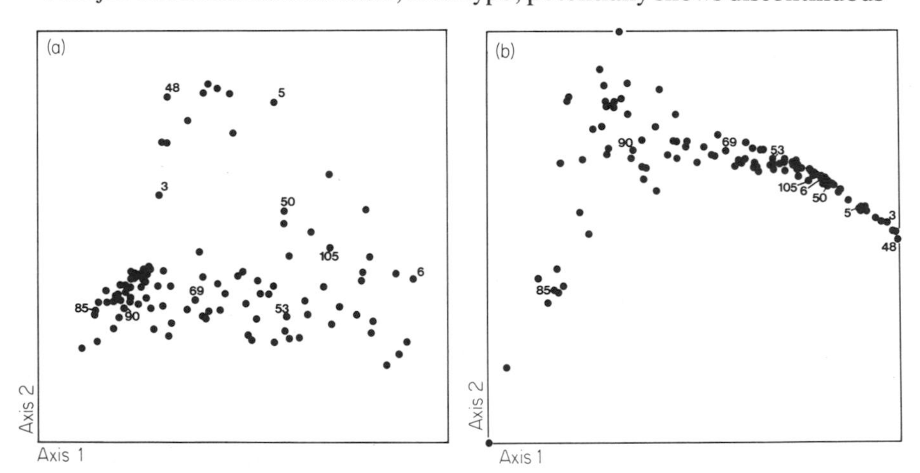

Fig. 1.8 Ordinations of seashore bryophyte and macrolichen vegetation employing the first two axes of (a) principal components analysis; (b) reciprocal averaging. Ten quadrats have been labelled to emphasize the distortion of the floristic sequence produced by principal components analysis. (Redrawn from Bates, 1975a, b.)

variation. The vegetation samples would be expected to fall into distinct clusters in the ordination corresponding to each rock type if rock type was an important feature in determining species' distributions. Bates (1978) used a form of Yarranton's (1966) plotless sampling technique to survey the bryophyte and macrolichen vegetation of four rock types (limestone, sandstone, basalt, peridotite) in western Scotland. The ordination of the sample grids, justifiable here because micro-environmental effects were not considered, resulted in a clear separation of the limestone from the other rock types on the first axis (Fig. 1.9a). The grid samples from non-calcareous rocks, however, did not fall into distinct groups but showed a gradation along axis 2, from ultrabasic rocks, through basalt to sandstone, thus indicating that levels of basic minerals present within the rocks were having a subtle influence on vegetation structure. Additional evidence for this hypothesis was obtained by ordination of rock samples on the basis of chemical data produced by leaching the samples with a chelating agent, EDTA (Fig. 1.9b). The first axis separated limestone samples from the remainder, mainly on the basis of high availability of Ca, Mg and K in limestone. The second axis could be interpreted mainly in terms of availability of heavy metals and gave a sequence, ranging from sandstone to igneous rocks, roughly corresponding to that observed in the floristic ordination (Fig. 1.9a).

In a similar study, Pentecost (1980) found that the bryophyte and lichen floras of rhyolite, dolerite and pumice-tuff rocks in North Wales showed little overlap in a reciprocal averaging ordination. Pentecost considered a number of physical and chemical differences between the rock types studied but concluded that the concentration of solutes in surface water,

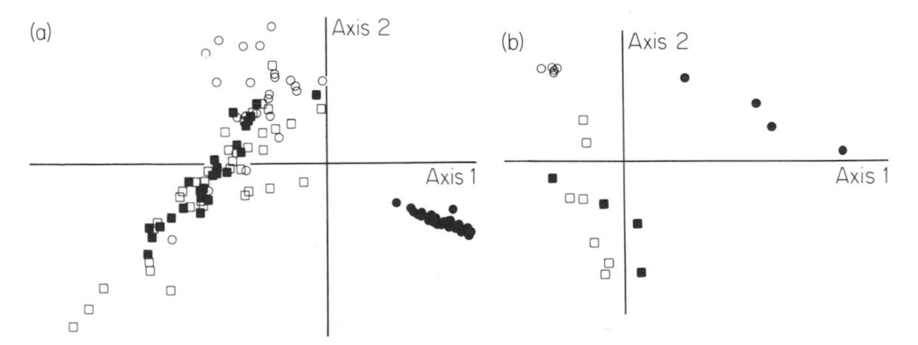

Fig. 1.9 A comparison of ordinations of floristic data and chemical leachate data from four Scottish rock types. (a) 'Step across' ordination (Williamson, 1978) of 121 bryophyte and macrolichen vegetation grid samples; (b) principal component ordination of 19 rock samples based on chemical analytical data from EDTA-leachates. Rock types: ●, Durness limestone; O, Torridonian sandstone; ■, basalt; □, ultrabasic rocks. (Redrawn from Bates, 1978.)

derived from the rocks or from animal excrement, was the most important factor in determining vegetation type.

Most of the above samples refer to floristic ordinations but, in some situations, it is useful to ordinate quadrat samples using environmental information collected from each site. The ordination of rock samples on the basis of chemical attributes (Bates, 1978), described above, is a simple example of this approach. Environmental ordination is most valuable as a means of replacing batteries of intercorrelated environmental variables by a few simple uncorrelated variables which may then be compared with species distributions in various ways (see Austin, 1972). There are several examples of the use of environmental ordinations to study higher plant communities (e.g. Goldsmith, 1973) but the approach has been little-used in bryophyte ecology. Bates and Brown (in press) found environmental ordination useful in a study of cryptogamic epiphytes on oak *(Quercus petraea)* and ash *(Fraxinus excelsior)*. Species cover and environmental variables were determined in several quadrats on each of 23 trees. An initial floristic ordination by reciprocal averaging produced an incomplete separation of quadrat samples from oak and ash trees (Fig. 1.10a) indicating that environmental control was by a factor, or factors, only loosely linked to phorophyte type. Bates and Brown attempted to identify the cause of the major pattern of epiphyte variation by calculating correlation coefficients between environmental variables and axis loadings (Table 1.3). In addition the environmental variables × quadrats data matrix was subjected to a principal components analysis to identify major trends of variation in the oak–ash environment. The first principal component axis gave a clear separation of quadrat samples from the two phorophyte species but variation along the axis was continuous rather than dis-

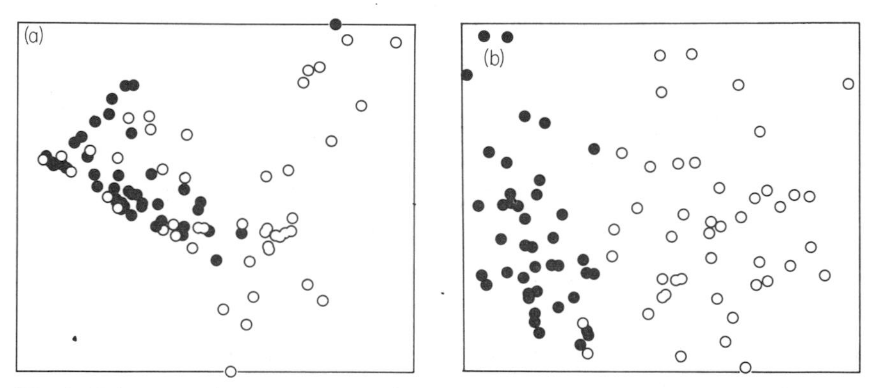

Fig. 1.10 A comparison of ordinations of quadrats from *Quercus petraea* (●) and *Fraxinus excelsior* (O) trunks employing (a) cover of epiphytic bryophytes and lichens (reciprocal averaging); (b) environmental variables (principal components analysis). (Bates and Brown, 1981).

Table 1.3 Spearman rank correlation coefficients (r_s) calculated between the first axis of a reciprocal averaging ordination and some environmental variables. 'Env. ord.' represents the variate obtained by a principal components analysis of the environmental data.

Environmental factor	r_s	Environmental factor	r_s
Tannin	−0.33†	Total Ca (bark)	0.31†
Bark pH	0.63‡	Total Mg (bark)	0.18*
Soluble Na (bark)	0.27†	Total N (bark)	−0.08
Soluble K (bark)	0.36‡	% unobstructed sky	0.02
Soluble Ca (bark)	0.17	Tree girth	0.39‡
Soluble Mg (bark)	0.24*	Inclination	0.05
Total Na (bark)	0.25*	Bark thickness	−0.06
Total K (bark)	0.38‡	Water-holding capacity	−0.15
Total Fe (bark)	−0.35‡	'Env. ord.'	0.26†

Significance levels: *, $p < 0.05$; †, $p < 0.01$; ‡, $p < 0.001$.

continuous (Fig. 1.10b). The major contributors to this axis of environmental variation were concentrations of several bark minerals and bark tannin concentration, the former being lower and the latter higher in oak compared to ash. Bates and Brown used the first principal component as an index of how 'oak-like' or 'ash-like' the environment was at each quadrat position and it is included as a separate environmental variable for correlation with the floristic ordination in Table 1.3. The relative sizes of the rank correlation coefficients (Table 1.3) indicate that variations in bark acidity are more likely to be the cause of epiphytic vegetation variation on oak and ash than those of other simple factors or a group of factors closely related to phorophyte identity.

1.5.4 Bryophyte geography

Phytogeography offers many distributional problems akin to those encountered in conventional ecology except that the 'quadrats' are geographical areas and the environmental factors are predominantly macroclimatic or historical rather than microclimatic and edaphic. Several bryologists have employed numerical approaches to aid the recognition of geographical assemblages of species or to provide a framework for comparison with gross environmental features (Lye, 1966; Proctor, 1967; Bryant *et al.*, 1973).

Proctor (1967) analysed the distribution of liverworts in the British Isles using the presence–absence records for Watsonian vice-counties published by the British Bryological Society (Paton, 1965). A classification of the 152 vice-counties was achieved by means of normal association analysis

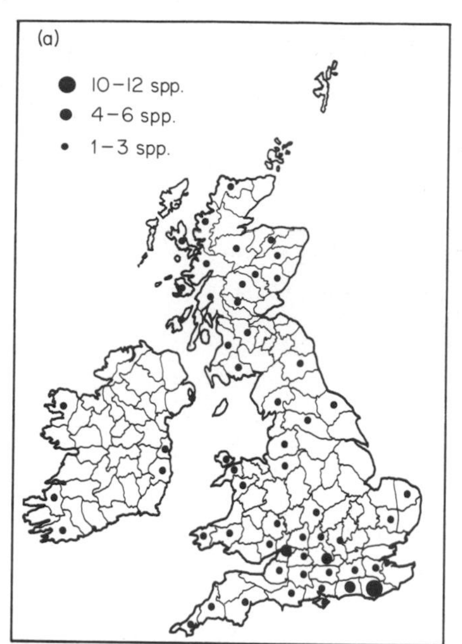
(a)
10–12 spp.
4–6 spp.
1–3 spp.

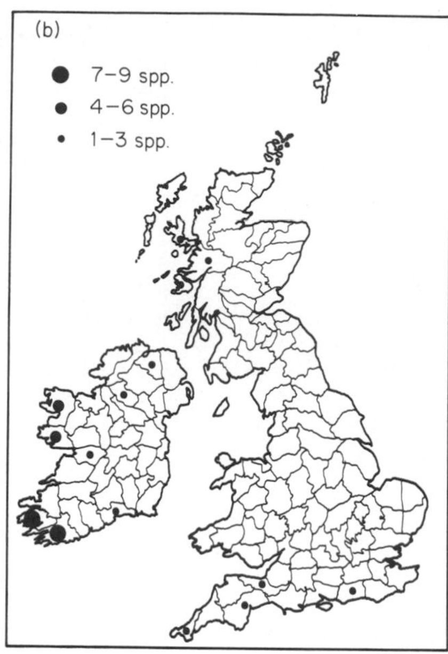
(b)
7–9 spp.
4–6 spp.
1–3 spp.

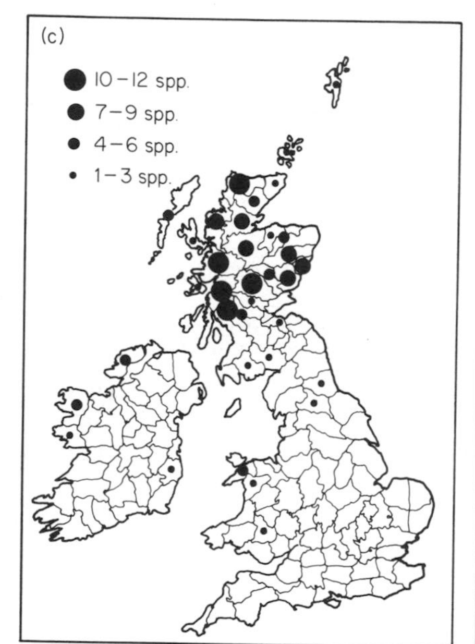
(c)
10–12 spp.
7–9 spp.
4–6 spp.
1–3 spp.

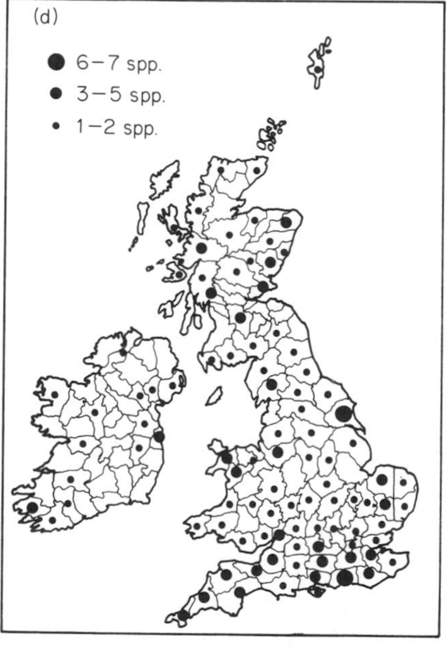
(d)
6–7 spp.
3–5 spp.
1–2 spp.

(Williams and Lambert, 1959). The 317 liverwort taxa were classified into 47 geographical groups by inverse association analysis (Williams and Lambert, 1961). Proctor noted that, on the basis of the first two divisions in the normal analysis (on *Anastrepta orcadensis* and *Solenostoma pumilum*), the vice-counties fell naturally into 'Highland' and 'Lowland' types. This major phytogeographical 'dividing line' corresponded closely with that obtained when a similar analysis was made of the moss vice-county records (Proctor, 1967). A very useful summary of the distribution centres of selected groups of species was produced by plotting maps showing the numbers of species of a particular inverse analysis class present within each vice-county (Fig. 1.11). In a further stage of the analysis, Proctor (1967) subjected the reduced data set, consisting of 25 normal vice-county groups×47 inverse species groups, to ordination by principal components analysis to clarify distributional relationships. The ordinations clearly demonstrated the importance of oceanicity and montanicity as separate, but partially correlated, factors influencing hepatic distribution patterns in the British Isles.

Many of the methods discussed in preceding sections could profitably be applied in studies of macro-distributional problems in bryology. It is to be hoped that further analysis of the type described by Proctor (1967) and Birks (1976) will be attempted as more detailed species distribution records become available from different regions.

1.6 EXPERIMENTAL APPLICATIONS OF MULTIVARIATE TECHNIQUES

The multivariate techniques of numerical classification and ordination have been used to summarize complex data in a number of studies related to bryophyte ecology.

Taxonomic problems in some difficult groups of bryophytes often arise because of uncertainties about whether variability is environmentally or genetically determined. In a number of cases, the application of multivariate analysis to complex biometrical data has aided interpretation (Lefebvre and Lennes, 1969; Rahman, 1972; Smith and Hill, 1975; Hill, 1976). Rahman (1972) undertook experiments in which taxa formerly included in the *Sphagnum subsecundum* aggregate were reciprocally transplanted into each other's habitats. Changes in morphology induced by the

Fig. 1.11 Vice-comital distributions of selected inverse association analysis groups of liverworts. (a) *Cephaloziella baumgartneri* group (12 representatives); (b) *Telaranea sejuncta* group (9 representatives); (c) *Mastigophora woodsii* group (12 reprsentatives); (d) *Anthoceros husnotii* group (7 representatives). (Redrawn from Proctor, 1967.)

new environment were followed within the framework of a principal component analysis ordination.

Austin (1977) proposed the wider use of ordination and related methods for the study of successional and other temporal changes occurring in vegetation. In the 'static' approach samples are collected from the various stages of an existing sequence at one point in time and ordinated to reveal their relationships to each other. The 'dynamic' approach is more flexible and may, for instance, involve the inclusion of a single site, scored on different occasions, in one ordination. Morton (1977) described an ordination of chalk grassland bryophyte communities in which the principal axes of variation were attributed to potential evaporation and higher plant shading and sheltering effects, respectively. The quadrats included samples from exclosures where sheep-grazing had been prevented for 7 years or some other grazing regime practised. The ungrazed quadrats were deflected, in comparison to the grazed controls, along both axes. The deflection indicated the occurrence of a more mesic type of bryophyte flora in the absence of grazing, probably due to protection from evaporation afforded by the lusher higher plant foliage. In this example, the ordination axes represent environmental factors but they provide a valuable reference system for studying changes in bryophyte communities provided by different management treatments.

Another application of multivariate methods where further developments are anticipated is in the identification of sources of mineral substances, including pollutants, in bryophyte tissues. Correlations between levels of elements may indicate a particular common source or physiological function (Garten, 1976). Merino and Garcia Novo (1975) provide an example of the application of ordination to a study of variations in Fe, Al, Mn, Cu, Zn, Ca, Mg, K, Na and P concentrations in the higher plant *Rosmarinus officinalis* L. and Hopke *et al.* (1976) and Saeki *et al.* (1977) have used a similar approach to identify sources of elements in urban aerosols and lichens, respectively. Bates (1978) applied principal components analysis to chemical data from a range of bryophytes collected from different Scottish rock types. Axes were obtained representing (1) unusual heavy metal-accumulating ability of *Andreaea rothii,* (2) the relative base status of the different rocks, and (3) physiological control of element levels.

1.7 CONCLUSIONS

The inclusion of many different aspects of bryophyte ecology within a scheme of 'Quantitative Approaches' is purely artificial but serves to highlight some important areas of study where deficiencies exist or more rigorous work is required. In this last section, brief consideration will be given to these points and attempt will be made to indicate areas where further developments may occur.

It is particularly difficult to decide whether the increasing use of sophisticated numerical analyses has led to a greater understanding of the ecology of bryophytes than would otherwise have been obtained. In many instances, it appears that a particular type of analysis has been applied almost arbitrarily, perhaps representing one of several possible alternative and equally valid approaches. Some workers, on the other hand have had strong convictions about the nature of community organization and have chosen their methods accordingly. A few may have been too enthusiastic about the mathematical properties of certain numerical techniques and have overlooked important biological considerations. One area for concern is the failure of many workers to follow up descriptive, hypothesis-generating field studies with experimental work designed to test their hypotheses. Proctor (1974) has questioned whether the routine application of phytosociological classification methods to plant communities could now be regarded as a valid scientific exercise. The same reasoning might also be extended to some studies where objective methods of ordination or classification have been applied but no follow-up work undertaken. Appropriate follow-up studies might take the form of field and glasshouse growth experiments (e.g. Morton, 1977), an examination of comparative physiology (e.g. Bates and Brown, 1974, 1975; Bates, 1976) or a detailed consideration of environmental factors (e.g. Bates, 1978; Pentecost, 1980).

Some areas where future applications of quantitative methods may be made have already been indicated and include experimental bryophyte ecology, bryophyte geography and the analysis of species interactions. It is also possible to predict some other fields where developments may occur by analogy with other, more established, topics in quantitative ecology.

Functional studies of particular bryophyte communities are long overdue. For instance, the manner in which nutrients are acquired, cycled and lost by some bryophyte-dominated communities is still poorly understood. It is likely that mathematical modelling techniques (see Jeffers, 1972b for references) will prove valuable in the analysis of such situations. A number of problems imposed by the dynamic balance between growth, death and decay of many bryophytes may also benefit from a quantitative, modelling approach (cf. Clymo, 1978; Jones and Gore, 1978). The construction of a simple mathematical model describing attachment of aquatic moss propagules to underwater substrata has recently been detailed by Glime *et al.* (1979).

The extent and intensity of competition within bryophyte communities has received very little attention. In sparsely vegetated habitats the existence of competition has been doubted (Yarranton, 1970) but in other situations strong evidence of interactions between species has been obtained (Rasmussen and Hertig, 1977; Watson, 1979). It is unlikely that many of the quantitative experimental approaches adopted for studying competition between higher plant species (e.g. de Wit, 1960; McGilchrist and Trenbath, 1971) will be available to bryophyte ecologists because of cultivation difficulties. However, alternative numerical methods, perhaps based on growth rates or productivities of individual species in naturally occurring mixtures and monocultures, could well be developed.

Demographic studies of bryophyte populations may offer considerable insight into both community organization and influence of environmental factors. Quantitative studies of the population dynamics of some *Polytrichales*-dominated moss communities have recently appeared (Collins, 1976; Watson, 1979) and it is probable that similar investigations of other bryophyte communities will soon be forthcoming.

Conventional, descriptive studies of particular bryophyte communities, however, are likely to remain of importance while large areas remain bryologically unknown. Even in well-worked countries such as Britain very little quantitative information is available for many bryophyte habitats. Workers who have applied objective, quantitative methods to study bryophyte vegetation soon discover that most 'information' resides in the common rather than the rare species of the community. Understandably, in the past bryophyte ecologists have tended to focus their attentions on the rarer species but many recent studies indicate that perhaps, after all, it is the common bryophytes which have the most interesting stories to tell.

ACKNOWLEDGMENT

I wish to thank Dr A. J. Morton for numerous discussions about the use of numerical methods and for reading and commenting on the manuscript.

REFERENCES

Austin, M.P. (1972), Models and analysis of descriptive vegetation data. In: Jeffers, J.N.R. *Mathematical Models in Ecology*, pp. 61–86. Blackwell Scientific, Oxford.

Austin, M.P. (1977), *Vegetatio*, **35**, 165–75.

Barkman, J.J. (1958), *Phytosociology and Ecology of Cryptogamic Epiphytes*. Van Gorcum, Assen.

Bates, J.W. (1975a), *J. Ecol.*, **63**, 143–62.

Bates, J.W. (1975b), *The Ecology and Physiology of Seashore Bryophytes*. Ph.D. thesis, University of Bristol.

Bates, J.W. (1976), *New Phytol.*, **77**, 15–23.

Bates, J.W. (1978), *J. Ecol.*, **66**, 457–82.

Bates, J.W. and Brown, D.H. (1974), *New Phytol.*, **73**, 483–95.

Bates, J.W. and Brown, D.H. (1975), *Oecologia (Berl.)*, **21**, 335–44.
Bates, J.W. and Brown, D.H. (1981), *Vegetatio*, **48**, 61–70.
Beals, E.W. (1965), *Oikos*, **16**, 1–8.
Becking, R.W. (1957), *Bot. Rev.*, **23**, 411–88.
Birks, H.J.B. (1973), *Past and Present Vegetation of the Isle of Skye. A Palaeoecological Study*. Cambridge University Press, London.
Birks, H.J.B. (1976), *New Phytol.*, **77**, 257–87.
Blackman, G.E. and Rutter, A.J. (1946), *Ann. Bot., Lond.* N.S., **10**, 361–90.
Bray, J.R. and Curtis, J.T. (1957), *Ecol. Monogr.*, **27**, 325–49.
Bryant, E.H., Crandall-Stotler, B. and Stotler, R.E. (1973), *Can. J. Bot.*, **51**, 1545–54.
Bunce, R.G.H. (1967), *Bot. Notiser*, **120**, 334–44.
Callaghan, T.V., Collins, N.J. and Callaghan, C.H. (1978), *Oikos*, **31**, 73–88.
Clymo, R.S. (1978), *Ecol. Stud.*, **27**, 187–223.
Cole, L.C. (1949), *Ecology*, **30**, 411–24.
Collins, N.J. (1976), *Oikos*, **27**, 389–401.
Culberson, W.L. (1955), *Ecol. Monogr.*, **25**, 215–31.
Curtis, J.T. and McIntosh, R.P. (1951), *Ecology*, **32**, 476–96.
Degelius, G. (1964), *Acta Horti Gotoburgensis*, **27**, 11–55.
Del Fabbro, A., Feoli, E. and Sauli, G. (1975), *G. Bot. Ital.*, **109**, 361–74.
Empain, A. (1973), *Lejeunia*, N.S., **69**, 1–58.
Empain, A. and Lambinon, J. (1974), *Bull. Soc. Bot. Fr.* (Coll. Bryologie), **121**, 257–64.
Fletcher, A. (1973a), *Lichenologist*, **5**, 368–400.
Fletcher, A. (1973b), *Lichenologist*, **5**, 401–22.
Foote, K.G. (1966), *Bryologist*, **69**, 265–92.
Garten, C.T. (1976), *Nature, Lond.*, **261**, 686–88.
Gauch, H.G., Whittaker, R.H. and Wentworth, T.R. (1977), *J. Ecol.*, **65**, 157–74.
Gilbert, O.L. (1971), *Trans. Br. Bryol. Soc.*, **6**, 306–16.
Gimingham, C.H. and Smith, R.I.L. (1970), Bryophyte and lichen communities in the maritime Antarctic. In: *Antarctic Ecology* (Holdgate, M.W., ed.), Vol. 2, pp. 752–85. Academic Press, London.
Gittins, R. (1969), The application of ordination techniques. In: *Ecological Aspects of the Mineral Nutrition of Plants* (Rorison, I.H., ed.), pp. 37–66. Blackwell Scientific, Oxford.
Glime, J.M., Nissila, P.C., Trynoski, S.E. and Fornwall, M.D. (1979), *J. Bryol.*, **10**, 313–30.
Goldsmith, F.B. (1973), *J. Ecol.*, **61**, 787–818.
Goodall, D.W. (1965), *J. Ecol.*, **53**, 197–210.
Greig-Smith, P. (1964), *Quantitative Plant Ecology*, 2nd edn. Butterworth, London.
Grime, J.P. and Lloyd, P.S. (1973), *An Ecological Atlas of Grassland Plants*, Arnold, London.
Hale, M.E. (1955), *Ecology*, **36**, 45–63.
Harris, G.P. (1971), *J. Ecol.*, **59**, 431–9.
Hill, M.O. (1973a), *J. Ecol.*, **61**, 225–35.
Hill, M.O. (1973b), *J. Ecol.*, **61**, 237–49.

Hill, M.O. (1973c), *Ecology,* **54,** 427–32.

Hill, M.O. (1976), *J. Bryol.,* **9,** 185–91.

Hill, M.O., Bunce, R.G.H. and Shaw, M.W. (1975), *J. Ecol.,* **63,** 597–613.

Hill, M.O. and Gauch, H.G. (1980), *Vegetatio,* **42,** 47–58.

Hoffman, G.R. (1971), *Bryologist,* **74,** 413–27.

Hoffman, G.R. and Boe, A.A. (1977), *Bryologist,* **80,** 32–47.

Hoffman, G.R. and Kazmierski, R.G. (1969), *Bryologist,* **72,** 1–19.

Hooker, T.N. and Brown, D.H. (1977), *Lichenologist,* **9,** 65–75.

Hopke, P.K., Gladney, E.S., Gordon, G.E., Zoller, W.H. and Jones, A.G. (1976), *Atmos. Env.,* **10,** 1015–25.

Hopkins, B. (1957), *J. Ecol.,* **45,** 451–63.

Ihm, P. and Van Groenewoud, H. (1975), *J. Ecol.,* **63,** 767–78.

James, P.W., Hawksworth, D.L. and Rose, F. (1977), Lichen communities in the British Isles: a preliminary conspectus. In: *Lichen Ecology* (Seaward, M.R.D., ed.), pp. 295–413. Academic Press, London.

Jeffers, J.N.R. (1972a), The challenge of modern mathematics to the biologist. In: *Mathematical Models in Ecology* (Jeffers, J.N.R., ed.), pp. 1–11. Blackwell Scientific, Oxford.

Jeffers, J.N.R. (1972b), *Mathematical Models in Ecology,* Blackwell Scientific, Oxford.

Jones, H.E. and Gore, A.J.P. (1978), *Ecol. Stud.,* **27,** 160–86.

Kershaw, K.A. (1957), *Ecology,* **38,** 291–99.

Kershaw, K.A. (1964), *Lichenologist,* **2,** 263–76.

Kruskal, J.B. (1964), *Psychometrika,* **29,** 1–27.

Lambert, J.M. (1972), Theoretical models for large-scale vegetation survey. In: *Mathematical Models in Ecology* (Jeffers, J.N.R., ed.), pp. 87–109. Blackwell Scientific, Oxford.

Lambert, J.M., Meacock, S.E., Barrs, S. and Smartt, P.M.F. (1973), *Taxon,* **22,** 173–6.

Lechowicz, M.J. and Adams, M.S. (1974), *Can. J. Bot.,* **52,** 55–64.

Lee, T.D. and La Roi, G.H. (1979), *Can. J. Bot.,* **57,** 914–25.

Lefebvre, J. and Lennes, G. (1969), *Taxon,* **18,** 291–9.

Lye, K.A. (1966), *Nytt Mag. Bot.,* **13,** 87–133.

McGilchrist, C.A. and Trenbath, B.R. (1971), *Biometrics,* **27,** 659–71.

McIntosh, R.P. (1967), *Bot. Rev.,* **33,** 130–87.

Malme, L. (1978), *Norw. J. Bot.,* **25,** 271–9.

Mead, R. (1971), *J. Ecol.,* **59,** 215–9.

Merino, J. and Garcia Novo, F. (1975), *Anal. Inst. Bot. Cavanilles,* **32,** 521–36.

Moore, J.J. (1962), *J. Ecol.,* **50,** 761–9.

Morton, M.R. (1977), *Ecological Studies of Grassland Bryophytes.* Ph.D. thesis, University of London.

Nagano, I. (1969), *J. Hattori bot. Lab.,* **32,** 155–203.

Noy-Meir, I. (1973), *J. Ecol.,* **61,** 753–60.

Omura, M. and Hosokawa, T. (1959), *Mem. Fac. Sci. Kyushu Univ.* Ser. E., **1,** 51–63.

Orlóci, L. (1975), *Multivariate Analysis in Vegetation Research,* Junk, The Hague.

Pakarinen, P. (1976), *Ann. Bot. Fenn.,* **13,** 35–41.

Paton, J.A. (1965), *Census Catalogue of British Hepatics,* British Bryological Society, Ipswich.

Pentecost, A. (1980), *J. Ecol.,* **68,** 251–67.

Perring, F. (1959), *J. Ecol.,* **47,** 447–81.

Perring, F. (1960), *J. Ecol.,* **48,** 415–42.

Petit, E. and Symonds, F. (1974), *Bull. Jard. Bot. Natn. Belg.,* **44,** 219–39.

Phillips, E.A. (1951), *Ecol. Monogr.,* **21,** 301–16.

Pielou, E.C. (1969), *An Introduction to Mathematical Ecology,* Wiley, New York.

Pike, L.H., Denison, W.C., Tracy, D.M., Sherwood, M.A. and Rhoades, F.M. (1975), *Bryologist,* **78,** 389–402.

Poore, M.E.D. (1955a), *J. Ecol.,* **43,** 226–44.

Poore, M.E.D. (1955b), *J. Ecol.,* **43,** 245–69.

Poore, M.E.D. (1955c), *J. Ecol.,* **43,** 606–51.

Poore, M.E.D. (1956), *J. Ecol.,* **44,** 28–50.

Proctor, M.C.F. (1962), *Devonshire Assoc. Trans.,* **94,** 531–54.

Proctor, M.C.F. (1967), *J. Ecol.,* **55,** 119–35.

Proctor, M.C.F. (1974), *J. Bryol.,* **8,** 135–7.

Rahman, S.M.A. (1972), *J. Bryol.,* **7,** 169–79.

Rasmussen, L. (1975), *Lindbergia,* **3,** 15–38.

Rasmussen, L. and Hertig, J. (1977), *Rev. bryol. lichén.,* **43,** 207–17.

Redfearn, P.L. (1960), *Rev. bryol. lichén.,* **29,** 235–43.

Richards, P.W. (1938), *Ann. Bryol.,* **9,** 108–30.

Roux, M. and Salanon, R. (1974), *Bull. Soc. Bot. Fr.* (Coll. Bryologie), **121,** 213–24.

Rutter, A.J. (1955), *J. Ecol.,* **43,** 507–43.

Saeki, M., Kunii, K., Seki, T., Sugiyami, K., Suzuki, T. and Shishido, S. (1977), *Env. Res.,* **13,** 256–66.

Scott, G.A.M. (1966), *Proc. N.Z. Ecol. Soc.,* **13,** 8–11.

Scott, G.A.M. (1972), *N.Z. J. Bot.,* **9,** 744–9.

Sheard, J.W. (1968), *Bryologist,* **71,** 21–8.

Siedel, D. (1976), *Flora, Jena,* **165,** 139–62.

Slack, N.G. (1976), *J. Hattori. bot. Lab.,* **41,** 107–32.

Slack, N.G. (1977), *Bull. N.Y. St. Mus. Sci. Surv.,* **428,** 1–70.

Smith, A.J.E. and Hill, M.O. (1975), *J. Bryol.,* **8,** 423–33.

Smith, R.I.L. (1972), *British Antarctic Survey Scientific Reports,* **68.**

Smith, R.I.L. and Gimingham, C.H. (1976), *Bull. Br. Antarctic Surv.,* **43,** 25–47.

Southwood, T.R.E. (1966), *Ecological Methods: with particular reference to the study of insect populations,* Methuen, London.

Stotler, R.E. (1976), *Bryologist,* **79,** 1–15.

Stringer, P.W. and Stringer, M.H.L. (1974a), *Bryologist,* **77,** 1–16.

Stringer, P.W. and Stringer, M.H.L. (1974b), *Bryologist,* **77,** 551–60.

Swan, J.M.A. (1970), *Ecology,* **51,** 89–102.

Turkington, R. and Harper, J.L. (1979), *J. Ecol.,* **67,** 201–18.

Van Groenewoud, H. (1976), *J. Ecol.,* **64,** 837–47.

Vitt, D.H. and Slack, N.G. (1975), *Can. J. Bot.,* **53,** 332–59.

Watson, E.V.(1960a), *J. Ecol.,* **48,** 397–414.

Watson, E.V. (1960b), *Trans. Proc. Bot. Soc. Edinb.,* **39,** 85–106.

Watson, M.A. (1979), *Ecology,* **60,** 988–97.
Whittaker, R.H. (1962), *Bot. Rev.,* **28,** 1–239.
Whittaker, R.H. (1965), *Science,* **147,** 250–60.
Whittaker, R.H. (1967), *Biol. Rev.,* **49,** 207–64.
Whittaker, R.H. (1973a), *Handbook of Vegetation Science. V. Classification and Ordination of Communities,* Junk, The Hague.
Whittaker, R.H. (1973b), Direct gradient analysis: techniques. In: *Ordination and Classification of Communities* (Whittaker, R.H., ed.), pp. 7–31. Junk, The Hague.
Williams, W.T. and Lambert, J.M. (1959), *J. Ecol.,* **47,** 83–101.
Williams, W.T. and Lambert, J.M. (1960), *J. Ecol.,* **48,** 689–710.
Williams, W.T. and Lambert, J.M. (1961), *J. Ecol.,* **49,** 717–29.
Williams, W.T., Lambert, J.M. and Lance, G.N. (1966), *J. Ecol.,* **54,** 427–45.
Williamson, M.H. (1978), *J. Ecol.,* **66,** 911–20.
Winner, W.E. and Bewley, J.D. (1978), *Oecologia (Berl.),* **35,** 221–30.
Wit, C.T. de (1960), *Versl. Landbouwk. Onderz. Ned.,* **66,** 1–82.
Yarranton, G.A. (1962), *Rev. bryol. lichén.,* **31,** 168–86.
Yarranton, G.A. (1966), *J. Ecol.,* **54,** 229–37.
Yarranton, G.A. (1967a), *Can. J. Bot.,* **45,** 93–115.
Yarranton, G.A. (1967b), *Can. J. Bot.,* **45,** 229–47.
Yarranton, G.A. (1967c), *Can. J. Bot.,* **45,** 249–58.
Yarranton, G.A. (1967d), *Lichenologist,* **3,** 392–408.
Yarranton, G.A. (1969), *J. Ecol.,* **57,** 245–50.
Yarranton, G.A. (1970), *Can. J. Bot.,* **48,** 1387–404.
Yarranton, G.A. and Beasleigh, W.J. (1968), *Can. J. Bot.,* **46,** 1591–99.
Yarranton, G.A. and Beasleigh, W.J. (1969), *Can. J. Bot.,* **47,** 959–94.
Yarranton, G.A. and Green, W.G.E. (1966), *Bryologist,* **69,** 450–61.

Chapter 2

Life-forms of Bryophytes

KARL MÄGDEFRAU

It is no use trying to define an organic structure in conceptual terms, nature does not create organs on this basis. All we can do is to try to provide a useful nomenclature for what exists in nature.

Translated from K. Goebel (1883)

2.1 INTRODUCTION

Unlike most of the higher plants, bryophytes are not found as single individuals but in groups of individuals which have characteristic features depending on their family, genus or species. These features enable the experienced bryologist to identify many genera and often even species from quite a distance. The shape and structure of the groups of individuals are determined by two different factors. The protonema produced from a spore forms one to several buds, each of which grows to become an 'individual'. The individuals are thus at the very outset part of an assemblage. Each individual (moss shoot) has a genetically fixed method of ramification, depending on species, genus or family, a particular 'growth-form' in the narrower meaning of the term. Assemblage of individuals and growth-form, modified by external conditions, together provide the characteristics which can be described as the 'life-form'. In the following pages we hope to show that the life-forms of bryophytes are very different in several plant formations and that this diversity is closely related to their life conditions.

2.2 BRYOPHYTE SOCIETIES

The plant formations of the world and the plant societies subordinate to them are made up almost exclusively of vascular plants. Only in tundra and on bogland do mosses grow so vigorously that they form an important part of the physiognomy of these formations. When describing and analysing plant societies mosses are therefore usually passed over or at most briefly mentioned as the lowest level ('moss level').

Bryophytes, like lichens, differ from vascular plants not only in height (vascular plants 10^2–10^4 cm, mosses and lichens 10^{-1}–10^2 cm) but also in important life processes (absorption of water and nutrients, water conduction, poikilohydry, resistance to dryness, heat and cold, position of compensation point, etc.). Bryophytes and lichens therefore form their own societies within vascular plant societies, 'a world on its own', as Goebel (1930) pertinently says.

Schimper, Lorentz, Molendo and Pfeffer established more than a hundred years ago that the various bryophyte species prefer certain substrates or even inhabit them exclusively, and that particular species are found together again and again on the different substrates, e.g. forest floor, bark, wood, siliceous rock, limestone (for more details see Mägdefrau, 1975).

Only a few years after Warming (1896) had coined the term 'plant community' ('Pflanzenverein'), Loeske (1901) applied this term to mosses, 'moss community' ('Moosverein'). Soon afterwards Quelle (1902) introduced the expression 'moss society' ('Moosgesellschaft'), which we shall use in the following pages. 'Bryo-coenology' was extended and given a more solid basis by Herzog, Hesselbo, Gams, Barkman, Koppe, Poelt, Philippi, etc. (see Mägdefrau, 1975). All recent publications, however, relate exclusively to the north temperate zone, whereas in the tropics, apart from Herzog's (1916) excellent description of the 'moss formations' of Bolivia and a few papers by P.W. Richards, we have still hardly gone beyond making lists of species.

The following summary of bryophyte societies will suffice for present purposes:

1 Forest societies
   (a) Societies on shady rocks (scio-epilithic)
   (b) Societies on the bark of living forest trees (scio-epiphytic)
   (c) Societies on leaves (epiphyllous)
   (d) Societies on decayed wood (epixylous)
   (e) Societies on the forest floor
2 Societies on free-standing trees (photo-epiphytic)
3 Societies on free-standing rocks (photo-epilithic)
4 Societies on bogland (fens, high moors)
5 Societies on open mineral soil
6 Societies in water (still water, watercourses and waterfalls)

2.3 THE LIFE-FORMS OF BRYOPHYTES

As we have already shown, the characteristics of bryophytes, the 'life-forms', consist of two components, the growth-form and the assemblage of individuals: both can be considerably modified by external factors.

2.3.1 Growth-form

The term 'growth-form' is defined by Meusel (1935) as the 'overall character of a plant that can only be determined by detailed morphological analysis'. Buchloh (1951) speaks in the same terms: 'The growth-forms are fundamentally established in the organization itself.' 'Growth-form' is thus a purely morphological term and therefore refers only to homologous structures. 'The morphologist places great value on situation and succession: to him function and adaptation are of secondary importance' (Wijk, 1932).

The terms 'acrocarpous' and 'pleurocarpous', introduced by Bridel in 1826 into the descriptive morphology of mosses, and which were used by Schimper (1860) to indicate basic systematic characteristics, relate not only to the position of the sporophyte on the gametophyte, but also cover variations in behaviour of the growth direction of the gametophyte. They are therefore also very important in any classification of growth-forms, as Meusel has shown.

In the case of mosses we can distinguish, by reference to Meusel, the following growth-forms:

(i) Orthotropic mosses. The stems stand up vertically from the substrate, gametangia positions and sporogonia are acrocarpous. Fertile plants continue their growth by lateral innovation.

(a) *Protonema mosses* (sensu latiore). The protonema produces short, orthotropic shoots, bearing gametangia, which wither after ripening of the sporogonium (comparable to hapaxanthous flowering plants) and are almost always annuals. Examples are the genera *Buxbaumia, Diphyscium* and, as extremes, *Ephemeropsis* and *Viridivellus.* Genera in which the caulonema is perennial and produces new buds and stems every year, e.g. *Ephemerum, Phascum, Pleuridium, Physcomitrium,* also belong here.

(b) *Turf mosses.* The sporogonium-bearing shoot forms new, orthotropic lateral shoots, which again bear archegonia and thus sporophytes. Innovation can be either basitonous (on the lower part of the parent shoot) or acrotonous (on the upper part of the parent shoot). Moss turfs of various densities and heights are therefore produced.

(ii) Plagiotropic mosses. The shoots lie more or less close to the substrate and can be differentiated into main and lateral shoots. Gametangia are borne on short lateral shoots (pleurocarpous).

(a) *Thread mosses.* There is little difference between main and lateral shoots. This growth-form is found in most of the Leskeaceae and in many *Amblystegium* and *Oxyrrhynchium* species.
(b) *Comb mosses.* This strong main shoot, lying close to or climbing over the substrate bears many simple or branched lateral shoots. Most Hypnaceae, Brachytheciaceae, Thuidiaceae, Amblystegiaceae, Meteoriaceae belong to this type.
(c) *Creeping-shoot mosses.* From creeping, rhizome-like main shoots branches arise, upright or standing away from the substrate. They are either single to slightly branched *(Leucodon, Antitrichia, Prionodon)* or simple to very branched in one plane (Neckeraceae, *Hypopterygium*) or branched like trees (Climaciaceae, Hypnodendraceae).

Of the Hepaticae the foliose orders (Sphaerocarpales, Jungermanniales) can be classified under the growth forms mentioned above (see Buchloh, 1951). Examples of orthotropic liverworts are: *Takakia, Haplomitrium*; of thread forms: many Lophocoleaceae, Lejeuneaceae; of comb forms: *Madotheca, Frullania, Lepidozia, Trichocolea,* and many others; of creeping forms: *Bryopteris, Lembidium.* The Hepaticae however contain another type (thalloid: Marchantiaceae, Metzgeriaceae) not present in the Musci.

2.3.2 The assemblage of individuals

It is rarely that we encounter bryophytes, as we do most flowering plants, as single individuals, e.g. *Buxbaumia.* Almost without exception a large number of individual shoots or thalli join together to form a species-specific or genus-specific 'assemblage of individuals'. This can come about by numerous buds developing on a protonema, by further shoots arising on subterranean rhizoid cords or a large number of spores germinating close to one another. These assemblages of individuals are of particular importance to various bryophyte life processes, particularly to their water economy (see Mägdefrau, 1935; Mägdefrau and Wutz, 1951).

2.3.3 The influence of external features

The appearance of the bryophytes as encountered in nature is influenced in a striking way by external factors, particularly by water and light. Indications to this effect are to be found even in the older literature, e.g. in *Bryologia Europapaea* by Bruch, Schimper and Gümbel (1836–66) and in Schimper (1864, published 1908). Strangely enough, only a few authors (e.g. Němec, Goebel, Fitting, Buch) have tried to examine these problems from the experimental angle. In the last section (p. 56) we shall again have something to say about the influence of external factors.

2.3.4 The life forms

By 'life-form' we mean, as did Warming (1896) who coined this term, the habit of a plant in harmony with its life conditions: this is made up of the components we have already mentioned, i.e. growth-form, assemblage of individuals and the influence of external factors.

The botanical literature does not seem to be very clear about the term *life-form*. Whilst many authors, e.g. Drude (1913), use the expression in the same sense as Warming, others, e.g. Raunkiaer (1934), narrow it down to overwintering forms, a concept that has no meaning in the tropics. Giesenhagen (1910) and Herzog (1916) use the term growth-form in the sense of life-form, according to the definition above. This confusion, which is perpetuated in the most recent literature, can be avoided if we define 'growth-form' in Meusel's (1935) sense as a strictly morphological term based on homologies, and 'life-form', with Warming (1896), in the ecological sense based on analogies.

There is extensive literature on the life forms of higher plants, from Alexander von Humboldt (1806) to Warming (1896) and Drude (1913) to Troll (1959). We would particularly refer the reader to the two-volume work *Die Lebensformen (The Life Forms)* by Koepcke (1971–1974), which deals with this subject comprehensively, covering the whole plant and animal kingdom. Other references dealing with life-forms in bryophytes include Amann (1928), Birse (1957, 1958), Düll (1969–70), Frey and Probst (1973, 1974), Gimingham and Birse (1957), Hamilton (1953), LeBlanc (1962) and Richards (1932).

In the following paragraphs we shall try, following Giesenhagen (1910) and Herzog (1916), and on the basis of our own observations in the tropics, to characterize the important life forms of the bryophytes.

(i) Annuals (Einjährige): The gametophyte stops growing once it has produced gametangia and dies after the sporogonium has ripened, without having first produced any regenerative shoots (Figs. 2.1a, 2.2a). These are always pioneer mosses on open mineral soils.
Musci: *Buxbaumia, Diphyscium, Phascum, Ephemerum, Discelium, Aloina.*
Hepaticae: *Sphaerocarpus, Calobryum, Riccia.*

(ii) Short turfs (Kurzrasen): The short shoots, hardly more than 1 cm high, stand close together and grow on after ripening of the sporogonia by means of (mostly sparsely and acrotonous) regenerative shoots. More or less closed, often very spreading turfs are thus formed which last for scarcely more than a few years (Figs. 2.1b, 2.2b). They grow on open mineral soils and on rocks.
Musci: *Trichostomum, Didymodon, Barbula, Ceratodon.*
Hepaticae: *Gymnomitrion, Marsupella.*

Fig. 2.1 Life forms of mosses. a, Annuals: *i, Diphyscium foliosum, ii, Phascum curvicolle.* b, Short turf: *Trichostomum brachydontium.* c, Tall turf: *Dicranum undulatum.* d, Cushion: *Grimmia pulvinata.* e, Mat: *Hookeria lucens.* f, Weft: *Thuidium delicatulum.* g, Pendants: *i, Papillaria deppei, ii, Phyllogonium viscosum.* h, Tail: *Prionodon densus.* i, Fan: *Neckeropsis undulata.* j, Dendroids: *i, Hypnodendron dendroides, ii, Rhodobryum roseum.* a*i* and *ii*×4; b–j×2/3 (from Mägdefrau, 1969).

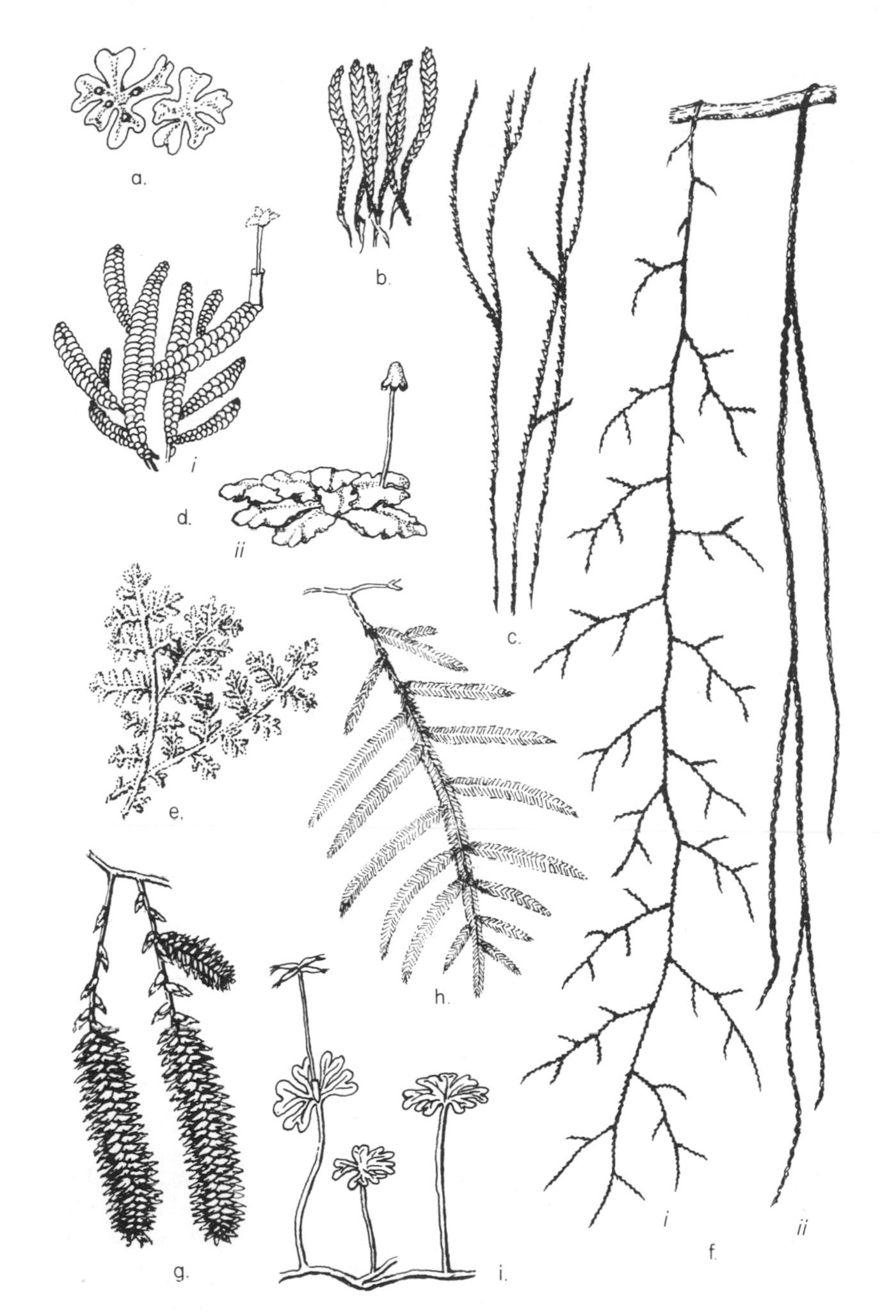

Fig. 2.2 Life forms of hepatics. a, Annual: *Riccia glauca*. b, Short turf: *Gymnomitrion coralloides*. c, Tall turf: *Herbertus sendtneri*. d, Mats: *i, Radula complanata, ii, Preissia quadrata*. e, Weft: *Trichocolea tomentella*. f, Pendants: *i, Frullania atrata, ii, Macrolejeunea pallescens*. g, Tail: *Schistochila appendiculata*. h, Fan: *Bryopteris fruticulosa*. i, Dendroid: *Hymnophytum flabellatum*. a, b, and d*i*×2; c and d*ii*×1; e–i×2/3.

(iii) Tall turfs (Hochrasen): The upright shoots, which are not branched or only slightly so, form turfs of considerable height (maximum up to 40 cm). The shoots grow on after gametangia formation or production of acrotonous regenerative shoots is continued (Figs. 2.1c, 2.2c). High turfs grow especially on forest floors in the temperate zones: they are able not only to hold water by capillary action but also to conduct it (see Mägdefrau, 1935, 1938).
Musci: Dicranaceae, Bartramiaceae, *Polytrichum, Dawsonia, Sphagnum, Tomentypnum, Drepanocladus* and *Rhytidiadelphus* species.
Hepaticae: *Herbertus, Isotachis, Plagiochila asplenioides* f. *major, Bazzania trilobata* f. *grandis.*

(iv) Cushions (Polster): Basal regenerative shoots are produced usually in considerable numbers on the upright shoots. The cushions therefore grow not only upwards but also extend sideways (Fig. 2.1d). If they are free-standing they are hemi-spherical in shape. If they are laterally inhibited (for instance in rock crevices), they become extraordinarily dense. Pleurocarpous mosses can also form cushions, their main axes remaining short and the lateral axes extending upwards *(Orthotrichum,* also *Meteorium, Leskea, Hypnum,* and *Brachythecium* species). They grow mainly on rocks and the bark of trees (photo-epilithic and photo-epiphytic) but also on the ground in the Arctic, Antarctic and in alpine regions of high mountains; here they are usually 'large cushions' ('Gross-Polster').
Musci: *Andreaea, Grimmia, Amphidium, Orthotrichum, Plagiopus.*
Hepaticae: no examples.

(v) Mats (Decken): Plagiotropic bryophytes, the main and lateral shoots of which lie close to the substrate and are attached to it by rhizoids (Figs. 2.1e, 2.2d). They grow on rocks and the bark of trees and, in the tropics, also on leaves (epiphyllous).
Musci: *Plagiothecium* species, *Homalothecium, Taxithelium.*
Hepaticae: *Lophocolea* and *Radula* species, Lejeuneaceae and most Marchantiaceae and Metzgeriaceae.

(vi) Wefts (Filze): Plagiotropic bryophytes, the main and lateral shoots of which grow loosely through one another and form a covering that is easy to lift from the substrate (forest floor, rotten trunks of trees), a new layer growing every year over that of the previous year (Figs. 2.1f, 2.2e). Together with high turfs, wefts form the main constituent of the mossy covering of the forest floors of temperate zones and are able to hold considerable quantities of rainwater by capillary action (Mägdefrau and Wutz, 1951).
Musci: many Hypnaceae, Brachytheciacae, Entodontaceae and Hylocomiaceae, *Thuidium* and *Rhacopilum* species.
Hepaticae: *Bazzania, Ptilidium, Trichocolea, Lepidozia.*

(vii) Pendants (Gehänge): Epiphytic, mostly plagiotropic bryophytes, the main shoots of which hang down from the branches and twigs of trees like beard mosses, whilst the lateral shoots remain short and stand out horizontally. Morphologically, the pendant mosses belong mainly to the same growth type as the comb mosses (Figs. 2.1g, 2.2f). The pendants grow most luxuriously in tropical cloud forests.
Musci: Meteoriaceae, Phyllogoniaceae, *Orthostichopsis, Pilotrichopsis, Calyptothecium* species.
Hepaticae: *Frullania* species (e.g. *F. atrata, F. convoluta, Macrolejeunea pallescens, Bryopteris flaccida, Mastigophora diclados, Schisma dicranum.*

(viii) Tails (Schweife): Bryophytes growing on trees and rocks, mostly shade-loving and of creeping habit. Their shoots stand out and are slightly branched or unbranched and usually radially leafed (Figs. 2.1h, 2.2.g).
Musci: *Prionodon, Leucodon, Spiridens, Cyathophorum, Lamprophyllum, Endotrichella.*
Hepaticae: *Schistochila* species, *Tylimanthus saccatus, Plagiochila* species (e.g. *P. bursata, P. nobilis*).

(ix) Fans (Wedel): Creeping mosses growing on a vertical base (trees, rocks), the shoots of which branch towards one another in the same plane, project horizontally to obliquely downwards and usually have flattened leaves (Figs. 2.1i, 2.2h).
Musci: *Thamnobryum, Echinodium,* Neckeraceae, Pterobryaceae (slightly branched: *Adelothecium, Lepidopilum).*
Hepaticae: *Bryopteris* and *Archilejeunea, Thysananthus* and *Plagiochila* sect. *Frondescentes* (e.g. *P. gigantea).*

(x) Dendroids (Bäumchen): Bryophytes growing on the ground, the negatively geotropic shoots of which bear at the top a tuft of large leaves or many lateral shoots. Their growth form groups them under the creepshoot mosses (Figs. 2.1j, 2.2i).
Musci: *Rhodobryum, Leucolepis, Climacium, Pleuroziopsis, Hypnodendron* (including *Mniodendron), Dendroligotrichum, Hypopterygium* species (transition to fan type).
Hepaticae: *Hymenophytum flabellatum, Pallavicinia connivens, P. crassifrons, Symphyogyna hymenophyllum, S. podophylla.*

These 10 life-forms cover only the important types of bryophytes, but by no means exhaust their diversity. It is often uncertain whether a moss or a moss assemblage should be classified under one or other life form. Sometimes one and the same type can include two different life forms, e.g.

Plagiomnium undulatum, the sterile shoots of which are fans, whilst the fertile shoots are classified under dendroids or *Bazzania trilobata* which, according to the environmental conditions, forms tall turfs or wefts.

This survey of life-forms does not take into account bryophytes living in water. The mosses found in ponds and lakes are mostly genera also found on land, e.g. *Sphagnum, Drepanocladus, Calliergon.* The growth-form does not change on transfer to water. The branches merely lengthen and the assemblage of individuals becomes looser. This also applies to many mosses in streams and waterfalls, e.g. *Cinclidotus.* The number of species which are strictly limited to water is very small. In quickly flowing waters there are two different life-forms, determined by the degree of adaptability to the stationary boundary layer (Prandtl layer) between rock and flowing water. The water flows over species which form dense cushions (e.g. *Brachythecium rivulare*) as the surface of the moss cushion is in the zone of the stationary boundary layer. The cushions are not very firmly fastened to the rock and can easily be removed. Loose moss assemblage, e.g. *Thamnobryum alopecurum* and *Platyhypnidium riparioides* project over the boundary layer into the quickly flowing water and are therefore very firmly attached to the rock (Schumacher, 1950). An interesting case is the hydroaerophytic moss *Hydropogon fontinaloides* on trees at the riversides of the Orinoco; it floats in the water like *Fontinalis* during the rainy season and hangs from the trees like *Usnea* during the dry season (Mägdefrau, 1973).

2.4 DISTRIBUTION OF LIFE-FORMS WITHIN BRYOPHYTE SOCIETIES

When considering bryophyte vegetation in different plant communities or in different habitats, considerable differences in the physiognomy of mosses are noticeable and this is due to the share the individual life-forms have in the bryophyte vegetation. Table 2.1 gives the percentage of the different life-forms in bryophyte vegetation for societies from the Arctic to the tropics (see also Horikawa and Ando, 1952). It is evident that annuals are limited to open mineral soils as found in fallow fields and sporadically also in the steppes and in pine forests. Short turfs are at home in the same habitats and on sunny rocks, whilst tall turfs are typical of damp to wet habitats (bogland, tundra, spruce forest floor). Cushions are nearly always found on free-standing rocks and trees (for the Páramo cushions, see Section 2.5), whereas wefts are decidedly shade-lovers (forest floors, roots of trees, rotten tree trunks, shady rocks) and so also are tails and fans. The epiphyllous bryophyte vegetation of moist tropical forests consists almost entirely of mats, the rhizoids of which often form special organs of attachment (Winkler, 1967). Pendants are a striking decoration in tropical cloud forests. The rare dendroid type prefers moist, shady habitats.

Table 2.1 Life forms of mosses in several plant communities (From Herzog, Loeske, Koppe, Poelt and own observations; for references see Mägdefrau, 1975).

	Number of species	Annuals	Short turfs	High turfs	Cushions	Mats	Wefts	Pendants	Tails	Fans	Dendroids
Shady limestone rocks	25		8	2		3	7		3	2	
Mountain trees	22		1			11	3		2	5	
Trees of cloud forests	30		2			1	4	10	7	6	
Forest soil (*Picea* forest)	35		6	8	1	4	15				1
Forest soil (*Pinus* forest)	52	3	25	5	1	6	12				
Exposed trees	18		2		9	5	1		1		
Exposed limestone rocks	13		9		2	2					
Exposed silicate rocks	13			1	9		3				
Steppe heath	50	10	28	1		6	5				
Fens	23			17			5				1
Páramos of Andes	13			2	7		4				
Moist tundra	25			21			4				
Mineral soil (fallow land)	22	8	10			3	1				

2.5 DEPENDENCY OF LIFE-FORMS ON ENVIRONMENTAL CONDITIONS

As in flowering plants, the life-forms of bryophytes are closely connected with two environmental factors: light and water.

Light inhibits elongation of the axes. The predominance of short turfs and cushions on brightly lit biotopes is directly connected with this fact. The presence hyaline hair-points (*Grimmia, Racomitrium*), which give these mosses a striking greyish appearance, is limited entirely to these habitats. The tails and fans however require a dim light. The fans hold their surfaces vertically to the direction of the light, i.e. horizontally to obliquely downwards, which is a phenomenon particularly noticeable in tropical rain forests and cloud forests.

The external, capillary conduction of water (Mägdefrau, 1935, 1938; Stocker, 1956) is important to those mosses that have sufficient quantities of soil water available to them, i.e. in bogland and in moist tundra. Here the life-form of the tall turfs predominates: because of their crowded shoots, dense foliage and the frequently occurring rhizoid weft these mosses show particularly high values for capillary water conduction. In the case of mats and wefts and also tails and fans capillary conduction plays no part in the natural habitat but capillary retention of water does, and the period of full activity, particularly of photosynthesis, is therefore extended considerably beyond the period of precipitation. This also applies to the short turfs and cushions in exposed habitats, although much less time is gained here than in the case of bryophytes growing on the soil.

Table 2.1 shows the large number of cushion mosses that are to be found in the Páramos of the tropical Andes and in the Arctic and Antarctic regions (e.g. Spitzbergen, Kerguelen, Macquarie Island). These are not the small cushions characteristic of *Grimmia* and *Andreaea* species but cushions of considerable height and spread. They are sometimes so firm (e.g. *Campylopus argyrocaulon*) that when they are trodden on the foot does not sink into them. Similar cushion forms are also found amongst the flowering plants of the Páramos (e.g. *Plantago rigida*). The ecological importance of these Páramo cushions is as yet unclear.

'Moss balls' ('Kugel-Moose') are a special form of cushion (in Iceland they are called 'Jöklamys' = glacier mice). They are cushions of *Andreaea* and *Grimmia* species which have been levered up from their substrate, probably due to the effects of frost, lie on the soil, and rolled hither and thither by the wind and therefore grow radially on all sides (a publication about moss balls by Beck and Mägdefrau is in preparation). Moss balls of this kind have been found on Kerguelen, where they were discovered in 1875, on Mount Kenya, in Iceland, on Jan Mayen Island, Amchitka Island and Norwegian mountains.

The striking habit of pendulous mosses, which are limited almost entirely to tropical cloud forests, is due to the action of water. The enormous elongation of the main axis is evidently determined by the growing tip being continually moistened by falling water and thus growing without interruption. Observations on European mosses close to waterfalls, (e.g. in the case of *Thuidium tamariscinum, Plagiothecium undulatum, Plagiomnium, Brachythecium rivulare, Rhytidiadelphus triquetrus*) support this explanation. Pendulous mosses grow on their substrate (twigs, leaves) at first as loose turfs and do not assume a hanging form until they reach the edge of their base (see Herzog, 1926, Figs. 31 and 60; Giesenhagen, 1910, Fig. 9). During (1979) has pointed out the relations between the life-forms and the life history ('life strategy') of bryophytes.

REFERENCES

Amann, J. (1928), *Bryogéographie de la Suisse,* Fretz Frères, Zurich.
Birse, E.M. (1957), *J. Ecol.,* **45,** 721–33.
Birse, E.M. (1958), *J. Ecol.,* **46,** 9–27.
Bridel, S.E. (1826–27), *Bryologia universa,* Barth, Leipzig.
Bruch, P., Schimper, W.P. and Gümbel, T. (1836–66), *Bryologia Europaea,* Schweizerbart, Stuttgart.
Buchloh, G. (1951), *Sitz. ber. Akad. Wiss. Heidelberg, math.-nat. Kl., Jg.,* **1951,** 211–79.
Drude, O. (1913), *Die Ökologie der Pflanzen,* Vieweg, Braunschweig.
Düll, R. (1969–70), *Mitteil. bad. Landesver. Naturk. Natursch.,* N.F. **10,** 39–138, 301–29.
During, H.J. (1979), *Lindbergia,* **5,** 2–18.
Frey, W. and Probst, W. (1973), *Bot. Jb. Syst.,* **93,** 404–23.
Frey, W. and Probst, W. (1974), *Bot. Jb. Syst.,* **94,** 267–82.
Giesenhagen, K. (1910), *Ann. Jard. bot. Buitenzorg,* **2.** Ser., Suppl. **3,** 711–89.
Gimingham, G.H. and Birse, E.M. (1957), *J. Ecol.,* **45,** 533–45.
Goebel, K. (1883), Vergleichende Entwicklungsgeschichte der Pflanzenorgane. In: *Handbuch der Botanik,* (Schenk, A., ed.), Vol. 3, pp. 99–432. Friedländer, Berlin.
Goebel, K. (1930), *Organographie der Pflanzen,* 2. Teil. 3, Aufl. Fischer, Jena.
Hamilton, E.S. (1953), *Bull. Torrey Bot. Club,* **80,** 264–72.
Herzog, T. (1916), *Bibliotheca bot.,* Heft **87.**
Herzog, T. (1926), *Geographie der Moose,* Fischer, Jena.
Horikawa, Y. and Ando, H. (1952), *Hikobia,* **1,** 113–29.
Humboldt, A. von. (1806), *Ideen zu einer Physiognomik der Gewächse,* Cotta, Tübingen.
Koepcke, W. (1971–74), *Die Lebensformen,* Bd. I–II. Goecke & Evers, Krefeld.
LeBlanc, F. (1962), *Can. J. Bot.* **40,** 1427–38.
Loeske, L. (1901), *Verh. bot. Ver. Prov. Brandenburg,* **22,** 75–164.

Mägdefrau, K. (1935), *Zeitschr. Bot.*, **29**, 337–75.
Mägdefrau, K. (1938), *Ann. Bryol.*, **10**, 141–50.
Mägdefrau, K. (1969), *Vegetatio*, **16**, 285–97.
Mägdefrau, K. (1973), *Herzogia*, **3**, 141–9.
Mägdefrau, K. (1975), *Acta hist. Leopold.*, **9**, 95–111.
Mägdefrau, K. and Wutz, A. (1951), *Forstwiss. Cbl.*, **70**, 103–17.
Meusel, H. (1935), *Nova Acta Leopold.* N.F., **3**, 123–277.
Quelle, F. (1902), *Göttingens Moosvegetation*, Eberhardt, Nordhausen.
Raunkiaer, C. (1934), *The Life Forms of Plants and Statistical Plant Geography*, Clarendon Press, Oxford.
Richards, P.W. (1932), Ecology. In: *Manual of Bryology*, (Verdoorn, F., ed.), pp. 367–95, Nijhoff, The Hague.
Schimper, K. (1908), Beih. bot. Cbl., **24**, II, 53–66.
Schimper, W.P. (1860), *Synopsis muscorum Europaeorum*, Schweizerbart, Stuttgart.
Schumacher, A. (1950), *Aus der Heimat*, **58**, 19–22.
Stocker, O. (1956), Wasseraufnahme, Wasserleitung und Transpiration der Thallophyten. In: *Handbuch der Pflanzenphysiologie*, (Ruhland, W., ed.), Vol. 3, pp. 1–172, 312–23, 514–21, Springer, Berlin.
Troll, C. (1959), *Jahresber. Ges. Fr. Univ. Bonn, Jg.*, **1958**, 1–75.
Warming, E. (1896), *Lehrbuch der ökologischen Pflanzengeographie*, Borntraeger, Berlin.
Wijk, R. van der (1932), Morphologie und Anatomie der Musci. In: *Manual of Bryology*, (Verdoorn, F., ed.), pp. 1–40, Nijhoff, The Hague.
Winkler, S. (1967), *Rev. bryol. lichénol.*, **35**, 303–69.

Chapter 3

Tropical Forest Bryophytes

T. PÓCS

3.1 INTRODUCTION

The tropical forests were known for a long period simply as an inexhaustible El Dorado of new bryophyte species and we did not know much about their ecology. Even in the 1950s Richards (1954), a known authority on bryophyte ecology, complained about the scantiness of our knowledge about the bryophyte communities in the tropical rain forests.

Papers on the subject include those of Massart (1898), Busse (1905), Pessin (1922) and Allorge *et al.* (1938) on epiphyllous communities; the classic work of Giesenhagen (1910) on the bryophyte types in rain forests; the studies of Seifriz (1924) on the altitudinal distribution of Javanese moss communities; Herzog's *Bryogeography* (1926) dealing briefly with rain forest bryovegetation, chapters in the *Manual of Bryology* devoted to the cenology (Gams, 1932) and to the ecology of bryophytes (Richards, 1932); the detailed study on the water relations of bryoepiphytes of Java by Renner (1933), and the various papers of Richards (1935, 1954), Stehlé (1943), Jovet-Ast (1949), Giacomini and Ciferri (1950), Horikawa (1950) and those of Thorold (1952, 1955) dealing with tropical and subtropical bryophyte communities.

From the 1950s onwards, the situation improved somewhat. A series of papers dealing with the epiphytic bryophyte communities in the subtropical forests of Japan are worth mentioning (e.g. Hattori and Kanno, 1956; Hattori *et al.*, 1956; Iwatsuki and Hattori, 1956, 1959, 1968, 1970 and the summarizing paper of Iwatsuki, 1960). In tropical Asia the most important work on the epiphytic communities is that of Tixier (1966a), which deals in detail with the composition of bryophyte synusia of the different vegetation belts of South Vietnam. In tropical America, Winkler (1967, 1971) studied the ecology of epiphyllous communities in upland and lowland rain forests. Fulford *et al.* (1970) described the liverwort communities of an elfin forest in Puerto Rico. Griffin *et al.* (1974) and Griffin (1979) dealt with altimontane bryophyte vegetation, while Steere (1970) observed the effect of ionizing caesium radiation on bryophyte communities in Puerto Rico. In Africa, the book of Hedberg (1964) on the afroalpine phanerogam vegetation contains data on bryophytes, and the monographic study of Johansson (1974) is worthy of mention. Although he deals only with the vascular epiphytic communities of the Nimba Mountains (West Africa), he discusses many general questions concerning the ecology of rain forest epiphytes. Petit and Symons (1974) analyse the bryophyte communities of forest plantations in Burundi. The foliicolous communities of Africa, Madeira and the Azores have been studied by Sjörgren (1975, 1978), in tropical West Africa by Olarinmoye (1974, 1975a, b) and in East Africa by Pócs (1974, 1978). The latter author also studied the role of bryophytes in different afromontane plant communities with special reference to water balance and humus accumulation in rain forests (Pócs, 1974, 1976a, b, c, 1980).

In the following account I will attempt to characterize the bryophyte communities of tropical forests and some adjoining vegetation types, illustrating it with examples based on my experiences in Vietnam (1963, 1965–66), East Africa (mostly in Tanzania) (1969–73, 1976) and Cuba (1978, 1979).

3.2 THE TROPICAL RAIN FOREST

3.2.1 The tropical rain forest as a bryophyte environment

The tropical rain forests (in the sense of Richards, 1932, 1957), which occupy or occupied large areas of the lower belt of the humid tropics (up to an altitude of 1500 m near the Equator, up to 600–700 m within the tropics of Capricorn and Cancer) are characterized in general by the complex structure of the canopy composed of tall, emergent trees (up to 50–60 m high), by a dominating, moderately high, closed layer ($\sim$25 m) and by many smaller trees (often palms, *Pandanus* and other pachycaulous types). The trees and shrubs are mostly evergreen – in seasonal rain forests the emergent trees may be deciduous. The canopy of tropical rain forests is seldom homogeneous but is usually composed of a large number of woody species. Lianas and epiphytes fill the gaps among tree crowns and branches and the biomass of the community is greater than that of temperate forests (5–600 000 kg ha^{-1} dry matter according to Rodin and Bazilivich, 1966). Climatic conditions comprise an average temperature higher than 20°C (up to 27°C) and an annual rainfall in excess of 1500 mm (up to 6000–8000 mm). Both climatic elements show relatively low seasonal fluctuations.

Fig. 3.1 Bryophyte microhabitats in tropical rain forests. 1, Bases of large trees; 2, upper parts of trunks; 3, macro-epiphyte nests; 4, bark of main branches; 5a, terminal twigs and leaves; 5b, bark of lianas, shrub branches and thin trunks; 6, *Pandanus* stems; 7, tree fern stems; 8, palm trunks and basal prop roots; 9, rotting logs and decaying wood; 10, soil surface and termite mounds; 11, roadside banks and cuttings; 12a, rocks and stones; 12b, submerged or emergent rocks in streams.

The rainy period is not interrupted by a dry season or seasons longer than 2–4 months (seasonal rain forests). Exceptionally, the dry season may be as long as 5 months in some seasonal rain forests (Longman and Jenik, 1974) if it is compensated for by 5000 mm or more rainfall during other parts of the year.

Tropical rain forests offer a large variety of habitats for bryophytes, because of their diverse microclimate and the variety of substrata available for small plants (Richards, 1954). The microhabitats of bryophyte communities (Fig. 3.1) are determined by the different amounts of direct or indirect light and heat radiation and by the availability of water and nutrients from direct rainfall, throughfall, stemflow, mist and dew and in the form of ground water. All of these are influenced by the relative humidity and by the physical and chemical character of substrata such as roots, stems and leaves of living or dead plants, underlying rock and natural or artificially disturbed soil surfaces. In general, the tropical rain forests are characterized by high temperatures, always adequate for bryophyte growth, and by a water supply from a variety of sources. In the interior of the rain forest water is continually present in adequate amounts accompanied by high relative humidity but near the upper surface of the canopy water is often deficient. The distribution of light shows a very uneven pattern in rain forests both vertically and horizontally and is often a limiting factor of bryophyte vegetation.

3.2.2 Microhabitats of bryophyte synusia in tropical rain forests

(a) Bases of large trees

Tree bases in tropical rain forest represent the most shady habitat for epiphytes, usually combined with the highest degree of humidity, especially where there are buttresses (see Fig. 3.2a, b) which sometimes reach as high as 8–10 m up the trunk. The large surface provides very suitable niches for corticolous epiphytic bryophytes (Fig. 3.2b), often associated with filmy ferns (Hymenophyllaceae). The same applies to trees with stilt roots and to the anastomosing aerial root systems of strangling hemiepiphytes, which develop into independent tree stems. Even on the smooth bases of large trees and on the top of main roots protruding above the soil surface there develops a similar sciophilous-hygrophilous bryophyte synusium. How far this synusium grows upwards on the trunk usually depends on humidity and the physical conditions of the bark. Under extremely wet conditions it might reach to the base of main branches, but usually it is restricted to the lowest 1–3 m and may often be absent, or poorly developed. In lowland rain forests, as indicated by Richards (1954) this synusium is the most luxuriously developed bryophyte community. Throughout the tropics, the initial phase of this community is

Fig. 3.2 (a) Buttressed base of a *Dracontomelon duperrianum* (Anacardiaceae) tree in the tropical rain forests of Vietnam with bryosynusia in its hollows.
(b) *Leucophanes glaucescens* between the buttresses of a *Pterospermum lancaefolium* (Sterculiaceae) tree in Vietnamese lowland rain forest.
(c) Characteristic upper trunk synusium with appressed mats of *Brachiolejeunea corticalis* (Lejeuneaceae/Ptychanthoideae) among crustaceous lichen thalli with orchid seedlings, Cuba. (d) Community with feathery bryophytes and with epiphytic orchids on the terminal twigs of lowland rain forest trees in Vietnam, Cuc Phüöng National Park.

characterized by the different mat- and weft-forming *Thuidium* species (*Thuidietum glaucini* associule of Iwatsuki (1960) in Japan; *T. glaucinoides, T. bonianum* in Vietnam; *T. gratum, T. tamariscellum* in Cuba; *T. involvens* in East Africa) in which minuscule turfs or solitary specimens of *Fissidens* species are intermixed with appressed mats of Lejeuneaceae. On more porous bark with a decomposing surface many *Leucobryum* species, *Leucophanes* (Fig. 3.2b) and *Calymperes* species occur. Indicators of the very shady, wet habitats of tree bases are the dendroid or pendulous Neckeraceae and Pterobryaceae species (in Asia many *Homaliodendron* and *Neckeropsis* species; in Africa *Porotrichum, Renauldia, Hildebrandtiella*; in America the ubiquitous *Neckeropsis disticha* and *N. undulata*). *Pinnatella* and *Lopidium* species are present on all continents. *Sematophyllum* and *Taxithelium* species are also common in this synusium. I observed about 100 species in Vietnam, 60 in East Africa and 50 in Cuba in the same habitat. Similar corticolous communities from rain forest tree bases are described by Richards (1954) from Guiana, Iwatsuki (1960) from southern Japan and Tixier (1966) from South Vietnam.

(b) The upper part of tree trunks

Above the hygrophilous-sciophilous tree base the trunks of rain forest trees offer a still relatively shady but much drier habitat (Fig. 3.2c). This part of the trunk is either bare or occupied by crustose lichens and by smaller or larger patches of appressed hepatics such as many of the *Frullania* and Lejeuneaceae (mainly the less hygrophilous Ptychanthoideae), smooth mat-forming or thread-like Sematophyllaceae, Hypnaceae and *Mitthyridium* species. Nearer to the ground, in more humid places, species of short turf-forming Dicranaceae (e.g. *Leucoloma*), *Calymperes* and feather-form or pendulous Pterobryaceae and Meteoriaceae occur. Especially typical of this habitat are some Pterobryaceae such as *Garovaglia* species in tropical Asia, *Jaegerina* species in Africa and *Pireella (Pterobyron)* species both in Africa and America. The dense mats of Macromitrioideae with creeping radial branches occur in this habitat and on branches as well.

(c) Branches and twigs

The branching system in the canopy of tropical rain forests is the first layer to receive precipitation but, at the same time, is the layer most exposed to direct radiation and to temporary desiccation. These environmental conditions are reflected in their macroepiphyte vegetation which often consists of succulent or at least partly water-absorbing and storing plants (epiphytic Cactaceae, succulent Polypodiaceae, Apocynaceae and many Orchidaceae with pseudobulbs). The relatively extreme water relations are partly compensated for by very favourable light conditions. The macroepiphytes mostly inhabit the main branches whilst the microepiphytes occur both on these and on the fully exposed terminal twigs. On the latter are found the specially adapted ramicolous microepiphytes, where the thinness of twigs is also a determining factor, adhering with their primary stems to the tiny twigs, while their secondary stems hang down or form pinnate branching systems. The ramicolous microepiphytes may be divided into three main groups:

(1) Those which inhabit the corticosphaera of the main branches. They do not differ much from the microepiphytes on the trunk but are exposed to more light and are therefore helio-xerophytes and include different dense mat-forming Macromitrioideae, Cryphaeaceae, Erpodiaceae, many Sematophyllaceae, appressed mats of *Frullania* and Lejeuneaceae throughout the tropics, *Erythrodontium* species in Asia and Africa and *Helicophyllum* in tropical America.

(2) Those which live on the terminal twigs, forming a dense rain absorbing system of feathery (e.g. *Ptychanthus striatus, Plagiochila, Bryopteris* species) or hanging mosses and liverworts (Meteoriaceae, Neckeraceae, Lejeuneaceae, *Frullania* subg. *Meteoriopsis*). This rain-absorbing, helio-hygrophilous, microepiphyte synusium develops in the high canopy only in the most humid types of tropical rain forest. In the seasonal and semi-decicuous types it develops only in the crowns of the second tree layer (Fig. 3.2d).

(3) Those which live in the lowest layers of the forest on thin trunks, on lianas and on stems of shrubs. This ramicolous synusium consists of scio-hygrophilous hanging or rough mat-forming species of *Plagiochila,* Meteoriaceae, Pterobryaceae, Neckeraceae, Phyllogoniaceae, Lejeuneaceae (Fig. 3.4d). The stem-embracing Rutenbergiaceae of East Africa (Fig. 3.4a) and Garovaglioideae in Asia-Oceania are especially adapted to this habitat.

The ramicolous synusia of tropical rain forests are described in detail: from Martinique and Guadeloupe by Jovet-Ast (1949) from Guiana by Richards (1954), from South Vietnam by Tixier (1966a). This synusium is typically rich in liverworts, especially Lejeuneaceae and *Frullania* species, in contrast to the trunk base synusium which is dominated by mosses.

(d) Epiphyllous communities

In the really wet types of tropical rain forests, where the air is continually water-saturated, epiphytic bryophytes occur not only on the bark of trees and shrubs but also on the surface of their living leaves. These are the epiphyllous bryophytes. Some lichens also occur on the lower surfaces of living leaves and are said to be hypophyllous, while the two groups together form the foliicolous epiphytes.

The epiphyllous life-form is most typical of the tropical rain forests. Even so they are not found everywhere, usually being restricted to the leaves of young trees, shrubs and of long-lived herbs of the lowest layer of forest, up to a height of 2–3 m and near streams. In the most humid types of rain forests they also occur in the canopy. According to Olarinmoye (1974) their growth rate is closely correlated with atmospheric humidity.

Epiphyllous liverworts occur in extratropical regions only amongst the most oceanic climatic conditions, e.g. in Japan (Kamimura, 1939), in China (Ch'en and Wu, 1964), in the southern Appalachians (Schuster, 1959), in the Macronesian islands (Allorge *et al.*, 1938; Sjörgren, 1975, 1978), in the southern Caucasus Mountains. (Pócs, unpublished), and even as far North as British Columbia (Vitt *et al.*, 1973).

Of equal importance with climatic factors is the nature of the leaves upon which epiphyllae occur. They must have a relatively long lifespan and, therefore, epiphyllae practically never occur on the leaves of deciduous trees or shrubs. They prefer evergreen, leathery or papyraceous (e.g. palm) leaves (Fig. 3.3a) but are often found on ferns, on leaves of monocot herbs, e.g. from the ginger family. They do not avoid hairy leaves, but leaves with 'unwettable', waxy surfaces are unsuitable (Richards, 1932). Some epiphyllous species show a preference for hard leaves of *Podocarpus* or *Encephalartos,* e.g. *Diplasiolejeunea* and *Leucolejeunea* species, whilst tiny *Aphanolejeunea, Cololejeunea* and Hookerioid mosses usually occur on filmy ferns (Pócs, 1978).

Although the high relative humidity that epiphyllous bryophytes require is often associated with shady localities this does not mean that they avoid light. In contrast, they never occur on the dark (and dry) lower leaf surface and, if humidity permits, they occur in open places such as cocoa or *Citrus* plantations, at streamsides or at the forest edge. Their water and nutrient supply is derived from precipitation. Rainfall in the tropics often contains more nutrients than in temperate regions (Jones, 1960) and rainwater falling through the canopy is enriched by exudates of leaves, excrement of caterpillars, etc. (Szabó and Csortos, 1975). Finally, the epiphyllous bryophytes usually live in a community on the leaf surface consisting of bryophytes, lichens, algae, fungi and small animals such as Tardigrada, Rotatoria and Acarina, forming the phyllosphere (Massart, 1898; Ruinen, 1961). The phyllosphere is able to fix free nitrogen probably due to the activity of its blue-green algal component (cf. also Olarinmoye, 1974).

Bryophytes, especially liverworts, play the most important role in the epiphyllous communities except in drier habitats where lichens can still occur. The liverworts, living in this habitat, are usually very specialized epiphytes, highly adapted to the epiphyllous conditions and most are obligate epiphyllae. For example, the African epiphyllous flora consists of about 170 liverwort species and 30 moss species and about 100 of the

Fig. 3.3 (a) Epiphyllous community of *Radula acuminata* on the leaflets of *Caryota monostachya* (Palmae) in Vietnam, Cuc Phüöng National Park. (b) Dr E.W. Jones, collecting bryophytes on a rain forest roadcut surface in the Uluguru Mountains, Tanzania, East Africa. (c) Bryophyte synusia of *Calymperes richardii* and *Octoblepharum albidum* on the trunk base of a planted coconut palm in the rain forest belt near Baracoa, Cuba. (d) *Lejeunea laetevirens* community on the proproots of a royal palm (*Roystonea regia*) in degraded rain forest near Santiago de Cuba.

liverworts are obligate epiphyllae, the rest facultative (Pócs, 1978). The obligate epiphyllous liverworts possess special attaching apparatus (rhizoid plates and paramphigastria) and adhesive exudates to establish themselves firmly on the leaf surface (Bischler, 1968; Winkler, 1967, 1970). Table 3.1 shows the number of species and distribution of epiphyllous liverworts. Many other epiphyllous genera are restricted to one continent e.g. *Cyclolejeunea* to tropical America. Africa has no endemic epiphyllous genera. Vegetative propagation by shoots and gemmae is also very widespread in epiphyllous liverworts and seems to be a compensation for the difficult substrate conditions.

Table 3.1 Distribution and number of species in genera of epiphyllous liverworts. Supplemented from Pócs (1978).

	Tropical America	Africa incl. all islands	Asia, Oceania and Australasia
Aphanolejeunea	17	8	7
Cololejeunea	40 (approx)	69	100 (approx.)
Colura	14	14	41
Diplasiolejeunea	19	28	8
Drepanolejeunea	34	18	37
Leptolejeunea	11	5	25
Lejeunea subgen. *Microlejeunea*	12	5	11
Radula (epiphyllous)	6	5	8

Among mosses the specially adapted types are far fewer. Obligate epiphyllous genera are *Ephemeropsis* (Nemataceae) in Asia and Australasia, most species of *Crossomitrium* (Daltoniaceae) in Latin America, but many species of *Cyclodictyon, Callicostella, Daltonia, Hookeriopsis* and *Lepidopilum* are common on living leaves. All the above mentioned belong to the Hookeriales. In addition, many Meteoriaceae also occur on leaves, usually spreading from the twigs or petioles. Very rarely, ferns and epiphytic bromeliads also germinate as epiphyllae, but they never reach maturity.

The number of species within one habitat varies from between 2–3 to 50–60. The average number of species per leaf in the Antilles is 6 (3–10) (Jovet-Ast, 1949), in South Vietnam 3–6 (Tixier, 1966a), in East Africa 8–9 (in some cases up to 25). This number varies with the age of phorophyte leaf. On young leaves only a few obligate epiphyllae are able to establish. Later more and more species and individuals appear (Richards, 1932; Olarinmoye, 1975a). The number of individuals can reach as many as 1745/100 cm^2 (Pócs, 1978). At this stage many facultative epiphyllae are able to establish in the community. In the final stage often only one or a few species dominate, suppressing the others until the leaf falls, ending the life of the epiphyllous community (Richards, 1932). Epiphyllous communities studied include those from South Vietnam (Tixier, 1966a), East Africa (Pócs, 1978), Sierra Leone (Harrington, 1967), Nigeria (Olarinmoye, 1974, 1975a), Guiana (Richards, 1932, 1954) and Colombia (Winkler, 1971).

(e) Specialized epiphyte communities on Pandanus and palm stems and on cultivated trees

The corticolous communities in tropical rain forests are not usually host-specific. Only the physical and chemical properties of the bark cause diversification in their composition. However, some trees with very specialized stem systems, such as pachycaulous monocot trees, possess specialized epiphyte communities or, at least, the vertical distribution of the communities on their stems is unusual. The most obvious case of this is the specialized community of tree fern stems which will be discussed under montane forests (p. 81). The stems of screw pines and palms, most common in lowland forests, are noteworthy.

(i) Pandanus stems. The very soft and somewhat spongy bark of *Pandanus* in the lowland rain forests of East Africa bears a very dense epiphyte cover formed by a few species of moss such as *Calymperes usambaricum, Thuidium gratum* and *Pinnatella oblongifrondea.* This community usually covers the stems from the base up to the branches and closely resembles the tree base communities in the same forest.

(ii) Palms. The very smooth, hard and therefore almost always dry bark of palm trees in the rain forests is usually poor in bryophytes. It resembles even at its lower part the communities of the uppermost trunk section of broad-leaved rain forest trees, dominated by lichens and ptychanthoid (holostipous) Lejeuneaceae and a few *Frullania* species. Under humid conditions, such as at lake shores or along streamlets, cushions of *Calymperes, Leucophanes* and *Octoblepharum* species are quite common on palm stems, especially on those which bear remains of petioles (Fig. 3.3c). The classic example is the oil palm *(Elaeïs guineensis)* native in Africa, which is always rich in epiphytes among the petiole stubs. Palms also offer another interesting microhabitat for bryophytes. Most of them have many pencil-like prop roots at the stem base which provide a shady, permanently moist niche in which humus accumulates. At the bases of royal palms, widespread in Cuba, there is almost always a community of *Lejeunea laetevirens* developed in this habitat, accompanied by the following species (see also Fig. 3.3d): *Cheilolejeunea rigidula, C. trifaria,*

Lejeunea phylliloba, L. pililoba, Cololejeunea cardiocarpa, Fissidens garberi, Neckeropsis disticha, Sematophyllum caespitosum, Octoblepharum albidum and *Taxithelium planum* (see Fig. 3.3d). This community also develops on planted coconut tree bases in rain forest environments and on palms in cultivated land in places of former forest area.

(iii) Cultivated trees. Tropical plantations often consist of different woody species (cocoa, coffee, tea, rubber, *Citrus,* cinnamon, cloves and forest plantations of *Eucalyptus, Acacia mearnsii, Aleurites, Cupressus* and *Pinus*) and many of them cultivated under a network of semi-natural or planted shade trees. With the exceptions of those with peeling bark (*Eucalyptus,* some of the *Pinus* species) or those which are planted too densely (tea), the conditions are suitable for epiphytic vegetation including bromeliads, orchids, bryophytes and lichens. Sometimes with the improved light conditions such a thick epiphytic vegetation develops on the branches of cultivated trees that it influences production in a negative way and is referred to as 'epiphytosis'. Where this occurs growers make efforts to remove the epiphytes regularly from the crop trees. According to Thorold (1952, 1953, 1955) the presence of some epiphytes, especially the epiphyllous liverworts on cocoa, might be used as an indicator of the need to apply fungicides against black pod disease (*Phytophtora palmivora*).

In semi-natural environments on plantations near natural forests the epiphytic communities of planted cocoa, coffee or *Citrus* trees are similar, although impoverished, to those of natural forest trees. Often planted trees do not have the variety of bryophyte habitats of native forest trees. However, it was observed in the East Usambara Mountains, Tanzania, in a clearing, that the planted mango and jacaranda trees had a very rich epiphytic bryovegetation comparable with that of the canopy branching system of forest trees nearby, probably due to the strong illumination. *Citrus* trees here and at many other places (e.g. in the Sierra del Escambray in Cuba) are especially rich in epiphyllae, probably due to the sticky, sugar-containing exudate on their leaf surfaces, caused by aphids, which is perhaps a suitable medium for attachment and germination of epiphyllous liverwort propagules. The frequency of some bryophytes is much higher on planted coffee, cocoa and on *Citrus* trees than in native trees; these include members of the Cryphaeaceae (e.g. *Acrocryphaea robusta* in Africa, *Cryphaea filiformis* and *Schoenobryum coffeae* in Cuba), of the Erpodiaceae (*Erpodium luzonense* in Vietnam, *E. beccarii* in East Africa and Mexico, *E. domingense* and *E. pringlei* in Cuba) and of Fabroniaceae (*Fabronia longipila* in East Africa and *F. wrightii* in Cuba). *Macrocoma tenue* is especially common on East African coffee. The corticolous community of roadside trees with dominant Erpodiaceae,

Fabroniaceae and *Acrolejeunea* species shows affinities with the xerophytic communities of savanna woodland trees (see p. 94). The bryovegetation of *Eucalyptus* and *Pinus* forest plantations is extremely poor due to the xeric microclimate, the continually peeling bark and the accumulation of dry undecomposed detritus. The situation is somewhat better in *Cupressus lusitanica* and *Acacia mearnsii* (black wattle) plantations of East and Central Africa, where at least the litter layer is more or less covered by bryophytes (Petit and Symons, 1974). Gilli (1975) observed a reasonably well developed corticolous community in the *Acacia mearnsii* plantations of Southern Tanzania. I observed a quite rich corticolous vegetation on planted *Juniperus excelsa* trees in East Africa, including *Erythrodontium, Schlotheimia, Macromitrium, Zygodon, Frullania* and *Lopholejeunea* species.

(f) Dead and decomposing wood

Fallen branches and even whole fallen trees, rotting logs and stumps are usually very common in undisturbed tropical rain forests. The cellulose and lignin more or less decomposed and softened by fungi absorbs much water and offers a physically and chemically specialized substrate for bryophytes. Akin to the more oceanic or montane beech and spruce forests in north temperate regions, this substrate, in the tropics, is densely covered by bryophytes. The lignicolous, hygrophilous and sciophilous rotting wood community of tropical rain forests consists of dense, thin, interwoven mat-forming or weft-forming bryophytes such as *Lophocolea, Chiloscyphus, Calypogeia, Callicostella, Cyclodictyon, Distichophyllum, Hookeriopsis* and other Hookeriaceae, *Thuidium,* Sematophyllaceae (e.g. *Trichosteleum*) and Hypnaceae (e.g. *Mittenothamnium, Ectropothecium* and *Vesicularia* species), feathery *Hypopterygium* and *Lopidium* species and smaller or larger cushion-forming Lepidoziaceae, *Leucobryum, Leucophanes, Campylopus,* Calymperaceae and *Rhizogonium spiniforme.*

(g) Disturbed soil and road-cut surfaces

As Richards (1954) has pointed out the undisturbed soil surface of tropical lowland rain forests is usually free from bryophyte vegetation. The reason for this is not really known and is often explained by the smothering effect of fallen sclerophyllous leaves forming a thin compact layer on the ground. Probably, the still unknown chemical and microbial processes occurring during decomposition of litter under hot tropic conditions also play a role. Where the litter layer is discontinuous as on earth banks, ant or termite hills, erosion gullies, ironstone outcrops, in animal (e.g. elephant) footprints or by human impact (road construction etc.) bryophytes immediately appear. On smaller earth banks and gullies the most common are *Fissidens* species (a very large number, especially in Africa), *Bryum, Callicostella, Campylopus, Hyophila, Hydrogonium, Hookeriopsis, Pogonatum* species and Lepidoziaceae (e.g. *Zoopsis* in America and Asia, *Sprucella* in Africa). These ground bryophytes in tropical rain forests grow in small cushions, short turfs or as separate shoots, the liverworts in closely attached mats.

Higher roadside banks and roadcuts are very interesting, because a more acidic and wetter section of soil is brought to the surface. Here, along with the large thalloid mats of *Marchantia* species and cushions of *Philonotis,* Polytrichaceae and different *Leucobryum* species, many interesting small hepatics form appressed or thread-like mats. These include *Calypogeia, Cephaloziella* and *Lophocolea.* Larger hookerioid mosses such as *Hookeria* and *Calyptrochaeta* also occur, and often otherwise epiphytic *Frullania* and *Radula* too, together with Lepidoziaceae (Fig. 3.3b).

(h) Rocks

Shady or half-shady rocks in the wet conditions of a rain forest are usually richly covered by bryophytes. The sharp distinction found in temperate regions, between corticolous and rupicolous species, or between the bryovegetation of basic and acidic rocks, is not apparent in the tropics. Many of the corticolous species occur on rocks and there are not many definite calcicole or calcifuge species in the rocky habitats. The richest in bryophytes are shady cliffs near streamlets where large amounts of filmy ferns (Hymenophyllaceae) also occur. On the limestone rocks of Vietnamese rain forest *Cyathodium balansae* (especially in cave mouths), *Trichostomum involutum, T. perinvolutum, Fissidens perpusillus, Pinnatella donghamensis, Horikawaea nitida*; in East Africa *Cyathodium africanum, Barbula lambarenensis, Tortella opaca* and *Haplohymenium pseudo-triste* are characteristic, although *Cyathodium africanum* also rarely occurs on other types of rock. In Cuba, *Cyathodium cavernarum, Tortella mollissima, Hypnella cymbifolia, Gymnostomiella orcuttii* seem to be typical. On siliceous rocks *Symphyogyna* species, *Dumortiera hirsuta,* many *Hookeriaceae*; in Asia *Fissidens nobilis*; in East Africa *Fissidens ovatus, Leucophanes rodriguezii*; in Cuba *Cephaloziopsis intertexta, Racopilum tomentosum* and *Taxiphyllum scalpellifolium,* seem to be characteristic. Many species of Meteoriaceae, Neckeraceae, Pterobryaceae, *Plagiochila* and Lejeuneaceae, generally regarded as corticolous, may occur in quantity on rocks.

3.3 MONTANE FORESTS

In the upper belt of the humid tropics, above the lowland and submontane tropical forests the montane forest zone develops. Its lower limit varies between 600–1500 m and the upper one between 2000–3000 m depending on the distance from the Equator and on the oceanity of climate. The mean annual temperature varies between 20 and 10°C, but it is still equable, with less than 10–15°C daily and seasonal variation of air temperature and without the occurrence of frost except in the uppermost belt near the altimontane forest line. The annual rainfall is usually higher than in the lower belt of the same region, varying between 2000–16000 mm, usually with an even yearly distribution, and dry seasons are compensated for by the frequent occurrence of mist.

Most of the montane rain forests are evergreen but, depending on the amount and yearly distribution of rainfall, it is possible to distinguish a range of types from the relatively dry microphyllous, hard-leaved forests (including Ericaceae and gymnosperms) to the broad-leaved, very humid mossy cloud forests (Nebelwald, bosque nulado). The structure of montane forest varies according to the temperature and nutritional conditions. At the lower part of the montane forest belt and on deep rich soils a relatively high forest develops, but its height (on average 20–30 m) never reaches that of the lowland and submontane forests. Higher up and on poor leached podsolic soils the canopy of montane forests is lower and more uniform, seldom exceeding 10–15 m. Finally, at the upper forest limit and often much lower too, with poor soil and rigorous climatic conditions, dwarf forests (4–6 m) of gnarled trees, often with umbrella-like very dense canopy develop. This so-called 'elfin forest' is present in all larger tropical mountain systems.

Due to the structural characters of montane forests it is not possible to distinguish as many different epiphytic habitats on any one tree as in lowland forests but, on the other hand, very different types of host plants, such as tree ferns and bamboos, ensure different epiphyte substrates. Finally, all habitats are much more dominated by bryophytes than in the lowland forest and the bryoflora is very rich in species. For example, in many montane forest types a thick carpet of many species of moss covers the soil. The bryovegetation of montane forests plays a much more important role in the water balance and humus accumulation than in the lowlands and in submontane forests (cf. p. 85), strongly influencing the microclimate.

3.3.1 Epiphytic communities

Seifriz (1924) subdivided the ever-wet rain forest according to mass vegetation of mosses and lichens on Mt. Gede in Java. His subdivisions coincide well with later literature data (Tixier, 1966a, 1966b; Steenis, 1972) and with my experience. The montane rain forests up to about 1600–2000 m are formed by relatively large trees where the tree bases have a similar bryovegetation to those of the submontane forests, although buttresses do

Fig. 3.4 (a) *Neorutenbergia usagarae* (Rutenbergiaceae) synusium on thin trunks in the montane rain forests of Uluguru Mountains, East Africa. (b) Pendulous type ramicolous Meteoriaceae: *Pilotrichella pilifolia* hanging down from branches in the rain forests of the Uluguru Mountains. (c) Pendulous Meteoriaceae (*Squamidium biforme*) hanging from twigs in the montane rain forests of Uluguru Mountains. (d) Ramicolous *Plagiochila* species on liana stem in the lower layer of a montane rain forest, Sierra Maestra, Cuba.

not develop at this altitude. The tree trunks are also, at least partly, covered by small cushion-forming or hanging-type bryophytes such as patches of *Leucobryum, Bazzania* and different Dicranaceae *(Leucoloma, Dicranoloma* and *Campylopus)* species. The branching system has a very rich bryophyte vegetation, mostly of pendulous Meteoriaceae, Pterobryaceae, Neckeraceae, feather-form, pinnate or dichotomous *Plagiochila, Porella, Lepidozia,* and dense mats of Orthotrichaceae *(Macromitrium, Macrocoma, Groutiella, Schlotheimia, Zygodon*) (Fig. 3.4b-d). This type of montane rain forest was studied in detail by Tixier (1966a) in South Vietnam.

In East Africa, where the greatest part of the African montane forests occur, the bryoflora of the old crystalline massifs, e.g. the Uluguru Mountains and those of the young tertiary volcanoes (e.g. Kilimanjaro) are very distinct, probably due to historical reasons (Pócs, 1975, 1976, 1980), even under similar climatic conditions.

The montane forests of the high East African volcanoes on the one hand lack the old relict elements found in Madagascar and other dissected parts of Gondwanaland but on the other hand have several endemic species. Typical epiphytes of the rain forests of Mt. Kilimanjaro and of the other volcanoes in East Africa include: *Cololejeunea fadenii, Colura berghenii, C. kilimanjarica, Leptoscyphus hedbergii, Marsupidium limbatum, Plagiochila barteri, P. subalpina, Porella hoehnelii, Syzygiella geminifolia, Acanthocladiella flexilis, Fissidens vogelianus, Hookeriopsis pappeana, Leucodon dracaenae, L. laxifolius, Mittenothamnium brevicuspis, Neckera platyantha, Pilotrichella cuspidata, Porotrichum ruficaule, P. subpennaeforme* and *Rhizofabronia perpilosa.*

In Latin America, Herzog (1916) described in detail the bryophyte vegetation of the montane rain forests of the Bolivian Andes, while Delgadillo (1979) analysed the distribution of the epiphytic mosses of Mexican *Liquidamber* forests. Epiphytes of similar montane forests, at an altitude of 900–1400 m in the Sierra Maestra mountains of Cuba, are listed below (only the more widespread species are given).

(i) Hepaticae. *Anoplolejeunea conferta, Ceratolejeunea cubensis, Cheilolejeunea rigidula, C. trifaria, Drepanolejeunea anoplantha, Herbertus juniperoideus, Microlejeunea ulicina, Omphalanthus filiformis, Plagiochila bursata* and *Radula kegelii.*

(ii) Musci. *Acroporium pungens, Adelothecium bogotense, Callicosta affinis, Fabriona wrightii, Groutiella apiculata, Homaliodendron decompositum, Isodrepanium lentulum, Leucobryum polakowskyi, Leucoloma albulum, L. serrulatum, Macromitrium cirrhosum, Meteorium illicebrum, Phyllogonium fulgens, Pireella angustifolia, Rhynchostegiopsis flexuosa,*

Schlotheimia torquata, Sematophyllum galipense and *Syrrhopodon prolifer.*

At higher altitudes, due to the greater amount and more even distribution of rainfall and the continuous mist cover, another type of montane rain forest occurs, the mossy forest or cloud forest (see Fig. 3.5a). In Java, this zone develops at 2400 m altitude (Seifriz, 1924) in Vietnam and the Philippines above 1700–2000 m (Tixier, 1966a, 1966b), in New Guinea above 2100–2400 m, whilst in more oceanic conditions (e.g. on the Solomon Islands in Melanesia) the forests are 'mossy' at as low an altitude as 700 m (Whitmore, 1966). Mossy forests in East Africa occur on the ranges near the sea, in general above 1800 m, while on the high volcanoes of the central plateau and rift valleys only above 2400 m. Their distribution is greatly influenced by the direction of rain-carrying winds, e.g. in the Uluguru Mountains they descend lower on the eastern slopes, while on Mt. Kilimanjaro they are extensive on the south-south-west slope and not present at all on the west, north and east slopes (Pócs, 1976c). On smaller islands (Steenis, 1972; Whitmore, 1966) because of cloud formation, mossy forests occur even lower. The same applies to 'inselbergs', where even dry woodland trees can be so rich in epiphytic mosses that they look like mossy montane forests (Pócs, 1976c, 1978). The descent of mossy forests and their asymmetric distributional pattern according to the direction of rain-carrying winds is described by Balázs (1973) and Gradstein (1979) with regard to the Galapagos Islands.

Besides the short account of Javanese mossy forests by Seifriz (1924) and Herzog's (1910) very spectacular and colourful description of the mossy forests of Sri Lanka, Richards' (1935) report on the mossy forests of Sarawak and Tixier's (1966a) study on the bryophyte communities of the South Vietnamese mossy forests are worth mentioning. The microclimate of mossy forests was analysed by Lötschert (1959) in Central America and by Pócs (1974) in East Africa. The mossy montane forests can be characterized by the very large amount of moss cover which is also present on the forest floor in the form of carpets and dense cushions. Tree trunks and branches, lianas and shrubs are covered in such a thick felt of bryophytes that the diameter of many thinner trunks is multiplied (Fig. 3.5b). There are no differences between the communities of tree bases and higher parts, only the thinner branches show some differences. Whilst with epiphytes in the lower montane belts pendulous types are the most frequent, in cloud forests turfs, mats and cushions are common, together with feathery forms and dendroid types on the branches. Species of *Syrrhopodon, Campylopus, Dicranoloma,* Macromitrioideae, many Meteoriaceae, Neckeraceae and Pterobryaceae, *Lepidozia, Bazzania, Herbertus, Plagiochila, Schistochila* and *Frullania* occur in Asia;

Fig. 3.5 (a) Interior of a montane mossy forest in the Uluguru Mountains at 2000 m altitude. Note the bamboo stem with moss balls at the nodes. (b) *Dicranoloma billardieri* cloak on a branch in the mossy elfin forest of Uluguru Mountains. (c) *Paraschistochila englerana* and *Lepidozia cupressina* on the trunk of an elfin forest tree, Uluguru Mountains. (d) *Chiloscyphus decurrens* and *Frullania serrata*, epiphytes on the branches of elfin forest trees in the Uluguru Mountains.

Hypnodendron, Braunfelsia and *Campylopodium* in Oceania; and *Leptodontium, Neckera, Pilotrichella* and *Adelanthus* species in Central America and in East Africa.

As already mentioned (p. 73), near the upper forest limit, and often much lower too, elfin forest develops, with a very dense canopy. Its occurrence at lower altitudes is explained by the 'telescoping effect' described by Steenis (1972) in which the continuous exposure to strong winds, cloud and mist formation and strong erosion result in the occurrence of an often peaty acid soil, poor in nutrients, which does not permit a taller forest to develop. Naturally, the bryophytic communities of these elfin woodlands differ from the other montane forests. Some of them are not rich in bryophytes, while others, because of higher rainfall or more regular mist formation, are very rich (cf. Seifriz, 1924; Pócs, 1980) and of the 'mossy' type. The microclimate of elfin forest was studied first by Baynton (1969) in Puerto Rico, in connection with its bryophyte vegetation, which was briefly described by Fulford *et al.* (1970). In East Africa Pócs (1974, 1976a-c) dealt with the ecology and plant communities of elfin forests, including bryophytes. Whilst the epiphytes on the trunks are more or less similar to those of the montane mossy forest, the epiphytes in the canopy of elfin forests are more exposed to direct radiation, wind and even low temperatures and occasionally frosts, than any other type of tropical rain forest. They usually form a quite closed ramicolous synusium together with small orchids (Fig. 3.5c, d).

Ericaceous forests, composed of giant heathers, usually occur on poor acidic soil (often peat). They are widespread in East Africa on the Ruwenzori and on the high volcanoes where they often form the forest line at about 3000 m. They should be regarded as a specialized kind of elfin forest and occur at the upper forest limit, or much lower on acidic rocks or on sandy or peaty soil within the montane forest belt. Their moss vegetation in Asia is described by Richards (1935, 1950) who underlines the striking differences between the bryovegetation of heath forests and broad-leaved montane forests. Their epiphytic vegetation in Vietnam is discussed by Tixier (1966a). Their corticolous species seem to show preference for the bark or twigs or ericaceous trees and shrubs, or at least to the better light conditions resulting from the more open canopy. The cushion-forming types are more common than in the broad-leaved elfin forests (Fig. 3.6a) otherwise the general physiognomy of moss vegetation is similar. In the misty environment of ericaceous forests on Mt. Kilimanjaro on the thin dead twigs of Ericaceae develops a community consisting of tiny, appressed or small cushion-forming bryophytes including *Anomalolejeunea pluriplicata, Drepanolejeunea physaefolia, Colura berghenii, C. kilimanjarica* and *Daltonia patula.* The composition of this community shows affinities with the one inhabiting bamboo stems at the same altitude.

Fig. 3.6 (a) Altimontane *Erica arborea–Hagenia abyssinica* forest ('giant heather') on the southern slope of Mt. Kilimanjaro, at 2800 m altitude, with large cushion-forming bryophytes. (b) Tree-fern stem-inhabiting *Rhizofabronia persoonii* var. *sphaerocarpa* community in the upper Mwere valley of Uluguru Mountains. (c) Older bamboo (*Arundinaria alpina*) stems with appressed mats of liverworts and with a few mosses on the nodes, in the Poroto Mountains, Southern Highlands of Tanzania, at 2100 m altitude. (d) Bamboo node inhabiting *Daltonia–Metzgeria–Radula* community on the south slope of Mt. Kenya at 2600 m altitude.

(a) Epiphyllous communities

The presence of foliicolous bryophytes at higher altitudes is restricted to the permanently wet areas or to the mist and dew affected habitats up to altitudes not subjected to night frosts or where the habitat is well sheltered from frost. Practically, it means an upper limit of 3000 m. Highest records of epiphyllous bryophytes are given in Table 3.2. Naturally the same and other epiphyllous bryophytes occur much higher too (up to 3900 m) but on other substrates. In East Africa epiphyllous bryophytes are common in the altimontane forests up to 2600 m if the conditions are suitable (Pócs, 1978).

Table 3.2 Maximum recorded altitudes of epiphyllae.

Cololejeunea pseudofloccosa: Sri Lanka, 2500 m (Pandé *et al.*, 1957); China, Yunnan, 2800 m (Chen and Wu, 1964).
Cololejeunea spinosa: Nepal, Himalaya, 2620 m (Hattori, 1975)
Cyrtolejeunea holostipa: Ecuador, Andes, 2700 m (Arnell, 1962)
Drepanolejeunea campanulata: Ecuador, Andes, 2800 m (Bischler, 1964)
Drepanolejeunea cupulata: Ecuador, Andes, 3000 m on *Herbertus* (Arnell, 1962)
Drepanolejeunea physaefolia: Africa, Mt. Kilimanjaro, 2900 m (Pócs, unpublished)
Drepanolejeunea tenera: Java, 2400 m (Herzog, 1930)
Drepanolejeunea tenuis: Java, 3060 m (Herzog, 1939 – epiphyll?) Vietnam, 2400 m (Pócs, 1968 – it grows together with *Lejeunea punctiformis, Cololejeunea planiflora, C. peraffinis, Apometzgeria pubescens*)

The Asian epiphyllous communities of montane forests are described from South Vietnam by Tixier (1966a), while in tropical America, epiphyllous communities of montane rain forests from El Salvador are detailed by Winkler (1967) who, besides describing the members of the communities, studied growth physiology and ecology of the foliicolous liverworts. He underlined the role of leaf exudates in the nutrition of epiphyllae, carried out field experiments to study their succession and established that the first leaf colonizers are usually obligate epiphyllae. He has also found a correlation between the growth rate of epiphyllae and the seasonal distribution of rainfall and studied the competition between different species. He established that slow-growing species, like *Diplasiolejeunea pellucida,* are often overgrown by more aggressive species such as *Odontolejeunea lunulata.*

(b) Specialized epiphyte communities

Tree ferns and bamboos are some of the most conspicuous tall plants of the wet montane rain forests in the tropics. Both offer peculiar opportunities for epiphytic bryophytes and usually have very specialized communities composed of species which are absent or very rare on other substrata. I have had the opportunity of studying their vegetation only in East Africa and Cuba and, therefore, the examples come from these localities, although I am sure that they also deserve more attention in other tropics.

(i) Tree ferns (Cyatheaceae). There seem to be various reasons why tree ferns harbour very specialized bryophyte vegetation. Firstly, their whole stem or at least the lower part is usually densely wrapped in a felt-like mass of adventive aerial roots, which often thicken the stem two or three times and are able to absorb a very large amount of water. Secondly, the remaining leaf bases and leaf scars form an intersecting pattern similar to a *Sigillaria* stem especially suitable for anchoring epiphytes. Thirdly, they always live in very wet conditions. Fourthly, the peat-like mass of the stem-covering adventicious aerial roots forms an acidic substrate – the pH of the wet stem surface is 4.5 and that of its moss cover, 4.0.

In the montane forests of tropical Africa the most characteristic moss species of tree fern stems are *Rhizofabronia perssonii* and *Leiomela africana.* Both form a short glossy emerald green turf, especially on leaf scars and in crevices among the remaining petiole bases. The close association of these two species on tree fern stems was observed by Richards and Argent (1968). The stems of *Rhizofabronia,* according to my experience, when moistened are able to lengthen their internodes by increasing the cell volume and to increase the water-absorbing capacity of the moss turf significantly. A third moss species, *Fissidens jonesii* occurs almost exclusively on tree fern stems (Pócs, 1980) but it is also epiphyllous in a few cases. Among the liverworts, *Lejeunea cyathearum* is the most characteristic species, often associated with *L. villaumei.* Fig. 3.6b shows the epiphytic community of a tall *Cyathea manniana* stem.

In Cuba I observed another pair of moss species, *Lepyrodontopsis trichophylla* and *Hymenodon aeruginosus,* which seems to be restricted mostly to the *Cyathea* stem habitat. These two species occur as frequently as *Rhizofabronia* and *Leiomela* do in Africa on tree fern stems, forming golden green turfs or pale green appressed mats.

(ii) Bamboos. The bamboos have their centre of evolution in tropical Asia where tall woody representatives are common in all vegetation belts as giant perennial tree-like plants or lianas. In tropical America the Bambuseae is represented only by small climbers but in Africa species are important vegetation builders. The giant African bamboo *(Arundinaria alpina)* forms dense thickets in the uppermost forest zone, replacing the elfin forest in many African mountain areas and forming the forest line in some cases, as on the Kimhandu peak of Uluguru Mountains or on Oldonyo Oldeani in the Massai Highland, between 2600–3000 m. Bam-

boos may form a continuous belt between 2200–3100 m within the broad-leaved evergreen montane forest zone as on Mt. Kenya or between the broad-leaved evergreen forests and *Erica* forest on the Ruwenzori. In other cases, they descend as low as 1600–2000 m and form mixed stands in the montane forest (Hedberg, 1951; Pócs, 1976c).

The bamboos, in contrast to trees, have very smooth dry woody stems, difficult to colonize by bryophytes. In addition, the life-span of bamboos in general is no more than 7–9 years. Despite this, under wet tropical montane conditions, bamboos have an epiphytic bryophyte vegetation. It consists partly of appressed types of Lejeuneaceae, *Radula* and *Frullania* which are able to creep adhering to the stems (Fig. 3.6c). Other types, which are much more common (e.g. *Daltonia* and *Metzgeria* species in East Africa and *Calymperopsis* in Hawaii), form small cushions, turfs, mats of wefts or even pendants attached to nodes and branch bases (Fig. 3.6d). In a wetter type of bamboo forest, as on the south slope of Mt. Kenya or in the Southern Highland of Tanzania these bryophyte colonies are very striking on the stem nodes of bamboo, looking like tennis balls.

3.3.2 Lignicolous communities

On fallen logs in the initial phase of decomposition usually hygro-sciophilous tree base communities occur, e.g. *Thuidium cymbifolium* and *Vesicularia demangei* in North Vietnam or *Thuidium glaucinum* associations in South Vietnam (Tixier, 1966a). At a more developed stage these species give way to more specialized lignicolous assemblages from Lepidoziaceae, *Odontoschisma,* Lophocoleaceae, Lejeuneaceae and *Riccardia* to Anthocerotales, Leucobryaceae, Calymperaceae, Hypnaceae, *Hypopterygium,* Hookeriaceae and Sematophyllaceae. These bryophytes are at least partly mycotrophic or are able in some other way to utilize the nutrients of the decomposed wood. Other species probably prefer the good water-absorbing and -retaining properties or the acidic properties of dead wood.

In Cuba, bryovegetation on decomposing wood (including decaying tree fern stems) appeared to be the richest in specialized lignicolous species. *Nowellia wrightii, Micropterygium trachyphyllum, M. carinatum,* and *Odontoschisma denudatum* are always present on rotten wood and also, from time to time, *Arachniopsis diacantha, Zoopsis antillana, Cephalozia caribbeana* and *Syrrhopodon prolifer.* Together with this very characteristic and constant combination of mostly lignicolous species, a number of corticolous and terricolous bryophytes also occur.

3.3.3 Terricolous communities

In the montane forest, terricolous bryophytes are not restricted to disturbed soil surfaces and a significant portion of the forest floor is covered by bryophyte mats and cushions. The topsoil is usually leached and acidophilous bryophytes, such as Polytrichaceae, *Dicranum, Dicranoloma,* Calymperaceae and Lepidoziaceae or, in the wettest types even *Sphagnum,* are common. The lower montane forest belt is not as rich in ground bryophytes, except under exceptional conditions, such as the elfin and heath forest at low altitude. In the mossy forests and high altitude ericaceous forests and elfin woodlands a more or less continuous bryophyte layer develops which, together with the epiphytic bryophytes, is very important in the water balance and humus formation of these forest types. In the montane forests of tropical Australasia the ground-covering carpets formed by *Dawsonia gigantea* and *Dawsonia papuana* are famous. Richards (1935, 1950) mentions, from the mossy forests of Borneo, *Sphagnum beccarii, S. sericeum, Mniodendron divaricatum, M. microloma* and *Leucobryum sanctum,* as the more common species. Table 3.3 shows the coverage of bryophytes on Bondwa Uluguru Mountains, tropical East Africa.

Table 3.3 Percentage cover of ground bryophytes in elfin forest on the top of Bondwa peak, Uluguru Mountains, tropical East Africa at an altitude of 2125 m (total bryophyte cover 45%).

	% Ground cover		% Ground cover
Bazzania borbonica	10	*Leucobryum isleanum*	1
B. nitida	3	*Metzgeria leptoneura*	2
Bryum leptoneurum	1	*Paraschistochila englerana*	1
Campylopus schroederi	0.1	*Pogonatum usambaricum*	2
C. subperichaetialis	0.1	*Rhodobryum keniae*	2
Cyclodictyon brevifolium	0.1	*Syrrhopodon stuhlmannii*	15
Dicranoloma billardieri	5	*Trachypodopsis serrulata*	1
Fissidens hedbergii	0.1	*Trichosteleum humbertii*	1
F. purpureocaulis	0.1	*Tylimanthus ruwenzorensis*	1
Kurzia verrucosa	5		

The vegetation of the ericaceous forest ground on the East African high mountains is very rich and usually covers the poor acidic soil in the form of a deep, continuous, ever-wet carpet of tall turf-forming or large cushion-forming bryophytes. I observed this community on Mt. Kilimanjaro, Mt. Meru and on the Rungwe volcano. The community is dominated by *Breutelia diffracta, B. perrieri, B. stuhlmannii, Campylopus jamesonii, Dicranoloma billardieri, Leptodontium luteum, Plagiochila ericicola,* and *Thudium matarumense,* and in more humid places by *Sphagnum davidii* and *S. pappeanum.*

In Cuba, on the soil surface of mossy forests (monte nublado) and elfin forests, a similarly rich bryophyte vegetation develops, usually dominated by large cushions of Lepidoziaceae, Dicranaceae and *Leucobryum* species.

3.3.4 Rocks

Rocky cliffs in the montane forest belt are usually rich in bryophytes. As already mentioned, there are many species, known as epiphytes, which grow on rocks as well. In East Africa, on the shady rock cliffs of the old granite massifs of Tanzania, as in the Uluguru, Nguru and Ukaguru Mountains, between 1500–2400 m altitude, a very interesting association of different *Saintpaulia* (African violet) species, *Elaphoglossum phanerophlaebium* and rupicolous bryophytes develops. Table 3.4 shows one example of such a community. I observed the same association on Lupanga peak in the Uluguru Mountains, at 2050 m, with the mass occurrence of *Aongstroemia vulcanica, Campylopus flageyi, Rhacocarpus purpurascens, Syrrhopodon spiralis* and *Sphagnum pycnocladulum.* In the Ukaguru Mountains, at 1540 m, a similar association occurs with *Saintpaulia pusilla, Streptocarpus schliebenii, Elaphoglossum phanerophlaebium, Selaginella abyssinica* and *Hymenophyllum splendidum* associated with such bryophytes as *Cephaloziella kiaeri, Fissidens holstii, Leucobryum perrotii* (codominant), *Metzgeria thoméensis, Syrrhopodon spiralis* (dominant) and *Wijkia trichocolea.*

On shady stream valley rocks of the same mountains *Dumortiera hirsuta* is associated with *Jamesoniella purpurascens, Symphyogyna podophylla, Ectropothecium regulare, Brachythecium atrotheca, Fissidens nitens, Hypopterygium mildbraedii* and *H. viridissimum* and *Leucoloma aspericuspis* with *Heterophyllium flexile.*

Table 3.4 Percentage cover of vegetation on shady, vertical or overhanging rock at 1750 m on the north slope of Bondwa peak, Uluguru Mountains. Up to 45% of the rock is covered by vascular plants, up to 90% by bryophytes.

Vascular Plants:			
Elaphoglossum phanerophlaebium	15	*Saintpaulia goetzeana*	20
Selaginella abyssinica	10	*Dorstenia* sp.	rare
Hepatics		Mosses	
Calypogeia sp.	rare	*Catagonium nitens*	1
Chandonanthus hirtellus	2	*Fissidens asplenioides*	60
Herbertus sp.	2	*F. nitens*	5
Pallavicinia lyellii	rare	*Leucobryum isleanum*	4
Plagiochila sp.	rare	*Sphagnum pycnocladulum*	3
		Syrrhopodon spiralis	10
		Wijkia cuynetii	3

The volcanic rocks in the ericaceous forest belt of Mt. Kilimanjaro have a different bryophyte vegetation. Firstly, the often porous basaltic rocks supply a more xeric and basic substrate; secondly, the history of young tertiary volcanoes is very different. Elements representing old Gondwanaland links in the Precambrian granite massifs are replaced by afromontane, afroalpine, afrosubalpine endemic species, such as *Atractylocarpus alticaulis, Kurzia irregularis, Mielichoferia cratericola, Rhabdoweisia africana* and *R. lineata*; or more widespread, even holarctic species such as *Ceratodon purpureus, Hymenostylium recurvirostre, Jungermannia sphaerocarpa, Leptodontium flexifolium* and *Pohlia elongata.*

In tropical America, Giacomini and Ciferri (1950) described an interesting cryptogamic association from the rock cliffs of mist forests of the Venezuelan Andes. The community is formed by *Cora pavonia* (Basidiolichenes) and by *Polytrichadelphus ciferrii.* Other bryophytes at the same locality include *Plagiochila tovarina, Frullania atrosanguinea* and *Pilopogon gracilis.*

In Cuba on shady granitic cliffs in the mountain rain forest belt some of the pendulous epiphyte species (e.g. *Omphalanthus filiformis, Phyllogonium fulgens, Pilotrichella flexilis, Papillaria imponderosa, Isodrepanium lentulum*) occur. Other epiliths such as *Pogonatum tortile, Fissidens asplenioides, Epipterygium wrightii* and *Mittenothamnium volvatum* are also terricolous. A small, but distinct group consists of obligate rupicolous species such as *Scapania portoricensis, Aongstroemia jamaicensis* or *Homalia glabella.* In shady wet crevices near springs and in stream valleys a more hygrophilous community is formed consisting of thalloid mats of *Dumortiera hirsuta, Monoclea forsteri, Riccardia* and *Symphyogyna* species, accompanied by *Trichocolea* species, *Jubula pennsylvanica* ssp. *bogotensis, Hygrolejeunea cerina, Radula antilleana, Hypopterygium tamariscinum* and by different Hookeriaceae.

3.4 THE BIOMASS OF BRYOPHYTES AND ITS SIGNIFICANCE IN THE WATER BALANCE OF TROPICAL RAIN FORESTS

The epiphytes in the northern temperate region do not usually play an important role in the ecosystem. Even in the oceanic part of Europe, Schnock (1972) estimated the epiphytic biomass to be only 355 kg ha^{-1} in oak forests, while Simon (1974) estimated the epiphytic biomass of a Hungarian oak forest under continental climatic conditions to be 41 kg ha^{-1}. On the other hand terrestrial mosses and lichens are important in soil fixation and humus accumulation of continental steppe communities (Simon, 1975; Simon and Láng, 1972; Láng, 1974; Orbán, 1977) where their biomass can be as much as 4000 kg ha^{-1}. Until recently (Pócs, 1980), there have been no attempts at estimating the biomass of tropical rain forest epiphytic bryophytes.

3.4.1 Analysis of two rain forest types

(a) Submontane rain forest

The sample area studied was a north-west-facing valley in the northern part of the Uluguru Mountains, Tanzania, at an altitude of 1415 m. On the flat bottom and gentle slopes of the valley is a very humid rain forest, with a 2–3 storey, up to 40 m high canopy. The forest type, described by Pócs (1974, 1976a) is evergreen, rich in epiphytic ferns, especially in nest epiphytes. The lower storey of the canopy is often formed by tree ferns, 6–10 m high. The understorey is very rich in ferns, more than 40 species occurring in the sample area.

The corticolous micro-epiphytes. Micro-epiphytes, mostly mosses, liverworts and filmy ferns, cover the bases of trunks in the sample area up to at least 4 m, often higher, forming an almost 10 mm thick carpet. Table 3.5 shows the biomass and interception capacity (amount of water held by freshly moistened plants relative to dry weight) of epiphytes on the trunk of a tree of 40 cm diameter, from a height of 110–120 cm. The girth at the lower edge of the sample was 126 cm, while at the upper edge it was 121 cm. Coverage of the 12 350 cm^2 was 60% by *Leucoloma* sp. and 15% by filmy ferns. The data clearly show why the water interception of mossy forests is much higher than in any other vegetation types in the tropics.

Table 3.5 Biomass and water interception capacity of epiphytes at a height of 110–120 cm on a trunk with a girth of 126–121 cm.

	Fresh weight (g)	Dry weight (g)	Water content (% dry weight)	Interception capacity (% dry weight)
The whole corticolous microphyte synusium	199.70	78.18	155	471

Although the vascular nest epiphytes are among the most striking features of tropical rain forests, their biomass is surpassed by that of smaller ramicolous bryoepiphytes. Nest epiphytes play an important role in aerial humus accumulation. Ramicolous microepiphytes (mostly bryophytes) which, with their feathery life-form cover the uppermost twigs in the canopy, proved to be the most effective interceptors of a rainwater, even exceeding the water capacity of the foliage of phorophyte trees. *Plagiochila* species in the sample plot are the most important ramicolous canopy rain-water interceptors. The same thing was found in lowland rain forests of Cuc Phüöng National Park in Vietnam. In really wet forests, these feather-like bryophytes, growing on twigs of pencil diameter or less, form a quite continuous intercepting layer which absorbs the rainwater falling on the canopy, retaining more than the phorophyte leaves.

An epiphyte biomass–water interception model based on Table 3.5 and other data was published by Pócs (1976b), and a modified and supplemented model is shown in Table 3.6.

The total interception in the submontane rain forest was estimated to be 25 578 litres per single rainfall per hectare. This amount, calculating 200 rainy days a year, represents 572 mm of rain, one fifth of the annual precipitation there. Hopkins (1960), in a submontane rain forest of Uganda over a long period, recorded 33.6% interception, which is not far from our estimation. Lundgren and Lundgren (1979) obtained very similar data (21–25%) during measurements carried out for two and a half years.

(b) Mossy elfin forest

The sample area was on the top of Bondwa peak, at 2120 m, on a gentle (15°) east-facing slope. The vegetation has been described by Pócs (1974). The elfin woodland is formed by 4–6 m tall trees with umbrella-shaped, dense canopy (up to 90% cover). The next layer consists of 1.0–2.5 m tall, half-woody Acanthaceae and there is a herb layer 10–60 cm high. The ground is covered by cushion- and carpet-forming bryophytes (up to 45% cf. p. 83) and sometimes by lichens. The sample plot had a rainfall of 3000 mm, well distributed throughout the year and an almost permanent cloud formation. Precipitation in the form of mist and dew compensates for short dry periods. The trunk and branching systems of trees are covered by a very thick (up to 15 cm) bryophyte and filmy fern carpet. The canopy twigs are interwoven so densely by bryophytes that their cover is almost continuous. Small epiphytic orchids, sometimes as many as 40–50 plants per square metre, occur among them. Table 3.7 is an epiphyte biomass-water interception model based on sample data from the elfin forest.

As can be seen, by far the best interceptor is the moss cover of the canopy, its interception capacity surpassing not only the other layers of the elfin forest, but also the whole interception ability of the submontane high forest (see Table 3.6), absorbing nearly 30 000 l water ha^{-1} during a single rain after a relatively dry period. If we add the thick bryophyte cloak of all trunks, lianas and the ground cover, with at least 250 rainy days a year, we reach a very high value and can understand the great significance of montane mossy forests on tropical watersheds. Here the presence of mossy forest cover in the cloud zone has a buffering role and eliminates the

Table 3.6 The epiphyte biomass – water interception model of a one hectare submontane rain forest plot.

		Fresh weight (kg)	Dry weight (kg)	Water content (% dry) weight)	Interception capacity	
					% dry weight	$l\ ha^{-1}$
I.	Phorophyte leaves	49 146	30 000	63.8	45.4	13 641
II.	Ramicolous microepiphytes in canopy	3075	1695	81.4	508.0	13 644
III.	Microepiphytes on tree trunks	200	78	156.4	434.0	339
IV.	Leaves of the macroepiphytes on tree trunks	183	23	695.6	45.4	73
V.	Root system of macroepiphytes on tree trunks	138	55	152.7	192.2	341
VI.	Leaves of nest epiphytes	195	34	473.5	45.4	15
VII.	Living root system of nest epiphytes	158	62	154.8	45.4	28
VIII.	Decomposed roots of nest epiphytes	107	36	197.2	334.9	122
IX.	Litter and detritus fallen in nests	36	17	111.8	130.4	22
X.	Humus accumulated in nests	308	130	136.9	271.6	353
X.	Grand total	53 547	32 130			28 578
II–X.	Total of epiphytes	4401	2130	106.6		14 937
II–III.	Total of bryophytic microepiphytes	3275	1773	84.7		13 983

Table 3.7 The epiphyte biomass – water interception model of the mossy elfin forest plot per hectare.

		Fresh weight (kg)	Dry weight (kg)	Water content (% dry weight)	Interception capacity	
					% dry weight	l ha^{-1}
I.	Microphyllous leaves of *Podocarpus milanjianus*	13 594	7488	81.5	67.5	5055
II.	Mesophyllous leaves of *Allanblackia* cf. *ulugurensis*	2753	595	363.0	173.3	1024
III.	Epiphytic orchids in the canopy	3305	834	296.2	145.0	1209
IV.	Epiphytic ferns (mostly Hymenophyllaceae) of the canopy	186	63	195.2	109.8	69
V.	Humus and detritus among the canopy epiphytes	4409	2453	79.3	193.0	4734
VI.	Microepiphytes on branches	13 419	6145	118.4	471.0	28 942
VII.	Microepiphytes on trunks	7062	2780	154.0	395.0	10 981
VIII.	Microepiphytes on roots	4103	1375	198.4	338.0	3958
IX.	Bryophyte cover on ground	2133	723	195.0	393.0	2841
I–X.	Grand total	50 964	22 456	126.9		58 813
III–VIII.	Total of epiphytes	32 484	13 650	138.0		49 893
VI–VIII.	Total of bryoepiphytes	24 584	10 300	138.7		43 881
VI–IX.	Total of bryophytes	26 717	11 023	142.7		46 772

detrimental effect of torrential rains, allowing the rainwater to filter gradually to the soil and, by keeping the environment moist during dry periods, regulating the level of agriculturally important watercourses. In the Uluguru Mountains, where the protecting cover of forest has been removed in large areas, sudden heavy rainfall has twice caused destructive large-scale landslides within the last ten years (Temple, 1972; Temple and Rapp, 1972; Lundgren and Rapp, 1974). Also significant in this respect is the demand for water from densely populated areas such as Dar-es-Salaam which depends upon the level of the Mgeta-Ruvu River which rises in the Uluguru Mountains.

Epiphytic humus accumulation is also very significant in the elfin forest, exceeding by more than ten times the aerial humus accumulation of the much taller and more complex submontane rain forest (see Tables 3.6 and 3.7). The humus in the elfin forest is not bound to epiphytic nests but gradually falls down, increasing the otherwise poor nutrient reserves of the soil. In the dwarf elfin forest, therefore, the canopy is not only the most important layer microclimatically (affecting light, temperature and water-regime) but is also the most active in humus formation.

3.4.2 Rainfall–epiphyte correlation theory

Analysing Tables 3.6 and 3.7 and comparing them with the corresponding climatological data (see the Gaussen–Walter diagrams in Pócs, 1974, 1976), there is an interesting correlation between the interceptive capacity of the epiphytic plants and the distribution of annual rainfall. The epiphyte biomass is proportional to annual rainfall amounts that exceed a monthly average of 100 mm. Tropical macrovegetation flourishes increasingly with increasing monthly rainfall up to 100–200 mm per month. Above about 2000–2500 mm rainfall a year there is no further 'improvement' and 'enriching' of vegetation. On the contrary (cf. Longman and Jenik, 1974), in perhumid regions, the productivity of forests is well below that of an average rain forest or even of a seasonal rain forest.

How does the surplus rainfall otherwise affect the tropical forest? Although the general physiognomy of the host community does not change much with further increase of rainfall, its epiphytic biomass and the interceptive capacity of epiphytes increases in proportion to the 'surplus' rainfall amount. In the two examples given here the surplus rainfall is 1100 and 1910 mm respectively. In other places in the same mountains there are submontane evergreen forests and elfin forests with much less precipitation and fewer epiphytes although tree species and ground flora are similar. The 'surplus' rainfall is utilized by the specially adapted epiphytes, which build up an active water-intercepting system. After intercepting the water, part is re-evaporated, part slowly drops or flows to the ground and part is retained by the epiphytes. In these three ways, they influence their environment by increasing the biomass and accumulating humus. My feeling is that the epiphytic life-form is not so much an adaptation to special light conditions but more a style of life initiated by the surplus rainfall and utilizing the other favourable climatic conditions of the tropics.

The epiphytic life-form seems to be, at least in most cases, a late adaptation which developed, not parallel to, but after the general development of rain forests since the Cretaceous. Vitt (1972) discusses this question in relation to the phylogeny of the genus *Orthotrichum*. The large angiosperm epiphyte families Bromeliaceae and Orchidaceae are at the end of evolutionary lines and highly specialized. There are no true gymnosperm epiphytes. There are many pteridophytes living as epiphytes but only one family, the Hymenophyllaceae, consists of highly specialized epiphytes and the most highly evolved families, the Aspleniaceae and Polypodiaceae, providing most of the rest. Among the bryophytes the most specialized epiphyllous Lejeuneaceae are not yet known from fossils and other members of the family are known only from the Eocene. They are obviously in the optimal phase of their evolution, producing many new species, highly adapted to the epiphyllous life-form by water-condensing, water-storing, anchoring and special vegetative reproductive organs. The same applies to a lesser extent to *Frullania* and to some Hookerioid mosses (cf. also Hutchinson in Tixier, 1966a and Schuster, 1969).

Summarizing, the epiphytic vegetation develops in the tropics to a great extent where the increased amount of evenly distributed rainfall does not result in further improvement of the host vegetation type. The quantity of epiphytic biomass and its interceptive capacity is in proportion to the otherwise ineffective rainfall. In consequence, the epiphytes, including epiphytic bryophytes, form a system in the tropical rain forest which has evolved, not to obtain more light, but to receive and to intercept more water from the rainfall, making it unnecessary to maintain direct physical and chemical contact with the soil. Adequate light (and nutrients from different sources) are essential for all epiphytes but light demand was not the factor which resulted in their evolution as it was thought for a long time. Epiphytism is an adaptation to utilize water resources which are generally surplus for the host plants.

3.5 BRYOPHYTE VEGETATION OF TEMPORARILY DRY TROPICS

In tropical areas, where the dry season or seasons are longer than 2–4 months, the water reserves of the soil are not able to compensate for the lack of adequate rainfall and the trees and shrubs reduce their evapo-transpiration surface by the way of leaf fall except for gymnosperms and succulents. After leaf-fall the interior of forest or bush is more exposed to radiation and drought. Even these conditions are tolerable for a limited amount of drought-specialized epiphytic vegetation including bryophytes. More deleterious in the dry forests and savanna are the regularly (annual, biennial or triennial) repeated bush fires which affect both terricolous and epiphytic bryophytes. In general the bryopopulations of sclerophyllous or deciduous tropical forests and woodland savannas are much poorer than those of the evergreen rain forests. In fire-affected types, the bryovegetation is restricted to ephemeral soil-inhabiting types (Ricciaceae, Ephemeraceae, Funariaceae) and to corticolous species in bark crevices of higher branches except where groundwater or mist affect stands. There are many types from the sclerophyllous evergreen forest to the semidesert bush with relatively rich to almost no bryovegetation.

3.5.1 Sclerophyllous evergreen forests

Lowland types are quite rare, due to extensive agriculture, and as a result of repeated fires most of them are already converted into open woodland and savanna types. In East Africa, among the corticolous species Frullaniaceae *(Frullania ericoides, F. spongiosa, F. diptera, F. variegata),* Fabroniaceae *(Fabronia pilifera, F. longipila, Schwetschkea schweinfurthii),* Erpodiaceae, Cryphaeaceae *(Forsstroemia producta, Cryphaea robusta)* and holostypous Lejeuneaceae *(Acrolejeunea emergens, Caudalejeunea* species) are more widespread, all belonging to the better-protected, appressed, interwoven or thread-like mat types of bryophytes. On fire-protected ground or on shady rocks *Fissidens* species (e.g. *Fissidens gumangense), Thuidium gratum* and *Trachyphyllum inflexum* are more widespread. In tropical America, e.g. in the sclerophyllous bush on the coral reefs of Cuba, Pottiaceae such as *Desmatodon sprengelii, Trichostomum involutum* are common, in the *Erpodiaceae* the commonest is *Erpodium domingense,* while on limestone rocks *Asterella elegans, Splachnobryum obtusum* and *Barbula cruegeri* occur.

In East Africa, in the montane types of sclerophyllous semi-dry forests the best known are the *Podocarpos gracilior–Juniperus excelsa–Olea chyrophylla* stands. In these high forests, as the result of cloud formation, a quite dense bryophyte vegetation develops on the trees, rich in species, but distinct from the wetter montane rain forests. The dominant moss species in the corticolous vegetation are *Leptodon beccarii, Pterogonium gracile, Forsstroemia producta, Macrocoma tenue, Herpetineuron toccoae, Calyptothecium hoehnelii, Porotrichum comorense, Neckera remota* and *N. submacrocarpa, Palamocladium sericeum* and *Rhizofabronia perpilosa* (the 'mouse moss' so-called by Sharp (in Bizot *et al.,* 1979) because of its characteristic small cushions on tree branches (Fig. 3.7a)). Among its liverworts *Frullania arecae, F. caffraria, F. obscurifolia, F. depressa, F. socotrana, F. truncatiloba, Plagiochila sinuosa* and other more drought-

Fig. 3.7 (a) 'Mouse moss', *Rhizofabronia perpilosa* community on large branches of a sclerophyllous, dry *Podocarpus gracilior–Juniperus excelsa–Hagenia* forest in West Kilimanjaro area, at 2600 m altitude. (b) Dry woodland in the Kiboriani Hills (Tanzania) at 1900 m altitude, with small epiphytic cushions of *Brachymenium leptophyllum* on the fire-protected parts of the deciduous *Dissotis bussei*. (c) Velloziaceae (*Xerophyta scabrida*) bush during the dry season on Mt. Mindu near Morogoro, Tanzania, at 500 m altitude. (d) Dew condensation from mist on *Brachymenium acuminatum* sporophytes densely covering the Velloziaceae stem, Kiboriani Hills.

resistant species are common. No epiphyllous synusia occur in the sclerophyllous montane forests and pendulous Meteoriaceae are common only in the uppermost zone near the forest line. Such types of sclerophyllous montane forests are common in Ethiopia, Kenya, and in Tanzania on Mt. Meru, Ngurdoto crater and Ngorongoro crater rim, West Kilimanjaro, further in the Southern Highland, Malawi and in Zimbabwe above 1800 m altitude up to the forest line. In Cuba, the *Pinus maestrensis* and *P. cubensis* forests in Oriente province belong to this type on granitic or on serpentine rocks. The latter are drier and relatively poor in bryophytes. In the *Pinus maestrensis* forests of the Sierra Maestra *Leucolejeunea xanthocarpa, Drepanolejeunea trigonophylla* and *Anoplolejeunea conferta* are the commonest inhabitants of pine bark. On the ground, *Plagiochila ekmanii, Kurzia capillaris, Syzygiella perfoliata, Campylopus lamellinervis* and *Bryum andicolum* form an often continuous bryophyte layer.

In the drier types of *Pinus cubensis* forests on serpentine rocks epiphytes occur sparsely, e.g. *Cheilolejeunea rigidula, Pycnolejeunea contigua, Octoblepharum albidum, Sematophyllum sericifolium* and *Groutiella apiculata.* On the chemically peculiar serpentine soil surface curious lichens often form a mass vegetation and the occurrence of *Fissidens imbricatus* and *F. palmatus* seem to be characteristic. Besides these, xerophytic mosses also occur. The same applies to the dry microphyllous bush vegetation on serpentine rocks.

3.5.2 Deciduous woodlands

Deciduous woodlands are widespread in seasonal tropical climates, where the dry season is longer than five months. In the canopy pinnate leguminous trees are usually dominant and a continuous grass layer of Gramineae and Cyperaceae is formed. Many of these woodlands are derivatives of former dry forests, resulting from regular burning and grazing over a long period of time. In south-east tropical Asia and south-east tropical Africa, and also in the New World, large areas are covered by deciduous woodlands. The best-known representative of this vegetation type is the African 'miombo', a woodland of *Brachystegia, Julbernardia* and *Copaifera* species. Bryophytes in fire-protected places are often richly represented by xerophytic species on the ruptured bark of miobo trees: *Fabronia longipila, Macrocoma tenue, Braunia captoclada, Rhachithecium perpusillum, Erythrodontium rotundifolium, Trachyphyllum species, Frullania ericoides, Acrolejeunea emergens, Mastigolejeunea carinata.* At sites of higher humidity the characteristic small cushion-forming *Brachymenium* species, first of all *Brachymenium angolense* and *B. acuminatum* and *B. pulchrum.* The fire-effected soil of miombo is bare and bryophytes are not common but at protected sites, termite mounds, in

Fig. 3.8 (a) *Asterella linearis* on the rocks in the dry, deciduous *Combretum–Pterocarpus* woodland of Uluguru Mountains, at 1200 m altitude. (b) Shaded granite streambed stones covered by a *Cololejeunea himalayensis–Erpodium biseriatum* community. Evergreen riverine forests surrounded by dry woodland, at 600 m altitude in the Uluguru Mountains, Tanzania. (c) Open giant groundsel (*Senecio cottonii*) stand at 4000 m altitude in the Barranco Valley of Mt. Kilimanjaro. Bryophyte cushions are present on the bare trunk parts without leaf cover. (d) *Tetraplodon mniodes* cushions developed on leopard dung, Barranco Valley on Mt. Kilimanjaro, 3900 m altitude.

gullies or on smaller stones many *Fissidens* species such as *F. sciophyllus, F. kegelianus* and *F. ventroalaris* occur. Roadcut surfaces are usually densely covered by *Brachymenium acuminatum, Microdus minutus, Philonotis hastata* and in some places by *Fossombronia husnotii.* Rocks in deciduous woodland have interesting thalloid liverwort vegetation, visible only during the rainy season, formed by *Riccia canescens* and other species like *Mannia capensis, Exormotheca pustulosa* and *Asterella linearis* (Fig. 3.8a) accompanied by such xerophytic mosses, as *Barbula indica, Gyroweisia latifolia, Hyophila potieri, Weisiopsis plicata, Anoectangium hanningtonii* and by *Trachyphyllum inflexum.*

On insolated larger rocks within the deciduous woodland zone of Africa, very characteristic xerophytic monocots, members of the Velloziaceae (*Xerophyta* and *Vellozia* species), form a vegetation type (Fig. 3.7c). These small branching shrubs have persistent fibrous leaf bases and this spongy mass readily absorbs water and even conducts it from the soil and ensures for epiphytes more favourable conditions, than should ensue from the dry environment. Therefore specially adapted orchids (e.g. *Polystachya tayloriana*) inhabit this niche together with lichens and bryophytes such as *Fissidens lacouturei, F. subobtusatus, F. gumangense, Fabronia abyssinica* and *Brachymenium acuminatum.* The rock surface among the shrubs is usually covered by poikilohydric *Selaginella* species (*S. dregei, S. mittenii*), *Octoblepharum albidum, Leptodontium viticulosoides* and *Barbula pertorquata.*

Savanna grasslands with scattered trees are extremely poor in bryophytes and in semi desert scrub there is practically no bryophytic vegetation except a few *Riccia* and *Fissidens* species in protected rock crevices.

3.5.3 Riverine forests

In the temporarily dry tropics riversides and other sweet groundwater-influenced vegetation types are the richest in bryophytes. In the riverine or gallery forests there usually occur many elements of more hygrophylous vegetation types, e.g. rain forest species, in the deciduous woodland belt of Central Africa (Pócs, 1981) although there are special 'riverside' bryophytes in arid zones such as *Micropoma niloticum* in the gallery wood of the Sudan desert (cf. Pettet, 1967); or *Loiseaubryum ephemeroides* along the Chari river in Chad)Bizot, 1976).

The most luxuriant type of African riverine forest is represented at Victoria Falls along the Zambezi River because of the continuous spray effect. The rich moss vegetation was first characterized by Sim (in Sim and Dixon, 1922). Although some parts of the forest around the falls are called 'rain forest', its bryovegetation is not comparable with that of true rain forests and epiphytic bryophytes are scanty. There are no epiphyllae at all and most of the species are concentrated on the shady rock surfaces and include *Fissidens dubiosus, Calyptothecium acutifolium, Mastigolejeunea rhodesica* and *Dicranolejeunea madagascariensis.* All over East Africa, *Stereophyllum radiculosum* seems to be a common element in riverine forests, together with *Hyophila acuminata, Thuidium varians* and *Fissidens helictocaulos.* More interesting corticolous species are *Erythrodontium rotundifolium, Groutiella laxotorquata* and *Rauiella subfilamentosa.* In mountain sites *Cololejeunea himalayensis* covers shady rock surfaces, associated with *Lejeunea ecklonii, Erpodium biseriatum, Dumortiera hirsuta* (Fig. 3.8b). On drier, more exposed rocks *Erpodium beccarii, Exormotheca pustulosa* and *Trachyphyllum inflexum* grow together. In West and Central Africa *Plagiochila africana, P. salvadorica, P. praemorsa, P. strictifolia, Frullania ericoides, Lopholejeunea abortiva, L. obtusilacera, Acrolejeunea emergens, Archilejeunea autoica, A. elobulata, A. linguaefolia, Cheilolejeunea newtonii* and *Lejeunea setacea* seem to be the characteristic epiphytic liverworts of gallery forests (Pócs, 1981).

In the lowland riverine forests of Cuba, I observed bryophytes in the same type of situation. These included a few bark-inhabiting species such as *Lejeunea pililoba, Cheilolejeunea adnata, C. trifaria, Frullania ericoides, Octoblepharum albidum* and *Calymperes richardii, Fissidens prionodes, Hyophila involuta,* and *Vesicularia amphibia* on limestone rocks at riverside, while on temporarily immersed soil surfaces *Tortula mniifolia* was common. Very rarely, one widespread epiphyllous species, *Cololejeunea cardiocarpa* also occurs.

3.5.4 Mist epiphytes

Cases of mist-induced epiphytic vegetation among otherwise dry conditions are known from several places. Mist and dew are forms of precipitation which become important during dry periods or when their amount exceeds other water sources. Even under desert conditions epiphytic bryophytes may develop in a mist-inducing oasis as was observed by Kassas (1956) in *Dracaena* stands in coastal Sudan.

In East Africa, the case of mist-affected, but otherwise dry, deciduous woodland is known from the top and ridges of inselbergs, places of frequent cloud formation even during dry season (Pócs, 1976a, c, 1980). Here, at altitudes between 800–1200 m, in a stand of deciduous *Brachystegia* trees, with an undergrowth of grasses, completely dry during the dry season, mist and dew allow a very rich development of epiphytic vegetation of both vascular and cryptogamic plants. On branches of *Brachystegia* and even of *Acacia* trees there are twelve fern species, two hanging *Lycopodiums* and several orchids accompanied by a very large number of bryophytes and lichens, mostly similar to the heliophilous, ramicolous species of submontane rain forests at the same altitude. The mist and dew do not affect the tree and herb growth, only the epiphytes, and often only the side of trees exposed to the moisture-carrying winds. The bryophytes occur in very large quantity, forming garlands of hanging Meteoriaceae and Pterobryaceae, feathery and pinnate forms, large cushions and dense mats. Wind-exposed soil and rock surfaces among the dry grass are also covered by bryophytes. Even more surprising are the mist epiphylls (Pócs, 1978). In closed dry deciduous forest, on hilltops in the very narrow strip of frequent cloud formation at 750–770 m altitude a relatively rich epiphyllous vegetation develops on the only evergreens: on the sclerophyllous leaves of *Encephalartos* sp. (*Cycadaceae*) and on the succulent leaves of a *Sanseviria* species. In Cuba, I observed an epiphyllous liverwort (*Cololejeunea cardiocarpa*) in a similar situation on the succulent leaves of an *Agave* species.

3.6 BRYOPHYTES IN THE WOODY COMMUNITIES ABOVE THE FOREST LINE

3.6.1 Ericaceous bush

In equatorial mountains, at altitudes where night frosts occur, the high or dwarf forest ends and gives place to shrubby vegetation. In East Africa and in many other tropics this is represented by ericaceous bushes, in East Africa between 3000 and 4000 m by a 1–2 m tall *Philippia* or by *Stoebe kilimandscharica* scrub. These are heathlike communities and the soil is more or less peaty or at least very poor in nutrients. The *Philippia* bushes sometimes bear epiphytes, as on Mt. Kilimanjaro, such as *Frullania capensis, Anomalolejeunea pluriplicata, Colura berghenii* and *Metzgeria convexa.* The ground layer, except when too dry, is often dominated by *Polytrichum* species (*P. piliferum, P. subformosum*), *Breutelia diffracta* and *Chandonanthus hirtellus* var. *giganteus.* On wet rocks *Lethocolea congesta* is common, while on dry acidic soil surfaces *Anomobryum filiforme, Ceratodon purpureus, Gongylanthus ericetorum, G. richardsii, Jungermannia abyssinica* and *J. pocsii* may be found.

3.6.2 Páramo vegetation

Páramo vegetation develops in humid equatorial regions above the level of or intermixed with microphyllous vegetation or ericaceous bush where there are marked diurnal variations in climate (relatively high day temperatures but with frost at night). It consists of pachycaulous trees with special adaptation to the diurnal change of temperature. The giant rosette plants extend tree growth well above the forest line and even above the general timberline and therefore offer special habitats for some epiphytes at an unusual altitude above sea level, between 3000 and 4500 m.

In tropical Asia true páramo is not known. In the South American Cordilleras many *Espeletia* species (Compositae) comprise this vegetation type. Their bryovegetation is briefly summarized by Griffin (1979). In tropical East Africa the giant groundsels (*Senecio* subgenus *Dendrosenecio,* Compositae) are the best known and most typical representatives of páramo trees. Although giant *Lobelia* species are also numerous they never form páramo woodland alone and epiphytes are not known on them. The *Senecio* stands are of different structure in the different African high mountains according to the climatic conditions. In the most humid, the Virunga volcanoes and Ruwenzori mountains they form a continuous belt of closed, 6–8 m high forests between 3800 and 4300 m. On Mt. Elgon and Mt. Kenya they form more open woodlands on the slopes, while on Mt. Kilimanjaro *Senecio* páramos are restricted to the valley bottoms except for the most humid south-south-west slope along Umbwe Route where its stands occupy large areas, intermixed with *Erica* and *Philippia*.

In the *Senecio* stands, most epiphytes live on those parts of the trunks where the dead, shrivelled leaves, which usually form a protecting cover on the stem, are shed and the bark is visible. Bryophytes usually form large cushions on *Senecio* trunk and branches and are most commonly present at the branching points (Fig. 3.8c). Almost all are xerophytes, resistant to severe insolation and frost exposure. Some are endemic afro-alpine species, others are disjunct temperate elements (see Table 3.8).

Hedberg (1964), studying the afro-alpine plant communities, dealt in detail with the composition of *Senecio* páramo woodlands and described their composition, including the bryophytes. It is clear from the literature that there are marked differences between the epiphytes of the humid *Senecio* stands of the Ruwenzori Mountains and those of the much drier Mt. Kilimanjaro stands.

The ground layer of *Senecio* páramos is very rich in bryophytes. Usually a mosaic of habitats is observable below the trees composed of spring-bog niches along streamlets and drier soil or rocky niches. On Mt. Kilimanjaro the dominant species of the spring-bog synusia are *Antitrichia curtipendula, Campylopus stramineus* and *Hygrohypnum hadbergii.* On dry soil and rocks the dominants are *Anastrophyllum auritum, Gynomitrion laceratum* and *Leptodontium tenerascens*.

In the *Philippia* and in páramo belts two further interesting communities are worth mentioning. Richards and Argent (1968) described the interesting bryovegetation of lava rock caves from Cameroon Mountain. There are very similar lava caves on Mt. Kilimanjaro, between 3300–3900 m, where on the inner walls in drier places *Rhizofabronia perpilosa, Cephaloziella* cf. *umtaliensis, Diplophyllum africanum* and *Gymnomitrion laceratum* occur; on wetter surfaces *Amphidium cyathicarpum, Pohlia afro-cruda, Blepharostoma trichophyllum, Lethocolea congesta* and

Table 3.8 Epiphytes of *Dendrosenecio* species in the Ruwenzori and Kilimanjaro Mountains. All from 3300–4500 m (mostly 3800–4200 m) altitude, well above the continuous forest line. Sources: Hau: Hauman, 1940; Hed: Hedberg, 1964; Klö: Klötzli, 1958; Lew: Lewinsky, 1978; BFLP: Lewinsky in Bizot, Friis, Lewinsky and Pócs, 1978; P: own records.

	Ruwenzori Mountains	Kilimanjaro Mountains
Hepaticae		
Anastrophyllum auritum	Hed	
Lophocolea molleri	Hed	
Lophocolea muricata	Hed	
Lophozia ruwenzorensis	Hed	
Plagiochila colorans	Hed	
Radula boryana	Hed	
R. meyeri	Hed	
Musci		
Antitrichia curtipendula	Hau, Hed	—
Bartramia afro-ithyphylla	Hed	—
Brachythecium afroglareosum	—	P
B. ramicola	Hed	—
B. vellereum	—	P
Breidleria africana	Hed	—
Bryoerythrophyllum rubrum	—	P
Bryum capillare s.l.	Hed	—
B. ellipsifolium	—	P
Campylopus stramineus	Hed	—
Drepanocladus uncinatus	Hed	—
Erythrophyllum papillinerve	Hau	—
Eurhynchiella decurrens	—	BFLP
Grimmia perichaetialis	Hed	—
Hedwigidium integrifolium	—	P
Hypnum cupressiforme	Hed	—
Leptodontiopsis fragilifolia	Hau, Hed	P, BFLP
Leptodontium sublaevifolium	Hed	—
L. tenerascens (only on decaying stems)	—	Klö
L. viticulosoides	—	P
Leucodon dracaenae	—	BFLP
Neckera macrocarpa	—	Klö
Orthotrichum affine	Hau, Lew	Lew, P, BFLP
O. arborescens	Lew	—
O. rupestre	—	Lew, P
O. speciosum	—	P, Lew
Tortula cavallii	Hau, Hed	BFLP, P
T. schmidii	—	BFLP, P
Rhacomitrium alare	—	P
Streptopogon erythrodontus	Hau	—
Zygodon erosus	Hau	—
Z. intermedius	Hed	—

Rhynchostegiella holstii form thin interwoven mats or short turfs. The presence of a few forest elements at this high altitude is ensured by the shady wet niche of the caves.

The other interesting community that occurs in several places (e.g. Barranco Valley) on Mt. Kilimanjaro, between 3600–4000 m, (Fig. 3.8d) is the very common coprophilous bryosynusium of the cushion-forming *Tetraplodon mnioides,* which always develops on old leopard dung composed of rock hyrax bones and hairs. This small mammal is the usual prey of leopards, and quite common at this altitude.

ACKNOWLEDGMENTS

The Author expresses his gratitude to Mme H. Bischler, M. Bizot, P.P. Duarte B., S.R. Gradstein, R. Grolle, E.W. Jones, Sra D. Reyes M., Mme S. Jovet-Ast, J.L. De Sloover, T.P. Tixier, J. Vána, C. Vanden Berghen, and other specialists for help in identifying bryophyte specimens and to Drs A.J.E. Smith and D.H. Vitt for the careful reading of the manuscript and for their valuable suggestions. He is also grateful to Ms G. Kis and to Mrs J. Forró for their technical assistance.

REFERENCES

Allorge, P., Allorge, V. and Persson, H. (1938), *Bol. Soc. Broteriana,* **13,** 211–31.
Arnell, S. (1962), *Svensk Bot. Tid.,* **56,** 334–50.
Balázs, D. (1973), *Galápagos,* Gondolat, Budapest.
Baynton, H.W. (1969), *J. Arn. Arb.,* **49,** 419–30.
Bischler, H. (1964), *Rev. bryol. lichénol.,* **33,** 15–179.
Bischler, H. (1968), *Rev. bryol. lichénol.,* **36,** 45–55.
Bizot, M. (1976), *Rev. bryol. lichénol.,* **42,** 843–55.
Bizot, M., Friis, I., Levinsky, J. and Pócs, T. (1978), *Lindbergia,* **4,** 259–84.
Bizot, M., Pócs, T. and Sharp, A.J. (1979), *J. Hattori bot. Lab.,* **45,** 145–65.
Busse, W. (1905), *Ber. dt. bot. Ges.,* **23,** 164–72.
Chen, Pan-chieh and Wu, Pan-cheng (1964), *Acta phytotax. sin.,* **9,** 213–76.
Delgadillo, C.M. (1979), *Bryologist,* **82,** 432–49.
Fulford, M., Crandall, B. and Stotler, R.E. (1970), *J. Arn. Arb.,* **51,** 56–69.
Gams, H. (1932), Bryo-Cenology (Moss Societies). In: *Manual of Bryology,* (Verdoorn, F., ed.), pp. 323–66, Nijhoff, The Hague.
Giesenhagen, K. (1910), *Ann. Jard. Bot. Buitenzorg,* suppl. **3,** 711–69.
Giacomini, V. and Ciferri, R. (1950), *Atti Ist. Bot. Univ. Pavia* Ser. 5, **9,** 211–17.
Gilli, A. (1975), *Feddes Rep.,* **86,** 233–52.
Gradstein, S.R. (1979), *Abstr. Bot. (Budapest)* **5,** Suppl. 3, 47–50.
Griffin, D. (1979), Briofitos y liquenes de los paramos. In: *El Medio Ambiente Páramo* (Salgado-Labouriau, M.L., ed.), Caracas.
Griffin, D., Breil, D.A., Morales, M.I. and Eakin, D. (1974), *Misc. bryol. lichenol.,* **6,** 174–5.

Harrington, A.J. (1967), Ecology of epiphyllous liverworts in Sierra Leone. Report submitted to the Royal Society.
Hattori, S. (1975), *Univ. Mus., Univ. Tokyo, Bull.*, **8**, 206–42.
Hattori, S., Iwatsuki, Z. and Kanno, S. (1956), *J. Hattori bot. Lab.*, **16**, 90–6.
Hattori, S. and Kanno, S. (1956), *J. Jap. Bot.*, **31**, 10–14.
Hauman, L. (1940), *Bull. Jard. Bot. Bruxelles*, **16**, 311–53.
Hedberg, O. (1951), *Svensk. bot. Tidskr.*, **45**, 140–202.
Hedberg, O. (1964), *Acta Phytogeogr. Suec.*, **49**, 1–144.
Herzog, T. (1916), *Biblioth. Bot.*, **87**, 1–347+T. 5–8.
Herzog, T. (1926), *Geographie der Moose*, Fischer, Jena.
Herzog, T. (1930), *Ann. Bryol.*, **3**, 126–49.
Herzog, T. (1939), *Ann. Bryol.*, **12**, 98–122.
Hopkins, B. (1960), *E. Afr. Agric. Forest J.*, **25**, 255–8.
Horikawa, Y. (1950), *Hikobia*, **1**, 1–6.
Iwatsuki, Z. (1960), *J. Hattori bot. Lab.*, **22**, 159–350.
Iwatsuki, Z. and Hattori, S. (1956), *J. Hattori bot. Lab.*, **16**, 83–90.
Iwatsuki, Z. and Hattori, S. (1959), *J. Hattori bot. Lab.*, **21**, 157–82.
Iwatsuki, Z. and Hattori, S. (1968), *J. Hattori bot. Lab.*, **31**, 189–97.
Iwatsuki, Z. and Hattori, S. (1970), *Mem. Nat. Sci. Mus. Tokyo*, **3**, 365–74.
Johansson, D. (1974), *Acta Phytogeogr. Suec.*, **59**, 1–129.
Jones, E.W. (1960), *Nature*, **188**, 432.
Jovet-Ast, S. (1949), *Rev. bryol. lichénol.*, **18**, 125–46.
Kamimura, M. (1939), *J. Jap. Bot.*, **2**, 52–63.
Kassas, M. (1956), *J. Ecol.*, **44**, 180–94.
Klötzli, F. (1958), *Ber. Geobot. Forsch. Inst. Rübel in Zürich* (1957), 33–59.
Láng, B. (1974), *Microkosmos*, **63**, 78–82.
Lewinsky, J. (1978), *Bot. Tidsskrift*, **72**, 61–85.
Longman, K.A. and Jenik, J. (1974), *Tropical Forest and its Environment*, Longmans, London.
Lötschert, W. (1959), *Bot. Stud.*, **10**, 1–88.
Lundgren, L. and Lundgren, B. (1979), *Geog. Ann. Svenska Sällsk. Antropol.*, **61A**, 157–178.
Lundgren, L. and Rapp, A. (1974), *Geog. Ann. Svenska Sällsk. Antropol.*, **56A**, 251–60.
Massart, J. (1898), *Ann. Jard. Buitenzorg*, **2**, 103–8.
Olarinmoye, S.O. (1974), *J. Bryol.*, **8**, 275–89.
Olarinmoye, S.O. (1975a), *Rev. bryol. lichénol.*, **41**, 457–63.
Olarinmoye, S.O. (1975b), *J. Bryol.*, **8**, 357–63.
Orbán, S. (1977), *Acta Bot. Acad. Sci. Hung.*, **23**, 167–77.
Pandé, S.K., Srivastava, K.P. and Ahmad, S. (1957), *J. Ind. Bot. Soc.*, **36**, 335–47.
Pessin, L.J. (1922), *Bull. Torr. Bot. Club*, **49**, 1–14.
Petit, E. and Symons, F. (1974), *Bull. Jard. Bot. Nat. Belg.*, **44**, 219–47.
Pettet, A. (1967), *Trans. Br. bryol. Soc.*, **5**, 316–31.
Pócs, T. (1968a), *Fragm. Florist. Geobot.*, **14**, 495–504.
Pócs, T. (1968b), *J. Hattori bot. Lab.*, **31**, 65–93.
Pócs, T. (1974), *Acta Bot. Acad. Sci. Hung.*, **20**, 115–35.
Pócs, T. (1975a), *Bull. Acad. Pol. Sci. Cl. II Sér. Sci. biol.*, **22**, 851–3.

Pócs, T. (1975b), *Boissiera,* **24,** 125–8.
Pócs, T. (1976a), *Boissiera,* **24,** 477–98.
Pócs, T. (1976b), *Boissiera,* **24,** 499–503.
Pócs, T. (1976c), *Acta Bot. Acad. Sci. Hung.,* **22,** 163–83.
Pócs, T. (1976d), *J. Hattori bot. Lab.,* **41,** 95–106.
Pócs, T. (1978), *Bryophytorum Bibliotheca,* **13,** 681–713.
Pócs, T. (1980), *Acta Bot. Acad. Sci. Hung.,* **26,** 143–67.
Pócs, T. (1982), *Boissiera,* **29** (in press).
Renner, O. (1933), *Planta Berlin,* **14,** 215–87.
Richards, P.W. (1932), Ecology. In: *Manual of Bryology* (Verdoorn, F., ed.), pp. 367–95, Nijhoff, The Hague.
Richards, P.W. (1935), In: Dixon, H.N., *J. Linn. Soc. Bot.,* **50,** 60–5.
Richards, P.W. (1950), In: Herzog, T., *Trans. Br. Bryol. Soc.,* **1,** 275–80.
Richards, P.W. (1954), *Vegetatio,* **5–6,** 319–27.
Richards, P.W. (1957), *The Tropical Rain Forest,* 2nd edn., Cambridge University Press, Cambridge.
Richards, P.W. and Argent, G.C.G. (1968), *Trans. Br. Bryol. Soc.,* **5,** 573–86.
Rodin, L.E. and Bazilevich, N.I. (1966), *Forest. Abs.,* **27,** 369–72.
Ruinen, J. (1961), *Plant Soil,* **15,** 81–109.
Schnock, G. (1972), *Bull. Soc. roy. Bot. Belg.,* **105,** 143–50.
Schuster, R.M. (1959), *Bryologist,* **62,** 52–5.
Schuster, R.M. (1969), *Taxon,* **18,** 46–91.
Seifriz, W. (1924), *J. Ecol.,* **12,** 307–13.
Sim, T.R. and Dixon, H.N. (1922), *South African J. Sci.,* **18,** 294–335.
Simon, T. (1974), *Acta Bot. Acad. Sci. Hung.,* **20,** 341–8.
Simon, T. (1975), *XII Int. Bot. Cong. Abst. Leningrad,* **1,** 87.
Simon, T. and Lang, B. (1972),
Sjörgren, E. (1975), *Svensk Bot. Tidskr.,* **69,** 217–88.
Sjörgren, E. (1978), *Mem. Soc. Brot.,* **26,** 1–283.
Steenis, C.G.G. van (1972), In: Amir Hamzah Moehamad Toha, *The Mountain Flora of Java,* pp. 1–56. Brill, Leiden.
Steere, W.C. (1970), Bryophyte Studies on the Irradiated and Control Sites in the Rain Forest at El Verde. In: *A tropical rain forest: A study of irradiation and ecology at El Verde, Puerto Rico. D.C.* (Odum, H.T., ed.), pp. D–213–225, U.S. Atomic Energy Commission, Washington.
Stehlé, H. (1943), *Caribbean Forest,* **4,** 164–82; **5,** 20–43.
Szabó, M. and Csortos, C. (1975), *Acta Bot. Acad. Sci. Hung.,* **21,** 419–32.
Temple, P.H. (1972), *Geog. Ann. Svenska Sällsk. Antropol.,* **54A,** 20–43.
Temple, P.H. and Rapp, A. (1972), *Acta svenska sällsk. antropol.,* **54A,** 157–93.
Thorold, C.A. (1952), *J. Ecol.,* **40,** 125–42.
Thorold, C.A. (1953), The control of black-spot disease of cocoa in the Western Region of Nigeria. In: *Report 1953 Cacao Conference,* pp. 108–115. Cocoa, Chocolate and Confectionary Alliance Ltd., London.
Thorold, C.A. (1955), *J. Ecol.,* **43,** 219–25.
Tixier, P. (1966a), *Flore et Végétation Orophiles de l'Asie Tropicale: Les Epiphytes du flanc méridional du Massif Sud Annamitique.* Published by the Author, Paris.

Tixier, P. (1966b), *Science Nature,* **78,** 28–37.
Vitt, D.H. (1972), *Nova Hedwigia,* **21,** 683–711.
Vit, D.H., Ostafichuk, M. and Brodo, I.M. (1973), *Can. J. Bot.,* **51,** 571–80.
Whitmore, T.C. (1966), *Guide to the Forests of the British Solomon Islands,* Oxford University Press, London.
Winkler, S. (1967), *Rev. bryol. lichénol.,* **35,** 303–69.
Winkler, S. (1970), *Rev. bryol. lichénol.,* **37,** 47–55.
Winkler, S. (1971), *Rev. bryol. lichénol.,* **37,** 949–59.

Chapter 4

Desert Bryophytes

G.A.M. SCOTT

4.1 INTRODUCTION

This is a tale of ignorance; my own and other people's. Knowledge of desert bryophytes is so incomplete, and the publications so fragmentary that I have had to rely largely on my own observations in Australia. Most of these have been made in or near the Big Desert in north-west Victoria, which is neither big (by world, or even Australian standards) nor a true desert.

The term 'desert', in the biological sense, implies a region where bryophytes (if there are any) will be subjected at least intermittently to severe drought, possibly during a time of great heat. In other words (Noy-Meir, 1972) 'water-controlled ecosystems with infrequent, discrete and largely unpredictable water inputs'. For the purposes of this chapter, I generally exclude the cold deserts of Antarctica, Siberia etc. (Logan, 1968), but include many, more temperate, regions and habitats which heat up as well as dry out during summer. Desert bryophytes, in this sense, have to tolerate great heat, by human standards, and great drought, by plant standards. As well as these two prime characteristics of the habitat there may also be associated secondary characteristics: excessive concentrations of unwanted and even toxic solutes e.g. sodium chloride, sodium sulphate; a concomitant shortage of important soil components, notably humus, phosphate and nitrate; and both the disturbing and erosive effects of substrate mobility. More intractable still, from the investigator's point of view, are the effects of the desert ecosystem as a whole (Ross, 1969). The areas over which these conditions obtain include southern

Note: Authorities for bryophyte names are not cited where they follow *Index Muscorum* and *Index Hepaticarum,* or are found in the reference cited. Those for flowering plants follow Willis (1970, 1973) and for lichens Weber and Wetmore (1972). Authorities for Australian *Riccia* names follow Na Thalang (1980).

North America and western and southern South America, North and South Africa and parts of the Mediterranean region, the Middle East, Australia, India and a belt stretching from Turkey across Afghanistan to Mongolia and China (Petrov, 1976). Fortunately, these are precisely the regions where bryophytes are less abundant so the task of covering this vast area is not as daunting as it might seem. There are many parts where bryophytes seem to be absent or at least so inconspicuous that they are not recorded (e.g. Kassas and Imam, 1959; Orshan and Zohary, 1962). There are floras of desert regions by Collenot *et al.* (1960), Flowers and Evans (1966), Jovet-Ast and Bischler (1970), McCleary (1959, 1968), Nash *et al.* (1977), Novichkova-Ivanova (1976), Pandé and Bhardwaj (1953), Pettet (1966a) and Townsend (1966).

4.2 HABITATS

The chief difficulty in studying the ecology of desert bryophytes, and one of the reasons for the scarcity of such work, is the discontinuity of the communities in time and space. Under the pressure of extreme desiccation, communities (of thallose liverworts especially) can disappear overnight, rolled up and shrivelled and sunk into the soil surface, only to reappear briefly again days later with a shower of rain or heavy dew (Cloudsley-Thompson and Chadwick, 1964). From year to year, the communities which will develop are as unpredictable as the rain that brings them to light, and although there are – on average – favourable and unfavourable patches of ground, the location of the richest communities shifts in response to trivial and unpredictable events. Where rainfall is too erratic (Pettet, 1966a) bryophytes will be virtually or totally absent. The cover of bryophytes is never great over large areas: Nash *et al.* (1977) recorded 0–0.27% cover in a part of the Sonoran desert and, although (as with all plants) cover must vary from 0–100% depending on the scale examined, the spatial discontinuities in dry lands are such that the regional average will always be very low. It therefore follows that random sampling of a large area, to achieve some kind of hypothetical average community, is impracticable and the only feasible method of study is partitioning into discrete habitats at an appropriate level of resolution. From my own experience I believe one can recognize some half dozen habitats in dry lands where bryophytes tend to grow, but there are bound to be more, even without further subdivision.

4.2.1 Rocks and rock crevices

From this category I exclude the temporary pools formed in rock basins, in which, for example, Jelenc (1957) has recorded *Riella notarisii* from Morocco. Danin (1978a) has shown that there is a fairly constant increase in species number with increasing area of sample in the Sinai desert, except for smooth-faced rock outcrops, which have a richer species list. This enrichment is brought about because 'the smooth surface of the hard rocks neither absorbs nor stores the rain water, and even after a weak shower water runs off and concentrates in the crevice . . . The smooth-faced rocks with a milder water regime are also a more heterogeneous habitat. Different sizes of soil pockets and crevices produce diverse water regimes and support many different species'. Although referring to vascular plants, his remarks apply equally to bryophytes. A method of investigating this variety of microhabitat in desert rock crevices has been shown by Kunkel (1975) for Gregory Canyon near Boulder, Colorado. Although hampered (or perhaps stimulated) by a lack of instrumentation, he assessed each of some 570 quadrats (10 × 10 cm) in terms of slope, abundance of surface water seepage, presence of crevices and degree of exposure to direct sunlight, as well as cover of each species. This represents a more sensitive extension of the habitat partitioning adopted by Davis (1951) for cliff vegetation and by Foote (1966) for rock outcrops in Wisconsin. Kunkel used an information measure to assess the degree to which individual species were restricted to particular micro-habitat types, i.e. to calculate the breadth of the ecological niche with respect to the various factors measured. All four ecological factors turned out to be of importance and the method permitted an assessment of the relative importance of each factor for each of the 14 principal moss species, as well as an overall quantification of ecological amplitude for each species. The general conclusions were that rock slope was the single most important factor (assessed in the categories: overhanging, vertical, steep, moderate, gentle) and that niche breadth increased in the different categories of slope with decrease of species richness – perhaps an indication of species expanding to fill the available space in the absence of competition.

In south-west Africa, with a summer wet season of between 100 and 500–800 mm year^{-1}, and a winter dry season lasting for six to eight months, Volk (1979) has recorded a rather rich flora of Marchantiales from clefts in rock outcrops, including *Plagiochasma* spp., *Targionia*, *Athalamia*, *Mannia*, *Oxymitra* and *Exormotheca pustulosa*.

Friedmann and Galun (1974) and Vogel (1955) give accounts of hypolithic algae in deserts, but mosses may occupy the same habitat – under quartz stones in the Namib desert, where they are protected from desiccation and receive sufficient light for growth (D.F. Gaff, personal communication).

It seems to be a general rule that bryophytes flourish best in the crevices where shade and moisture may be more frequently available. Danin (1972) observed that high run-off from rock faces converted the crevices and soil pockets into nearly mesic habitats, and observed that 'even weak showers of 1 mm led to small floods concentrated in the crevices'. McCleary (1968) found the same feature in Arizona. On fully exposed rock surfaces colonization, initially at least, is in minute fissures where protection and water supply are likely to be enhanced, on a miniature scale, but increasing heat and dryness of the habitat tend to favour lichens rather than bryophytes.

4.2.2 Sand dunes

There is no essential difference between the conditions of inland desert sand dunes and those of coastal sand dunes, as considered by Moore and Scott (1979), except for the general absence of salt spray inland. Rather, the same conditions have to be withstood but sometimes in greater intensity: mobility of substrate, high levels of solar radiation and temperature, high evaporation rates and, correlated with these, extreme desiccation. Of these, mobility is the one with which bryophytes are least able to cope. Moore and Scott (1979) showed that *Barbula torquata* and *Tortula princeps* on shore dunes in Australia could tolerate burial to depths of 4 cm in the field, growing up through the overlying sand; and these are major species in the inland semi-arid regions also. Similar depths have been recorded by Gimingham (1948) for *Barbula fallax* and *Bryum pendulum* ('about an inch') and, in the laboratory, by Birse *et al.* (1957): 4 or 5 cm for *Ceratodon purpureus, Bryum pendulum, Brachythecium albicans* and *Pohlia annotina,* and up to 6 or 7 cm for *Polytrichum juniperinum* and *P. piliferum.*

No Australian species are likely to tolerate much greater depths of burial than that, nor have they any mechanism to withstand excavation by wind, although the soil crusts and vegetation mats in which they participate decrease the chances of erosion greatly, and excavated plants are normally capable of renewed growth in their displaced position, provided conditions are congenial. It is probably true of all deserts that really mobile sands cannot be colonized except by vascular plants. Danin (1978b) says that, in the Negev desert, 'in mobile sands no such algae and mosses occur'. Small size and low growth rates are themselves sufficient to ensure this.

Once sand dunes have reached the level of stability with which mosses can cope, the power of subsequent stabilization becomes very great. Both Birse and Gimingham (1954) and Moore and Scott (1979) have found that the erect growth forms (turfs) predominate in such pioneer conditions. The abundance of rhizoid production by mosses is a main contributor to the process of stabilization and Moore and Scott (1979) have proposed the ratio of dry weights of gametophore/rhizoids as a measure of progress in colonization. In desert dunes, the growth period when adequate moisture is available will be, if not spasmodic, at least greatly interrupted and it is during the hottest and driest periods, when surface sand mobility is greatest, that mosses are most prone to burial and erosion. At such times, they may be the main agents binding the surface together, since ephemeral flowering plants will have withered and gone. Eventually, continual build-up of humus in the sand will lead to more or less permanent stabilization, and the development of a perennial vascular flora. Danin (1978b) illustrates 'biological mounds' in the Negev, initiated by *Stipagrostis scoparia* (Trin. et Rupr.) de Wint., which ceased growing when the sand was stabilized and then became colonized by algae and mosses. Mosses thus act as the consolidators of sand dunes, not as primary colonists. Given the 'resurrection-plant' capabilities of dune mosses, their success in this role is evidently precariously determined by the balance between adequate moisture for growth and the intensity of erosion/burial by the agency of wind in the dormant season.

4.2.3 Seepage areas and ponds

Where there is a more persistent source of moisture, ensuring considerable periods of growth, either along lines of seepage or in shallow basins which accumulate temporary pools, the moss and liverwort flora becomes more conspicuous. In the Sudan, Pettet (1966a) has recorded that bryophytes develop where there is a regular seepage of water: several species of *Riccia, Tortula khartoumensis, Barbula unguiculata, Micropoma nilotica, Funaria* and *Bryum*. From the seepage areas round dams in south-west Africa, Volk (1979) has recorded *Riccia cavernosa* (nitrophilous) *R. runssoriensis,* and *R. volksii*; similarly species of *Riccia* are a major component of the normal flora in Australia round long-term but temporary tanks and dams in the desert areas, especially *R. cavernosa* and *R. crystallina.* These are usually confined to clay soils which stay moist for at least several weeks, for example round the edges of temporary lakes, or on river flood plains, even where these may go for several years without receiving water. The spores therefore remain viable in the dry soil, although it is unknown for how long. Many species of *Riella* also occur on flooded ground (Jelenc, 1957).

A comparable habitat is found on sandy or silty river banks, which are frequently colonized by *Riccia* spp. in arid areas such as Turkey (*R. frostii*; Jovet-Ast, 1960) and Sudan where Pettet (1966b) has recorded how the female plants are greatly influenced by subtle changes in microtopography, growing in the very shallow depressions between slight undulations of the bank surface. In Arizona, McCleary (1968) found sparse *Amblystegium varium* and *Sphaerocarpos texanus* on desert river banks.

Where pools develop and persist for some time, an underwater flora may develop: the sole bryological component always seems to be *Riella,* found especially where the water is brackish or strongly saline but also where it is fresh or nearly so (e.g. *R. parisii*; Jelenc, 1957).

4.2.4 Soil crusts

A common feature of desert soils throughout the world is a greater or lesser development of a soil crust (Jackson, 1957). Initially, in sandy soils, this is a thin platy structure of sand grains formed, according to Fuller (1974), by the beating of rain-drops, cemented together chemically, primarily by carbonate but also by silica. Purely chemical processes may intensify this crust feature to form a hard pan layer – of geological but little biological consequence. Alternatively, the crust may become stabilized biologically to give a softer and more permeable layer which acts as a basis for further colonization. In this kind of crust formation, the earliest stage – in Australia – is commonly infiltration and covering with a weft of filamentous blue-green algae, together with the finest roots and root hairs of the nearest trees and bushes. Roots of trees in arid regions may travel 50 m or more from the trunk so that feeding rootlets may be a significant component of soil crust formation even in soils where trees and bushes are very few and far between. To this algal/root crust, lichens are added; with the capability for nitrogen fixation, by some species (Cameron and Fuller, 1960; Rogers *et al.,* 1966; Rogers and Lange, 1971; Shields *et al.,* 1957) they are important forerunners of the next stages in crust stabilization, by bryophytes. In the northern Negev, Danin (1978b) records similar crusts of algae – mainly blue-greens – followed by mosses in the most stable conditions. Similar crusts have also been described by Schwabe (1960) from Chile, by Vogel (1955) from South Africa and Fletcher and Martin (1948) from Arizona.

Over most desert soils there are neither crusts nor plants (Forman and Dowden, 1977; Shields *et al.,* 1957) and the formation of crusts leading to permanent colonization is the exception rather than the rule. It is on these stabilized crusts, however, especially where there is some shade from full sun, that the bryophytes of arid regions reach their greatest development, with what can be a rich flora of both mosses and liverworts (as well as lichens) in Australia (Table 4.1). The stabilized surface is often more or less glazed by a skin of blue-green algae, which is perennial and persists, even if cracked, during the hottest season. This skin greatly reduces sand mobility and helps to ensure that the plants growing on it do not become engulfed in blown sand. The moss flora of the region is accordingly very rich in minute ephemeral species. Within this crust microtopography is evidently important as both cause and effect. For example, patches of *Lecidea decipiens* in the arid Australian mallee form raised scabs on the soil around which cracks tend to develop during the driest weather and in the shelter offered by such cracks there can develop the minute leafy liverwort *Cephaloziella arctica.* The effects of microtopography on drainage are even more influential. Very slight, wide, concavities in an

Table 4.1 Bryophyte-dominated flora of humus-rich, consolidated soils in Wyperfeld National Park, Victoria. Black Box floodplains. August 1974. Figures are percentage frequency in 1×1 cm microquadrats, averaged over 8 blocks of 100 each.

Mosses		Liverworts		Lichens	
Barbula torquata	38	*Fossombronia* c.f. *intestinalis*	40	*Collema coccophorum*	29
Tortella calycina	37	*Riccia limbata*	14	*Siphula coriacea*	18
Triquetrella papillata	12	*Fossombronia* c.f. *wondraczekii*	11	*Cladonia* sp. (squamules)	17
Desmatodon convolutus	11	*Fossombronia leucoxantha*	7	*Lecidea decipiens*	16
Gigaspermum repens	10	*Asterella drummondii*	6	*Parmelia pulla*	14
Eccremidium pulchellum	10	*Riccia* c.f. *vesiculosa*	4	*Dermatocarpon lachneum*	6
Bryum caespiticium	8	*R. marginata*	3	*Parmelia convoluta*	5
Funaria apophysata	7	*R. nigrella*	3	*P. amphixantha*	4
Bryum pachytheca	3	*Cephaloziella arctica*	+	*Cladia aggregata*	4
Goniomitrium enerve	1	*Lethocolea squamata* (Tayl.) Hodgs.	+	*Diploschistes scruposus*	3
Bryobartramia novae-valesiae	+			*Heterodea beaugleholeii* Filson	1
Eccremidium arcuatum	+				
Bryum argenteum	+				
Ephemerum cristatum	+				
Pleuridium nervosum	+				
Aloina sullivaniana	+				
Crossidium geheebii	+				
Tetrapterum cylindricum	+				

apparently flat surface may be sufficient to accumulate and retain extra organic matter and water to a significant extent but the difference in level is almost immeasurable and can only be detected by drainage patterns after rain.

4.2.5 Saline habitats

Many desert soils in which bryophytes grow are moderately rich in minerals, especially salt (sodium chloride), lime (calcium carbonate) and gypsum (sodium sulphate) but this feature becomes extreme in some localities where these substances crystallize out on the soil surface. Surprisingly, even these, apparently totally inhospitable habitats may have a distinctive sparse flora of both mosses and liverworts. In salt pans, *Riella* spp. are once again almost specific and Jelenc (1957) describes *R. helicophylla* and *R. numidica* as strict halophytes in North Africa, the former growing in water with a salt content of up to 98 g l^{-1}.

In Australia, *R. halophila* Banwell used to grow in the highly saline Pink Lake at Dimboola (north-west Victoria) where salt was at one time commercially extracted. It grew round the shore, elevated a centimetre or two on sticky clay lumps projecting above the lake surface, together with *Funaria salsicda*. In about 1977 the lake flooded, raising the water level by about half a metre and killing off not only all the bryophytes but the dense thicket of *Melaleuca halmaturorum* trees round the shores. They have not yet returned. It seems as if the *Riella* must have grown in a low-salt microhabitat on top of the emerging blocks of clay, relying on occasional fresh water leaching by rain water to reduce its salt stress; flooding with saline water killed it. The physiology of this species would be of great interest to investigate. *Carrpos sphaerocarpos* is likewise characteristic of saline areas in north-west Victoria but in this case usually in dry salt pans, again in heavy soils, under the shade of *Arthrocnemon* bushes, apparently a similar habitat to that in which Wolff (1968) recorded *Tortella flavovirens* in Greece. The bryophyte flora of salt pools and salt pans in Australia and elsewhere is under-collected and poorly known.

In gypsum-rich soils the flora may be rather similar. In Wyperfeld National Park (Victoria) and elsewhere in Australia, on these *copi* soils where the surface is often quite white with gypsum crystals, a dense colony of *Carrpos* and *Riccia albida* and some lichens and small vascular plants has been recorded (Table 4.2) but the quantities visible in any year depend on the extent of the previous winter's rain. In Mesopotamia *Tortula fiorii* occupies a similar habitat (Schiffner, 1913).

4.2.6 Epiphytes and commensals

It may seem surprising to speak of epiphytes in deserts, since they are usually thought of as being characteristic of habitats with high humidity.

Table 4.2 Vegetation of gypsum-rich soils on Copi Plains, Wyperfeld National Park, Victoria. August 1978. Figures are % frequency out of 26 quadrats of ½ × ½ m.

Liverworts		Dicotyledons	
Riccia albida	58	*Crassula colorata*	88
Carrpos sphaerocarpos	27	*Erodium cicutarium*	73
		Medicago minima	46
		Marrubium vulgare	23
Mosses		*Spergularia* sp.	23
		Ajuga australis	19
Desmatodon convolutus	77	*Myosotis australis*	15
Bryum sp.	19	*Toxanthes* sp.	8
Funaria apophysata	8	*Helipterum corymbiflorum*	8
Barbula torquata	8	*Helipterum pygmaeum*	8
Gigaspermum repens	8	*Harmsiodoxa blennodioides*	8
Aloina sullivaniana	4	*Plantago indica*	4
		Gnaphalium involucratum	4
		Urtica urens	4
Lichens		*Hypochoeris radicata*	4
		Brassica tournefortii	4
Fulgensia subbracteata	92	*Sonchus asper*	4
Lecidia decipiens	73		
Diploschistes scruposus	35		
Dermatocarpon lachneum	31	Monocotyledons	
Toninia coeruleonigricans	31		
Aspicilia sp.	19	*Vulpia myuros*	92
Synalissa symphorea	12	*Pentaschistes airoides*	54
Collema coccophorum	12	*Anguillaria dioica*	27
Caloplaca sp.	4	*Vittadinia triloba*	15
Parmelia sp.	4	*Bromus rubens*	15
		Danthonia setacea	4

No doubt in true deserts they will be absent, but in the dry north-west of Victoria in open Black Box woodland (*Eucalyptus largiflorens*) it is not uncommon to find small patches of a mixture of *Tortula papillosa* and *T. pagorum* at the very base of the trunks where there is shade from the canopy and a concentration of the occasional rains by rain-tracks down the trunks. The epiphytes thus receive more than their expected share of rain when it comes. It seems as if *T. pagorum* increases proportionately in the mixture with increasing aridity, but this requires proof. A somewhat similar effect of increased shade and moisture is found in the same region under the bristling spikes round the rim of Porcupine Grass tussocks (*Triodia irritans*) where, on sandy soil, there is an added luxuriance to the typical flora of mosses and small vascular plants including the minute *Ophioglossum coriaceum.* These effects, which are analogous to the microclimatological effects of crevices in desert rocks, promoting increased shelter and moisture, are no doubt common in arid regions.

4.3 BIOLOGICAL FEATURES

As well as their physiological characteristics which allow them to cope with living in arid environments, desert bryophytes show a number of anatomical and biological features associated with their way of life, although few of these are peculiar to desert plants.

4.3.1 Moss leaf movements and structures

It is a very general feature of arid region perennial mosses that their leaves change position markedly between the dry and imbibed states and there can be no doubt that this is associated with a reduction in water loss and/or damage from sunlight. In some cases as in *Triquetrella papillata* there is a comparatively simple change, on drying out, from having leaves spreading to having them tightly pressed up against the stem in the erect (strict) position. This is accompanied by a considerable shrinkage of the lamina, narrowing of the angle between the two sides of the leaf and increased rolling-up of the recurved margin down each side. In other species such as *Barbula crinita* and *Tortula princeps,* as well as these movements the whole leaf winds helically round the stem, so that the dried leaf is then protected from both insolation and desiccation by the crowns of bristling hair points provided by crowded shoots in the turf; in *Barbula torquata* this rolling up can be very tight at the stem apex and the whole plant becomes encased in a kind of outer armour formed by the shining and dark coloured abaxial surfaces of the midribs. In *Desmatodon convolutus,* where colonies are more open and sparse, so that the sides of individual shoots are not protected by their neighbours, this is so pronounced that the plant often almost disappears in the soil and all that is left exposed is a flattish disc of coiled-down leaves. Most described species showing this adaptation are in the Pottiaceae and are provided with surface papillae which no doubt greatly accelerate the uptake of water when it becomes available. They can soak out in no more than minutes, if not seconds.

Although the function of leaf movements and hair points is generally accepted as being a reduction of damage by water loss and insolation, these features can have another, quite different, effect. It is noticeable, on the sand hills in Wyperfeld National Park, that *Tortula princeps* is especially prominent under the perimeter of the canopy of Black Box (*Eucalyptus largiflorens*) where there is a considerable fall of dead leaves from the trees but the dense slightly raised patches of the moss, often 15–30 cm across, are surprisingly free of litter. Class experiments at Monash (W. Fellows, personal communication) have shown that, when dead eucalyptus leaves are laid on the surface of the colony, the moss movements on wetting and drying carry the leaves, supported on the hair-points, over to the side of the colony where they then drop off the edge. The process is analogous to hearth rugs, laid on a pile carpet, 'walking' across a room; it relies on all the moss leaves being twisted in the same direction when dry. A total of 55 mm travel was recorded in one case, over 16 wetting/drying cycles; although impeded by projecting sporophytes and by a tendency for litter to rotate rather than travel, there seems no doubt that *T. princeps* is able to grow in the rain-drip of the eucalypt canopy partly as a result of being able to shed fallen litter in this way. For this mechanism to operate the moss shoots have to be erect and their leaves need hair points. Comparable tests with *Tortella calycina* (erect) and *Triquetrella papillata* (prostrate) which lack hair points, showed maximum movements of only 6 mm in 10 cycles and 4 mm in 5 cycles respectively, which may be too small to be of significance. It is likely that *Barbula crinita* too would show this feature, but it has not been tested. It might be thought that the soaking out of the moss turf is so infrequent in an arid region that the mechanism would scarcely be advantageous but, at least in and after the rainy season, condensation and/or dew is sufficient for these mosses to unfurl regularly every night. It is likely that drips of condensation from the overlying canopy will also contribute to the nightly water supply and the possibility of actual absorption of water vapour from the air, aided by condensation on the hair points, must not be discounted since Walter (1936) has recorded that dew and even fog are important sources of water for desert angiosperms in south-west Africa. Even though total quantities may be small (Milthorpe, 1960), Kappen *et al.* (1975) consider the same sources 'decisive for lichen growth'. Moore (1980, unpublished) has shown that, of 11 species of sand dune mosses examined, all were capable of absorbing sufficient moisture from prolonged exposure to a saturated atmosphere to be able to resume some photosynthesis after desiccation (see below). In areas where there is an accretion of wind-blown sand on top of mosses another ability, that of flexing the leaves back strongly when moistened, appears to be an important part of the soil-trapping and soil-stabilizing mechanism, the movements allowing the moss to swim-up, as it were, through the overlying sand.

Other desert mosses have hyaline or partly hyaline leaves. This is striking in the case of *Gigaspermum repens* and *Bryum argenteum* where no doubt the dead, hyaline, upper parts of the leaves act as a protective covering over the green living parts of the plants, like a glasshouse, protecting them against desiccation and possibly also mechanical damage. The total photosynthetic capacity is thereby reduced, but the photosynthetic efficiency is very high, permitting maximum use of the available water and light (R.D. Seppelt, personal communication).

Another common feature in arid-zone species is the presence of lamellae, filaments or granules or other outgrowths on the adaxial surface of leaves: *Aloina, Crossidium, Desmatodon, Pterygoneurum* mostly have these outgrowths. Although such structures are frequently thought of as increasing the photosynthetic surface of the leaves, in these cases they may well be more significant as sun shades, giving thick and opaque coverings when curled in over the stem apex and, later, over the developing sporophyte. Thick leaves ('doppelschichtige') were noted by Schiffner (1913) to be remarkably common in Mesopotamian mosses. Frequently, too, arid zone mosses contain large quantities of lipid material e.g. *Tortula oleaginosa* (Stone, 1978). This is likely to be of adaptive significance, but the function is not known.

4.3.2 Liverwort movements and structures

A corresponding mechanism to moss leaf movements is found in those thallose liverworts which roll up on desiccation into a tubular form so that the dorsal photosynthetic surface is hidden and the ventral surface and/or ventral scales, intensely pigmented with anthocyanin, is the only one exposed. The pigmentation is usually dense enough to appear black. This is typical of *Asterella drummondii, Targionia hypophylla* L. and many species of *Riccia* such as *R. nigrella* and *R. limbata.* In *Fossombronia* spp. which are surprisingly common in arid soils in Australia, the whole uppermost surface of the fronds may be tinted with anthocyanin, although never as heavily as the ventral surfaces and scales in Marchantiales. The same is true of *Cephaloziella arctica,* a minute liverwort which is probably abundant throughout the arid regions of Victoria, usually coloured a dark brown or crimson and wholly invisible to the naked eye on black or brown soils. Even with a hand lens it is scarcely detectable but careful observation of the soil surface with a binocular microscope will commonly reveal it.

What appears to be an alternative response to the rigours of desert life is found in *Riccia* spp. where the scales are hyaline and large, as in *R. marginata,* or the upper surface partly covered, especially in the dry state, by a cheval-de-frise of long bristles, as in *R. crinita*; these are likely to have protective effects in relation to desiccation, insolation and blown dust, but their biological consequences are yet to be worked out. The collapsed epidermal cells of *R. lamellosa* probably have a similar effect.

In *R. caroliniana* (Na Thalang, 1980) this hyaline protective layer – no doubt corresponding to the moss leaves mentioned above – extends through the whole dorsal part of the thallus, and the green photosynthetic stratum occurs below, thus achieving a structure like that of *Exormotheca.* In the leafy liverwort *Lethocolea squamata* (Tayl.) Hodgs., which has a very long marsupium penetrating the soil (this too may be thought of as a desiccation-avoiding feature), the leaves normally are green, flat and spreading, when growing in moist forest habitats. It is, however, also common in arid Australia and there the stems are sunk partly in the soil and the leaves tend to be vertically orientated (set on edge) closely imbricated, concave and silvery-hyaline. In this condition, the green growing and photosynthetic parts are protected inside the sheath provided by the hyaline parts.

Riccia albida in Australia is unusual not only in its apparent preference for calcium-rich habitats and for being extremely drought tolerant but also in having the upper surface heavily impregnated with minerals. There are thick deposits of calcium carbonate (possibly also calcium sulphate) crystals within the uppermost walls of the surface cells. This white incrustation may well be hygroscopic and must also influence the absorption of solar radiation and consequent desiccation resistance, but no experimental data are available.

4.3.3 Underground parts and perennation

Very little is known about this topic even in mesophytic habitats, less still in deserts. A few species may have underground stems (e.g. *Gigaspermum repens*) from which buds can undoubtedly arise. More commonly the underground parts are only rhizoids. In the case of mosses it is certainly possible for new gametophore shoots to arise from perennating underground rhizoids and Moore and Scott (1979) have listed *Barbula torquata, Tortella calycina* and *Tortula princeps* among those for which they have recorded regeneration from rhizoids. It is likely that this perennating ability is often enhanced by the production of tubers (rhizoid-gemmae) but the only example I know of is *Tortula desertorum* (Novichkova-Ivanova, 1976). In liverworts, tubers seem to be rarer than one might expect in deserts and the normal vegetative method of persistence of thallose liverworts is certainly by the shoot apex which can often withstand great desiccation. The more extreme the desert conditions, however, the more probable it becomes that perennation is largely by spores, but the proportion of new plants arising from this cause has never been investigated.

4.3.4 Reproduction

At first sight it seems surprising that desert bryophytes should be so fertile. Although it is impossible to prove, they seem to be distinctly more fertile than the average in more mesophytic habitats. Since sexual reproduction involves the swimming of a male gamete into the archegonium in a film of liquid water one might have expected such events to be rather rare in arid habitats, especially with short-lived plants, since there the successful production of sporophytes involves moisture at two distinct periods, one for the growth of the gametophyte and the second time for fertilization. It may be partly that the tiny cleistocarpous mosses on arid soils are just not seen except when their relatively large capsules are present but that will not explain the abundance of sporophytes in large mosses and liverworts. *Riccia* spp., for example, seem to produce spores as the rule rather than the exception, but these species are most commonly synoicous and it is true that the autoicous *Asterella drummondii,* of comparable size, rather seldom produces sporophytes in the driest part of its range. Apogamy, although known for mosses, has never been thought of as anything but a bizarre abnormality and there is no evidence for its being a normal way of life in any desert species. The likeliest explanation of fertilization in desert archegoniate plants seems to me to be condensation/dews. Assuming sufficient rain in winter to produce the growth of a gametophyte for a few weeks, there is still likely to be sufficient moisture in the soil to produce a surface film, by condensation at night, especially as nights in the desert can often be very cold. For deserts where the rain falls in summer this explanation is less likely but I have no knowledge of their bryophytes.

Asexual propagation is rare, and the drier the habitat the rarer it is. In marginally dry situations, *Tortula papillosa* and *T. pagorum* produce abundant gemmae; *Eccremidium pulchellum* has shoots which fragment rather readily and *Lethocolea squamata* (Tayl.) Hodgs. produces discoid gemmae (although I have never actually seen these produced in really dry conditions). Despite the well known propensity of bryophytes to regenerate from casual fragments it seems from my own observations that, in dry regions, this is less important than spore production and that the latter must be the main method for perennation. It seems probable that there will be mechanisms for producing heterogeneous spores so that, like seeds, a false start, by too brief a rain to allow full growth of the gametophyte, will not use up all the store of spores in the soil. Although highly probable, such a mechanism has not yet been demonstrated. The spores in many cases are very large, of the order of 100 μm or more, and may be multifaceted in some mosses *(Gigaspermum, Goniomitrium)* although others such as *Crossidium* and *Aloina* have small spores, nearer 10 μm. No doubt each of these will have its own particular biological niche determined by the rate of growth and the probability of successful re-establishment but we have almost no information, at even the most elementary level, on such topics. A great many of the most distinctive mosses of arid areas (e.g. *Goniomitrium*) seem to fall into During's 'annual shuttle life strategy' (1979), with a short life span (at least of the leafy shoot) and large spores. Those with small spores *(Crossidium)* may fit in his 'fugitive life strategy' but we do not yet know.

These small mosses of dry areas are often either gymnostomous or cleistocarpous with sessile or nearly sessile capsules, and the spores are released by collapse and breakdown of the capsule wall and dispersed by wind and (when it comes) water. In the case of *Goniomitrium* the foot of the sporophyte is only feebly attached to the gametophyte and the whole capsule, still completely enclosed in its large calyptra, often falls off and is blown away as a miniature tumbleweed. *Bryobartramia,* also with a giant calyptra, may do the same.

4.4 DROUGHT PHYSIOLOGY

Of the rather numerous investigations of bryophyte drought physiology (e.g. Lange, 1955; Abel, 1956; Stocker, 1960; Hinishiri and Proctor, 1971; Dilks and Proctor, 1976; see also review by Bewley, 1979) few deal with desert plants. The most recent investigation has been by Moore (1980, unpublished) on 12 species, of which 9 are sand dune species and 4 or 5 are common on arid soils in Australia (Moore and Scott, 1979): *Barbula torquata, Tortula princeps, Tortella calycina, Bryum argenteum, Triquetrella papillata.* All of them are desiccation-tolerant rather than

Table 4.3 Poikilohydric behaviour of arid region mosses (from Moore, 1980, unpublished).

Species	Drought period	Time elapsed before net photosynthesis (hr)	Time elasped before net photosynthetic gain (hr)
Barbula torquata	1 day	immediatc	immediate
	2 days	immediate	immediate
	1 week	immediate	immediate
	2 weeks	0.6	1.0
	6 weeks	0.6	1.0
	24 weeks	15.0	26.0
	32 weeks	17.4	34.0
	52 weeks	17.1	38.0
	72 weeks	16.5	42.0
Tortula princeps	1 day	immediate	immediate
	2 days	0.45	1.2
	1 week	8.7	17.0
	4 weeks	11.5	23.0
	24 weeks	14.4	34.0
	32 weeks	11.0	31.0
Tortella calycina	4 weeks	20.5	48.5
Triquetrella papillata	1 week	immediate	immediate
	4 weeks	2.2	8.5
	8 weeks	4.6	38.0
Bryum argenteum	2 weeks	5.3	13.0

Species in order from most to least drought-tolerant.

desiccation-resistant. That is, they lose water readily but are capable of surviving desiccation (for various periods according to species) without damage and of rapidly resuming normal metabolic activity when rehydrated. Using an infra-red gas analyser and oxygen electrodes, Moore showed that all these species were drought-tolerant and capable of rapid recovery to full photosynthetic activity after prolonged – sometimes very prolonged – drought. It is not a simple matter to compare their performances in this respect since it depends on a number of conditions (and their interactions): rates of drying and subsequent rehydration, level of dehydration reached, duration of desiccation, number of cycles of drying and re-wetting, temperature, light and humidity conditions before and during desiccation, prior history of the individual plants and the experimental conditions. Net assimilation rate of different species at different water contents cannot be directly compared since their water contents will differ at any given water potential, because of structural differences between species. Nevertheless, in broad terms, Moore showed that all those species were capable of: (a) photosynthesis at water contents well below 100% Relative Water Content, (b) surviving long periods without water and (c) rapid resumption of photosynthesis on rehydration after desiccation. Of these species, the extreme example is *Barbula torquata* which survived a year and a half of ambient temperature and humidity in an unheated glasshouse without water (Table 4.3). Slow drying and hardening-off by previous experience of short droughts seem to be important factors in conditioning mosses to withstand desiccation.

There are no comparable studies that I know of on desert liverworts although several mesophytic species have been investigated (e.g. Höfler, 1942). Many of the plants seem to have a desiccation-avoidance strategy, and most of the plant body dies off in drought, leaving only the protected apex alive but not dehydrated. *Riccia albida* may, however, be an exception to this.

Bryophytes in general seem to have excellent potential for life in deserts because of their naturally poikilohydric water régimes (Kallio and Kärenlampi, 1975, p. 394): 'In addition to stress resistance to temperature extremes and water deficiency, there is an ability for rapid reactivation without a lag phase. Both lichens . . . and mosses . . . are able to take advantage of very short periods of temperature suitable for photosynthesis'.

ACKNOWLEDGMENTS

I am grateful to Mr G. Kunkel, Mr W. Fellows and Dr C. Moore for permission to quote from their unpublished work, to Dr R.D. Seppelt, Dr W.A. Weber, Dr A. Danin and Dr I. Herrnstadt for comments, the

National Parks Service for permission to collect and Miss C. Cargill for assistance with the literature. The ARGC and ABRS provided research grants thanks to which some of my investigations were made possible.

REFERENCES

Abel, W.O. (1956), *Sber. Akad. Wiss. Wien. Math—Naturw. Kl.* Abt. I., **165**, 619–707.

Bewley, J.D. (1979), *A. Rev. Plant Physiol.*, **30**, 195–238.

Birse, E.M. and Gimingham, C.H. (1954), *Trans. Br. bryol. Soc.*, **2**, 523–31.

Birse, E.M., Landsberg, S.Y. and Gimingham, C.H. (1957), *Trans. Br. bryol. Soc.*, **3**, 285–301.

Cameron, R.E. and Fuller, W.H. (1960), *Proc. Soil Sci. Soc. Am.*, **24**, 353–6.

Cloudsley-Thompson, J.L. and Chadwick, M.J. (1964), *Life in Deserts*, Foulis, London.

Collenot, A., Dubuis, A. and Faurel, L. (1960), *Bull. Soc. Hist. Nat. Afr. N.*, **51**, 235–54.

Danin, A. (1972), *Notes R. Bot. Gdn. Edinb.*, **31**, 437–40.

Danin, A. (1978a), *Vegetatio*, **36**, 83–93.

Danin, A. (1978b), *Flora, Jena*, **167**, 409–22.

Davis, P.H. (1951), *J. Ecol.*, **39**, 63–93.

Dilks, T.J.K. and Proctor, M.C.F. (1976), *J. Bryol.*, **9**, 249–64.

During, H.J. (1979), *Lindbergia*, **5**, 2–18.

Fletcher, J.E. and Martin, W.P. (1948), *Ecology*, **29**, 95–100.

Flowers, S. and Evans, F.R. (1966), The flora and fauna of the Great Salt Lake region, Utah. In: *Salinity and Aridity. New approaches to old problems*, Boyko, H. (ed.), pp. 367–93. Junk, The Hague.

Foote, K.G. (1966), *Bryologist*, **69**, 265–92.

Forman, R.T.T. and Dowden, D.L. (1977), *Bryologist*, **80**, 561–70.

Friedmann, E.I. and Galun, M. (1974), Desert algae, lichens, and fungi. In: *Desert Biology* (Brown, G.W. Jr., ed.), Vol. II, pp. 165–212. Academic Press, New York and London.

Fuller, W.H. (1974), Desert Soils. In: *Desert Biology* (Brown, G.W. Jr., ed.), Vol. II, pp. 31–101. Academic Press, New York and London.

Gimingham, C.H. (1948), *Trans. Br. Bryol. Soc.*, **1**, 70–2.

Hinishiri, H.M. and Proctor, M.C.F. (1971), *New Phytol.*, **70**, 527–38.

Höfler, K. (1942), *Ber. dt. bot. Ges.*, **60**, 94–107.

Jackson, E.A. (1957), *J. Aust. Inst. agric. Sci.*, **23**, 196–208.

Jelenc, F. (1957), *Rev. bryol. lichénol.*, **26**, 20–50.

Jovet-Ast, S. (1960), *Rev. bryol. lichénol.*, **26**, 67–8.

Jovet-Ast, S. and Bischler, H. (1970), *Rev. bryol. lichénol.*, **37**, 265–87.

Kallio, P. and Kärenlampi, L. (1975), Photosynthesis in mosses and lichens. In: *Photosynthesis and Productivity in Different Environments* (Cooper, J.P., ed.), pp. 393–423. Cambridge University Press, New York.

Kappen, L., Lange, O.L., Schulze, E.-D., Evenari, M. and Buschbom, V. (1975), Primary productivity of lower plants (lichens) in the desert and its physiological basis. In: *Photosynthesis and Productivity in Different Environments* (Cooper, J.P., ed.), pp. 133–43. Cambridge University Press, New York.

Kassas, M. and Imam, M. (1959), *J. Ecol.*, **47**, 289–310.
Kunkel, G.P. (1975), Unpublished *Microhabitats and Phytosociological Studies of Cryptogamic Chasmo- and Lithophytes.* M.Sc. Thesis, University of Colorado.
Lange, O.L. (1955), *Flora, Jena,* **142,** 381–99.
Logan, R.F. (1968), Causes, climates, and distribution of deserts. In: *Desert Biology* (Brown, G.W. Jr., ed.), Vol. I, pp. 21–50. Academic Press, New York and London.
McCleary, J.A. (1959), *Bryologist,* **62,** 58–62.
McCleary, J.A. (1968), The biology of desert plants. In: *Desert Biology* (Brown, G.W. Jr., Ed.), Vol. I, pp. 141–94. Academic Press, New York and London.
Milthorpe, F.L. (1960), The income and loss of water in arid and semi-arid zones. In: *Plant-water Relationships in Arid and Semi-arid Conditions.* UNESCO Arid Zone Research. XV., pp. 9–36.
Moore, C.J. (1980), Unpublished *Factors determining the Spatial Distribution of some Coastal Sand Dune Mosses.* Ph.D. Thesis, Monash University.
Moore, C.J. and Scott, G.A.M. (1979), *J. Bryol.,* **10,** 291–311.
Nash, T.H., White, S.L. and Marsh, J.E. (1977), *Bryologist,* **80,** 470–9.
Na Thalang, O. (1980), A revision of The Genus *Riccia* (Hepaticae) in Australia. *Brunonia,* **3,** 61–140.
Novichkova-Ivanova, L.N. (1976), *Bot. Zh.,* **61,** 1168–79.
Noy-Meir, I. (1972), *A. Rev. Ecol. Syst.,* **4,** 25–51.
Orshan, G. and Zohary, M. (1962), *Vegetatio.,* **11,** 112–20.
Pandé, S.K. and Bhardwaj, D.C. (1953), *Palaeobotanist,* **1,** 368–81.
Petrov, M.P. (1976), *Deserts of the World,* Wiley, New York.
Pettet, A. (1966a), *Trans. Br. Bryol. Soc.,* **5,** 316–31.
Pettet, A. (1966b), *Trans. Br. Bryol. Soc.,* **5,** 332–7.
Rogers, R.W. and Lange, R.T. (1971), *Oikos,* **22,** 93–100.
Rogers, R.W., Lange, R.T. and Nicholas, D.J.D. (1966), *Nature,* **209,** 96–7.
Ross, M.A. (1969), *Proc. Ecol. Soc. Aust.,* **4,** 67–81.
Schiffner, V. (1913), *A. K. K. naturh. Hofmus. Wein.,* **27,** 1–34.
Schwabe, G.H. (1960), *Öst. bot. Z.,* **107,** 281–309.
Shields, L.M., Mitchell, C. and Drouet, F., (1957). *Am. J. Bot.,* **44,** 489–98.
Stocker, O. (1960), Physiological and morphological changes in plants due to water deficiency. In: *Plant-water Relationships in Arid and Semi-arid Conditions.* UNESCO Arid Zone Research. XV., pp. 63–104.
Stone, I.G. (1978), *J. Bryol.,* **10,** 117–24.
Townsend, C.C. (1966), *Trans. Br. Bryol. Soc.,* **5,** 136–41.
Vogel, S. (1955), *Beitr. Biol. Pflanz.,* **31,** 45–135.
Volk, O. (1979), *Mitt. bot. München.,* **15,** 223–42.
Walter, H. (1936), *Jb. Wiss. Bot.,* **84,** 58–222.
Weber, W.A. and Wetmore, C.M. (1972), *Beih. Nova Hedwigia.,* **41,** 1–137.
Willis, J.H. (1970, 1973), *A Handbook to Plants in Victoria,* Vol. I. Monocotyledons. Vol. II. Dicotyledons. Melbourne University Press, Melbourne.
Wolff, W.J. (1968), *Vegetatio,* **16,** 95–134.

Chapter 5

Bryophyte Vegetation in Polar Regions

R.E. LONGTON

5.1 INTRODUCTION: THE PHYSICAL BACKGROUND

No universal agreement exists as to the limits of the Arctic and Antarctic, but these areas are generally regarded as tundra in the broad sense of 'treeless regions beyond climatic timberlines' (Webber, 1974). The present account considers the low and high Arctic (Fig. 5.1) as defined by Bliss (1979), and the sub-Antarctic and Antarctic zones (Fig. 5.2) of Greene (1964a) and Holdgate (1964, 1970). These regions include most of the polar tundra, but exclude certain lands such as the Falkland Islands where the absence of arboreal vegetation may be attributable to oceanic as well as polar features of the climate.

The regions under discussion are diverse in topography and climate. Arctic lands encircle a polar ocean (Fig. 5.1), and the terrain ranges from folded mountains, often high and imposing, to rugged uplands on igneous Precambrian rocks and extensive flat-bedded plains and plateaus. Except in Greenland, glaciation is now localized and confined to the mountains. The principle Antarctic land mass is centred over the Pole and is surrounded by ocean (Fig. 5.2), which reaches a minimum width of c. 1000 km between the Antarctic Peninsula and Cape Horn. Islands, both those fringing the continent and others widely scattered in the ocean, though small in area, are highly significant in terms of Antarctic terrestrial biology, as over 98% of the continent lies buried in ice. Even South Georgia, a sub-Antarctic island at latitude 54°S, is heavily glaciated. Thus, vegetation is restricted to the almost universally rugged terrain of coastal regions and the islands, and to nunataks penetrating the inland ice sheet.

Differences between Arctic and Antarctic in the extent of glaciation, and distribution of land and sea, result in comparable variation in climate. Short, cool summers are the most consistent feature but, latitude for latitude, the summers are colder in the southern hemisphere than in the

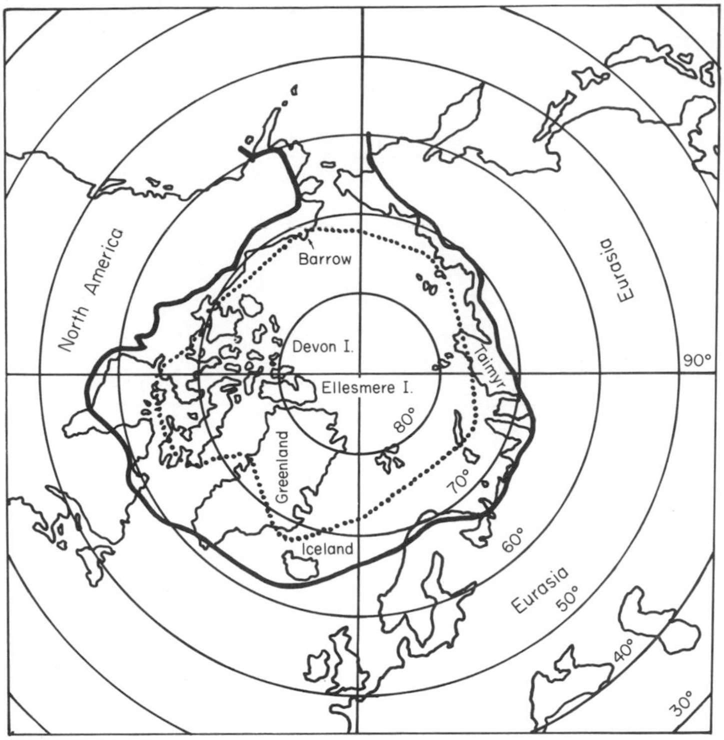

Fig. 5.1 Northern circumpolar region showing the southern boundaries of the low Arctic (solid line) and high Arctic (dotted line). After Bliss (1979).

north (Table 5.1). Mean monthly air temperatures in mid summer exceed 10°C at some low Arctic sites. They are normally above 3°C in the high Arctic, but remain below this value in the polar desert region of the USSR (Alexandrova, 1970) and throughout the Antarctic zone. Means exceed 0°C for one or two months in summer near the west coast of the Antarctic Peninsula and on Antarctic islands in the South Atlantic, i.e. in the maritime Antarctic of Holdgate (1964), but are below freezing throughout the year elsewhere in the Antarctic zone, i.e. in continental Antarctica. Prolonged winter frost and snow cover is a feature of most polar climates, but oceanic influences reduce the annual temperature range in the maritime Antarctic (Table 5.1), and on sub-Antarctic islands, some of which have mean monthly air temperatures above 0°C throughout the year.

Precipitation is moderate to low (Table 5.1), with the notable exception of the sub-Antarctic islands. Moderate precipitation, low evaporation rates, impeded drainage due to permafrost, and extensive spring run-off combine to produce waterlogged conditions over extensive areas of the low Arctic plains. Water is also freely available in many maritime Antarctic

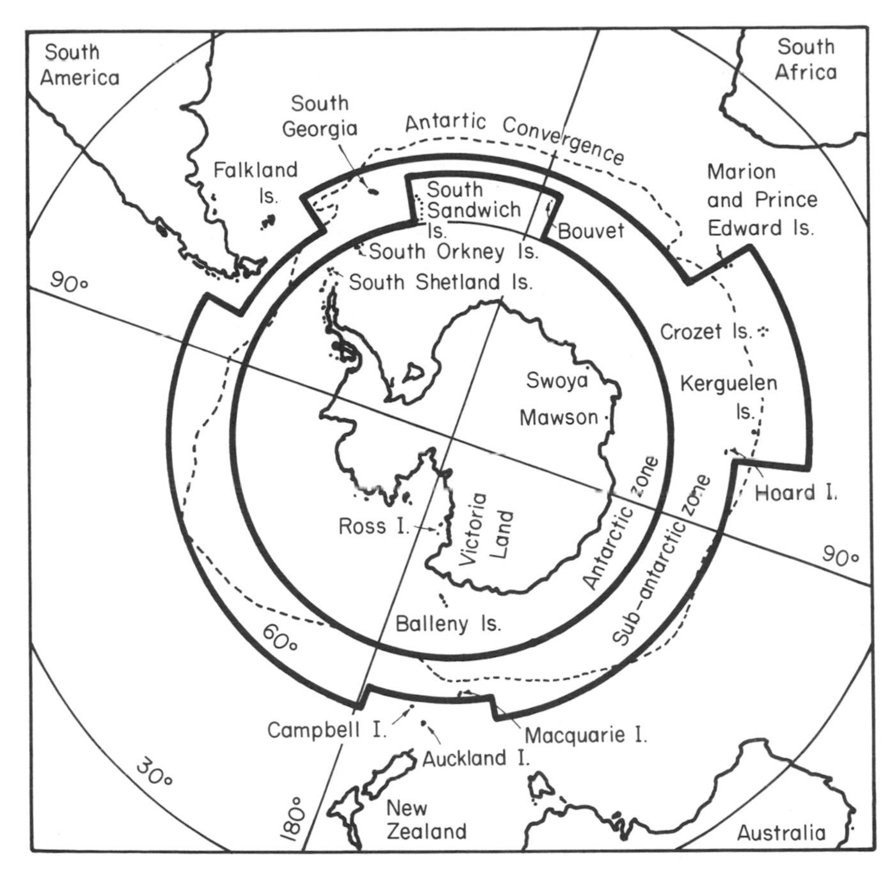

Fig. 5.2 Southern circumpolar region showing the boundaries of the sub-Antarctic and Antarctic zones. After Greene (1964a).

Table 5.1 Air Temperatures and precipitation in polar regions. Data from Anonymous (1978), Bliss (1979), Holdgate (1964), Thompson (1967).

	Mean air temperature, warmest month (°C)	Mean air temperature, coldest month (°C)	Mean annual precipitation (cm rainfall equivalent)
High Arctic	1 to 6	−43 to −27	7 to 20
Low Arctic	6 to 11	−33 to −24	10 to 64
Sub-Antarctic	3 to 8	−2 to 2	93 to 258
Maritime Antarctic	1 to 3	−11 to −8	c. 40
Coastal sites in continental Antarctica	−7 to 0	−40 to −26	16 to 43

habitats. In contrast, aridity is a major limiting factor for biological activity throughout much of the high Arctic and in continental Antarctica. In the latter the influence of low precipitation is enhanced by the absence of rain, and by sub-freezing air temperatures throughout the summer. Aridity and cold are commonly accentuated by strong winds, while cloud cover also depresses ground level temperatures, particularly at oceanic localities.

Except on sub-Antarctic islands, which lie at only moderate latitudes, the maximum intensities of incoming solar radiation are lower in polar than in temperate regions, but total daily amounts of solar radiation in mid summer are similar due to increased day length (Gates, 1962). This regime may favour efficient energy utilization by plants, as relatively little radiation is received at intensities far above saturation levels for photosynthesis (Warren Wilson, 1966).

Extensive pleistocene glaciation, combined with harsh climatic conditions and a consequent predominance of physical weathering processes in more recent times, have resulted in the prevalence of immature, primarily mineral soils which may be poor in available nutrients, particularly nitrogen, except in coastal areas (Tedrow, 1977). However, upward water movement results locally in toxic concentrations of salts at the surface in arid regions. The soils are poorly stratified, with little incorporation of organic matter, although extensive superficial peat deposits occur in the wetlands. Permafrost is extensive, except on sub-Antarctic islands. Soil instability, resulting from solifluction and frost-induced phenomena, imposes further limitations on the establishment of closed plant cover, but presents conditions exploitable by pioneer communities. Only locally do soils approach the degree of development normal in temperate regions.

Arctic phanerogamic vegetation can be considered under three headings: wetlands, mesic communities and polar desert. Wetlands occupy level or gently sloping ground with impeded drainage, and support well-developed swards of graminoids over-topping mosses and herbs such as *Polygonum viviparum* and *Saxifraga hirculus.* A variety of vegetation types occurs under mesic conditions, including further graminoid-dominated communities, dwarf shrub heaths, and associations of cushion plants, notably *Dryas integrifolia,* with mosses and lichens. Taller scrub formed by species of *Betula* and *Salix* occurs locally. Mosses, lichens, perennial herbs and suffrutescent perennials such as *Salix arctica* are also conspicuous, with plant cover either continuous or discontinuous. In contrast, bare ground predominates in the polar deserts, where cover of scattered vascular plants and cryptogams totals less than 15%. The proportion of ground occupied by closed vegetation decreases northwards. Thus, wetland and mesic communities predominate in the low Arctic, while much of the high Arctic is occupied by polar desert. The transition is gradual, however, as polar desert is extensive in elevated or exposed habitats in the low Arctic, and occasional stands of luxuriant vegetation extend almost to the northern limits of land.

Much of the vegetation at low altitudes on sub-Antarctic islands is physiognomically similar to that in the low Arctic. Dwarf shrub heaths are absent, but suffrutescent perennials in the genus *Acaena* are abundant, both in almost pure stands and in association with grasses. Mires and bogs are also widespread, but a number of distinctive southern hemisphere vegetation types also occur, notably herbfield and coastal tussock grassland (Jenkin and Ashton, 1970). Open communities reminiscent of Arctic deserts are widespread under exposed conditions and at higher elevations, where large cushion-forming flowering plants are another distinctive feature.

The sub-Antarctic and Antarctic zones are clearly demarcated, as only two native species of vascular plants occur in the latter. Cryptogamic communities are variable, extensive and often luxuriant in the maritime Antarctic, but are strikingly depauperate in the continental region. Antarctic vegetation is sometimes referred to as polar desert by analogy with that in the high Arctic. This term is appropriate for continental Antarctica, but its application to the maritime Antarctic is misleading in view of the favourable conditions of water availability.

Bryophytes are highly significant in polar ecosystems in terms of aerial cover, species richness, phytomass and production. An indication of the numbers of species involved is provided by the following figures, all of which will be subject to revision, probably upward, following further taxonomic study: Arctic Alaska, 415 mosses and 135 hepatics (Steere and Inoue, 1978; Steere, 1978a); Northern Ellesmere Island, Canadian high arctic, 166 mosses and 43 hepatics (Brassard, 1971a, 1976; Schuster, 1959); sub-Antarctic island of South Georgia, 111 mosses and 27 hepatics (Greene, 1973; Grolle, 1972); Antarctic zone, 85 mosses and 14 hepatics (Greene, 1968; Grolle, 1972; Steere, 1961) Phytogeographical relationships and possible origins of polar bryophyte floras are discussed in the above papers and in Robinson (1972), Schofield (1972) and Vitt (1979).

British and other national scientific activities in the Antarctic since IGY, and IBP investigations in the Arctic, have recently begun to elucidate the ecological relationships of polar bryophytes. Community studies have received detailed attention in the maritime Antarctic, while some aspects of bryophyte autecology are further advanced in the Arctic. Relationships between the physical environment and the distribution of cryptogamic life-form types stand out particularly clearly in the Antarctic in the absence of extensive vascular plant cover, and they have been embodied in a hierarchical classification of the vegetation as discussed on p. 139. No comparable classification of Arctic or sub-Antarctic bryophyte vegetation has yet been formulated. This may reflect either the greater emphasis

Table 5.2 Bryophyte life-forms in polar regions (after Gimingham and Birse, 1957).

Life-form	Description
Small cushion	Systems with main shoots radiating in dome-shaped colonies: each colony < 5 cm in diameter. Branches adopting the same direction of growth as the main shoots. High spatial density of shoots and branches resulting in compact colonies.
Large cushion	Systems with main shoots radiating in dome-shaped colonies: many colonies > 5 cm in diameter. Branches adopting the same direction of growth as the main shoots. Spatial density of shoots and branches normally lower, and the colonies frequently looser, than in small cushions.
Short turf	Systems with main shoots parallel and erect. Branches erect, of indeterminate growth. Colonies loose or compact, > 2 cm tall.
Tall turf, branches erect	Systems with main shoots parallel and erect. Branches erect, of indeterminate growth. Colonies loose or compact, > 2 cm tall.
Tall turf, branches divergent	Systems with main shoots parallel and erect. Branches divergent, of determinate growth. Colonies loose or compact, > 2 cm tall.
Carpet	Systems with distal portions of shoots parallel and erect or ascending, derived from a prostrate, basal region of partially denuded stems. Colonies loose or compact.
Mat	Systems of prostrate or ascending shoots, interweaving to form compact colonies.
Weft	Systems of long, often robust, erect ascending shoots, interweaving and forming loose colonies.
Canopy	Systems of sympodial shoots, at first stoloniferous, later becoming erect. Erect portion dendroid: unbranched and bearing scale leaves below, with abundant branches above, the branches bearing photosynthetic leaves and forming loose canopies.

placed on cryptogamic community ecology by students of Antarctic botany, or intrinsic difficulties arising from the wide range of species and vegetation types present in the Arctic. Steere (1978b) doubts whether it is feasible to define formal bryophyte communities in the Arctic because of the gradual and subtle changes in species composition which occur in response to gradients in water availability and other environmental variables. In the present account, Arctic and sub-Antarctic vegetation is described under the informal headings of wetland, mesic and dry ground communities. Attention is paid to the contribution of bryophytes both to vegetation with a significant vascular plant component and to essentially cryptogamic communities. It is hoped that this account will assist in the development of a comprehensive classification of polar vegetation which takes due account not only of flowering plants, but also of bryophytes and the other major groups of cryptogams. Two possible approaches have been outlined by Bliss (1979) and Longton (1979a).

An attempt is also made to describe relationships between bryophyte life-form and environment in the Arctic and sub-Antarctic, in order to complement the Antarctic work, the life-form types considered here being outlined in Table 5.2. It should be noted that certain terms such as bryophyte carpet are commonly used less precisely than defined in Table 5.2, while many accounts of polar vegetation make no mention of life-form. The present assignment of life-form to species in areas not studied by the author must therefore be regarded as tentative. The term life-form is employed in preference to growth form, as used in recent Antarctic literature (Gimingham and Smith, 1970; Longton, 1979a), for reasons discussed elsewhere in this volume by Mägdefrau (p. 49).

5.2 BRYOPHYTE VEGETATION IN THE ARCTIC

The widespread notion that bryophytes are abundant in Arctic vegetation is only partially true, and applies particularly to wet, low Arctic tundra. Steere (1978b) describes the landscape in the coastal plain province of Alaska as 'an almost unbroken expanse of bog, marshy meadows and fens' in which 'bryophytes – and particularly *Sphagnum* – may become the dominant and continuous vegetation over large areas, usually associated with grasses, carices and *Eriophorum*'. However, the abundance of bryophytes, as of flowering plants, varies broadly within the Arctic in relation to water availability, on both a local and a regional basis. Thus, considerable differences in abundance and species composition occur between the dry summits of ice wedge polygons and the wet troughs between them (Steere, 1978b), while the high Arctic polar desert in places contrasts strongly with wetlands in the low Arctic in 'the extremely restricted development of the bryophyte vegetation' (Schuster, 1959).

The contribution of bryophytes to the plant cover of the Truelove Lowland region of Devon Island is considered by Vitt and Pakarinen (1977) and other reports on the Canadian IBP Tundra Biome investigations in Bliss (1977a). Truelove Lowland supports more extensive closed vegetation than is common at its latitudes of 76°N (Bliss, 1977b), and a representative selection of wetland, mesic and polar desert communities is developed in the Lowland and surrounding plateau region. In addition, Brassard (1971b), Holmen (1955) and Steere (1978a) have listed bryophyte associations from high Arctic localities in northern Ellesmere Island, Peary Land, north Greenland, and Arctic Alaska respectively, while comprehensive accounts of bryophyte vegetation have been presented for Iceland (Hesselbo, 1918), northern Swedish Lapland (Mårtensson, 1956) and south-east Spitzbergen (Philippi, 1973), the last based on methods of the Zurich–Montpellier School. The following account is based largely on the papers cited above, supported by the present author's observations on northern Ellesmere Island and in low Arctic localities on the west coast of Hudson's Bay. Prominence is given to the diverse, well-documented bryophyte vegetation of Truelove Lowland which is summarized in Table 5.3.

5.2.1 Wetlands

Bryophytes are abundant in the Arctic wetlands. Four different meadow communities are recognized in Truelove Lowland (Table 5.3), each of which comprises stands of graminoids and other vascular plants rooted in an almost closed bryophyte understorey. Significant breaks in the moss cover occur principally on hummocks or disturbed areas. The dominant mosses generally assume the tall turf life-form, and include pleurocarps such as *Drepanocladus* spp. as well as *Cinclidium arcticum* and other acrocarps. Other life-forms are represented more sparingly: for example, *Seligeria polaris* forms short turfs on small stones in relatively open areas of the frost boil meadow type. Over 40 species of bryophyte have been recorded in the meadow communities of Truelove and the nearby Sparbo-Hardy Lowlands of Devon Island. *Riccardia pinguis* was the only hepatic present at high frequency, while *Sphagnum* was represented only by the local occurrence of *S. orientale*.

A comparable bryophyte stratum, comprising largely tall turfs of peat-forming mosses, both acrocarpous and pleurocarpous, occurs in wetlands throughout the Arctic. Wetlands reach their most extensive development in the low Arctic where, as in Alaska, species of *Sphagnum* are prominent. *Lophozia* spp. and other hepatics are also frequent in low Arctic wetlands, both as isolated stems among mosses, and in small mats on eroding peat surfaces. However, isolated stands of sedge meadow extend in damp hollows to beyond 82°N in both Ellesmere Island and Peary Land, and here the bryophyte layer is represented principally by a *Drepanocladus brevifolius* community described by Brassard (1971b) and Holmen (1955). A detailed account of mire vegetation in the Torneträsk area of Lapland is provided by Sonesson (1967, 1969, 1970a, 1970b) and Sonesson and Kvillner (1980).

5.2.2 Mesic communities

Five mesic communities have been described from the Truelove Lowland region (Table 5.3), including graminoid-moss tundra with the vascular plant cover dominated by *Hierochloe alpina* and *Luzula confusa*, dwarf shrub heath with abundant *Cassiope tetragona*, and three felfield types dominated by *Dryas integrifolia* and other cushion-forming species. Mosses form an important component of all these vegetation types, which differ from most of the wet meadows in the greater abundance of lichens.

Table 5.3 Plant communities in the vicinity of Truelove Lowland, Devon Island, Canadian High Arctic.

Habitat – plant community types	Topographic position	Plant cover	Examples of representative species		
			Vascular plants	Bryophytes	Lichens
Wetland communities					
Wet sedge-moss meadow	By lowland streams and ponds: on peat often > 30 cm deep	Homogeneous vascular plant cover and abundant mosses	*Carex stans* (D) *Hierochloe pauciflora* (C)	*Calliergon giganteum* (D – tall turf) *Drepanocladus revolvens* (D – tall turf) *Meesia triquetra* (D – tall turf)	—
*Hummocky sedge-moss meadows	Level ground in lowlands: on peat often 10–15 cm deep	Vascular plants and mosses abundant. Hummocky relief due to differential growth of the latter	*Carex stans* (D) *Salix arctica* (C) *Saxifraga hirculus* (C)	*Cinclidium articum* (D – tall turf) *Drepanocladus revolvens* (D – tall turf) *D. brevifolius* (C – tall turf)	—
*Frost boil sedge-moss meadow	Level ground in lowlands: little or no peat formation	Frost boils cover c. 50% of ground, their surfaces usually covered by blue-green algae, mosses and scattered sedges, with *Seligeria polaris* on small stones. Continuous plant cover in intervening areas	*Eriophorum triste* (D) *Carex membranacea* (D) *Carex arctica* (C) *Polygonum viviparum* (C)	*Drepanocladus revolvens* (D – tall turf) *Campylium arcticum* (C – tall turf) *Ditrichum flexicaule* (C – tall turf) *Seligeria polaris* (C – short turf)	—

Table 5.3 Continued. Plant communities in the vicinity of Truelove Lowland, Devon Island, Canadian High Arctic.

Habitat – plant community types	Topographic position	Plant cover	Examples of representative species		
			Vascular plants	Bryophytes	Lichens
Hummocky graminoid meadow	Coastal, recently emerged areas	Graminoids and bryophytes abundant	*Arctagrostis latifolia* (D) *Dupontia fischeri* (D)	*Orthothecium chryseum* (C – tall turf) *Drepanocladus revolvens* (C – tall turf) *Cinclidium arcticum* (C – tall turf) *Calliergon giganteum* (C – tall turf) *Meesia triquetra* (C – tall turf)	—
Tidal salt marsh	10–30 m wide strip along coast	Continuous turf: mosses and lichens a very minor component	*Puccinellia phryganoides* (D)	—	—
Mesic and dry ground communities					
Raised centre polygons	Beach ridge margins and dry meadows	Scattered vascular plants, bryophytes and lichens	*Alopecurus alpinus* (D) *Dryas integrifolia* (D) *Salix arctica* (D)	*Aulacomnium turgidum* (C – tall turf) *Ditrichum flexicaule* (C – tall turf) *Polytrichun alpinum* (C – tall turf)	*Cladonia poccilum* (C) *Certraria nivalis* (C) *Lecanora epibryon* (C)
*Dwarf shrub heath-moss	Among rock outcrops	Cover averages 37% for vascular plants and 14% for bryophytes	*Cassiope tetragona* (D) *Dryas integrifolia* (C) *Saxifraga oppositifolia* (C)	*Racomitrium lanuginosum* (D – mat) *Hypnum revolutum* (C – mat) *Ditrichum flexicaule* (C – tall turf) *Hylocomium splendons*	*Certraria nivalis* (C) *Thamnolia vermicularis* (C) *Dactylina arctica* (C)

				(C – tall turf)	
Graminoid-moss	Near summit of rocky slopes	Bare rock gives greatest cover. Cover of cryptogams exceeds that of vascular plants	*Hierochloe alpina* (D) *Luzula confusa* (D)	*Rhacomitrium lanuginosum* (D – mat) *Hylocomium splendens* (C – mat) *Polytrichum juniperinum* (C – tall turf)	*Alectoria ochroleuca* (D) *Umbilicaria vellea* (D)
*Cushion plant moss	Mid to lower slopes of raised beaches and other lowland areas	Plant cover approaches 100% with mosses and lichens contributing 30–50%	*Dryas integrifolia* (D) *Carex rupestris* (C) *Saxifraga oppositifolia* (C) *Salix artica* (C)	*Tortella arctica* (D – tall turf) *Oncophorus wahlenbergii* (D – tall turf) *Schistidium holmenianum* (D – tall turf)	*Certraria nivalis* (C) *Lecanora epibryon* (C)
*Cushion plant lichen	Crests and upper slopes of raised beaches, limestone pavement and rock outcrops: little winter snow cover, well drained and relatively warm in summer	Vascular plant cover c. 20% moss and lichen cover c. 40–45%	*Carex nardina* (C) *Dryas integrifolia* (C) *Salix arctica* (C) *Saxifraga oppositifolia* (C)	*Distichium capillaceum* (C – short turf) *Encalypta rhaptocarpa* (C – short turf) *Hypnum bambergeri* (C – mat or carpet) *Myurella julacea* (C – mat or short turf)	*Alectorea pubescens* (D) *Rhizocarpon geographicum* (C) *Thamnolia subliformis* (C) *Umbilicaria arctica* (C)
Polar desert communities					
*Moss-herb (polar desert)	Dry plateau areas. Plant cover best developed in troughs of polygons, and between and under frost shattered rock fragments	Total plant cover 2–7%: mosses predominant	*Draba corymbosa* (M) *Papaver radicatum* (M) *Saxifraga oppositifolia* (M)	*Distichium capillaceum* (C – short turf) *Ditrichum flexicaule* (C – short turf) *Hypnum bambergeri* (D – mat or carpet) *Mnium lycopodioides* (M – single stems among rocks)	*Alectorea nigricans* (M) *Pertusaria dactylina* (M) *Polyblastia bryophila* (M)

Table 5.3 Continued. Plant communities in the vicinity of Truelove Lowland, Devon Island, Canadian High Arctic.

Habitat – plant community types	Topographic position	Plant cover	Examples of representative species		
			Vascular plants	Bryophytes	Lichens
Bryophyte-dominated communities					
Herb-moss snowbed	Small areas in lee of rocks: snow persists until July	Bryophyte cover exceeds that of vascular plants	*Oxyria digyna* (C) *Ranunculus sulphureus* (C) *Saxifraga cernua* (C)	*Bryum cryophyllum* (D – large cushion) *Bryum pseudotriquetrum* (D – tall turf or large cushion) *Cinclidium arcticum* (D – tall turf or large cushion) *Philonotis fontana* var. *pumila* (D – tall turf or large cushion)	—
Scorpidium scorpioides community	Submerged in shallow calcareous pools	Almost continuous moss cover	—	*Scorpidium scorpioides* (D – mat or carpet) *Scorpidium turgescens* (C – mat or carpet) *Calliergon giganteum* (C – mat)	—
Schistidium alpicola community	Submerged in swiftly-flowing streams	Discontinuous moss cover	—	*Schistidium alpicola* var. *rivularis* (C – small cushion) *Blindia acuta* (C – short turf)	—
Aplodon wormskjoldii	Dung of musk ox	Continuous moss cover	—	*Aplodon wormskjoldii* (C – tall turf) *Bryum* spp.	—

				(tall turf) *Tetraplodon* spp. (tall turf) *Voitia hyperborea* (tall turf)	
Desmatodon–Funaria community	Disturbed ground around lemming burrows	Discontinuous moss cover	—	*Desmatodon heimii* var. *arctica* (D – short turf) *Desmatodon leucostoma* (D – short turf) *Funaria polaris* (C – short turf)	—
Community of small acrocarpous mosses	Disturbed ground on sides of ice wedges	Discontinuous moss cover	—	*Ceratodon purpureus* (C – short turf) *Dicranella crispa* (C – short turf) *Leptobryum pyriforme* (C – short turf) *Psilopilum cavifolium* (C – short turf) *Stegonia latifolia* (C – short turf)	—
Racomitrium lanuginosum community	Dry sites among rocks	Extensive cover of mosses and lichens	—	*Racomitrium lanuginosum* (D – mat)	*Rhizocarpon* spp. (C) *Umbilicaria vellea* (D)

Additional, undescribed bryophyte communities occur, particularly on rock outcrops.
*Major communities in terms of area occupied.
D = dominant, C = common, M = minor component
Data from Bliss *et al.* (1977), Muc and Bliss (1977) and Vitt and Pakarinen (1977), with additional information from D.H. Vitt (personal communication).

Mosses are particularly abundant in the graminoid tundra, dwarf shrub heath, and in a closed cushion plant-moss community (Table 5.3), and the range of life-forms is wider than in the wetlands. Tall turfs are again represented, for example by *Polytrichum juniperinum* in the graminoid tundra, but short turf- and mat-forming species are also prominent, particularly in the drier habitats. Examples include mats of *Racomitrium lanuginosum,* a dominant moss in both dwarf shrub heath and graminoid tundra, and *Encalypta rhaptocarpa,* a short turf-forming species occurring commonly in the cushion plant-lichen community.

A similar range of life-forms occurs in association with *Cassiope* and *Dryas* elsewhere in the Arctic. Abundant tall turfs of *Aulacomnium* spp. occur in *Cassiope* heath on northern Ellsemere Island where a varied bryophyte assemblage, including mats of *Abietinella abietina* and the hepatic *Arnellia fennica,* with short turfs of *Encalypta alpina,* occurs among hummocks of *Dryas integrifolia* (Brassard, 1971b). In Peary Land, mats of *Hypnum revolutum* predominate in the bryophyte component of *Cassiope* heath, with *Aulacomnium turgidum* present only in the moister places (Holmen, 1955). Mats of *Hylocomium splendens* with tall turfs of *Aulacomnium turgidum* and *Tomenthypnum nitens* are reported as the dominant bryophytes in a *Dryas*–sedge–moss tundra in the Taimyr region (Matveyeva *et al.*, 1974).

At some low Arctic sites, the dwarf shrub heathlands support bryophyte communities resembling those of heaths in boreal and temperate regions. In Iceland, the weft-forming mosses *Pleurozium schreberi* and *Hylocomium splendens* are prominent, as well as tall turfs of *Dicranum* spp. with mats of such leafy hepatics as *Frullania tamarisci, Lophozia* spp. and *Ptilidium ciliare* (Hesselbo, 1918). In addition, the canopy-forming moss *Climacium dendroides* occurs in association with wefts of *Hylocomium* and *Rhytidiadelphus* spp. and a wide range of other bryophytes on mesic, grassy slopes. Weft-forming mosses are also abundant in birch woodland, the presence of which indicates that not all Icelandic lowland vegetation is truly Arctic in character.

5.2.3 Polar desert

The dry polar desert soils support felfields dominated by cushion-forming flowering plants and mosses, and other open communities comprising scattered vascular plants and cryptogams among which short, compact life-forms again generally predominate. Mats or carpets of *Hypnum bambergeri,* and short turfs of *Distichium capillaceum* and *Ditrichum flexicaule,* are among the characteristic mosses of the dry plateau areas around Truelove Lowland (Table 5.3), while Brassard (1971b) lists a wide range of small acrocarpous mosses as constituents of a *Bryum–Encalypta* community on dry plains of sand, silt or clay in northern Ellesmere Island.

Similar communities occupy dry ground in the low Arctic. In Iceland, large stones on dry rocky flats in montane regions support scattered mats, small cushions and short turfs of such mosses as *Andreaea rupestris, Dicranoweisia crispula, Racomitrium fasciculare* and *Schistidium apocarpum*, and a similar assortment of life-forms is represented in an open, pioneer community on porous substrata in the extensive young lava fields (Hesselbo, 1918).

5.2.4 Bryophyte-dominated communities

The bryophyte assemblages so far considered form an integral part of plant communities also comprising vascular plants and often lichens. Additional bryophyte communities may be recognized in habitats such as rock faces where significant vascular plant cover has failed to develop, or in small areas where the habitat differs markedly from the norm for a major community, e.g. on muskox dung within a sedge meadow. Examples occur locally in sheltered sites where snow persists until well into July. As well as an unusually short growing season with freely available water, such sites are characterized by shelter from wind, well developed, protective winter snow cover, and often an accumulation of sand or organic debris blown into the hollows over the winter snow. Distinctive assemblages of both vascular plants and bryophytes may be found under such conditions. Mårtensson (1956) has noted, however, that many bryophytes which are characteristic of late snow areas in the low alpine belt of the Torneträsk area occur more widely in the high apline region.

Late lying snowbeds on Truelove Lowland support luxuriant, large cushions and tall turfs of *Brynum cryophyllum, Philonotis fontana* var. *pumila* and other mosses, in which are rooted scattered herbaceous flowering plants (Table 5.3). The hepatic *Anthelia juratzkana* is characteristic of late snowbeds throughout the Arctic, for example in Peary Land (Holmen, 1955) and on Spitzbergen (Philippi, 1973), and it is one of the few hepatics to assume dominance in Arctic plant communities. Holmen noted, however, that a covering of sand deposited by melting snow restricts the bryophyte vegetation of some late snow areas: *Polytrichum alpinum* and *Timmia austriaca* were considered two of the mosses best able to overcome this stress, and *P. alpinum* is also dominant in some snow bank vegetation on northern Ellesmere Island (Brassard, 1971b).

Vegetation resembling the snowpatch community of Truelove Lowland in the abundance of large cushion-forming mosses such as *Bryum cryophyllum* and *Philonotis* spp. is widely distributed on stony or marshy ground by permanently flowing rivers and streams (Brassard, 1971b; Hesselbo, 1918; Holmen, 1955). Bryophytes are only sparsely developed on rocks submerged in streams in some high Arctic regions (Holmen, 1955), but communities with abundant *Hygrohypnum* and *Schistidium* spp. have been reported from Truelove Lowland (Table 5.3) and elsewhere, particularly by waterfalls (Brassard, 1971b; Holmen, 1955; Steere, 1978b). *Scapania* and *Fontinalis* are among other genera characteristic of such habitats in low Arctic regions (Hesselbo, 1918; Mårtensson, 1956). *Scorpidium scorpioides* and other mat or carpet-forming pleurocarpous mosses, including *S. turgescens* and species of *Calliergon* and *Drepanocladus*, are widely distributed in shallow calcareous ponds.

Boulders, cliffs and other rocky habitats at high Arctic sites are commonly too dry to support an abundance of bryophytes. Rapid weathering of sedimentary rock is another limiting factor (Brassard, 1971b). Vitt and Pakarinen (1977) note that small cushion-forming acrocarps are rare in calcareous conditions on Devon Island, but that such species as *Amphidium lapponicum, Andreaea rupestris, Dicranoweisia crispula* and *Grimmia torquata* are common on acidic rock outcrops. Similar communities occur in other high Arctic sites (Brassard, 1971b; Holmen, 1955). In contrast, small mat- and turf-forming species appear to predominate in the open vegetation found in sheltered rock crevices. Examples include the *Cyrtomnium hymenophylloides* community of northern Ellesmere Island, where associated species include *Fissidens arcticus, Isopterygium pulchellum, Pohlia cruda* and *Myurella* spp., and an assemblage of *Leskeella nervosa, Seligeria* spp. and other mosses from Alaska (Brassard, 1971b; Steere, 1978b). Cliff vegetation is best developed at oceanic low Arctic sites where many communities resemble those of temperate montane regions (Hesselbo, 1918; Mårtensson, 1956).

The Arctic is notable for the number of bryophyte communities associated with substrata enriched in nutrients by animal activity. Most striking is the assemblage comprising members of the Splachnaceae, with *Bryum* spp. and other mosses at later stages in succession (Webster and Sharp, 1973), found on faeces, bone and other animal remains. It occurs throughout the Arctic, and a wide range of splachnaceous mosses may be represented. Brassard (1971b) regards six such species, and a comparable number of other acrocarpous mosses, as characteristic of an *Aplodon wormskjoldii* community in northern Ellesmere Island. Seventeen members of the Splachnaceae have been recorded in Alaska (Steere, 1978a), but not all are restricted to animal remains. In Scandinavia, the spores of splachnaceous mosses characteristic of dung have been shown to differ from those of related species growing on soil or as epiphytes, and the attraction of insects to coprophilous species, possibly by odours or secretions from the enlarged and brightly coloured hypophysis, has been demonstrated experimentally. These and other factors which suggest the evolution of entomophily within the Splachnaceae have been discussed by Koponen (1978) and Koponen and Koponen (1978); see also Chapter 9, pp. 312–315.

Bird perches and the mouths of lemming burrows also support distinctive bryophyte communities, composed, at least in the high Arctic, of short turf-forming acrocarps. Several of the species associated with lemming burrows are apparently Arctic endemics, e.g. *Desmatodon leucostoma* and *Funaria polaris* (Steere, 1978b; Vitt, 1975). In contrast, the cosmopolitan *Bryum argenteum* is among the mosses colonizing sites habitually frequented by birds in several areas. Mårtensson (1956) has noted that certain mosses with southern distribution patterns extend to alpine sites in the Tornaträsk area only around bird perches.

Short turf-forming acrocarps are also prominent in successional communities on ground disturbed by ice action and other factors. Examples include a community on the sides of ice-wedge polygons on Truelove Lowland (Table 5.3), and an assemblage of such cosmopolitan weeds as *Bryum argenteum, Ceratodon purpureus, Funaria hygrometrica* and *Marchantia polymorpha* on ground disturbed by man and other animals in Alaska (Steere, 1978b).

A final community type which should be mentioned is the moss 'heath', generally dominated by species of *Racomitrium*, notably *R. lanuginosum*. The latter is widespread throughout the Arctic: it is a frequent component of dwarf shrub heath and other vascular plant communities, and may become dominant in dry rocky sites, as on Truelove Lowland (Table 5.3). Extensive areas of *Racomitrium* 'heath' are characteristic of certain islands where porous volcanic soils occur in an oceanic climate characterized by strong winds, frequent precipitation, and generally high relative humidity with frequent fog. The combination of strong winds and porous substrata restricts vascular plant cover, and bryophytes, whose colonies can absorb and retain precipitation, and even water from a saturated atmosphere (Tallis, 1959), predominate. Thus, large tracts of stony ground on Iceland support a closed *Racomitrium lanuginosum* 'heath', often with only scattered phanerogams and a small admixture of other bryophytes (Hesselbo, 1918), and a similar community on exposed hillsides forms the principal vegetation of Jan Mayen. A second bryophyte community, in which *Drepanocladus uncinatus* may predominate (E.V. Watson, personal communication), is developed extensively in parts of Jan Meyen thought to experience particularly frequent precipitation and mist (Russell and Wellington, 1940), and similar communities dominated by *D. uncinatus* and *Racomitrium* spp. occur on Spitzbergen (Philippi, 1973).

5.3 BRYOPHYTE VEGETATION IN THE MARITIME ANTARCTIC

5.3.1 Classification

A vegetation classification for the maritime Antarctic has been developed by Holdgate (1964), Gimingham and Smith (1970) and Longton (1967). In this hierarchical scheme (Table 5.4), the basic unit, the sociation, is recognized by aerial cover of the dominant or codominant species, and sociations are grouped within associations, defined by the floristic similarity of the component sociations and the high constancy of a group of characteristic species, following Poore (1962). Associations are included within subformations on the basis of life-form of the community dominants, and two formations are recognized according to the dominance of vascular plants or cryptogams.

The classification has been generated subjectively and there is intergradation between the vegetation units at all levels. However, quantitative analysis of quadrat data from Signy Island, South Orkney Islands (Fig. 5.2), provided strong support for a number of the units recognized (Smith and Gimingham, 1976). The classification has been applied successfully in describing vegetation at several maritime Antarctic localities (Allison and Smith, 1973; Lindsay, 1971; Longton and Holdgate, 1979; Smith and Corner, 1973), and notably in Smith's (1972) detailed account of Signy Island. More recently, the classification has been extended to parts of continental Antarctica (Longton, 1973a; Nakanishi, 1977; Seppelt and Ashton, 1978). Literature on Antarctic plant communities is reviewed by Longton (1979a), who lists the sociations so far recognized.

5.3.2 Vascular plant and lichen communities

Vegetation developed under the cold oceanic conditions prevailing in the maritime Antarctic is almost entirely cryptogamic. The two native species of flowering plants only become abundant locally in small stands of the grass and cushion chamaephyte formation (Table 5.4). In contrast, vegetation dominated by mosses, lichens and, locally, algae is extensive and often luxuriant. The crustaceous lichen subformation (Table 5.4) is widely distributed on boulders and rock faces, particularly in dry exposed situations and on sea cliffs, but also around pools and in other wet habitats. *Buellia, Caloplaca, Placopsis, Xanthoria,* with *Verrucaria* low on marine cliffs, are among the prominent genera. Bryophytes are generally rare or absent. Mosses, particularly small cushion-forming species of *Andreaea, Dicranoweisia* and *Grimmia,* are more prominent in some stands of the fruticose and foliose lichen subformation. These communities, in which species of *Himantormia, Umbilicaria* and *Usnea* generally predominate, cover extensive areas on cliffs, scree slopes and other exposed, rocky habitats. Small moss cushions are most frequent as associates of the macrolichens on the moister rocks, and the green alga *Prasiola crispa* may become abundant where water trickles over rock surfaces.

Table 5.4 Outline of a classification of vegetation in the Antarctic zone (After Longton, 1979a).

Antarctic nonvascular cryptogam tundra formation

Crustaceous lichen subformation
Buellia – Lecanora – Lecidea association: 2 sociations
Caloplaca – Xanthoria association: 10 sociations
Placopsis contortuplicata association: 3 sociations
Verrucaria association: 5 sociations

Fruticose and foliose lichen subformation
Usnea – Umbilicaria association: 15 sociations

Short moss turf and cushion subformation
Andreaea association: 6 sociations
Bryum antarcticum – B. argenteum association: 4 sociations
Bryum inconnexum association: 5 sociations
Ceratodon association: 3 sociations
Pohlia nutans association: 2 sociations
Pottia austro-georgica association: 1 sociation
Sarconeurum glaciale association: 1 sociation
Tortula – Grimmia antarctici association: 6 sociations

Tall moss turf subformation
Campylopus association: 2 sociations
Polytrichum alpestre – Chorisodontium aciphyllum association: 6 sociations
Polytrichum alpinum – Pohlia nutans association: 7 sociations

Bryophyte carpet and mat subformation
Brachythecium association: 1 sociation
Calliergidium austro-stramineum – Calliergon sarmentosum – Drepanocladus uncinatus association: 6 sociations
Cephalozia badia association: 1 sociation
Cryptochila grandiflora association: 2 sociations
Marchantia association: 1 sociation

Moss hummock subformation
Brachythecium austro-salebrosum association: 1 sociation
Bryum algens – Drepanocladus unciantus association: 3 sociations

Alga subformation
Prasiola crispa association: 1 sociation
Nostoc association: 1 sociation

Snow alga subformation
Chlamydomonas nivalis association: 2 sociations

Antarctic herb tundra formation
Grass and cushion chamaephyte subformation
Deschampsia antarctica – Colobanthus quitensis association: 4 sociations

5.3.3 Short moss turf and cushion subformation

Mixed communities comprising fruticose and foliose lichens, with small cushions of *Andreaea* spp. and other mosses, in places intergrade with essentially bryophytic vegetation in the *Andreaea* association of the short moss turf and cushion subformation (Table 5.4). The latter is characteristic of stones, gravel and sand on level ground or gentle slopes, principally on acidic substrata, at sites covered by winter snow. Small cushion-forming mosses predominate, principally *Andreaea depressinervis, A. gainii, A. regularis* and locally *Racomitrium austro-georgicum.* Mats of *Drepanocladus uncinatus* are also widespread, and small areas on Signy Island support an *Andreaea*-hepatic sociation in which *Barbilophozia hatcheri, Caphaloziella varians, Herzogobryum teres, Hygrolembidium isophyllum* and *Pachyglossa dissitifolia* form mixed mats up to 50 cm in diameter (Smith, 1972). Bryophyte cover may be high, particularly at the moister sites where moss cushions coalesce to give a closed sward. Lichens are present and become abundant in drier areas. They include epipetric species such as *Usnea antarctica,* and other, principally crustose forms, which grow epiphytically on the mosses. Solifluction and frost heave may disrupt the moss cover, and Smith (1972) describes hillsides on Signy Island where the *Andreaea*–lichen sociation occurs as narrow strips parallel with the slope, on coarse, relatively stable substrata, which alternate with strips of mobile clay almost devoid of vegetation.

The *Andreaea* association is commonly replaced on calcareous substrata by the *Tortula–Grimmia* association. The dominant mosses again assume the small cushion life-form, but short turf-forming species of *Encalypta, Pohlia, Pottia* and other genera may also be present, as well as mats of *Brachythecium* and *Drepanocladus* spp. Species of *Amblystegiella, Cephaloziella* and *Pseudoleskea* have been recorded as scattered stems among the more abundant mosses, while lichens become frequent only in the drier sites. The *Tortula–Grimmia* association is particularly well developed on relatively dry sandy or gravelly soils around local outcrops of marble on Signy Island. Here, moss cover may reach 100% in small stands of vegetation supporting several species not recorded on the prevalent acidic substrate (Smith, 1972).

The remaining associations within the short moss turf and cushion subformation are all characteristic of particularly arid habitats or, in the case of the *Pottia austrogeorgica* association, occur as temporary colonizers of unstable soils. Those communities dominated by species of *Bryum* or *Sarconeurum* (Table 5.4) occur primarily in continental Antarctica, while the *Ceratodon* and *Pohlia nutans* associations are well represented on porous, volcanic substrata in the maritime Antarctic, notably on Deception Island, South Shetland Islands, and on the South

Sandwich Islands. For example, volcanic ash and scoria on level ground and gentle slopes on Candlemas Island support extensive turfs of *Pohlia nutans,* which form a crust *c.* 1 cm deep. Much of the moss is moribund and colonized by a species of *Lepraria* and other crustose lichens, which may be so frequent as to give the vegetation a whitish-green colour when viewed from a distance. The *Pohlia nutans–Lepraria* sociation comprises species-poor, short, open turfs of *Polytrichum* spp., only 1–2 cm high and frequently intermixed with *Pohlia nutans,* being the principal associates.

Short turf- and small cushion-forming mosses are also characteristic of rock crevices, together with mats of both mosses and hepatics. On Signy Island, crevices on acidic and on base-rich rocks usually support mosses characteristic of the *Andreaea* and *Tortula–Grimmia* associations respectively, while additional bryophytes include species of *Bartramia, Dicranoweisia, Isopterygium, Plagiothecium, Barbilophozia, Metzgeria* and *Pachyglossa* on siliceous rocks, and of *Amblystegiella, Barbula, Didymodon, Pseudoleskea* and *Sarconeurum* under base-rich conditions (Smith, 1972). These communities have not yet been formally grouped into sociations and associations.

5.3.4 Tall moss turf subformation

The tall moss turf subformation (Table 5.4) is characteristic of mesic substrata, often on gentle to moderately steep rocky slopes. Of particular interest are communities in the *Polytrichum alpestre–Chorisodontium aciphyllum* association which form raised banks commonly overlying 1–2 m of ombrogenous peat, and extending continuously for as much as 50 m. The surface of the banks is hummocky or undulating, and the underlying peat shows permafrost below 20–30 cm. Uptake of water from melting ice above the permafrost layer is considered important in irrigating the shoot apices. *C. aciphyllum* prefers wetter conditions than *P. alpestre,* and is often dominant in shallow banks on level ground, while on the drier slopes either species may occur singly or in mixed turf. Lichen encrustation is widespread on the higher, more windswept parts of the banks.

Smith (1979) has pointed out that peat in the *Polytrichum–Chorisdodontium* banks is unusual in being largely unhumified, and in developing under conditions that are neither waterlogged nor anaerobic. He attributes its formation to permafrost, and to slow decomposition in the active layer due to a very low pH and an unusually poor microbial flora. Decomposition of material above the permafrost has been estimated as only 2% per year (Baker, 1972). The growth of *P. alpestre* in the Antarctic and elsewhere is compared by Longton (1974b, 1979b).

Polytrichum alpinum is a frequent component of many vegetation types in the maritime Antarctic, but only locally becomes dominant. In particular, the *P. alpinum* association replaces the *P. alpestre–*

Chorisodontium aciphyllum association on recent volcanic substrate on the South Sandwich Islands and on many of the South Shetland Islands. On these loose-textured soils *P. alpinum* forms circular colonies up to 1 m in diameter and 20–30 cm in height, but deep peat deposits do not occur. *Polytrichum piliferum, Chorisodontium aciphyllum,* a tall form of *Pohlia nutans* and mats of *Drepanocladus uncinatus* have all been recorded as codominant with *Polytrichum alpinum.* Several species of *Campylopus* may be abundant in stands of the *Campylopus* association, but this vegetation type is confined to the vicinity of fumaroles on the South Sandwich Islands.

5.3.5 Bryophyte carpet and mat subformation

Bryophyte carpets have so far been recorded widely only in the maritime Antarctic. They comprise pleurocarpous mosses, or less commonly leafy hepatics, occurring as closely packed, almost parallel, erect shoots connected to a basal prostrate layer of partially denuded stems (Gimingham and Smith, 1970; Longton, 1967). Some species typically form carpets in wet habitats, and mats under drier conditions. Communities in the *Calliergidium austro-stramineum – Calliergon sarmentosum – Drepanocladus uncinatus* association are widespread, and may cover extensive areas in permanently moist or wet habitats on level or gently sloping ground at low altitudes, particularly on seepage slopes receiving melt water throughout the summer. The carpet-forming mosses commonly occur in continuous stands overlying soligenous peat up to 15 cm deep. Other life-forms are seldom abundant, although mats of *Cephaloziella varia* may occur on the surface of the mosses. Lichens are rare except for crustose species on stones. A striking feature of many moss carpets are 'fairy rings' up to 20 cm in diameter composed of white, moribund shoots, associated with fungal infection (Longton, 1973b).

Carpets of an unidentified species of *Brachythecium* occur in similar habitats on the South Sandwich Islands, where the fumaroles support a range of hepatic-dominated communities also placed in the bryophyte carpet and mat subformation (Longton and Holdgate, 1979). Luxuriant carpets of *Cryptochila grandiflora,* and locally of *Triandrophyllum subtrifidum,* occupy finely divided ash and scoria around many of the fumarole vents. Tall turfs of *Pohlia nutans* and species of *Campylopus* are frequent associates, while crustose lichens are abundant on projecting stones. Shallow, compact mats of *Cephalozia badia* were recorded around a single fumarole. Conversely, a species of *Marchantia* is widespread in the fumarole vegetation, and becomes abundant locally both there, and at well insolated sites in the South Orkney Islands not influenced by current volcanic heat and moisture.

5.3.6 Moss hummock subformation

The margins of melt water streams, wet rock ledges subjected to dripping water, and other flushed sites in contact with moving water form the typical habitat for the remaining major terrestrial bryophyte vegetation type, the moss hummock subformation (Table 5.4). The life-forms represented have been described by Smith (1972) as 'a tall compact cushion (*Bryum algens*), a tall loose cushion (*Brachythecium austro-salebrosum*), or a deep undulating carpet (*Drepanocladus uncinatus*)'. The large cushions may attain heights of 20 cm, and frequently coalesce to form a closed, hummocky stand. Communities in the moss hummock subformation are widely distributed, but are seldom extensive, and may intergrade with those dominated by carpet-forming mosses as distance from flowing water increases.

5.3.7 Pattern and succession

Cryptogamic vegetation is extensive at favourable locations in the maritime Antarctic, as indicated in Table 5.5 for Signy Island, which has an area of 17 km^2 and a maximum elevation of 281 m. It can be seen that 30% of the land surface is exposed from snow and ice in summer. Some 13% supports closed bryophyte vegetation, principally in the tall moss turf, and bryophyte carpet and mat subformations. Additional closed vegetation occurs within the short moss turf and cushion subformation, and less than 26% of the exposed land surface lacks either open or closed macroscopic plant cover. Table 5.5 also emphasizes the limited role of phanerogams.

The relationships between topography and life-form of the dominant cryptogams are indicated in outline in Fig. 5.3, which represents diagramatically the distribution of major vegetation types across an idealized, well-vegetated valley. Carpet-forming mosses cover large areas on the lower slopes and valley floor, particularly in areas below late snow beds, with large cushion mosses occurring locally by streams. Mesic, often steeper, rocky slopes support banks of tall turf-forming acrocarps, while short moss turfs and small cushions occur principally on more exposed, elevated slopes of sand, gravel and scree. Lichen communities occupy boulders, cliffs and even more windswept slopes of scree and gravel, with crustose lichens particularly prominent on coastal cliffs. The highest, most exposed cliffs and scree slopes are largely devoid of vegetation.

A more detailed discussion of pattern within and between bryophyte communities on Signy Island has been presented by Smith (1972) who recognized four major types, namely environmental, zonal, morphological and sociological pattern. Of these, environmental pattern is the most widespread. It develops in response to interacting, and sometimes intersecting gradients in several environmental factors, and results in a

Table 5.5 Extent of major vegetation types on Signy Island, South Orkney Island (Data from Tilbrook, 1970).

	Area occupied (km^2)	% total snow-free surface
Crustaceous lichen subformation or bare ground	4.25	26
Fruticose and foliose lichen and short moss turf and cushion subformations		
Cover >50%	3.5	21
Cover <50%	5.0	30
Tall moss turf subformation	1.0	6
Bryophyte carpet and mat and moss hummock subformations		
Cover >50%	1.25	7.5
Cover <50%	0.75	4.5
Grass and cushion chamaephyte subformation	0.0001	<0.001
Biotically formed organic soil	0.65	4.0
Penguin nests and seal wallows	0.1	0.6
Other bird nests	0.0005	<0.005

mosaic of communities in which different cryptogams assume dominance. An example discussed by Smith occurs on low knolls in coastal districts (Fig. 5.4). The thin mantle of sandy and gravelly soil on the windswept summits of the knolls supports a sociation dominated by the fruticose lichen *Usnea antarctica.* In more sheltered locations below the summit, subject to shallow snow accumulation in winter, this gives way to a community dominated by other lichens and *Andreaea depressinervis,* while slopes receiving deeper snow accumulation, and moist level ground at the foot of the knoll, support stands of *Drepanocladus uncinatus* and *Polytrichum alpinum.* The gradual nature of the transition between the three communities, and the wide ecological amplitude of certain of the species, is emphasized in Fig. 5.4, though the mosaic of communities associated with environmental pattern is not well illustrated by the transect data.

Less complex, zonal patterns of plant distribution in response to environmental gradients are not frequent on Signy Island, and are best exemplified by coastal lichen communities. However, clear zonal patterns are associated with the gradients of temperature and moisture around fumaroles on the South Sandwich Islands (Longton and Holdgate, 1979), which give rise to concentric distributions of plant communities. An example is described in Table 5.6.

Morphological pattern is associated with the life-form of individual species and is commonly exhibited only by species of *Polytrichum,* notably

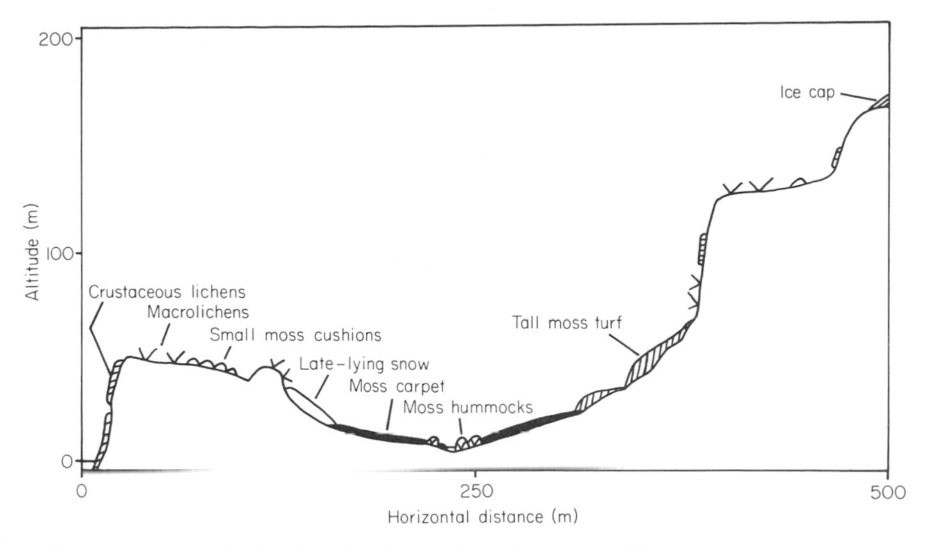

Fig. 5.3 Diagram indicating the distribution of the major life-form types across an idealized valley in the maritime Antarctic.

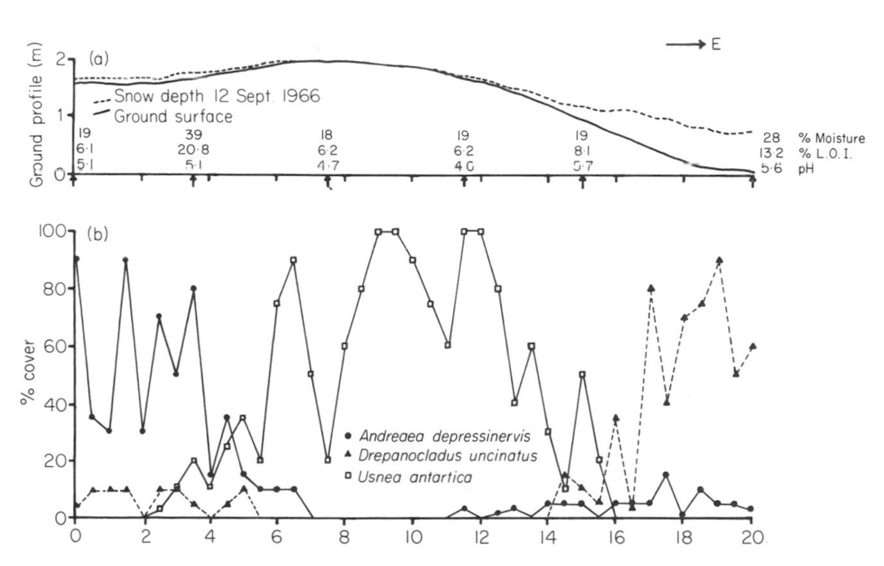

Fig. 5.4 Relationships between ground profile, soil data and winter snow cover (a) and % cover of the principal species (b) along a transect across a low knoll on Signy Island, South Orkney Islands. Arrows indicate positions where soil data were obtained: L.O.I. is loss on ignition. Redrawn from Smith (1972).

Table 5.6 Zonation of vegetation around a fumarole on Bellingshausen Island, South Sandwich Islands. (From Longton and Holdgate, 1979.)

Zone	Approximate distance from vents	Dominant species	Notes
1	0–30 cm	Algae and a small dicranoid moss	Glistening areas of algae, together with abundant turfs of a small dicranoid moss, scattered stems of *Bryum sp.* and taller turfs of *Campylopus spiralis*.
2	30–150 cm	*Campylopus spiralis*	Turfs of *Campylopus spiralis* dominant. *Cryptochila grandiflora* and *Marchantia berteroana* frequent. *Polytrichum alpinum* and *Pohlia nutans* recorded as associates.
3	150–270 cm	*Cryptochila grandiflora*	*Cryptochila grandiflora* increased in frequency to form a luxuriant, closed, bright orange stand. *Campylopus spiralis* and *Marchantia berteroana* abundant in central areas, dying out further from the vent, where *Psilopilum antarcticum* recorded as an associate. Basidiomycete fruiting bodies scattered throughout.
4	270–390 cm	*Cryptochila grandiflora* and mosses	*Cryptochila grandiflora* became drier and dark reddish brown in colour, being interspersed with large areas of *Racomitrium crispulum* and other mosses including *Drepanocladus uncinatus* and *Pohlia nutans*. Basidiomycetes persisted.
5	390–570 cm	*Cryptochila grandiflora* and *Usnea antarctica*	*Cryptochila grandiflora* very dark brown and dry, forming an open community with *Usnea antarctica*. *Psilopilum antarcticum* frequent. *Polytrichum alpinum* and *Racomitrium crispulum* recorded as associates.
6	>570 cm	*Usnea antarctica*	Abundant *Usnea antarctica* on bare stones, with less common cushions and turfs of *Dicranoweisia* sp., *Pohlia nutans* and *Polytrichum alpinum* as well as small mats of *Lophocolea secundifolia* and *Cephaloziella* sp.

Temperature readings 2.5 cm deep in the vegetation on a cloudy day in late summer ranged from 38°C at the junction between zones 1 and 2 to 7°C in zone 6.

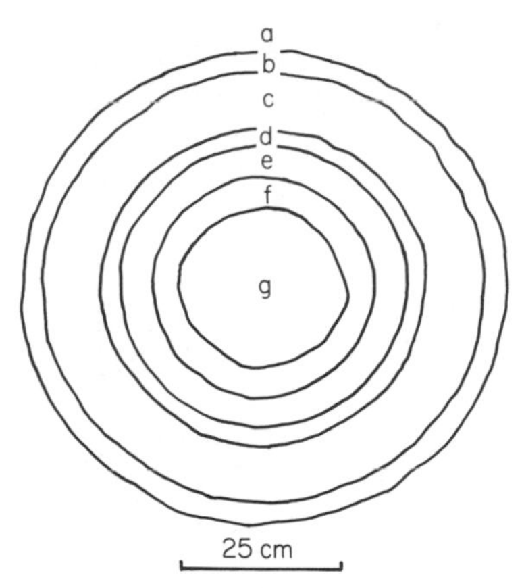

Fig. 5.5 Diagram of a circular colony of *Polytrichum alpinum* on Signy Island, South Orkney Islands, exhibiting morphological pattern. Redrawn from Smith (1972). (a) Area outside *Polytrichum* colony; (b) Scattered young shoots of *Polytrichum* arising from rhizomes; (c) Living, green *Polytrichum* 1–5 cm tall; (d) Dying, brown *Polytrichum*; (e) Moribund, black *Polytrichum*; (f) Thin layer of decaying *Polytrichum* peat colonized by bryophytes and lichens; (g) Peaty soil with occasional bryophytes, often heavily encrusted with lichens.

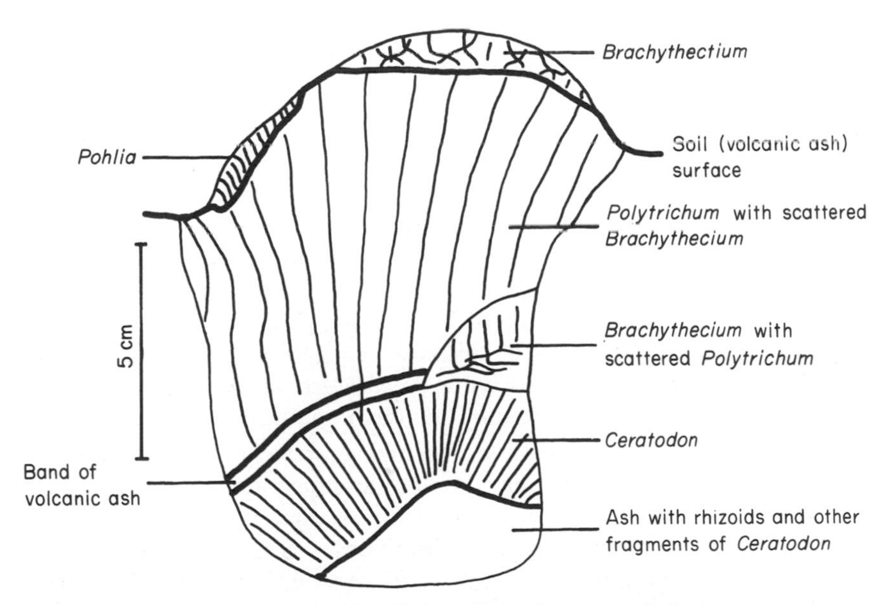

Fig. 5.6 Profile from a stand of the *Polytrichum alpinum* sociation on Candlemas Island, South Sandwich Islands. Redrawn from Longton and Holdgate (1979).

P. alpinum. On level ground, *P. alpinum* tends to form circular stands which gradually expand as underground rhizomes grow outwards from the periphery and give rise to aerial shoots. Translocation of photosynthate from older shoots along rhizomes to young growing points has been demonstrated in Arctic *P. alpinum* (Collins and Oechel, 1974). On Signy Island individual colonies may reach 1 m or more in diameter but peripheral expansion of the larger examples is usually accompanied by progressive death of plants from the centre outwards followed by colonization of the decaying *Polytrichum* remains by other species (Fig. 5.5). In some cases *P. alpinum* may eventually re-invade the central region so that a cyclic succession is established.

A somewhat different pattern of cyclic succession involving *P. alpinum* was demonstrated on a slope of volcanic ash on Candlemas Island by observations on the living vegetation combined with analysis of the shallow peat deposits beneath (Longton and Holdgate, 1979). Here, *P. alpinum* forms large hummocky turfs up to 15 cm in diameter, associated with similar turfs of *Pohlia nutans* and mats of *Brachythecium* and *Drepanocladus* spp. The larger turfs are subject to lichen encrustation and ultimately to severe wind erosion which leaves areas of bare mineral soil. A cushion-forming species of *Ceratodon* is the principal colonizer of the bare areas, and shoots of *P. alpinum* are common in the larger cushions. The peat profiles suggest that re-establishment of *P. alpinum* frequently occurs in areas initially invaded by *Ceratodon* (Fig. 5.6).

Sociological pattern as understood by Smith (1972) arises as a result of successional processes involving the invasion of one community by species characteristic of a second. An example is provided by the banks of *Chorisodontium aciphyllum* and *Polytrichum alpestre,* whose development has been discussed by Collins (1976) and Smith (1972, 1979). It was suggested that the banks may be initiated either through colonization by *P. alpestre* of cushion-forming mosses on slopes, or by *C. aciphyllum* invading carpet-forming mosses on moist level ground followed by the establishment of *P. alpestre* on the higher, relatively dry areas of the *Chorisodontium* turfs. Eventually, peat formation results in the upper parts of the banks being blown clear of snow in winter, thus initiating cyclic succession involving the mosses and epiphytic lichens, which become established following erosion of the surface layer of moss.

Sociological pattern of the type occurring where *P. alpestre* and *C. aciphyllum* are actively invading communities dominated by other mosses appears to be surprisingly uncommon in Antarctic vegetation. Smith (1972) noted that the cryptogamic communities on Signy Island are seldom affected by competition from adjacent communities, and that once established, most appear to be climax units showing little evidence of seral succession. This view is supported by radiocarbon dates ranging from several hundred to 5000 years for the base of *Polytrichum–Chorisodontium* banks (Smith, 1979). Similarly, analysis of peat profiles beneath the living vegetation on Candlemas Island suggested that all the major bryophyte community types had developed independently of each other. Thus there is some evidence that succession in Antarctic cryptogamic vegetation may be primarily cyclic rather than directional (Longton and Holdgate, 1979).

5.4 BRYOPHYTE VEGETATION ON CONTINENTAL ANTARCTICA

Vegetation in continental Antarctica contrasts strongly with that in the maritime Antarctic in its sparsity, and in the restricted diversity of bryophyte species and life-form types. Bare rock, scree and gravel predominate in unglaciated areas. Crustose, fruticose and foliose lichens are all locally abundant. Bryophyte vegetation is restricted to the short moss turf and cushion subformation, which occurs most extensively on sandy and gravelly soils and is seldom well developed on rock. Perhaps surprisingly, the range of algal communities is wider than in the maritime Antarctic, particularly in semi-terrestrial vegetation in and beside melt water streams. It is possible that low temperatures prevent the growth of carpet and hummock-forming mosses and that algae flourish due to reduced competition from bryophytes. Vegetation is best developed at localized coastal sites but mosses, lichens, algae and also micro-organisms and invertebrates occur on inland nunataks as far as 84°S to 87°S (Cameron *et al.*, 1971; Janetschek, 1970; Wise and Gressit, 1965).

The dominant mosses may vary between sites. Thus communities in the *Bryum antarcticum–B. argenteum* and *Sarconeurum* associations have been reported from Cape Hallet (Rudolph, 1963) and McMurdo Station (Longton, 1973a) in southern Victoria Land (Fig. 5.2), while the *Bryum inconnexum* and *Ceratodon* associations are represented at Swoya Station on East Ongul Island (Matsuda, 1968; Nakanishi, 1977). The moss communities recently reported from Mawson Station (Seppelt and Ashton, 1978) are dominated by *Bryum algens,* and thus recognition of a *B. algens* association, additional to those listed in Table 5.3, may become necessary. It is possible, however, that the apparent diversity of *Bryum*-dominated vegetation results in part from differences in taxonomic interpretation.

Variation in community structure is also evident. At McMurdo Station *Bryum argenteum* and *Sarcoenurum glaciale* typically form widely scattered short, compact turfs nestling among stones or with their surfaces flush with the surrounding substrate. Other species form small cushions which may give continuous cover over several square metres in particularly favourable sites. At Swoya Station the moss vegetation reaches 5–9 cm in depth, and Matsuda (1968) has estimated the age of the larger *Bryum inconnexum* cushions as c. 100 years on the basis of rhizoid banding. Development of the moss cover at Mawson begins with small cushions, less than 1 cm in diameter, which in some cases have living shoots all over the surface and are blown around by the wind (Seppelt and Ashton, 1978). Larger colonies become sedentary and anchored to the ground by rhizoids. Complex vegetation may be built up by coalescence of the cushions and by young colonies growing over older ones, often of the same species. The older cushions become convoluted and longitudinal sections reveal cores of sand, gravel and small stones, features attributed by Seppelt and Ashton to differential frost action resulting from the relatively high water content of the moss cushions and the substrate immediately below. Cushions with similar mineral cores, and others with hollow centres, have been observed in strands of *Bryum* and *Ceratodon* spp. on Deception Island in the maritime Antarctic, and again attributed to frost action (Longton, 1967). The older colonies of many mosses in continental Antarctica become colonized by lichens or more locally (Matsuda, 1968) by algae.

5.5 BRYOPHYTE VEGETATION IN THE SUB-ANTARCTIC

The sub-Antarctic as understood here (Fig. 5.2) comprises a series of isolated islands characterized by rugged terrain and a cool, oceanic climate (Table 5.1). Descriptions of the vegetation have been provided for Macquarie Island (Taylor, 1955), Marion and Prince Edward Islands (Huntley, 1971) and South Georgia (Greene, 1964b) and these, combined with the author's observations on South Georgia, form the basis of the present account.

Tall turf-forming mosses predominate in the bryophyte stratum of most phanerogamic vegetation at low altitudes, at least on South Georgia. The extensive wetlands support communities dominated by graminoids, which physiognomically resemble comparable Arctic vegetation. On South Georgia, *Calliergon sarmentosum, Drepanocladus uncinatus* and *Tortula robusta* are abundant in the continuous bryophyte understorey in large areas of mire dominated by the sedge *Rostkovea magellanica.* Stands of *Sphagnum fimbriatum* up to 20 m in diameter are locally frequent in *Rostkovia* mire, while conspicuous, circular colonies of *Philonotis acicularis* are interspersed among other mosses at some flushed sites. Comparable communities, although without *Sphagnum,* occur on the other islands and wetland vegetation may comprise principally bryophytes with only scattered vascular plants. For example, on Marion and Prince Edward Islands, some flushes support an assemblage of mosses dominated by *Breutelia integrifolia* and *Bryum laevigata,* while the leafy hepatic *Blepharidophyllum densifolium* assumes dominance in lowland bogs.

Mesic sites in the lowlands support grassland and herbfield, the most conspicuous component of the latter being large forbs such as *Cotula plumosa* or suffrutescent perennials in the genus *Acaena.* The vascular plants reach heights up to 25 cm. Herbfields are to some extent the counterpart of heaths, from which they differ principally in the absence of ericoid shrubs. Heaths occur in the southern temperate zone but not in the sub-Antarctic. The most widespread mesic communities on South Georgia form an intergrading series ranging from swards of *Acaena magellanica* through mixed stands in which *A. magellanica* and the caespitose grass *Festuca contracta* are both abundant to areas of *Festuca* grassland. Stands of *A. magellanica* occur principally on steep scree slopes, and the creeping, woody stems are usually embedded in a closed turf of *Tortula robusta.* On more gently sloping ground, where *F. contracta* is abundant, *T. robusta* is largely replaced by other large acrocarps, particularly *Polytrichum alpinum*: other prominent mosses include *Chorisodontium aciphyllum, Conostomum australe* and *Psilopilum trichodon.* The bryophyte layer in *Festuca* grassland is generally discontinuous, as the mosses are interspersed among macrolichens, notably *Cladonia* spp., and areas of bare ground.

It is noteworthy that the association of *Acaena magellanica* and *Tortula robusta* on South Georgia may be found both on scree slopes and in wet flushes. Similarly, Huntley (1971) records *A. magellanica* as dominant in flushes and in some mesic sites on Marion and Prince Edward Islands, with *Drepanocladus uncinatus* and *Rhyncostigium brachypterygium* the principal bryophyte associates in both cases.

Tussock grassland dominated by species of *Poa* is a distinctive feature of coastal sites in south temperate and sub-Antarctic regions, where it may form an almost continuous zone along shorelines, and ascend to at least 750 m altitude on cliffs subjected to wind-blown spray. *Poa flabellata* is the principal component on South Georgia where it forms tussocks reaching 1.0 m in diameter and 1.5 m in height, each comprising a stool bearing a dense crown of leaves. Tall turf-forming mosses, particularly *Chorisodontium aciphyllum, Polytrichum alpestre* and *P. alpinum,* are abundant in the more open stands. *P. alpestre* locally forms extensive banks resembling those in the maritime Antarctic, particularly where *Poa flabellata* has been overgrazed by introduced herbivores.

A wider range of bryophyte life-form types is represented in essentially cryptogamic communities of rock and scree, which predominate in South Georgian vegetation at elevations above 250 m. Here, the exposed mineral soils support open felfields in which short turfs, small cushions and mats formed by species of *Andreaea, Dicranoweisia, Grimmia, Polytrichum* and

Racomitrium are prominent, in association with lichens and scattered vascular plants. Small cushion-forming mosses, including species of *Andreaea, Blindia* and *Dicranoweisia* are widespread on rock faces, while sheltered crevices and ledges support a further range of cushions, turfs and mats. Genera represented here include *Bartramia, Brachythecium, Dicranum, Distichium, Drepanocladus, Pohlia, Polytrichum* and *Racomitrium.*

Concerning vascular plants, the South Georgian felfields frequently support small cushions of *Colobanthus quitensis* but lack the large compact cushions of *Azorella selago* which are a conspicuous feature of comparable vegetation on some of the other islands. On Macquarie, Marion and Prince Edward Islands, individual cushions of *A. selago* may exceed 50 cm in diameter and they frequently support epiphytic mats of *Racomitrium lanuginosum* and cushions of *Ditrichum strictum.* The latter is dominant in the more exposed felfields on these islands, where its cushions may become dislodged by the wind and develop as spherical, mobile 'moss balls' up to 8 cm in diameter. This process, and successional relationships in the felfield communities, have been discussed by Ashton and Gill (1965), Huntley (1971) and Seppelt and Ashton (1978).

Finally, it may be noted that, as in other areas, particularly rich and luxuriant bryophyte assemblages are to be found on sub-Antarctic islands on rocks in the spray zone of waterfalls and in other permanently wet habitats.

5.6 BENTHIC COMMUNITIES

Attention should also be drawn to the bryophyte communities occurring submerged at depths up to 10 m in polar lakes. Such vegetation appears to be widespread in both Arctic and Antarctic regions (Light and Heywood, 1973, 1975; Mårtensson, 1956; Persson, 1942; Welch and Kalff, 1974), particularly under oligotrophic conditions where little incoming radiation is absorbed by phytoplankton (Priddle, 1980a, b). Individual moss stems may reach 40 cm in length in these benthic communities, whose luxuriance in continental Antarctica contrasts strongly with the sparsity of the terrestrial vegetation (Light and Heywood, 1975). Genera recorded submerged in polar lakes include *Bryum, Calliergon, Campylium, Dicranella, Distichium, Drepanocladus, Fontinalis* and *Marsupella.* In several cases, the submerged plants showed larger leaves and longer internodes than terrestrial material of the same species (Priddle, 1979). A detailed study of benthic bryophytes on Signy Island has shown that positive net photosynthesis occurs during all except the three darkest months of the year, although the lake surfaces are frozen for 8–12 months annually. It was confirmed that the mosses have unusually low light compensation points, and an estimate of 40 g m^{-2} was obtained for net annual production in closed stands (Priddle, 1980a, b). This relatively low figure suggests that the luxuriance of some benthic communities may result from stability rather than rapid growth.

5.7 ENVIRONMENTAL RELATIONSHIPS OF POLAR BRYOPHYTE VEGETATION

5.7.1 Temperature

One of the most striking conclusions to be drawn from recent biological investigations in the Arctic and Antarctic is that polar environments are diverse in the extreme, and that generalizations concerning environmental relationship of polar biota should therefore be undertaken with caution. The regions considered here include some having mean temperatures in mid summer around 10°C and freely available water in many habitats, others where mean summer temperatures remain below 3°C with water again freely available, and yet others which experience extremes of both cold and aridity. Differences in these and other climatic, edaphic and biotic variables are reflected in the diversity of tundra vegetation.

The influence of environment on the distribution of plant life-forms has long been recognized. Thus, the timber-line marking the southern boundary of the Arctic as understood here corresponds closely with the mean summer position of Arctic frontal systems (Bryson, 1966), and is no doubt determined largely by the contrast in summer climate to the north and to the south of the fronts. Of the various factors involved, low summer temperature has traditionally been regarded as the most significant. Similarly, low summer temperature is thought to be the most important single factor determining the essentially cryptogamic facies of Antarctic vegetation. This is due to the relatively mild winters, the free availability of water and mineral nutrients in many maritime Antarctic habitats, the restriction of the two native phanerogams to sheltered, well insolated sites, and the poor performance of vascular plants in experimental introductions to the Antarctic (Edwards, 1979; Longton, 1979a; Longton and Holdgate, 1967).

That temperature is important in restricting the number of species able to colonize polar regions is also demonstrated by the floras of sites influenced by geothermal heat. Many hot springs on Iceland and Greenland are surrounded by concentrically zoned vegetation comparable with that around fumaroles on the South Sandwich Islands (Table 5.6), and support a wide range of mosses and hepatics not recorded elsewhere on these islands (Halliday *et al.*, 1974; Hesselbo, 1918; Lange, 1973). On the South Sandwich Islands, also, 19 bryophytes have been recorded only from the fumaroles. Several, including *Campylopus* spp. and *Cryptochila grandiflora* (Table 5.6), are locally abundant at these highly localized sites of volcanic heat and moisture but are unknown elsewhere in the Antarctic zone. Summer temperature again appears to be a key factor (Longton and Holdgate, 1979).

The influence of temperature on bryophyte distribution within the polar regions is difficult to assess, since temperature seldom varies in isolation from other factors. Thus, increasing latitude is generally accompanied by decreases in both temperature and water availability. It may be noted, however, that some arid sites in northern Ellesmere Island have, for their latitude, both unusually warm summers and rich floras of vascular plants and of bryophytes (Brassard, 1971b; Savile, 1964). Concerning aspect, the luxuriant banks of tall turf-forming mosses in the maritime Antarctic are characteristic of north- and north-west facing slopes which benefit most strongly from radiant heating. Similarly, Steere (1978b) notes a contrast between the floras of some north- and south-facing slopes in Alaska, the former comprising principally of Arctic mosses, while the latter includes many species of temperate affinity. Again, however, the drier nature of the south- compared with the north-facing slopes make interpretation difficult.

Bryophyte microclimates throughout the polar regions are more favourable for plant growth and survival than indicated by the air temperature data in Table 5.1. Relationships between summer temperatures and carbon assimilation in mosses are explored in Collins (1977), Longton (1974a) and Oechel and Sveinbjörnsson (1978).

5.7.2 Life-form and water relations

Among bryophytes, it is clear that relatively compact turfs, mats and small cushions predominate in polar regions, with carpets also widespread in the Antarctic. The looser life-forms described by Gimingham and Birse (1957) from temperate regions, notably the weft and the canopy (Table 5.2), are of local occurrence only in the low Arctic. The latter types are typical of sheltered, humid habitats under trees and shrubs, and would hardly be expected to prosper under open, windswept polar conditions, especially as wefts have few anchoring rhizoids and are easily disrupted by wind. Schofield (1972) noted that species such as *Hylocomium splendens*, which form wefts in boreal forest regions, normally occur as mats in the Arctic, where many other pleurocarps occur as tall turfs. Among species assuming the latter life-form, the acrocarps can best be described as 'tall turfs branches erect', and the pleurocarps and *Sphagnum* spp. as 'tall turfs, divergent branches' (Table 5.2).

Habitat relationships of the life-form types have been better documented in the Antarctic than in the Arctic, but several parallels between the two regions may be drawn, and it is clear that water availability is of utmost importance in determining distribution patterns in each case. Thus small, compact cushions and short turfs predominate in particularly dry, exposed habitats, both on rocks and on soil. These are the major bryophyte life-forms in continental Antarctica and in the Arctic polar desert, being locally abundant also on porous, volcanic substrata in areas of higher precipitation in Iceland and in the maritime Antarctic. In addition, they are among the most frequent pioneers in successional sequences on moraines and disturbed ground.

Small acrocarpous mosses are also well represented in some mesic habitats in the Arctic, in which mats and tall turf-forming acrocarps are also abundant. The latter predominate under similar conditions in the Antarctic, where mats are less common than might be expected. Tall turf-forming mosses, particularly pleurocarps and *Sphagnum,* form a high proportion of the bryophyte stratum beneath the graminoids of wetland communities in the low Arctic and on South Georgia. Carpet-forming mosses largely replace the turfs on seepage slopes in the maritime Antarctic. The family Amblystegiaceae is well represented in both types of habitat, and a careful comparison of these carpet and turf-forming pleurocarps would be of interest. It is possible that the carpet form (p. 144) results in part from the pressure of winter snow on essentially erect-growing but weak-stemmed mosses lacking protection and support from vascular plants. Finally, it is of interest that large, loose cushions appear to be characteristic of stream banks and other sites influenced by flowing water in both northern and southern polar regions, while the contribution of hepatics to the vegetation cover is disproportionately low in relation to the numbers of species recorded.

Aggregation of bryophyte shoots into colonies may substantially enhance capillary uptake and retention of externally held water, and thus indirectly affect the internal water content of the plants. By comparing rates of water uptake and loss in individual shoots and in portions of intact colonies representing several common mosses on Signy Island, Gimingham and Smith (1971) were able to demonstrate that the most widespread life-form types differ in their degree of influence on water relations in a manner corresponding with their habitat preferences. Thus, small compact cushions from dry habitats showed the greatest effect of life-form in reducing evaporation rates, and also possessed a marked ability to hydrate rapidly when provided with water either at the base or at the upper surface. These effects were least marked in the case of carpets and large cushions from wet habitats. Anomalous results were given by *Polytrichum alpinum,* which extends into dry habitats despite forming tall, loose turfs with a relatively weak capacity for retaining external water. This may be attributable to the partially endohydric nature of *P. alpinum* which, compared with ectohydric species, may have greater access to soil water and higher internal resistance to water loss due to cuticularization and ability to fold leaf margins over lamellae and leaves against stems in dry conditions.

Mosses of wet meadows in the low Arctic may benefit from close to optimal conditions of hydration throughout the summer, as a result both of precipitation and freely available ground water (Oechel and Sveinbjörnsson, 1978; Vitt and Pakarinen, 1977), and the same may apply in many habitats in the sub-Antarctic and maritime Antarctic (Collins, 1977). In contrast, the scattered mosses of the polar deserts are largely dependent upon melt water derived from ground ice and late-lying winter snow. Under the arid conditions prevailing in these areas it is doubtful whether resistance to the loss of externally held water as a result of compact life-form is sufficient to prevent periodic desiccation of the predominantly ectohydric mosses, which have little resistance to the loss of internal water. Bryophytes are generally regarded as poikilohydrous, however, with an ability to withstand a considerable degree of cytoplasmic dehydration (Hinshiri and Proctor, 1971). The adaptive value of compact growth forms may thus principally lie in extending periods when the plants are sufficiently hydrated to achieve significant rates of net photosynthesis in habitats with a low and intermittent supply of water.

5.7.3 Light

Polar bryophytes appear to be well adapted to the prevailing summer conditions of long days or continuous illumination at relatively low intensities as the light intensities required for compensation and saturation of net photosynthesis are depressed at low temperatures (Longton, 1974a; Oechel and Sveinbjörnsson, 1978). Field experiments have confirmed that positive net assimilation is sustained for 24 h per day in some Alaskan mosses under favourable conditions in early summer (June) when the lightest 'nights' occur (Oechel and Collins, 1973). Maximum light intensities during summer may have an inhibitory effect on photosynthesis in some Alaskan mosses (Oechel and Sveinbjörnsson, 1978), and direct insolation may lower chlorophyll content and alter the chlorophyll a:b ratio in *Bryum antarcticum* in southern Victoria Land (Rastorfer, 1970). Otherwise, there is little evidence that either continuous illumination in summer, or lack of shade from vascular plants, have seriously deleterious effects on polar byrophytes. Kallio and Valanne (1975) have shown that rates of photosynthesis in temperate and boreal forest bryophytes may be reduced by cultivation in continuous light as compared with alternating light and darkness, due to reduction in chlorophyll content and to ultrastructural modifications. However, these effects were not well marked in the Arctic moss *Dicranum elongatum*. In *Racomitrium lanuginosum,* Arctic and temperate strains showed contrasting responses suggestive of ecotypic adaptation to polar conditions in the former. Both

Arctic and boreal forest populations of *Pleurozium schreberi* showed reduced net photosynthesis after cultivation in continuous light, however, and this could be one factor responsible for the restricted distribution of *P. schreberi*, and possibly other mosses, in the Arctic.

5.7.4 Wind and snow

The effects of wind and snow on polar vegetation are powerful, and they may be inter-related as wind largely controls the distribution of snow cover. Wind is important through its cooling and desiccating effects. It also acts as an agent of erosion, particularly in winter when particles of sand and hard snow driven by strong winds can cause severe abrasion of frozen vegetation. Even thin snow cover gives protection from these effects, while deeper snow provides insulation from minimum temperatures and repeated freezing and thawing (Longton, 1979b; Matsuda, 1968). Thus many polar bryophyte communities are characteristic of habitats protected by winter snow (Smith, 1972), but conversely bryophyte vegetation decreases with winter snow cover in some areas (Holmen, 1955). Late-lying snow banks are of great importance in providing moisture for surrounding vegetation in summer although unusually prolonged snow may damage the plants below. Light summer snowfall accumulating in sheltered hollows may also be significant in the water relations of mosses in arid regions.

Effects of wind and snow on bryophyte distribution patterns are most evident in the high Arctic and in continental Antarctica, where snow cover is discontinuous in winter, and water of extreme scarcity. Thus, the scattered short turf- and cushion-forming mosses of the latter region have, at several stations, been reported principally from the vicinity of late-lying snow on lee slopes (Seppelt and Ashton, 1978). The important influence of wind and snow on moss distribution is indicated by the observation that near Swoya Station (Fig. 5.2), where the winds are predominantly from the north-east, the mosses are most frequent on south-west-facing slopes (Matsuda, 1968), despite the more favourable summer temperature regime that may be expected under conditions of stronger insolation on slopes of northerly aspect.

5.7.5 Edaphic and biotic factors

While the distribution of bryophyte life-form types appears to be controlled largely by water relations, habitat preferences at the level of genera and species are strongly influenced by pH and other chemical features of the substrate. Many bryophytes show a strong affinity for either acidic or base-rich conditions, for reasons discussed in Longton (1980). The preferences of the *Andreaea* and the *Tortula–Grimmia* associations of the short moss turf and cushion subformation for siliceous and calcareous substrata respectively in the maritime Antarctic have already been noted

(p. 142). The effect may be extremely localized, as cushions of *Grimmia* spp., *Tortula* spp. and other calcicoles occur around deposits of limpet shells left by gulls on otherwise acidic soil. Steere (1978b) gives similar examples of distribution patterns in lithophytic bryophytes in Alaska apparently being determined by the calcium content of the rocks, and the distribution of *Sphagnum* spp. and other bryophytes in the wetland communities on peat varies according to the base status of the ground water (Mårtensson, 1956).

The general level of the major nutrients in maritime Antarctic soils is unnlikely to be seriously limiting for plant growth, due to the marine influence and the presence of large bird and seal colonies (Allen and Heal, 1970). Similarly at Barrow, Alaska, Oechel and Sveinbjörnsson (1978) found that watering with a dilute nutrient solution resulted in no increase in moss growth or photosynthesis.

In contrast, many soils of inland regions in the high Arctic and in continental Antarctica are strongly deficient in certain nutrients. This applies particularly to available nitrogen, due to the limited extent of microbial activity (Tedrow, 1977). Nitrogen fixation by blue-green algae associated with tundra mosses has recently been demonstrated (Alexander, 1975) and could be important in the early development of the soils. Other elements may occur in high or even toxic concentrations in polar desert soils due to lack of leaching or in extreme cases to prevalence of upward water movement. Steere (1978b) noted that Arctic desert soils are apt to be calcareous due to the low rate of leaching and that this is reflected in the bryophyte flora. Ugolini (1970, 1977) has suggested that there is 'an inverse relationship between pedologic development and life succession' in continental Antarctica because of the high salt content of the most highly developed soils.

The distribution of bryophyte communities associated with lemming burrows and other sites of animal activity in the Arctic (p. 138) could well be controlled in part by concentrations of available nitrogen. Similarly, the most luxuriant vegetation on inland nunataks in continental Antarctica occurs in the vicinity of seabird colonies (Llano, 1965), and again availability of nitrogen could be the key factor. It is not always clear, however, to what extent such distributions reflect nitrophily in the species concerned, as opposed to low nitrogen levels in areas not under strong biotic influence. Experimental studies have shown that a wide range of nitrogen sources is capable of supporting *Bryum algens* on Ross Island (Schofield and Ahmadjian, 1972).

In conclusion, it may be stated that Arctic and Antarctic bryophytes survive, and often thrive, in environments marked by severity not only of temperature but also of several other major factors. Space constraints do not permit a detailed consideration of recent studies on the adaptations of bryophytes to polar environments, or of the contribution of bryophytes to polar ecosystems, to be presented here. Reviews of these topics may be found on Oechel and Sveinbjörnsson (1978), Longton (1980) and Smith (in press).

ACKNOWLEDGMENTS

This account is an outgrowth from work supported by the British Antarctic Survey, United States Antarctic Research Programme, National Research Council of Canada and University of Manitoba Northern Studies Committee, to whom grateful acknowledgment is made. I also thank Dr D.H. Vitt for comment and criticism of a large section of the manuscript.

REFERENCES

Alexander, V. (1975), Nitrogen fixation by blue-green algae in polar and subpolar regions. In: *Nitrogen Fixation in Free-Living Micro-Organisms* (Stewart, W.D.P., ed.), pp. 175–88. Cambridge University Press, London.

Alexandrova, V.D. (1970), The vegetation of the tundra zone in the USSR and data about its productivity. In: *Proceedings of the Conference on Productivity and Conservation in Northern Circumpolar Lands, Edmonton, Alberta,* 1969 (Fuller, W.A. and Kevan, P.G., eds), pp. 93–114. International Union for Conservation of Nature and Natural Resources, Morges, Switzerland.

Allen, S.E. and Heal, O.W. (1970), Soils of the maritime Antarctic zone. In: *Antarctic Ecology* (Holdgate, M.W. ed.), Vol. 2, pp. 693–6. Academic Press, London and New York.

Allison, J.S. and Smith, R.I.L. (1973), *Br. Antarct. Surv. Bull.*, **33, 34,** 185–212.

Anonymous (1978), *Polar Regions Atlas.* Central Intelligence Agency, Washington, D.C.

Ashton, D.H. and Gill, A.M. (1965), *Proc. R. Soc. Victoria,* **79,** 235–45.

Baker, J.H. (1972), *Br. Antarct. Surv. Bull.,* **27,** 123–9.

Bliss, L.C. (1977a), *Truelove Lowland, Devon Island, Canada: a High Arctic Ecosystem,* University of Alberta Press, Edmonton.

Bliss, L.C. (1977b), Introduction. In: *Truelove Lowland, Devon Island, Canada: a High Arctic Ecosystem* (Bliss, L.C., ed.), pp. 1–11. University of Alberta Press, Edmonton.

Bliss, L.C. (1979), *Can. J. Bot.,* **57,** 2167–78.

Bliss, L.C., Kerik, J. and Peterson, W. (1977), Primary production of dwarf shrub heath communities, Truelove Lowland. In: *Truelove Lowland, Devon Island, Canada: a High Arctic Ecosystem* (Bliss, L.C., ed.), pp. 217–24. University of Alberta Press, Edmonton.

Brassard, G.R. (1971a), *Bryologist,* **74,** 282–311.

Brassard, G.R. (1971b), *Bryologist,* **74,** 233–81.

Brassard, G.R. (1976), *Bryologist,* **79,** 480–7.

Bryson, R.A. (1966), *Geogr. Bull.,* **8,** 228–69.

Cameron, R.E., Lacy, G.H., Morelli, F.A. and Marsh, J.B. (1971), *Antarct. J. U.S.*, **6,** 105–6.

Collins, N.J. (1976), *Br. Antarct. Surv. Bull.*, **43,** 85–102.

Collins, N.J. (1977), The growth of mosses in two contrasting communities in the maritime Antarctic: measurement and prediction of net annual production. In: *Adaptations within Antarctic Ecosystems* (Llano, G.A., ed.), pp. 921–33. Smithsonian Institution, Washington, D.C.

Collins, N.J. and Oechel, W.C. (1974), *Can. J. Bot.*, **52,** 355–63.

Edwards, J.A. (1979), *Br. Antarct. Surv. Bull.*, **49,** 73–80.

Gates, D.M. (1962), *Energy Exchange in the Biosphere,* Harper and Row, New York.

Gimingham, C.H. and Birse, E.M. (1957), *J. Ecol.*, **45,** 433–45.

Gimingham, C.H. and Smith, R.I.L. (1970), Bryophyte and lichen communities in the maritime Antarctic. In: *Antarctic Ecology* (Holdgate, M.W., ed.), Vol. 2, pp. 752–85. Academic Press, London and New York.

Gimingham, C.H. and Smith, R.I.L. (1971), *Br. Antarct. Surv. Bull.*, **25,** 1–21.

Greene, S.W. (1964a), Plants of the land. In: *Antarctic Research* (Adie, R.J., Priestley, R. and Robin, G. de Q., eds), pp. 240–53. Butterworths, London.

Greene, S.W. (1964b), *Br. Antarct. Surv. Sci. Rep.*, **45.**

Greene, S.W. (1968), *Rev. bryol. lichénol.*, **36,** 132–8.

Greene, S.W. (1973), *Br. Antarct. Surv. Bull.*, **36,** 1–32.

Grolle, R. (1972), *Br. Antarct. Surv. Bull.*, **28,** 85–95.

Halliday, G., Kliim-Nielsen, L. and Smart, I.H.M. (1974), *Meddr. Grønland,* **199** (2), 1–49.

Hesselbo, A. (1918), The Bryophyta of Iceland. In: *The Botany of Iceland* (Rosenringe, L.K. and Warming, E., eds), Vol. 1, pp. 395–676. Frimodt, Copenhagen.

Hinshiri, H.M. and Proctor, M.C.F. (1971), *New Phytol.*, **70,** 527–38.

Holdgate, M.W. (1964), Terrestrial ecology in the maritime Antarctic. In: *Biologie Antarctique* (Carrick, R., Holdgate, M.W., and Prévost, J., eds), pp. 181–94. Hermann, Paris.

Holdgate, M.W. (1970), Vegetation. In: *Antarctic Ecology* (Holdgate, M.W., ed.), Vol. 2, pp. 729–32. Academic Press, London.

Holmen, K. (1955), *Mitt. Thuringischen Bot. Ges.*, **1,** 96–106.

Huntley, B.J. (1971), Vegetation. In: *Marion and Prince Edward Islands* (van Zinderen Bakker, J.M., Winterbottom, J.M. and Dyer, R.A., eds), pp. 98–160. Balkema, Cape Town.

Janetschek, H. (1970), Environments and ecology of terrestrial arthropods in the high Antarctic. In: *Antarctic Biology* (Holdgate, M.W., ed.), Vol. 2, pp. 871–85. Academic Press, London and New York.

Jenkin, J.F. and Ashton, D.H. (1970), Productivity studies on Macquarie Island vegetation. In: *Antarctic Ecology* (Holdgate, M.W., ed.), Vol. 2, pp. 851–63. Academic Press, London and New York.

Kallio, P. and Valanne, N. (1975), On the effects of continuous light on photosynthesis in mosses. In: *Fennoscandian Tundra Ecosystems* (Wielgolaski, F.E., ed.), pp. 149–62. Springer-Verlag, New York, Heidelberg and Berlin.

Koponen, A. (1978), *Bryophytorum Bibliotheca,* **13,** 535–67.
Koponen, A. and Koponen, T. (1978), *Bryophytorum Bibliotheca,* **13,** 569–77.
Lange, B. (1973), *Lindbergia,* **2,** 81–93.
Light, J.J. and Heywood, R.B. (1973), *Nature,* **242,** 535–6.
Light, J.J. and Heywood, R.B. (1975), *Nature,* **256,** 199–200.
Lindsay, D.C. (1971), *Br. Antarct. Surv. Bull.,* **25,** 59–83.
Llano, G.A. (1965), The flora of Antarctica. In: *Antarctica* (Hatherton, T., ed.), pp. 331–50. Methuen, London.
Longton, R.E. (1967), *Phil. Trans. R. Soc.,* **B252,** 213–35.
Longton, R.E. (1973a), *Can. J. Bot.,* **51,** 2339–46.
Longton, R.E. (1973b), *Br. Antarct. Surv. Bull.,* **32,** 41–9.
Longton, R.E. (1974a), *Bryologist,* **77,** 109–127.
Longton, R.E. (1974b), *J. Hattori Bot. Lab.,* **38,** 49–65.
Longton, R.E. (1979a), *Can. J. Bot.,* **57,** 2264–78.
Longton, R.E. (1979b), *Bryologist,* **82,** 325–67.
Longton, R.E. (1980), Physiological ecology of mosses. In: *The Mosses of North America* (Taylor, R.J. and Leviton, A.E., eds), pp. 77–113. Pacific Division, American Association for the Advancement of Science, San Francisco.
Longton, R.E. and Holdgate, M.W. (1967), *Phil. Trans. R. Soc.,* **B252,** 237–50.
Longton, R.E. and Holdgate, M.W. (1979), *Br. Antarct. Surv. Sci. Rep.,* **94.**
Mårtensson, O. (1956), *Kungl. Svenska Vetenskapsakademiens Avhandlinger I Naturskyddsarenden,* **15.**
Matsuda, T. (1968), *Jap. Antarct. Res. Exped. Sci. Rep.,* **E29.**
Matveyeva, N.V., Polozova, T.G., Blagodatskykh, L.S. and Dorogostaiskaya, E.V., (1974), A brief essay on the vegetation in the vicinity of the Taimyr Biogeocoenological Station, International Tundra Biome Translation, I.B.P.
Muc, M. and Bliss, L.C. (1977), Plant communities of Truelove Lowland. In: *Truelove Lowland, Devon Island, Canada: A High Arctic Ecosystem* (Bliss, L.C., ed.), pp. 143–54. University of Alberta Press, Edmonton.
Nakanishi, S. (1977), *Antarct. Rec.,* **59,** 68–96.
Oechel, W.C. and Collins, N.J. (1973), Seasonal patterns of CO_2 exchange in bryophytes at Barrow, Alaska. In: *Proc. Conf. Primary Production Processes, Tundra Biome* (Bliss, L.C. and Wielgolaski, F.E., eds), pp. 197–203. Int. Biol. Programme, Dublin.
Oechel, W.C. and Sveinbjörnsson, B. (1978), Primary production processes in Arctic bryophytes at Barrow, Alaska. In: *Vegetation and Production Ecology of an Alaskan Arctic Tundra* (Tieszen, I., ed.), pp. 269–98. Springer-Verlag, New York, Heidelberg and Berlin.
Persson, H. (1942), *Bot. Notiser,* **1942,** 308–24.
Philippi, G. (1973), *Moosflora und Moosvegetation des Freeman-Sund-Gebietes (Sudost-Spitzbergen).* Franz Steiner Verlag, Wiesbaden.
Poore, M.E.D. (1962), The method of successive approximation in descriptive ecology. In: *Advances in Ecological Research* (Cragg, J.B., ed.), Vol. 1, pp. 35–66. Academic Press, London and New York.
Priddle, J. (1979), *J. Bryol.,* **10,** 517–29.
Priddle, J. (1980a), *J. Ecol.,* **68,** 141–53.

Priddle, J. (1980b), *J. Ecol.*, **68**, 155–66.
Rastorfer, J.R. (1970), *Bryologist*, **73**, 544–56.
Robinson, H.E. (1972), *Antarct. Res. Ser.*, **20**, 163–78.
Rudolph, E.D. (1963), *Ecology*, **44**, 585–6.
Russell, R.S. and Wellington, P.S. (1940), *J. Ecol.*, **28**, 153–79.
Savile, D.B.O. (1964), *Arctic*, **17**, 237–56.
Schofield, W.B. (1972), *Can. J. Bot.*, **50**, 1111–33.
Schofield, E. and Ahmadjian, V. (1972), *Antarct. Res. Ser.*, **20**, 97–141.
Schuster, R.M. (1959), Hepaticae. In: *The Terrestrial Cryptogams of Northern Ellesmere Island* (Schuster, R.M., Steere, W.C. and Thomson, J.W., eds.) pp. 15–71. *Natl. Mus. Can. Bull.*, **164**, Ottawa.
Seppelt, R.D. and Ashton, D.H. (1978), *Aust. J. Ecol.*, **3**, 373–88.
Smith, R.I.L. (1972), *Br. Antarct. Surv. Sci. Rep.*, **68.**
Smith, R.I.L. (1979), Peat-forming vegetation in the Antarctic. In: *Classification of Peat and Peatlands* (Kivinen, E., Heikurainen, L. and Pakarinen, P., eds), pp. 58–67. International Peat Society, Helsinki.
Smith, R.I.L., Terrestrial plant ecology of the sub-Antarctic and Antarctic. In: *Antarctic Ecology* (Laws, R.M., ed.). Academic Press, London (in press).
Smith, R.I.L. and Corner, R.W.M. (1973), *Br. Antarct. Surv. Bull.*, **33, 34,** 89–122.
Smith, R.I.L. and Gimingham, C.H. (1976), *Br. Antarct. Surv. Bull.*, **43,** 25–47.
Sonesson, M. (1967), *Bot. Notis*, **120,** 272–96.
Sonesson, M. (1969), *Bot. Notis*, **122,** 481–511.
Sonesson, M. (1970a), *Opera Botanica*, **26.**
Sonesson, M. (1970b), *Bot. Notes*, **123,** 67–111.
Sonesson, M. and Kvillner, E. (1980), *Ecol. Bull. (Stkh.)*, **30,** 113–25.
Steere, W.C. (1961), A preliminary review of the bryophytes of Antarctica. In: *Science in Antarctica Vol. 1, The Life Sciences in Antarctica*, pp. 20–33. National Academy of Science, Washington, D.C., Publication 839.
Steere, W.C. (1978a), *Bryophytorum Bibliotheca*, **14.**
Steere, W.C. (1978b), Floristics, phytogeography and ecology of Arctic Alaskan bryophytes. In: *Vegetation and Production Ecology of an Alaskan Arctic Tundra* (Tieszen, L., ed.), pp. 141–67. Springer-Verlag, New York, Heidelberg and Berlin.
Steere, W.C. and Inoue, H. (1978), *J. Hattori bot. Lab.*, **44,** 251–345.
Tallis, J.H. (1959), *J. Ecol.*, **47,** 325–50.
Taylor, B.W. (1955), *Aust. Natl. Antarct. Exped. Rep.*, **B2.**
Tedrow, J.C.F. (1977), *Soils of the Polar Landscapes*, Rutgers University Press, New Brunswick, New Jersey.
Thompson, H.A. (1967), *The Climate of the Canadian Arctic*, Dominion Bureau of Statistics, Ottawa.
Tilbrook, P.J. (1970), The terrestrial invertebrate fauna of the maritime Antarctic. In: *Antarctic Ecology* (Holdgate, M.W., ed.), Vol. 2, pp. 886–96. Academic Press, London and New York.
Ugolini, F.C. (1970), Antarctic soils and their ecology. In: *Antarctic Ecology* (Holdgate, M.W., ed.), Vol. 2, pp. 673–92. Academic Press, London and New York.

Ugolini, F.C. (1977), The protoranker soils and the evolution of an ecosystem at Kar Plateau, Antarctica. In: *Adaptations within Antarctic Ecosystems* (Llano, G.A., ed.), pp. 1091–110. Smithsonian Institution, Washington, D.C.
Vitt, D.H. (1975), *Can. J. Bot.*, **53**, 2158–97.
Vitt, D.H. (1979), *Can. J. Bot.*, **57**, 2226–63.
Vitt, D.H. and Pakarinen, P. (1977), The bryophyte vegetation, production and organic components of Truelove Lowland. In: *Truelove Lowland, Devon Island, Canada: A High Arctic Ecosystem* (Bliss, L.C., ed.), pp. 225–44. University of Alberta Press, Edmonton.
Warren Wilson, J. (1966), *Ann. Bot.*, **30**, 383–402.
Webb, R. (1973), *Br. Antarct. Surv. Bull.*, **36**, 61–77.
Webber, P.J. (1974), Tundra primary productivity. In: *Arctic and Alpine Environments* (Ives, J.D. and Barry, R.G., eds), pp. 445–73. Methuen, London.
Webster, H.J. and Sharp, A.J. (1973), *Am. Biol. Soc. Bull.*, **20**, 90 (Abstr.).
Welch, H.E. and Kalff, J. (1974), *J. Fish. Res. Board Can.*, **31**, 609–20.
Wise, K.A.J. and Gressit, J.L. (1965), *Nature*, **207**, 101–2.

Chapter 6

Alpine Communities

PATRICIA GEISSLER

6.1 INTRODUCTION

The term 'alpine' was originally applied only to the Alps, the mountain range in central Europe folded in the Tertiary. The word is derived from the celtic expression 'alp' meaning 'height' and has been incorporated into all European languages. Nowadays it is often used in a general sense for elevated areas both inside and outside Europe. At medium latitudes climatic conditions are comparable on different continents. In the northern hemisphere these areas also show floristic relationships. Boreal mountain floras originated from the arcto-tertiary flora which included many of our present species. During the last glaciation these mountains were in direct communication with the arctic vegetation as is still the case for the Rocky Mountains, the Scandinavian mountains and thc Urals.

The tropical alpine environment, however, differs considerably. The change of seasons is replaced by pronounced diurnal changes. The vegetation, composed of respectively different floristic elements, shows different adaptation patterns.

A number of observations on alpine bryophytes have been made in the European Alps and the excellent description of the alpine environment by Schröter (1926) is still considered as a classic. The conditions of the nival belt have been investigated by Braun-Blanquet (1913). Early papers on alpine moss-vegetation such as that of Pfeffer (1871) present adequate observations on relations between species and habitat. In most alpine phytosociological treatments bryophytes have not been dealt with as adequately as in Scandinavia (e.g. Samuelsson, 1917). Exceptions are Frey (1922) who describes the dynamics of the associations of the upper subalpine and alpine belt in a central Switzerland beginning with the succession of lichen and bryophyte pioneer communities, Allorge (1925) and Gams (1927) in his survey of all vegetation types of a limited area in the

Rhone Valley from 440 m to almost 3000 m. A careful study on the altitudinal distribution and pH-range of the moss-vegetation in a central Swiss valley was carried out by Greter (1936). Giacomini (1940) described some bryophyte associations from the Lombardian Alps. Oostendorp-Bourgonjon's (1976) observations confirm the conclusions of Amann (1918) but, since Amann (1928), little attention has been paid to the ecology of bryophytes in the Alps. The results of the more recent research of Dahl (1956) or Gjaerevoll (1956) on Scandinavian moss-rich vegetation can only be compared with alpine conditions in a limited way as the differences in the floristic composition of the field layer, and in climate due to the higher latitude, result in changes in bryophyte behaviour.

In phytogeography, the term alpine defines the vegetation belt above the timber-line as does the term arctic in higher latitudes. The term subalpine corresponds in classical literature to the coniferous zone below the timber-line, including the shrub zone at its upper limit. Human activity has artificially depressed the timber-line in most parts of the Alps to enlarge the belt of natural meadows and pastures. In the central Alps the timber-line has risen because of the more favourable climatic conditions due to the 'Massenerhebungseffekt' (the effect of a mountain massif displacing climate and biological limits to higher altitudes than would otherwise be expected). Vertical differences in relatively short horizontal distances diversify the vegetation pattern because of the influence of elements of the neighbouring vegetation belts.

6.2 CLIMATE

The climate above the timber-line is characterized by a decrease in temperature and a shorter growing season, the principal common features of arctic and alpine climate. In contrast to arctic conditions, where the differences between day and night and summer and winter temperatures are less marked, summer temperatures in the Alps can be higher, especially the maxima, due to the lower latitude. Daily freezing at night in spring limits growth. Soil surfaces and the air near the soil are, in contrast to the low temperatures 2 m above soil, considerably warmer. Intensive insolation produces a favourable microclimate for low-growing plants. On the other hand, reradiation at night is also higher and with the low night temperatures respiration is decreased (Billings, 1974). Radiation with cloudy sky is higher than in the lowlands and productivity is scarcely affected under cloudy conditions in comparison with direct insolation (Turner and Tranquillini, 1961). In the arctic regions, light is less rich in u.v. radiation and the air is less thin.

Above the timber-line the climate in the Alps is always more or less humid, even in the continental inner valleys, while in the Arctic continental conditions predominate. The period when precipitation occurs as snow is prolonged. The growing period above the timber-line lasts about 3 to 4 months and is gradually reduced towards the snow-line. The snow-cover is influenced by wind and by relief and this causes the particular mosaic of melting areas in spring and their characteristic vegetation. Under snow cover, the temperatures rarely fall below freezing point and sometimes assimilation takes place if enough light penetrates the snow. Snow also protects against wind action and desiccation as there is almost no evaporation.

Permafrost may occur in the Alps above 2500 m and down to 2300 m on slopes facing west to north-east. Solifluction phenomena are observed as terraces, and polygons are caused by the freeze-thaw cycles. Continuous swampy regions due to low evaporation over frozen ground are not found in the Alps. Wind drying exposed ridges and transporting snow to late-lying snow beds should not be ignored as an ecological factor. Snow-grinding polishes the rocks which then have fewer cavities for colonization and it damages the plants although this may also be an effect of frost–drought (Turner, 1968).

The main differences in the environmental characteristics of an arctic and an alpine station in North America (European data are similar) are pointed out by Billings and Mooney (1968). Climatological investigations in the Alps have been carried out by Turner (1971) and Turner *et al.* (1975) with regard to afforestation projects.

6.3 ENVIRONMENTAL EFFECTS

Ecological experiments on bryophytes have been carried out in recent years on arctic and antarctic regions, e.g. Pakarinen and Vitt (1974), Greene and Longton (1970) but not in the Alps. Adaptation to the tundra climate means, above all, tolerance of cold temperatures. The results of research on the temperature factors apply equally to alpine bryophytes. The effects of the cold climate also concern life-forms. As in higher plants these are mainly perennial prostrate forms, dense tufts and cushions. Vegetative propagation by means of gemmae, bulbils or fragments is very common. At least two seasons are necessary for a reproductive cycle, and spore maturation often takes place after two winters and is even possible under snow. The lack of sexual organs makes identification difficult and the gametophytic structures of many taxa are very variable phenotypically because of varying environmental conditions. Culture experiments would provide information on the degree of phenotypic variation of alpine taxa.

Beside the temperature factor, many other climatic conditions such as radiation and precipitation, are more favorable in the Alps than in polar regions and, therefore, competition from phanerogams is much greater. In some vegetation types, bryophytes play a less important rôle than in the Arctic. This applies especially to alpine meadows over a well developed soil. Nevertheless, bryophytes dominate in wet as well as in dry habitats and not only in the pioneer vegetation. Mixed samples on unstable soils provide better possibilities for recolonization on moraines, solifluction slopes or screes. On rocks, competition with lichens or, on irrigated cliffs, between hepatics and blue-green algae, takes place. Succession corresponds to pedogenesis but on raw soil the first arriving diaspore is the one that prevails, perhaps a flowering plant before a lichen. Mosses enhance eolic or alluvial humus accumulation.

6.4 CLASSIFICATION

Gimingham and Smith (1970) clearly point out two approaches to vegetation classification, one with a complete analysis of sample stands and following statistical evaluation, the other on selective criteria such as physiognomy, floristic composition or dominance. The former is regarded as more objective. In regions with a limited number of species so-called subjective methods, based on field experience, are also often successful.

The most important factor affecting the distribution of bryophytes is the microtopography with its corresponding microclimate. Slope and neighbouring mountains influence snow cover, melting pattern, soil moisture and radiation (Barry and Van Wie, 1974). These topo-climatological effects work also at the single stand level. Bryophytes interact only at the uppermost soil surface and are particularly sensitive to micro-environmental changes. Closed vegetation types may present a homogenous phanerogamic layer with heterogenous cryptogamic synusiae. These problems occur less frequently in open vegetation which is widely distributed in the alpine levels and even more towards the snow-line.

Phytosociological approaches to the description of alpine bryophyte communities may follow the example of Ochsner (1954). He showed the importance of bryophytes in alpine plant communities with regard to the orders of the Braun–Blanquet system. From the bryological point of view it seems better to investigate the correlation between habitat, which is often so small that it cannot be treated using the Braun–Blanquet method, and bryofloristic composition. The units obtained may then occur in different formations. The most suitable method depends on the aim of the research and the required scale.

The main habitats for alpine bryophytes are, on the one hand, wet sites: open water with successional stages towards mire vegetation (in the upper alpine belt, replaced by snow-beds) and communities of running water, springs and streams; on the other hand, dry sites such as rock faces, boulders and screes. In these two ecological extremes bryophytes play a predominant role. Nevertheless, they should not be ignored in mesophytic vegetation where they occupy an appropriate niche in meadows, pastures and heaths. The following classification is based on habitats defined primarily by the water factor, secondarily on the substratum.

6.5 HYDROPHILOUS COMMUNITIES

6.5.1 Communities of standing water

Few bryophytes are found in open water. Most alpine lakes freeze in winter down to depths below which plants grow. Only *Drepanocladus exannulatus* (or *D. aduncus* or *D. fluitans*) in diverse modifications, and sometimes *Calliergon sarmentosum,* are found submersed or floating but always in relation to the neighbouring fen communities (Fig. 6.1). Many of the shallow ponds dry out in autumn. A very rare hepatic, *Riccia breidleri,* probably a true alpine endemic, occurs on mud in temporary pools on siliceous or schistose ground never completely destitute of calcium. Some nutrients will be derived from the cattle of surrounding pastures. About ten localities scattered over the whole alpine range (Jovet-Ast, 1977) are now known for *R. breidleri* but, as this species is easily overlooked when it is dry or submerged, further stations are likely to be discovered.

6.5.2 Mire communities

Fens reach their upper altitudinal limit in the lower alpine belt. Single species may occur up to 2500 m but productivity in the alpine belt is too low for peat formation above 2300 m in the central Alps. Therefore only fragments of bog-communities are found just below the timber-line. Bog-species such as *Dicranum undulatum, Cephalozia connivens* or *Mylia anomala* may occur in alpine stations on sufficiently acid mor. Hummock-building sphagna such as *Sphagnum fuscum, S. magellanicum, S. rubellum,* reach their altitudinal limit far below 2000 m. Relief and only slightly raised hummocks do not exclude the influence of mineral water. Similarly, in northern Fennoscandia, large bogs are absent and bog vegetation appears in scattered patches in mixed mires. In spite of the cold local climate in a bog, bog sphagna demand minimum temperatures and a sufficiently long growth period, conditions not found above 800 m in the northern and 1500 m in the central Alps.

Closest to open water in the *Eriophorum scheuchzeri* belt, species other than *Drepanocladus exannulatus* are only occasionally found. In the description of the *Drepanocladi exannulati–Eriophoretum scheuchzeri* belt (Rybníček and Rybníčková, 1977), however, patches of snow-bed vegetation arc found in a variant of this association. In the adjacent *Carex nigra*

Fig. 6.1 (a) Alpine lake in the Mont Blanc massif (2150 m). In the foreground, a spring with *Philonotis seriata*; on the far slope, spring fens with *Nardia compressa*. (b) Zonation of snowbed communities in central Switzerland at 2650 m. Around the outlet of the lake, pioneer communities with *Marsupella brevissima*; at the lake shore, *Polytrichum sexangulare*; above the path, *Salicetum herbaceae*; on the mountain ridge in the foreground, *Caricetum curvulae*.

community a significant part of the surface is covered by bryophytes. Under acid conditions, *Drepanocladus exannulatus, D. aduncus, Calliergon sarmentosum, C. stramineum, Aulacomnium palustre, Dicranum bonjeanii, Gymnocolea inflata, Scapania undulata* or *S. paludosa* dominate. Among the peat-mosses, *Sphagnum nemoreum, S. palustre, S. teres* and *S. subsecundum* may be encountered but only in small quantities. *Sphagnum compactum* is widely distributed on wet slopes and at the edge of springs and mires. Species of springs such as *Philonotis seriata, P. tomentella, Bryum pseudotriquetrum* and *Dicranella palustris* are found at sites with water outlets.

In fens with more neutral water (pH 5–6) the moss layer is composed of *Drepanocladus revolvens, Sphagnum warnstorfii, Tomenthypnum nitens* and some arctic-alpine relict species such as *Meesia triquetra, Cinclidium stygium* and *Paludella squarrosa* (Geissler and Zoller, 1978). Main distribution of this fen-type is below the timber-line but isolated sites are known up to 2240 m. *Cinclidium* is also sometimes found beside open water above 2500 m within stands of *Carex rostrata*. The ecology of some of these glacial relics has been investigated by Rybníček (1966) in the Bohemian–Moravian highlands. Water analysis showed that *Calliergon trifarium* and *Scorpidium scorpioides* prefer permanently flooded rich basic habitats while *Paludella* and *Cinclidium* require less basic water and a low water table. Moving away from calcareous fen habitats the bryophytic layer gets floristically less rich. It is dominated by *Drepanocladus revolvens, Campylium stellatum* and *Cratoneuron commutatum* but less abundantly. Tufa formation may be observed rarely in the upper alpine belt. Yerli (1970) observed that lateral movement of ground water is as important as the water hardness for biogenous precipitation of carbonate. The coefficient of these two factors combined is significantly different for fens with *Carex davalliana* on slopes from fens with *Tomenthypnum* and *Trichophorum caespitosum* on level terrain in the pre-alps of western Switzerland at about 1500 m. Further study is needed to see whether these results can be applied to the calcareous fens of other regions. Few hepatics are present in these fens rich in Cyperaceae – the ubiquitous *Aneura pinguis, Pellia endiviifolia* and *Lophozia collaris, Philonotis calcarea, P. tomentella, Bryum pseudotriquetrum, Oncophorus virens, Fissidens adianthoides, Aulacomnium palustre, Calliergon stramineum* or *Dicranum bonjeani* indicate more peaty or wetter habitats in the communities belonging to the *Caricion davallianae* which has its main distribution range between 1500 and 2000 m. Stands with *Catoscopium nigritum, Ablyodon dealbatus, Meesea uliginosa* and *Bryum pallens* together with *Bryum pseudotriquetrum* and *Crato-neuron commutatum,* found up to 2670 m, may be linked to a rare relict vegetation belonging to the nordic alliance *Caricion bicoloris-atrofuscae* in associations with *Carex maritima* on glacier alluvions and along streamsides (Richard and Geissler, 1979). Similar vegetation types are described from north Norway (Dierssen, 1977) on slightly basic to slightly acid damp gravel or sandy river banks. However, the aforementioned mosses occur nowadays with decreasing frequency down to pre-alpine areas where they are found in tufa springs with *Cratoneuron commutatum, Aneura pinguis* and *Hymenostylium recurvirostre.*

6.5.3 Snowbed communities

The vegetation type in sites where drainage is slow in the upper alpine belt were recognized by Wahlenberg (1813) as 'loci uliginosi frigidi a nive serius relicti'. In these places, snow cover is prolonged compared with the surrounding vegetation and the growing season varies from 1 to 3 months, differing from one year to another. It is assumed that the melting pattern remains the same every year but the depth of snow-cover does not only depend on the precipitation in the form of snow but also on wind and its velocity from different sectors of the compass. It may happen that the snow-cover persists for two or even more winters and except for the very short snow-free season, conditions are rather favorable. Temperatures under snow-cover rarely drop below freezing point and frost danger occurs only on summer nights. The melting snow on which nutrients are accumulated by the wind provides a sufficient water supply. These vegetation types are not restricted to the bottoms of depressions or snow-beds in the strict sense but may cover large areas on slopes where snow persists. There are few phanerogams adapted and able to grow, flower and produce seed in such a short time. They are common only in the earliest snow-free areas. Bryophytes are also favoured by the fact that they are less disturbed by solifluction phenomena.

Succession in calcareous and in siliceous snow-beds proceeds in different ways. Initial snowbeds on calcareous substrata consist of coarse sand or stones with a high permeability and scattered vegetation of phanerogams and cryptogams. The latter are represented by *Brachythecium glaciale, Pohlia commutata, Tayloria froelichiana, Sauteria alpina, Asterella lindenbergiana* the wide-spread species *Drepanocladus uncinatus, Tortula alpina, Lescuraea incurvata, Distichium inclinatum* and *Polytrichum juniperinum.* A characteristic species of snowbeds, *Anthelia juratzkana,* is also found on calcareous soil. With colonization by such phanerogams as *Salix retusa, Luzula alpina-pilosa* and *Sibbaldia procumbens* humus formation occurs. Well developed snowbed communities over calcareous ground are difficult to distinguish from those on siliceous soils.

Initial stages on acid ground are characterized by liverwort communities. Because of the taxonomic difficulties they have not been well described in a phytosociological context. *Anthelia juratzkana* is mixed with

Marsupella brevissima (the rare *M. condensata* and *M. boeckii* may also be found there), *Cephalozia ambigua* and *C. bicuspidata, Lophozia sudetica, Nardia breidleri, Pleuroclada albescens, Scapanias* of the Section *Curtae (S. helvetica, S. scandica, S. praetervisa), Moerckia blytii*

Fig. 6.2 (a) *Polytrichum sexangulare*; in the background Galenstock (same locality as Fig. 6.1b). (b) Snowbeds at 2700 m in the Grisons, Switzerland (Val da Fain) in a stony *Curvuletum*. In the centre, *Polytrichum sexangulare* with patches of *Pohlia commutata*, surrounded by *Salix herbacea*. In the foreground on the bare soil, *Marsupella brevissima* and *Anthelia juratzkana*.

and often Cyanophyceae. The vegetation often shows the polygon pattern of permafrost soil. *Kiaeria starkei, K. falcata* and *Pohlia commutata* may also be found there (Fig. 6.1), while *Polytrichum sexangulare* has a tendency to dominate stands. In the *Polytrichetum sexangularis* (Fig. 6.2a), patches of the above mentioned hepatics are only met with where the humus-layer is rather thin.

All these species also occur in the best-known snowbed association, the *Salicetum herbaceae* which benefits from the longest snow-free time (Fig. 6.2b). Quantitatively, the bryophyte layer is less important but many more species are represented, e.g. *Conostomum tetragonum* on low humus and species normally more frequent on heaths such as *Barbilophozia floerkei, Dicranum fuscescens, Bartramia ithyphylla, Lophozia wenzelii* and *Polytrichum piliferum.* The distribution of these species is correlated with various microtopographical conditions which are difficult to analyse. Sometimes it is hard to distinguish snowbed and spring habitats when both vegetation types occur mixed together.

6.5.4 Spring communities

Alpine spring communities can hardly be compared with those of the lowlands. At lower altitudes, bryophytes have less competitive capacity and spring associations are dominated by flowering plants. In the subalpine and alpine belt spring habitats are well defined though the characteristic vegetation is only found as a small strip along a rivulet. The number of flowering plant species is restricted in alpine springs whereas the bryophyte layer plays a predominant rôle. The floristic composition depends not only on the altitude but also on the limnological spring type, on the type of outlet of the spring water, its speed and, in relation to these two factors, the temperature. Possible desiccation and snow cover also have their effects. The results of the investigations of Geissler (1976) mainly in the eastern part of the central Alps showed that the pH of the spring water is less important in determining the composition of the vegetation than the above mentioned factors, in contrast with the current view of alpine phyto-sociologists.

The most frequent alpine spring type is characterized by a rather fast running outlet in a well-defined basin. The running water is warmer than its surroundings in winter and melts the snow early in spring or even keeps them snow-free for a larger part of winter. *Cratoneuron commutatum* (mainly the ssp. *falcatum*) is present in calcareous as well as in siliceous regions and may occur even with the pH of the spring water as low as 4.6. It has to be remembered that, on the one hand, carbonates are present, although in low quantities, in gneissic rocks and that this low concentration is sufficient for the growth of *Cratoneuron* and, on the other hand, other competitive species may be lacking. *Philonotis tomentella* is found over the whole pH-range but the only vicarious species are *Philonotis seriata* and *P. calcarea*. *Bryum pseudotriquetrum* with allied forms is ubiquitous in wet places, *Campylium stellatum* and *Aneura pinguis* occur constantly with *Cratoneuron*, the *Philonotis* species, *Oncophorus virens* and *Dicranella palustris*. The spring water is always rich in oxygen, rather cold (4–8°C in summer) and some nitrophilous species may be introduced by grazing cattle, e.g. *Desmatodon latifolius, D. muticus, Tayloria longicollis* and *Rhizomnium punctatum*. In the field layer, nitrophily is more evident with the presence of *Cirsium spinosissimum, Poa alpina* or *Ligusticum mutellina*. They grow abundantly on cattle-trodden ground but show optimal growth in snow-patches of screes.

Bryum schleicheri has long been considered a characteristic species of silicicolous alpine springs. This may be true for the relict stations in paleozoic mountains such as the Massif Central, Vosges or Black Forest from which carbonates have been leached, but as long ago as 1918 Amann stated that *Bryum schleicheri* was a preferentially calcicole species. The pH values of its habitat range between 4.4 and 7.4 around an average value of 6. It grows on bare gravel soil as well as in nutrient- and humus-rich springs, but seems to be sensitive to desiccation. Hepatics are rather rare in calcareous springs, only the ubiquitous *Scapania irrigua, Lophozia bantriensis* or *Preissia quadrata* occurring in appropriate habitats. Patches of the pioneer vegetation in springs and flushed rocks, represented by *Blindia acuta, Racomitrium sudeticum* and *Jungermannia sphaerocarpa*, may remain on stones even in well developed spring communities.

Two more or less neutrophilous species, *Bryum weigelii* and *Cratoneuron decipiens*, require ecological investigation. The latter is sometimes found scattered in mires or less acid springs and eutrophic fens. It would be interesting to compare its alpine habitat with that in northern Europe where it seems to be calciphilous. *Bryum weigelii* also grows in wet peat-rich fens as well as on mineral- to humus-rich soil in springs. The problem of the distinction between calcicolous and calcifugous species is discussed by Richards (1932). Several experiments done since need confirmation under natural conditions.

As in calcareous fens tufa formation has hardly been observed in the alpine belt. It is suggested that precipitation of calcium carbonate is impeded by the low temperatures and perhaps also by the high speed of the running water.

In the upper alpine belt over gravel soils and in slowly running streamlets the vegetation is composed of *Pohlia wahlenbergii* var. *glaciale* (Fig. 6.3a), *P. ludwigii* in large bright-green tufts. Additional species have their optima in snow-beds or screes; *Pohlia wahlenbergii* is also frequent in spring communities with faster running water. Along glacier streams, on variably irrigated, sometimes overflowing, sometimes completely dry, sandy soil other Bryaceae occur: *Pohlia filum, P. commutata, P. bulbifera, Bryum blindii, B. badium,* together with *Aongstroemia longipes* (rare) and *Philonotis tomentella; Brachythecium glareosum, Polytrichum piliferum* and *Racomitrium canescens* with lichens (*Stereocaulon, Cladonia*) dominant on higher and drier stands.

Fig. 6.3 (a) Irrigated scree, with *Pohlia wahlenbergii,* Fuorcla Tschüffer, 2830 m, Grisons, Switzerland. (b) Mosaic of alpine vegetation types composed of snowbeds, siliceous rocks and heaths on the ridge, Gimel, central Switzerland.

Fig. 6.4 (a) Spring with *Marsupella emarginata* and *Scapania uliginosa* at 2420 m. (Val Viola, Grisons, Switzerland. (b) Spring in a block-scree at 2700 m. (Val, Languard, Grisons, Switzerland) with *Hydrogrimmia mollis* beside the melting snow. In front, snowbed with *Polytrichum sexangulare*.

A second spring type in addition to the 'rheocrenes' with fast-running water are the 'helocrenes' – the terms are defined by Thienemann (1925). They are characterized by spring-rivulets slowly trickling over the source vegetation (Fig. 6.4a). These spring-fens are well represented in the siliceous parts of the Alps with a typical vegetation consisting mainly of hepatics together with *Philonotis seriata* and *Dicranella palustris*. *Scapania undulata* is nearly always present but this species has a very wide ecological range and is also common in mountains of medium altitude and in streams, as is also the case with *Nardia compressa*. *Nardia compressa* is, in the Alps, restricted to peaty spring fens where there is hardly any water movement and it is often associated with species of snow beds. The spring water is probably too slow to accelerate snow-melting. *Scapania uliginosa* grows in deep, sometimes overflowing, spring-fens. The typical composition of alpine spring fens consist of dominant *Scapania undulata, Philonotis seriata, Jungermannia obovata, Dicranella palustris, Drepanocladus exannulatus, Calliergon sarmentosum; Blindia acuta* and *Racomitrium sudeticum* over stony micro-habitats and, rarely, also representatives of other spring types like *Bryum pseudotriquetrum, Rhizomnium punctatum* and *Cratoneuron commutatum*. *Marsupella emarginata* dominates in springs with a thin humus layer and *M. sphacelata* on drier more gravelly habitats and also on irrigated rocks. Associations with the two latter species may be regarded as a stage of succession towards spring fens with *Scapania undulata, S. uliginosa* or *Nardia compressa* but, more often, they represent the outer belt in the zonation of a spring fen.

These hepatics may, together with *Scapania irrigua* or *S. subalpina*, encroach into the *Cratoneuron–Philonotis* spring type. As a characteristic feature they show vinous-red or fuscous colouration of the cell walls. As the water slowly flows over the sometimes almost blackish mats it is considerably warmer, often exceeding the air temperature though it may only be 4°C at its outlet. The dark colour may represent an adaptation to the higher intensities of solar radiation and work as a u.v. filter. These hepatic spring fens occur with decreasing frequency up to 2500 m. It is possible that they are related to oceanic climate types in the Alps but then it would be difficult to explain their presence in the continental Oetztal (Austria) where the well developed spring fens also contain a northern species, *Harpanthus flotovianus*, absent from the western part of the Alps.

In the upper alpine belt, outlets of water may be found at the base of screes. As the phanerogamic flora is rather poorly developed this vegetation type has been ignored for a long time. The spring water is very cold, scarcely above freezing point and often as rich in sandy suspensions as glacier water, the major part issuing from permafrost soils or block glaciers (Fig. 6.4b). The two characteristic cryptogams, often covering large areas of siliceous rock, are the lichen *Dermatocarpon rivolorum* and the moss

Hydrogrimmia mollis. Other species that come from the surrounding boulders include *Lescuraea incurvata, L. radicosa, Racomitrium sudeticum, Tritomaria polita, Dicranoweisia crispula* and *Dichodontium pellucidum.* Species such as *Brachythecium glaciale, Kiaeria falcata* and *Philonotis tomentella* may encroach from snowbeds. There may also be such representatives of stream communities as *Hygrohypnum dilatatum, Schistidium rivulare* and *Cratoneuron commutatum. Hydrogrimmia* is absent from calcareous rocks where the dominant species are *Lescuraea incurvata* and *Tortula alpina.* Water flow may be very irregular, the course often changing with movements in the scree and drying in autumn, but most of the species do not survive drought over several seasons.

Hydrogrimmia is distributed from 2000 m upwards with increasing frequency up to the limits of plant life with the appropriate water supply. It has never been found with capsules in the Alps, although it seems to fruit commonly in the Rocky Mountains. Its lower limit is probably determined less by increase of water temperature, which can exceed 20°C on a slowly irrigated exposed rock on a bright summer day at 2500 m, but rather by competition.

6.5.5 Communities of running water

The floristic composition of alpine stream communities is considerably different from that of lowland streams. *Fontinalis antipyretica* rarely exceeds 1500 m, no *Cinclidotus* species have been collected above 1200 m and *Rhynchostegium riparioides* becomes rare above 1600 m. Ecological differences from lowland streams are expressed in lower temperatures and very fast running water. Speeds of over 2 ms^{-1} may occur and even be far exceeded by snow-melt over a short horizontal distance. Adaptation is required for attachment under such mechanical stress. At the very bottom of the river bed the running speed is slower in comparison with the surface. At the base of the long floating mats, which is the typical life-form of alpine streams, sediment accumulates between the stems. Mechanical damage is rather frequent and the lower part of the stems may lack leaves; in *Hygrohypnum dilatatum* the leaf lamina may be eroded and in *Cratoneuron commutatum* often only leaf midribs remain on the stem. Capsules are only found at the margins of mountain streams where the influence of fast running water is less dominant.

Associations with *Hygrohypnum dilatatum, Schistidium rivulare* and *Cratoneuron commutatum* are widespread in siliceous regions. *Philonotis seriata, Brachythecium rivulare* and *Blindia acuta* may also be present but *Hygrohypnum ochraceum* and *H. molle* are rarer species. This community is distributed from 1800 m up to over 2700 m. In most cases, in winter, water runs under the snow cover accumulated at the bottom of valleys, but it does not erode the snow up to the surface. Mosses are always attached to stones and rocks, mainly in places with the highest turbulance and maximum oxygenation. The temperature depends on the distance from the spring and may rather quickly reach 8–10°C in summer due to the rapid movement of the water. In calcareous regions the communities of springs in block-screes are less frequent and are composed of *Hygrohypnum luridum* and *Cratoneuron commutatum* and sometimes also *Philonotis calcarea, Brachythecium rivulare* and *Bryum schleicheri.* Colonization of calcareous rocks may be impeded by their smooth surfaces which offer few cavities and crevices for rhizoid attachment. Decomposition of the rock and detritus formation are also slower than with granitic or gneissic rocks.

In the upper subalpine belt the arctic species, *Jungermannia exsertifolia* ssp. *cordifolia,* grows in silt-free brooks at its relict stations scattered over the alpine range from the Dauphiné to Austria. Constant associates are *Scapania undulata, Hygrohypnum dilatatum, H. smithii, Philonotis seriata* and *Blindia acuta* in the rather fast streams. In the centre of its distribution belt the arctic *Jungermannia* is dominant in relation to the accompanying species. The distribution area is related to the presence of spring fens with leafy hepatics in more oceanic climatic regions of the Alps.

6.6 XERIC AND MESIC COMMUNITIES

6.6.1 Saxicolous communities

Rocks are another habitat where bryophytes show strong competitive capacities. In humid and wet habitats low temperatures and short growing season are the main limiting factors. On rocks, however, the extremes are more pronounced. Exposed rock faces do not benefit from the protection of a snow cover and are subjected to minimal winter temperatures as well as overheated surfaces, which may even reach 60°C in summer. Effects of ice and wind do not facilitate colonization. Phanerogams only grow in fissures where initial soil formation has taken place whereas little irregularities on the rock surface are sufficient for bryophytes, some of which are attached by a ramified and sometimes rhizome-like system of rhizoids which makes it difficult to detach the tufts. In *Andreaea* the protonema becomes encrusted with dust and sand and cushions and mats accumulate detritus. The rhizoid tomentum retains capillary water and impedes evaporation and reradiation. Temperatures inside the moss mats are often considerably higher than those of the air and this has a positive effect as shelter from strong wind as has been shown by Holdgate (1964) in the Antarctic.

For saxicolous communities the chemical nature of the rock is an important factor because the bryophytes are in direct contact with the substratum. Two distinct habitats are provided by basic and by acidic rocks but many gneissic or granitic rocks contain such calcareous inclusions as calcite and some schists may be composed of carbonate and non-carbonate sediments. Calcareous water may irrigate rocks. In contrast with ferns and flowering plants there is no particular serpentine bryophyte flora, the species being more or less widespread and basipholous. Amann (1928) and Meylan (1924) analysed statistically the Swiss bryoflora of which about two-thirds occurs in the Alps. Their conclusions are indicated in Table 6.1.

Table 6.1 Proportions of calcifuge, calcicole and indifferent bryophytes in the Swiss alpine flora.

	Mosses	Hepatics
Calcifuge species	41.5%	70%
Calcicole species	27.4%	10%
Indifferent species	31.1%	20%

Distinction between rock-face and scree as a habitat for mosses is less important than for flowering plants. In any case they are chomophytes, able to survive with a minimal layer of humus. Rock-faces have less snow-cover or none at all in winter while snow is often accumulated at the base of screes. The upper parts of scree are unstable and move downwards with avalanches but, on boulders, succession takes place according to the local climatic conditions with the decomposition of the rock. Important factors for bryophyte colonization are exposure, inclination, pattern of rock decomposition and, above all, humidity. These factors determine the ecological stress of temperature amplitude, radiation, wind, snowblast effects and drought resistance.

These sites, and the successions on them, have been adequately classified by Frey (1922), especially with regard to the lichens which are the most important constituents of the saxicolous flora. The mosses of dry rocks extend to the uppermost localities where bryophytes have been collected. A list of these stations is given by Pitschmann and Reisigl (1954). These species often have a large ecological amplitude and include *Grimmia donniana* (4230 m, Monte Rosa), *Tortella fragilis* (3800 m, Matterhorn and *Polytrichum piliferum* (3740 m, Oetztal). Hepatics are rather rare at high altitudes and less resistant to drought. They ascend to 3740 m (*Gymnomitrium coralloides* in the Oetztal). The vegetation of calcareous mountain tops is composed of *Grimmia anodon, Pseudoleskeella catenulata, Schistidium apocarpum, S. trichodon* and *S. atrofuscum* together with depauperate cushions of more nitrophilous species of *Encalypta, Bryum argenteum* or *Tortula mucronifolia.* Nitrogen provision by birds is very important (Poelt, 1955). In more humid clefts *Ctenidium procerrimum* grows together with *Hypnum vaucheri, H. bambergeri, H. revolutum* and other basiphilous species. In the same habitat on siliceous rocks, the characteristic species are *Grimmia alpestris, G. incurva, G. montana, G. sessitana* and *Dicranoweisia crispula,* both on vertical rocks and on the tops of boulders which may be snow-free in winter.

Wet rocks show a greater diversity especially the siliceous rocks with a richer hepatic flora. *Andreaea nivalis* communities bear resemblances to the communities of springs in block-screes and are more frequent in the eastern part the Alps. They have been described by Krajina (1933) from the Tatra. A widespread community of wet rocks is the *Andreaea rupestris* association described by Frey (1922) with *Gymnomitrium concinnatum, Oreoweisia torquescens, Racomitrium sudeticum, Amphidium lapponicum, Dryptodon patens, Kiaeria falcata, Andreaea rothii* ssp. *frigida* and *Bryum muehlenbeckii,* often situated on rocks covered by humus-rich vegetation above. The presence of *Marsupella alpina, M. emarginata, Bryum muehlenbeckii* or *Andreaea crassinervia* indicates continuously trickling water. On rocks beside spring fens *Marsupella sphacelata, Scapania undulata, Campylopus schwarzii* or *Cephaloziella grimsulana* may colonize the wet rock surface.

Cavities in screes are occupied by shade-loving species such as *Molendoa hornschuchiana, Schistostega pennata, Timmia* species or *Cyrtomnium hymenophylloides.* The tops of boulders are often visited by birds and are a well-known site for lichenologists. They are, however, poor in bryophytes being covered by two ornithocoprophilous lichens (*Ramalina, Caloplaca*). The rock habitats offer a wide spectrum of very variable ecological conditions. Space is never limiting and many rare or poorly known species occupy particular niches and their autecology is often unknown.

Special attention may be drawn to the group of bryophytes found on heavy-metal containing rocks, the so-called 'copper-mosses'. In central Europe they are only found within the alpine zone. An excellent discussion on this topic is given by Persson (1956) and see also Chapter 12, pp. 459–468. On copper, iron or zinc ores in Europe four mosses and three hepatics are found. Their tolerance to high concentrations of copper sulphate has been investigated by Url (1956) who showed that the two most common and exclusively copper-mosses, *Mielichhoferia mielichhoferi* and *M. elongata,* tolerate a concentration of 0.1 mol. They are known from several localities in the High Alps, *M. mielichhoferi* even being collected at 3480 m on the Theodulhorn. The genus *Mielichhoferia* is also widespread in the Andes on ores. *Merceya ligulata,* the only European representative of a genus with three other species from Asia, North and South America, and *Gymnocolea acutiloba* also grow exclusively on ore deposits with a very low pH. The others, *Grimmia atrata, Cephaloziella phyllacantha* and

C. massalongoi are less exclusive but nevertheless rare species. Other species may be facultative associates and tolerate some higher concentrations of heavy metals than is usual but the copper concentration of the rock surface is often lower than that found on the detritus around old mines. The copper concentration tolerated by the copper-mosses is lethal to phanerogams and on such sites bryophytes have no competition. Shacklette (1961) investigated the bryoflora from other mineral deposits but the species found on chalcopyrite, gypsum and pyrites have a wide ecological amplitude. They include *Oligotrichum hercynicum, Calypogeia muelleri, Cephalozia bicuspidata, Nardia scalaris, Pleurocladula albescens, Dicranella subulata, D. cerviculata, Racomitrium fasciculare.* These minerals are not toxic to phanerogams and, in the Alps the gypsum flora is similar to a limestone flora.

6.6.2 Coprophilous communities

Coprophilous communities are abundantly developed in alpine regions while they seem to be absent from the lowlands in central Europe. Most bryophytes disappear with excess nitrogen concentrations resulting from manuring but many Splachnaceae are adapted to living on humus patches produced by decomposition of animal excrement or carcases. The only epiphyte of the family, *Tayloria rudolphiana,* occurs on branches of old maples and beeches probably on pellets of birds of prey (Gams, 1932), in some localities between 900 and 1600 m in the northern Alps. Although Greter (1936) could not confirm this observation, a nitrogen supply from birds excrement is evident. The coprophilous mosses normally occupy a single habitat, but it may happen that several species occur mixed together. They are more frequently found in humid climates or at least with increased local humidity in hollows or nooks on mountain slopes, in pastures or near exits of marmot burrows. The dung is always old and humified and not easily recognizable as such.

Tayloria species may be found on rich wet soil in springs or snow-beds, but *Splachnum* species are exclusively coprophilous. The genus *Tetraplodon* prefers stomach pellets of birds of prey or owls. The means of propagation of *Voitia nivalis,* known from some stations in the Alps and one in Tien-shan, is not yet clear. The capsules are cleistocarpic and, perhaps breaking off with the seta, are dispersed in the following year to another growing place by wind or the feet of ruminants. It is found between 2500 m and 3000 m in shelters or trails of sheep, chamois or ibex, often on rather dry cliff ledges.

6.6.3 Terricolous communities

On soils which are, according to the climatic conditions, optimally developed, the vegetation is, at first glance, poor in bryophytes. Meadows and swards develop in the most favourable sites in the alpine belt; they have moderate snow-cover, sufficient insolation, continuous water flow, no extreme wind-effect; the limiting factors are determined by the general definition of alpine climate. The vegetation tends to develop towards the climax communities of the alpine belt. Phanerogamic vegetation shows distinct patterns according to the soil reaction, development and texture, moisture, snow-cover and so forth. The climax vegetation type is referred to in current manuals (e.g. Ellenberg, 1979).

The bryophyte flora of these stands does not correlate with those divisions based on higher plants. Bryophytes do not compete directly with the rooting plants as, for example, in fissures or in alluvions of ponds. However, they may represent a significant amount of the biomass whilst always occupying particular niches. From bare soil thermophilous and photophilous elements are reported, often composed of cosmopolitan or southern species. More hygrophilous species occur at protected sites such as holes, quarries or under tufts or sedges. Stony places amidst a homogeneous phanerogam community are inhabited by epilithic mosses. Floristically rich and interesting are the vegetation mosaics met with on stony humus-rich slopes (Fig. 6.3b) which are very difficult to investigate in a phytosociological or ecological context. Rarities such as *Plagiobryum demissum* may be found on humus edges in grass heaths. On very steep slopes in humous hollows compact cushions of *Oreas martiana* are found in the eastern Alps. Birds often perch on prominent stands and nitrify the humus. *Oreas* is also reported from eastern Asian mountains and Alaska. *Campylopus schimperi, Paraleucobryum albicans, Bartramia ithyphylla, Anoectangium aestivum, Tetraplodon urceolatus* are other species found in such sites up to the limits of plant life near the snowline.

Dwarf shrub heaths show relationships with the subalpine forest and shrub communities at the timber-line wth hygrophilous elements like *Hylocomium splendens, H. pyrenaicum, Pleurozium schreberi, Dicranum muehlenbeckii, Barbilophozia floerkei, B. lycopodioides, Leucobryum glaucum, Sphagnum girgensohnii, S. compactum* etc., and hummocks with *Aulacomnium palustre.* Towards wind-spent mountain-ridges mosses are in direct competition with lichens. Only *Racomitrium lanuginosum* or *R. canescens* cover larger non-calcareous surfaces mixed with *Polytrichum juniperinum, Rhytidium rugosum*; on calcareous sites *Stegonia latifolia, Encalypta rhaptocarpa* and *Desmatodon latifolius* are found. From southern France, Hébrard (1973) described the associations of the moss layer in these communities.

The moss flora of the alder forests, with tall herbaceous vegetation on wet north-exposed steep slopes extending from the subalpine to the middle alpine belt, is composed of hygrophilous and nitrophilous mainly pleurocarpous species of streams or wet places near cattle-trodden ground. These species include *Brachythecium rivulare, B. reflexum, B. salebrosum, Plagiothecium denticulatum, Rhizomnium punctatum, Drepanocladus uncinatus, Campylium stellatum, Ctenidium molluscum, Rhodobryum roseum* and *Lescuraea radicosa.*

6.7 CONCLUSIONS

Alpine climate is defined by low temperature, high precipitation and short growing periods. Another limiting factor is often shortage of nitrogen. Ecological research, investigating the reactions of a bryophyte to any one of these particular alpine environmental factors has not yet been carried out in the European Alps although in the lowlands there has been research into indicator species of pollution. There, abiotic factors have been measured and their relation to plant distribution analysed. While the alpine phanerogamic vegetation may be rather well described by the Braun–Blanquet method with the same characteristic combinations of species reappearing under similar ecological conditions, the size of bryophyte communities make such a treatment more difficult. It is not easy to work on vegetation pattern when the ecological variation of the species involved is not yet well known. Ecological behaviour may not be the same throughout the distribution area of a species, especially with regard to competition with other species. Obviously, many taxa have been described from the Alps but for some of them it has yet to be proved that they are taxonomically sound. There is an excellent field for future research in the study of the correlations between morphological variation, ecological modification and habitat factors. Simulation of alpine conditions in culture experiments in the lowlands is difficult. Further essential taxonomic work is still needed and there are considerable gaps in the floristic knowledge of the alpine bryoflora. Filling in these gaps may clarify the distribution and ecology of many so-called rare species and give us a better general view of the alpine bryoflora.

REFERENCES

Allorge, P. (1925), *Veröff. Geobot. Inst. Rübel Zürich,* **3,** 108–26.

Amann, J. (1918), *Flore des Mousses de la Suisse,* Imprimeries Réunies, Lausanne.

Amann, J. (1928), *Mat. F. Cryptog. Suisse,* **6,** (2), 1–437.

Barry, R.G. and Van Wie, C.C. (1974), Topo- and microclimatology in alpine areas. In: *Arctic and Alpine Environments* (Ives, J.D. and Barry, R.G., eds), pp. 78–83. Methuen, London.

Billings, W.D. (1974), Arctic and alpine vegetation: plant adaptations to cold summer climates. In: *Arctic and Alpine Environments* (Ives, J.D. and Barry, R.G., eds), pp. 403–43. Methuen, London.

Billings, W.D. and Mooney, H.A. (1968), *Biol. Rev.*, **43**, 481–529.
Braun-Blanquet, J. (1913), *Neue Denkschr. Schweiz. Natf. Ges.*, **48**, 1–347.
Dahl, E. (1956), *Skr. Norske Vidensk.-Akad. Oslo, Mat.-Naturvidensk. Kl.*, **3**, 1–374.
Dierssen, K. (1977), *Mitt. Flor.-soz. Arbeitsgem. N.F.*, **19/20**, 297–312.
Ellenberg, H. (1979), *Vegetation Mitteleuropas mit den Alpen in ökologischer Sicht*, Ulmer, Stuttgart (2. Auflage).
Frey, E. (1922), *Mitt. Naturf. Ges. Bern*, **6**, 1–195.
Gams, H. (1927), *Beitr. Geobot. Landesaufn. Schweiz*, **15**, 1–760.
Gams, H. (1932), *Ann. Bryol.*, **5**, 51–68.
Geissler, P. (1976), *Beitr. Kryptogamenfl. Schweiz*, **14** (2), 1–52.
Geissler, P. and Zoller, H. (1978), *Candollea*, **33**, 299–319.
Giacomini, V. (1940), *Att. Ist. Bot. Lab. Critt. Univ. Pavia*, s.4, **12**, 1–139.
Gimingham, C.H. and Smith, R.I.L. (1970), Bryophyte and lichen communities in the maritime Antarctic. In: *Antarctic Ecology* (Holdgate, M.W., ed.), Vol. 2, pp. 752–85. London.
Gjaerevoll, O. (1956), *Kongel. Norske Vidensk. Selsk. Skr. (Trondheim)*, **1956**, 1- 405.
Greene, S.W. and Longton, R.E. (1970), The effects of climate on Antarctic plants. In: *Antarctic Ecology* (Holdgate, M.W., ed.), Vol. 2, pp. 786–800. London.
Greter, P.F. (1936), *Die Laubmoose des obern Engelberger Tales*, Stiftsdruckerei, Engelberg.
Hébrard, J.P. (1973), *Rev. bryol. lichénol.*, **39**, 1–41.
Holdgate, M.W. (1964), Terrestrial Ecology in the Maritime Antarctic. In: *Biologie Antarctique/Antarctic Biology* (Carrick, R., Holdgate, M.W. and Prévost, J., eds), Paris.
Jovet-Ast, S. (1977), *Rev. bryol. lichénol.*, **43**, 465–72.
Krajina, V. (1933), *Beih. Bot. Centralbl.*, **50** (3), 774–957.
Meylan, C. (1924), *Beitr. Kryptogamenfl. Schweiz*, **6**, **1**, 1–318.
Ochsner, F. (1954), *Vegetatio*, **5/6**, 279–91.
Oostendorp–Bourgonjon, C. (1976), *Nova Hedwigia*, **28**, 637–47.
Pakarinen, P. and Vitt, D.H. (1974), *Can. J. Bot.*, **52**, 1151–61.
Persson, H. (1956), *J. Hattori bot. Lab.*, **17**, 1–18.
Pfeffer, W. (1871), Neue Denkschr. Allg. Schweiz. Ges. Gesammten Naturw., **24**, **(5)**, 1–142.
Pitschmann, H. and Reisigl, H. (1954), *Rev. bryol. lichénol.*, **23**, 123–31.
Poelt, J. (1955), *Feddes Repert. Spec. Nov. Regni Veg.*, **58**, 157–79.
Richard, J.-L. and Geissler, P. (1979), *Phytocoenologia*, **6**, 183–201.
Richards, P.W. (1932), Ecology. In: *Manual of Bryology* (Verdoorn, F., ed.), pp. 367–95, Nijhoff, The Hague.
Rybníček, K. (1966), *Folia Geobot. Phytotax.*, **1**, 101–19.
Rybníček, K. and Rybníčková, E. (1977), *Folia Geobot. Phytotax.*, **12**, 245–91.
Samuelsson, G. (1917), *Nova Acta Regiae Soc. Sci. Upsal.*, (ser. 4), **4**, 1–252.
Schröter, C. (1926), *Das Pflanzenleben der Alpen*, Raustein, Zürich.
Schacklette, H.T. (1961), *Bryologist*, **64**, 1–16.
Thienemann, A. (1925), Die Binnengewässer Mitteleuropas. *Die Binnengewässer*, Vol. 1, pp. 1–255. Schweizerbart, Stuttgart.

Turner, H. (1968), *Wetter und Leben,* **20,** 192–200.
Turner, H. (1971), *Ann. Met. N.F.,* **5,** 275–81.
Turner, H., Rochat, P. and Streule, A. (1975), *Mitt. Eidg. Anst. Forstl. Versuchswesen.,* **51,** 95–119.
Turner, H. and Tranquillini, W. (1961), *Mitt. Forstl. Bundes-Versuchsanstalt Mariabrunn,* **59,** 69–103.
Url, W. (1956), *Protoplasma,* **46,** 768–93.
Wahlenberg, G. (1813), *De vegetatione et climate in Helvetia septentrionali inter flumina Rhenum at Arolam observatis.,* Orell & Füssli, Zürich.
Yerli, M. (1970), *Veröff. Geobot. Inst. Rübel Zürich,* **44,** 1–119.

Chapter 7

Epiphytes and Epiliths

A.J.E. SMITH

7.1 INTRODUCTION

Epilithic or saxicolous bryophytes may be defined as those growing directly on the surface of rock, and in this chapter do not include aquatic species; epiphytic or corticolous species are those growing on the bark of living trees and shrubs. In the literature, authors dealing with epiphytes usually adhere strictly to the above definition, probably because of the well-defined nature of the habitat, although a few (e.g. Barkman, 1958) include species on dead tree stumps and logs. On the other hand, and this makes the interpretation of some data difficult, the term saxicolous is often taken (e.g. Yarranton, 1967a) to include plants growing on soil or detritus overlying rock as well as those occurring directly on rock surfaces. In the account of epiliths in this chapter the strict definition is followed as far as possible. There has been considerably more work done on epiphytes, especially of a phytosociological nature, than on epiliths. There are probably several reasons for this – the epiphytic habitat is much more discrete and the number of species involved is usually much fewer (see, for example, Table 7.1).

With both epiphytes and epiliths there are obligate and facultative species. Obligate epiphytes are usually restricted to bark and very rarely occur on other substrates. Similarly, obligate epiliths are usually only found on rock. Facultative species are regularly found on more than one substrate type.

7.2 SEXUALITY AND HABITAT

Before dealing with the two groups there are data on the reproductive behaviour of epiphytes and epiliths in Britain that are worth considering. From Table 7.1 it may be seen that the proportions of monoecious obligate epiliths and epiphytes are greater than might be expected when compared with mosses in general. This suggests that the tendency toward homozygosity that may result from monoecism may be of selective advantage in species that are adapted to specific habitats. Conversely, the proportions of dioecious species among facultative epiliths and epiphytes is higher than might be expected suggesting that the heterozygosity resulting from dioecism is advantageous in species that are catholic in their choice of habitat.

It is also of interest to note that twelve of the thirteen dioecious obligate epiphytes have vegetative propagules, usually in the form of multicellular gemmae. These include *Dicranum flagellare, D. montanum, Orthotrichum lyellii* and *Habrodon perpusillus*; the exceptional species being *Leptodon smithii*.

Table 7.1 Numbers of monoecious and dioecious British mosses.

	Monoecious	Dioecious
All species	287 (41.5%)	494 (58.5%)
Obligate epiliths	68 (53.1%)	60 (46.9%)
Facultative epiliths	19 (23.7%)	61 (71.3%)
Obligate epiphytes	20 (60.6%)	13 (39.4%)
Facultative epiphytes	15 (34.1%)	29 (65.9%)

A further curious feature of the reproductive behaviour of some corticolous mosses is that the peristome of the capsule opens when it is moist so that spore dispersal will occur during or shortly after rain or heavy dew. This phenomenon has been observed in corticolous species of the Leucodontaceae and Cryphaeaceae and in such genera as *Forstroemia, Pylaisia* and *Neckera* (Patterson, 1953). This behaviour has been noted only in epiphytic species although many epiphytes exhibit the more usual opening of the peristome on drying. It is difficult to see the selective advantage of spore dispersal during wet periods unless this is associated in some way with spore germination in which case it might be expected to occur in epiliths as well.

7.3 EPIPHYTES

During the past thirty years there has been intensive work on epiphytic bryophytes in Japan by Horikawa, Hosokawa, Iwatsuki, Nakanishi and their co-workers but, elsewhere, the studies have been less systematic. The most detailed account of cryptogamic epiphytes is Barkman's (1958) *Phytosociology and Ecology of Cryptogamic Epiphytes*. Jones (1959) says of this work 'Every conceivable aspect of everything relating to epiphytes is considered in turn with great thoroughness. . .' Over the past twenty years there has also been much done on the effect of atmospheric pollution on epiphytic bryophytes (see Chapter 12). Most of the work on epiphytes is of a descriptive nature and, except in Japan, there has been little experimentation.

Epiphytic bryophytes may be obligate or facultative epiphytes. In Britain there are some 33 obligate epiphytes and about 44 facultative epiphytes (see Table 7.1). Twenty-four facultative and twelve obligate epiphytes have been reported from the Adirondak Mountains, New York State (Slack, 1975). Examples of British obligate epiphytes include *Tortula laevipila, T. papillosa, Zygodon conoideus, Orthotrichum* and *Ulota* spp., *Leskea polycarpa, Cryphaea heteromalla* and *Pylaisia polyantha.* Such species are only very rarely found on other substrates. Facultative epiphytes include *Bryum capillare, Dicranum* spp., *Isothecium myosuroides, Cirriphyllum crassinervium, Hypnum cupressiforme* s.l., *Metzgeria* spp., *Plagiochila* spp., and *Frullania* spp. Some species that are epiphytes in one geographical area are not so in another. Thus, the following species occur on the bark of softwood and hardwood trees up to a height of two metres in the United States: *Campylium chrysophyllum, Fissidens cristatus, Thuidium delicatulum, Rhodobryum roseum, Hylocomium brevirostre, Ctenidium molluscum, Lepidozia sylvatica, Blepharostoma trichophyllum, Hypnum imponens, Nowellia curvifolia* and *Bartramia pomiformis* (Billings and Drew, 1938). These species only exceptionally occur on tree trunks in Europe. On the other hand, such species as *Ulota crispa* and *Ptilidium pulcherrimum* are epiphytic throughout their range. The majority of pioneer species on trees are obligate epiphytes, whilst most later colonizers and succession climax species are facultative epiphytes (Herzog, Koskinen, cited in Barkman, 1958).

7.3.1 The tree as a habitat

Whilst the majority of workers have confined themselves usually to the lowest two metres of trees a few (e.g. Richards, 1938; Hale, 1952; Pike *et al.*, 1977 and many Japanese workers) have studied the epiphytes of whole trees. There are usually three or four well-marked regions recognized: the tree base (mostly to one metre above soil level); the trunk and the crown with the latter being divided into two parts, large branches, and small branches and twigs. Study of the whole tree presents obvious difficulties which may be overcome either by the examination of fallen trees (e.g. Hale, 1952) or by using climbing techniques (e.g. Pike *et al.*, 1977). The tree base is colonized mostly by facultative epiphytes including species from the surrounding ground; the trunk and larger branch species may be facultative or obligate epiphytes whilst those of the smaller branches and twigs are usually pioneer obligate epiphytes.

Trees present a wide variety of habitat types for bryophytes, these depending upon light intensity, relative humidity and atmospheric pollution, which in turn are influenced by geographical location and the proximity of other trees, and upon the physical and chemical nature of the bark of the host tree or phorophyte.

7.3.2 Epiphytic communities

The most comprehensive accounts of epiphytic communities are by Barkman (1958), Iwatsuki (1960) and Iwatsuki and Hattori in a series of nineteen papers, the most recent being 1970. Other papers on the topic published since Barkman (1958) are Beals (1965), Engle (1960), Gough (1975), Hoffman (1971), Hoffman and Boe (1977), Hoffman and Kazmierski (1969), LeBlanc (1963), Nakanishi (1966), Pike *et al.* (1975), Rasmussen (1975), Slack (1976, 1977) and Sjögren (1961).

(a) Classification of communities

Barkman (1958) used an hierarchical system of noclemature to describe cryptogamic epiphytic communities in the Netherlands and adjacent regions. The basic unit is the association and associations are grouped into alliances and these into orders and there is a system of rules governing nomenclature akin to the *International Code of Botanical Nomenclature*. In the area concerned Barkman recognized 40 associations grouped into 10 alliances and four orders. The taxa are usually defined on the basis of faithful species and ecology. Iwatsuki (1960), using a similar system of community classification, recognized 33 associations and 10 alliances from the five major forest types in Japan.

Whether it is possible to produce a really satisfactory unit of classification for epiphytic bryophytes is open to question. The hierarchical system of the Zürich–Montpellier school as utilized by Barkman (1958) and Iwatsuki (1960) is unsatisfctory as it is to some degree subjective and attempts to categorize variation in community composition in an artificial taxonomic manner and is too inflexible, particularly where governed by nomenclatural rules. Further, it is the life-forms of the species making up the communities, not their taxonomic identity, that is important ecologically. A greater understanding of the ecology of epiphytic communities might be gained if these were analysed in terms of life-forms as well as species composition.

Phillips (1951) pointed out that 'The fact that many species appear in associations other than the ones they dominate and are even characteristic constituents of those other associations, shows that the dividing line between associations is not sharp but that many of them intergrade'. Further, Hale (1952) says that 'When variation in abundance with height is analysed quantitatively the delimitation of associations becomes a questionable procedure'.

Nevertheless, within a limited geographical area, it is useful to have associations based on the degree of constancy and cover of each species as this allows for comparisons both within and between different areas. It should be borne in mind, however, that although the name of an association is based upon that of one or more species, those species are not necessarily dominant or even present. The significance of this is indicated by a comment of Hoffman and Kazmierski (1969) who point out that species associations, because of different effects of competition, may be better indicators of microclimatic conditions than presence or absence of particular species which may be widely distributed. There is also an observation, relevant in this context, by Rasmussen and Hertig (1977) that 'It is often mistakenly assumed from "normal" association analyses that species under certain ecological circumstances are closely associated with each other. In many cases, the truth is that they are only associated with the ecological conditions in a given habitat, but otherwise show high interspecific competition . . .'.

If the species composition of an association is the result of competition then, clearly, life-forms are highly important. Japanese authors have made a point of indicating the life-forms of epiphytic bryophytes (pp. 199–201). If other authors had done the same it would be possible to make more meaningful comparisons, both between different host species and between different geographical areas.

(b) Examples of epiphytic communities

As might be expected there is vertical zonation in epiphytic bryophyte communities and the following examples will suffice to illustrate this. Life-forms, where known, are based on those of Gimingham and Birse (1957), Horikawa and Nakanishi (1954) and Mägdefrau (see Chapter 2).

In the Killarney oakwoods, south-west Ireland, there are three associations (or climax associules) (Richards, 1938). The first association occurs on tree bases up to 1–2 m and is dominated by *Hylocomium brevirostre, Thuidium tamariscinum* (both wefts), *Isothecium myosuroides** and *I. myurum* (both rough mats) and the upward limit is determined by humidity. Trunks and larger branches have an association dominated by *Isothecium myosuriodes* (rough mat) with other species such as *Plagiochila spinulosa, P. punctata* (tall turfs) and *Hymenophyllum wilsonii*; this association is intolerant of high winds and bright illumination. The climax association of twigs and small branches is dominated by *Ulota crispa* (small cushions) and *Frullania tamarisci*

*Although the individual shoots of *Isothecium myosuroides* are denroid the overall effect is that of a rough mat.

(smooth mat). This latter association is also found on the branches of the shrub *Vaccinium myrtyllus* and on trees of *Ilex aquifolium* the latter two associations occur though they are not very well defined.

On *Fraxinus excelsior* and *Fagus sylvatica* in northern Jutland, Denmark (Rasmussen, 1975) on the first 2.5 m of the trunks there are three inter-related associations, discernable using random sampling and statistical techniques. Nearest the ground is an association composed of *Homalia trichomanoides* (fan), *Lophocolea heterophylla* (rough mat), *Brachythecium rutabulum, Plagiomnium undulatum* (wefts) and *Isothecium myosuroides* (rough mat). Higher up the trunk is an association, the most important constituents of which are *Metzgeria furcata* (smooth mat or thread-like), *Homalothecium sericum, Leucodon sciruoides* (rough mats), *Neckera pumila* (fan), *Frullania dilatata* (smooth mat), *Orthotrichum lyellii* and *Zygodon viridissiumus* (small cushions); in intermediate situations is an association composed of *Porella platyphylla, Radula complanata, Pylaisia polyantha* (smooth mats), *Antitrichia curtipendula* (weft), *Isothecium myosuroides, Pterogonium gracile* (rough mats), *Neckera complanata* (fan) and *Bryum capillare* (cushion). *Hypnum cupressiforme* (smooth mat) occurs in all three associations. The species of the intermediate association and of the other two associations are not mutually exclusive. If an analysis using only the five most common species is carried out (Rasmussen and Hertig, 1977) there is a distinction into two communities, one near the tree bases consisting of *Hypnum cupressiforme, Isothecium myosuroides* and *Neckera complanata* and one at the upper levels composed of *Homalia trichomanoides, Hypnum cupressiforme, Neckera complanata* and *Metzgeria furcata*. Clearly, the inter-relationships of species within communities even in a small area of relatively uniform environmental conditions are highly complex.

The only systematic studies on the physiological ecology of epiphytic bryophytes have been carried in Japan (see Hosokawa *et al.*, 1964) in the *Fagus crenata* forest of Mt. Hiko, Kyushu, south-west Japan and it is useful to know something of the communities involved. The epiphytic vegetation is described by Omura *et al.* (1955). They recognized five associations, which they referred to as epilia, on *Fagus crenata*. At or near the tree bases where light intensity is low and relative humidity high the predominant species are *Thuidium cymbifolium, Hylocomium cavifolium, Bryhnia novea-angliae* (all wefts), *Homaliodendron scalpellifolium* (feather) and *Thamnobryum sandei* (dendroid). On the trunks, and occasionally the bases of main branches, are *Pterobryum arbuscula, Anomodon giraldii, Neckera yezoana, Dolichomitra cavifolia* (all dendroid), *Macrospiriella scabriseta* (tall turf) and *Metzgeria conjugata* (thalloid mat). The third association, which occurs mainly on the undersides of the primary branches, is composed of lichens. On the upper parts of trunks and on branches the species that occur are *Boulaya mittenii, Macromitrium prolongatum, Okamuraea hakoniensis* (all rough mats), *Frullania moniliata* (smooth mat), *Dicranum fragiliforme, Dicranodontium denudatum* (both short turfs) and five lichen species. On the ultimate branches the small cushion-forming *Ulota crispa* is the only bryophyte and is associated with several lichen species.

Two examples illustrate variation within an association and in the distribution of associations under different climatic conditions. Omura and Hosokawa (1959), studying the *Thuidium cymbifolium–Homaliodendron scarpellifolium* association at the base of *Fagus crenata* trees on Mt. Hiko, south-west Japan; found that there was variation in species composition and cover within and between individual associations with location of the tree, height from the ground and aspect and considered that these were probably related to humidity and light intensity.

In a study of bryophyte–macrolichen communities of *Pseudotsuga menziesii* at eight sites in an area approximately 100 × 100 km on the Olympic Peninsula, Washington, USA (Hoffman and Kazmierski, 1969) it was found that community structure was related to climatic factors. Six associations occur and their vertical distribution depends on climate and aspect. Taking the north-east side of trunks as an example, under mesic conditions a community dominated by *Hypnum circinale, Dicranum fuscescens* and *Scapania bolanderi* extends well above 4 m. As conditions become drier this community extends less far up the trunks and gives way, at first above and then right to the tree bases, to a community composed of *Hypnum circinale, Dicranum fuscescens* and the lichen *Lepraria membranacea.* As conditions become more xeric this community in turn gives way to one of *Hypnum circinale* and the lichens *Lepraria membranacea* and *Sphaerophorus globosus.* Under the most xeric conditions the communities are composed entirely of lichens.

On a wider geographical basis there may be similarities between associations in different regions particularly if the associations are not treated as rigid inflexible units. For example, some of the associations of epiphytes in Michigan, USA (Phillips, 1951) resemble those in Virginia, USA, southern Norway, Switzerland and northern Italy, Poland and the Rhône Valley. The *Neckera pennata* association has parallels in Tennessee, USA and various parts of Europe. The *Ptilidietum californicae* association and certain associations of the *Dicrano–Bazzanion* in Japan are somewhat similar to those of boreal Europe and North America (Iwatsuki, 1960). Similarly, associations of the Japanese *Ulotion asiaticae* may have affinities with associations of the *Ulotion crispae* of Barkman (1958) and *Ulota crispa* communities in North America.

On the other hand, although statistically delimited associations on *Fraxinus excelsior* and *Fagus sylvatica* in northern Jutland, Denmark

(Rasmussen, 1975) do bear resemblance to Barkman's (1958) *Ulotetum bruchii, Neckereto–Isothecietum myosuroides* associations and *Homalion* sub-alliance (categories of the Zürich–Montpellier school), there are too many differences to relate the type of communities arrived at using different methods of definition.

A possibly more realistic approach is, rather than making comparisons on the basis of species composition, to make them on the basis of life-form composition since it is the life-form spectra of the communities that is ecologically important, not their taxonomic constitution.

7.3.3 Succession

To study colonization and succession it would be necessary to place permanent quadrats on young trees and observe these over a period of very many years. The alternative is the study of communities on trees of differing ages within a more or less uniform environment. This was done in Michigan, USA (Phillips, 1951). On young trees pioneer associations are *Frullania* spp., sometimes *Ptilidium pulcherrimum* or *Radula complanata* (which may follow *Frullania*), *Orthotrichum sordidum* and *Ulota crispa,* the latter two requiring protection, such as knot holes, for establishment. These are followed by pleurocarpous moss associations. Depending on environmental factors the association may be, from xerophytic to less xerophytic, *Homomallium adnatum, Pylaisia selwynii, Leucodon sciuroides, Porella platyphyloides, Neckera pennata* or *Anomodon minor.*

Whilst the pioneer species vary from place to place, primary colonizers are usually algae or crustose lichens, followed by foliaceous lichens or mosses and the latter by fruticose mosses (Barkman, 1958). In Europe, pioneer bryophytes include *Orthotrichum, Ulota* and *Zygodon* spp. (usually requiring protection for establishment), *Hypnum cupressiforme, Neckera pumila, Frullania* spp. and *Metzgeria furcata* (Barkman, 1958; Grubb *et al.,* 1969; Rasmussen, 1975). Martin (1938) reports *Pylaisia polyantha* as a pioneer in western Scotland but this is probably an error for *Hypnum cupressiforme* var. *resupinatum.* Which particular species arrives first and how succession proceeds varies with the situation of the tree species and the microclimate. Thus, Grubb *et al.* (1969) report that *Ulota crispa* is a primary colonizer in Britain that is suppressed by later species, especially *Hypnum cupressiforme* which in turn is succeeded by *Dicranum scoparium.* At Fontainbleau, France, succession follows a different pattern (Doignon, cited in Barkman, 1958). On oaks, after 15–20 years, foliaceous lichens appear, after 20–25 years, *Frullania dilatata,* then *Orthotrichum affine* and *O. schimperi,* then *Hypnum cupressiforme* and *Orthotrichum lyellii* : *Ulota crispa* appears after 25 years, *Anomodon viticulosus, Porella platyphylla* and *Neckera* after 35 years, *Leucodon sciuroides* and *Zygodon* after 40 years. On beech twigs at Fontainbleau

(Doignon, cited in Barkman, 1958) the first bryophytes to appear are *Ulota bruchii* and *Hypnum cupressiforme* succeeded by *Frullania dilatata* and the lichen *Parmelia asperata*. In woods in Amsterdam (Rynders, cited in Barkman, 1958), the first bryophytes to establish on *Salix alba* and *Sambucus nigra* after 16–17 years are *Frullania dilatata* followed two years later by *Ulota bruchii*. In the Killarney oakwoods, south-west Ireland (Richards, 1938), the tree base pioneer is *Isothecium myosuroides* which may be invaded but not replaced by woodland floor species. On trunks and larger branches *Isothecium* is again a pioneer, possibly preceded by lichens. It may invade the *Ulota–Frullania* association of larger branches but not of smaller branches or twigs where the *Ulota–Frullania* association is the climax. In that habitat the pioneers are usually *Ulota crispa* and *Frullania germana* and later species that come in include *Frullania tamarisci, Plagiochila spinulosa* and *P. punctata*.

Except perhaps in the tropics and subtropics it would seem that the bryophyte communities on mature tree trunks are climax communities. Barkman (1958) points out that in the tropics and subtropics the cryptogamic epiphytic communities are merely a stage in a succession that will culminate in a phanerogamic epiphyte community but that this climax is never reached in temperate climates as the trees do not live long enough for sufficient accumulation of 'soil' to allow the establishment of higher plants to any extent. The implication is, therefore, that in such regions the cryptogams do not form climax communities. This seems a rather extreme approach especially as, except in very moist areas, there is no indication that any further stage in the succession is likely to occur even on very old trees. This is supported by the observations of Hosokawa and Omura (1959) on *Fagus crenata* trees on Mt. Hiko, Kyushu, Japan. Analysis of data gathered from quadrats at various heights on the trees over a period of five years indicated the presence of a climax community.

7.3.4 Life-forms

It is only in Japan that large-scale studies have been carried out on the distribution of epiphytic bryophyte life-forms. The life-forms used are based on the categories of Gimingham and Robertson (1950) and added to by Horikawa and Nakanishi (1954) and Iwatsuki and Hattori (1956). The detailed phytosociological accounts of such authors as Hattori, Hosokawa, Iwatsuki, Nakanishi and their co-workers contain tables in which the life-forms of the various species are indicated.

Iwatsuki (1960) discusses the life-forms from the five main Japanese forest types. Table 7.2 shows the vertical distribution of life-forms in the forest alliances *Vaccinietum Pinetum pumilae* (alpine forest), *Abieton mariesi* (sub-alpine coniferous forest), *Fagion crenatae* (montane deciduous hardwood forest), *Tsugion sieboldii* (montane coniferous

Table 7.2 Distribution of main life-forms on tree base, trunk and crown of trees of main Japanese forest types. Modified from Iwatsuki (1960).

Forest type														Life-form
Vaccinieto–Pinctum pumilae		*Abietion mariesii*			*Fagion crenatae*			*Tsugion sieboldii*			*Shiion sieboldii*			
Base	Stem	Base	Trunk	Crown	Base	Trunk	Crown	Base	Trunk	Crown	Base	Trunk	Crown	
–	–	–	–	×	–	–	–	–	–	–	–	–	–	Small cushions
–	–	×	–	–	×	×	×	–	×	×	–	–	×	Tall turfs
×	×	×	×	×	–	–	–	×	–	–	–	–	–	Short turfs
–	–	–	–	–	–	–	–	–	–	–	–	×	–	Feather forms
–	–	–	–	–	×	×	–	×	×	–	×	×	–	Dendroid forms
–	–	–	–	×	–	×	×	–	×	×	–	–	×	Rough mats
×	×	×	×	–	–	×	×	×	×	×	×	×	×	Smooth mats
×	×	×	×	–	–	–	–	–	×	–	–	–	–	Thread-like forms
×	×	–	–	–	×	–	–	×	–	–	×	–	–	Weft forms
–	–	–	–	×	–	–	×	–	×	×	–	–	×	Appressed forms
–	–	–	–	–	–	–	–	–	×	×	–	–	×	Pendulous forms

forest) and *Shiion sieboldii* (lowland broad-leaved evergreen forest). Clearly there is variation in life-form with vertical distribution on the tree, with forest type and with altitude. Generally, small cushions are characteristic of tree crowns in sub-alpine and montane regions; short turfs usually occur on tree bases in similar regions. Tall turfs arising from creeping rhizomes are found on tree trunks in humid montane areas. Feather forms are shade- and moisture-requiring and are found mainly on the lower trunks in montane deciduous forests; dendroid forms grow on trunks and tree bases in similar habitats and also in evergreen broad-leaved forests. Of the mat forms, rough mats are light-requiring and drought-resistant and occur in tree crowns in montane areas; smooth mats develop on bases of conifers and also trunks of montane deciduous trees; thread-like forms occur on trunks and tree bases in most areas of Japan. Appressed forms (especially species of *Frullania* and Lejeuneaceae) are light-requiring and are most frequent in southern and lowland parts of Japan. Wefts occur on tree bases and pendulous forms on trunks and branches in warm humid localities. Although there have been no such detailed studies in Europe it is clear from field observations that the distribution of life-forms is essentially similar in both geographical areas, although pendulous forms are lacking in Europe.

7.3.5 Environmental factors

It is not really possible to treat individual environmental factors in isolation, as the composition and structure of epiphytic communities results from the interaction of all the factors. The complexity of the situation is illustrated by the observations of Coleman *et al.* (1956) who carried out a statistical study of the distribution of epiphytes on the Olympic Peninsula, Washington, from 43 species of conifers, hardwoods and shrubs at 20 stations with annual rainfall varying from about 38 cm to 310 cm. They concluded that the presence of a given epiphyte on a given host at a given station was dependent upon the surface characteristics of a given host reacting with suitable climatic conditions to produce suitable conditions on the surface of the host. Establishment of the epiphyte depended on the presence of a suitable host and the opportunity for a given epiphyte to arrive at a given site.

It is not even possible to state that any one factor or factors are more or less important than others as there is no means of isolating effects which may vary with circumstances, although Culberson (1955), studying epiphytes in Michigan, considered that it is the nature of the bark that is of greatest importance but superimposed upon this is the effect of microclimate. Olsen (1917) n the other hand considers that age of trees is the most important factor. From studied of *Populus deltoides* Hoffman and Boe (1977) concluded that it is bark texture that is important in determining the structure of epiphytic communities rather than bark pH and moisture content. Bearing in mind that it is not possible to consider any one environmental factor in isolation, they can be dealt with under two main headings, the nature of the bark of the host tree and microclimate.

(a) The bark

(i) Host specificity. Whilst most authors consider that the species of the host tree is of major importance in determining the nature of the epiphytic community there is one claim to the contrary. Phillips (1951) points out that similar associations in different geographical areas occur on quite different tree species indicating that the tree species is not an important factor. He considers that it is microclimatic conditions that are significant and that optimum moisture conditions will enable an epiphyte to colonize more or less any tree as moisture will compensate for such inimicable features as low pH, smooth bark, youth of the tree and exposure. He considers restriction to particular tree species as being characteristic of drier areas. However, it is probably only in tropical montane cloud forest that such optimum climatic conditions prevail (see Chapter 3, pp. 97–8) so that the argument is not relevant. Further, it may well be that different tree species in different geographical areas which have similar epiphytic communities have similar bark types or that the combination of one bark type under one set of climatic conditions produces an environment as compatible to a particular epiphytic association as the combination of a somewhat different pair of variables in another geographical area. From the fact that different associations occur on different tree species it can only be concluded that the nature of the bark and, therefore, the identity of the host species is of great importance in determining the composition and structure of epiphytic communities and this thesis is supported by numerous observations (e.g. Beals, 1965; Culberson, 1955; Hale, 1955; Iwatsuki, 1960; Martin, 1938; Rasmussen, 1975; Slack, 1976). A few examples will demonstrate this.

In Argyll, western Scotland (Martin, 1938) the number of bryophytes recorded from different tree species varies. Nineteen taxa were recorded from *Quercus robur*, 19 from *Fraxinus excelsior*, 11 from *Betula pendula*, 8 from *Acer pseudoplatanus*, 9 from *Fagus sylvatica* and 9 from conifers. The frequency of species on different trees also varies. In Argyll, *Dicranum scoparium* is most common on *Quercus* and *Betula* and is absent from *Fraxinus* and *Acer*; *Homalothecium sericeum* is most common on *Fraxinus* and *Acer* and absent from *Quercus* and *Betula*.

In northern Wisconsin, at breast height, *Anomodon minor* is most common on *Acer saccharum*, and *Tilia americana*, *Dicranum montanum* on *Pinus stobus*, *Tsuga canadensis* and *Betula lutea*, *Orthotrichum* spp. on

Populus spp., *Ptilidium pulcherrimum* on *Pinus strobus* and *Ulota crispa* on *Betula lutea* and *Populus* spp. (Culberson, 1955).

Slack (1976) found in the Adirondack Mountain Preserve, New York State that there was a considerable degree of phorophyte specificity. *Neckera pennata, Porella platyphlloidea, Anomeodon rugelii, Pylaisia intricata, Pseudoleskeella nervosa, Radula complanata* and *Leucodon brachypus* are restricted to *Acer saccharum; Frullania asagrayana, Dicranum montanum, Ptilidium pulcherrimum, Hypnum pallescens* and *Dicranum viride* are restricted to *Betula alleghaniensis*; only *Frullania eboracensis* and *Platygyrium repens* are common to both tree species.

In the coniferous forest of Mt. Honokawa, Shikoku, Japan (Iwatsuki and Hattori, 1966) *Leucodon okamurae* is common on *Tsuga sieboldii* trunks but rare on those of *Abies firma. Brotherella yokohamae, Clastobryella kusatsuensis, Dicranum hamulosum, Hypnum tristo-viride, Bazzania fauriana, Chandonanthus hirtellus* and *Odontoschisma grosseverrucosum* are more abundant on *Tsuga* whilst *Haplohymenum longinerve, Hypnum plumaeforme, Macromitrium gymnostomum, Macvicaria ulophylla, Okamuraea plicata, Pilotrikopsis dentata* and *Trocholejeunea saudvicensis* are more abundant on *Abies*. Similar differences were found between the epiphytic communities of the two conifer species in another part of Japan, Wariiwa, Nichinan, Honshu (Iwatsuki and Hattori, 1956).

If a trunk is inclined this has an effect on the distribution of epiphytes. Barkman (1958) suggests that with inclined trunks, rain water will trickle to the lower side so that epiphytes will grow all round the trunk. When there is an epiphyte cover on the upper side this will absorb rain, depriving the lower side of water. In England, Pitkin (1975) found that trickling of rainwater to the under side does not appear to occur and epiphytes rarely if ever colonize the underside of inclined trunks. On inclined trunks of *Fraxinus excelsior* and *Fagus sylvatica* (Rasmussen, 1975) species diversity is higher when the angle of inclination is 5°–15° but at greater angles of inclination the number of species drops, with *Hypnum cupressiforme* and *Homalothecium sericeum* becoming dominant on the upper side. Probably, these two species have a greater competitive ability on an inclined than on a vertical trunk.

(ii) Physical nature of the bark. The only attempt to determine the hardness of bark, a feature said by many authors to be important, was by Culberson (1955) who investigated hardwood and softwood trees in northern Wisconsin. He found no direct correlation between epiphyte communities and hardness of bark although softwood and hardwood trees differed with respect to this feature and in their epiphyte communities.

Data on the microtopography of bark is conflicting. Olsen (1917) says that small acrocarpous mosses such as *Orthotrichum* and *Ulota* spp. are typical of smooth bark although other authors (e.g. Barkman, 1958) say they occur on rough bark and in knot holes and leaf scars. According to Barkman (1958) *Neckera pumila, Pterigynandrum* and *Leskea* occur on smooth bark. The stability of the bark is obviously important and Gough (1975) considers that exfoliation of the bark is the most important factor determining the distribution of epiphytes communities in Boulder County, Colorado, USA.

In Japan (Iwatsuki, 1960), associations characteristic of smooth bark, for example the *Ulo–Frullanietum* associations extend down the trunks of smooth barked trees such as *Betula grossa,* but on other kinds of trees are usually restricted to the crowns. It also appears that xerophytic associations in Japan tend to occur on smooth bark and hygrophilous ones on soft-barked trees.

(iii) Moisture and pH. Estimates of bark water content vary, depending upon the methods of estimation and of expressing the results. Reports of bark pH are also variable although differences may be due to the degree of atmospheric pollution and amount of rainfall. Staxang (1969) relates differences in bark pH of deciduous trees in Sweden to atmospheric pollution. The pH of bark of *Quercus robur* and *Q. petraea* in North Wales and Devonshire, areas of high rainfall and low pollution, varies from 4.9 to 5.2 whilst from sites in the vicinity of Oxford, where rainfall is lower and pollution higher, the range is 3.4 to 4.0 (Pitkin, 1973b). Factors affecting bark pH are discussed in Chapter 11 (p. 412).

An example of the importance of bark moisture capacity and pH is provided by studies on the moisture content of the bark of *Liriodendron tulipiferam, Fagus grandifolia, Fraxinus americana* and *Tsuga canadensis* in an area of eastern Tennessee (Billings and Drew, 1938). These showed that the field moisture content of the bark of the three angiosperms was higher by a factor of two to seven than that of *Tsuga.* In all four species, moisture content was greater at a height of one metre than at 1.5 m and that in turn was greater than at 2 m. The moisture content on the south side, the direction from which the rain came, was higher than on the north side. The *Tsuga* had a bark pH of 4.28 to 4.96, the other species a pH of 5–6. There is a marked difference in the bryophyte associations of the angiosperms and the gymnosperm. Every specimen of the former studied had communities of *Neckera pumila, Anomodon attenuatus* and *A. rostratus* and many had *Campylium chrysophyllum, Brachythecium oxycladum* and *Fissidens cristatus,* none of which was found on *Tsuga.* Conversely, communities of *Ulota crispa, Dicranum fulvum, Bazzania denudata, B. trilobata, Hypnum reptile* and *Mnium hornum* were exclusive to the bark of *Tsuga.* These differences were interpreted in terms of differences in the moisture content and pH of the various barks as other environmental conditions were similar. Vertical zonation on the trunks is explicable in terms of the upward decrease in moisture content. This latter is also remarked on by Rasmussen (1975). On *Fraxinus excelsior* in Denmark there is a decrease in the number of species upwards, whilst on *Fagus sylvatica* there is no distinct trend. This correlates with the data (Rasmussen, 1975 citing Sjogren) that *Fraxinus* trunks are moister near the base than above, whereas the moisture content of *Fagus* bark is more uniform.

The effect of pH on corticolous bryophytes in culture is shown in Table 7.3. These data agree with field observations (Pitkin, 1973a, b) and may account at least in part for the distribution patterns of bryophytes in Britain where low pH may be related to high atmospheric pollution.

Table 7.3 Growth of epiphytes on agar cultures at various pH's over a period of twelve weeks (Pitkin, 1973b).

pH	*Hypnum cupressiforme*	*Cryphaea heteromalla*	*Isothecium myosuroides*
3	Many shoots dead.	Dead.	Many shoots dead.
4	Dark green, shoots spindly.	Some shoots dead, others dark green.	Growth largely protonemal, plants dark green.
5	Robust, shoots dark green.	Vigorous shoot growth, shoots dark green.	Robust, shoots dark green.
6	Very robust, shoots dark green.	Robust, shoots dark green.	Robust, shoots dark green.
7	Robust, shoots pale.	Shoot growth, shoots pale.	Robust, shoots dark green.
8	Poor growth, shoots yellow/green.	Poor growth, shoots pale yellow.	Poor growth, shoots yellow/green.

An example where the occurrence of epiphytes is related to bark pH and air pollution comes from the observations in a region of high atmospheric pollution in the Aalborg–Nørrendly area of north Denmark (Johnsen and Søchting, 1976). The species, *Tortula latifolia, T. subulata, Leskea polycarpa, Orthotrichum diaphanum, Tortula virescens* and *Zygodon virudissimus* only occur near to a cement plant which releases calcareous dust, those species with the narrowest range being listed first. The dust raises the bark pH and clearly this allows the establishment of species that would otherwise be susceptible to the degree of atmospheric pollution in the area concerned.

(iv) Chemical nature of the bark. It is often suggested that the chemical nature of the bark is important in determining the composition of epiphytic communities (e.g. Martin, 1938; Iwatsuki, 1960), although little evidence has been given until recently in support of this. Barkman (1958) divides

trees into three groups on the basis of bark electrolyte content. Those with eutrophic bark, including *Acer* spp., *Sambusus nigra,* and *Prunus avium,* those with mesotrophic bark, including *Quercus* spp., *Fagus sylvatica, Fraxinus excelsior* and *Salix alba,* and those with oligotrophic bark such as *Betula, Picea* and *Abies*. Several associations are restricted to eutrophic and mesotrophic barks. Facultative epiphytes that occur on eutrophic bark, if also facultative epiliths, occur on basic rocks whilst similar bryophytes of oligotrophic bark occur on siliceous rocks.

Table 7.4 shows the results of culture experiments with corticolous mosses using bark extracts. These agree well with field observations, especially the growth of *Lophocolea heterophylla* on *Picia abies* as it is often the only bryophyte on spruce in Britain (Pitkin, 1973b). This suggests that the chemical nature of the bark is indeed a factor affecting growth of epiphytic bryophytes.

Table 7.4 Growth on water extracts of different barks after 13 weeks (g organic carbon per plant) (from Pitkin, 1973b).

	Fraxinus excelsior		*Quercus* sp.		*Sambucus nigra*	*Picea abies*
Height above ground	60 cm	120 cm	60 cm	120 cm		
Cryphaea heteromalla	2.25	1.13	0.43	0.37	2.31	0.28
Hypnum cupressiforme	3.10	1.06	1.88	1.27	2.12	0.72
Lophocolea heterophylla	3.14	0.92	2.37	1.15	1.86	2.40

In experiments with gemmae of *Ulota phyllantha* (Pitkin, 1973a, b) it was shown that these would not germinate on bark discs taken from 10 m up a *Quercus petraea* trunk but grew readily on discs from beneath the epiphyte cover at the foot of the tree. The gemmeae germinated on discs from 10 m up the tree after these had been washed in running water for one month suggesting the presence of a leachable inhibitor or inhibitors in the bark. If such leachable substances occur in bark this could explain in part the epiphyte cover extending further up a tree in wetter areas and supporting Phillip's (1951) argument that ample moisture will negate the effect of deleterious aspects of the environment.

(b) Microclimate: light intensity and atmospheric humidity

Light intensity and relative humidity are inversely correlated, the former being at its maximum at the tree crown and the latter at the tree base. This had been suggested by numerous authors (e.g. Olsen, 1917; Richards, 1938) and has been confirmed by Omura *et al.* (1955). The effect of humidity, which is related to rainfall, is marked. This is indicated by growth rates of *Hypnum cupressiform* and *Isothecium myosuroides* described on p. 210. In another instance, in mesic areas of the Olympic Peninsula, Washington, with annual rainfall in the region of 300 cm, up to 17 epiphytic bryophytes occur on the trunks of *Pseudotsuga menziesii*; at the other extreme, with an annual rainfall of about 60 cm the number of bryophyte species is three (Hoffman and Kazmierski, 1969). Further, as mentioned on p. 197 vertical zonation is also markedly affected by rainfall and consequent relative humidity. In Argyll, Scotland, in exposed areas, bryophyte cover is less dense, zoning less marked and vertical extent reduced as compared with protected areas with higher relative humidity (Martin, 1938).

An illustration of the effect of height on cover by particular species is given by Slack (1976) from the Adirondack Mountain Preserve. Over a height range from 0 to 2 m, the mean percentage cover of *Neckera pennata* rises from 0% at 0 m to 27% at 1 m dropping to 6% cover at 2 m: *Anomodon rugelii* rises from 0% cover at 0 m to 8% at 0.5 m and drops to 0 at 1 m; *Ptatygyrium repens* rises from 0% cover at 1 m to 20% at 2 m; *Pylaisia intricata* shows a similar pattern to *Platygyrium.*

Whilst there are no quantitative data available it is amply evident from field observations that there is a close correlation between rainfall and the luxuriance of epiphyte cover. There is, however, one observation that suggests that another source of moisture may be important. In the Transvaal, South Africa (Jacobsen, 1978), over a three year period, the growth of epiphytic mosses at Bloemfontein and at Woodbush was 45 and 47 gm respectively per 10×10 m plot although the rainfall was 2.5 times greater at the latter site than the former. Although no meteorological data were available it was considered that mist was an important contributory factor to this phenomenon at Bloemfontein.

There is a relationship between vertical distribution, rate of evaporation and desiccation resistance. Generally, species in the crown are more resistant to desiccation than those at the tree base (Hosokawa and Kubota, 1957; Hosokawa *et al.*, 1964). For example, *Ulota crispula* and *Boulaya mittenii* which occur in crowns are more drought-resistant than tree base species such as *Homaliodendron scalpellifolium, Hylocomium cavifolium* and *Thuidium cymbifolium.* The resistance of crown species to desiccation is greater in the winter than in summer. This may be due partly to increased humidity when the phorophyte *(Fagus crenata)* is in leaf and to the greater physiological activity in the summer. this correlates with observations of Tagawa (1959) who found that the evaporation potential in tree crowns is greater in winter than in summer.

There is also a relationship between compensation point and vertical distribution. This was shown experimentally by Hosokawa and Odani (1957) as is indicated in Table 7.5. Clearly, the light requirements of species affects their vertical distribution. Light saturation curves (Miyata

Table 7.5 Vertical zonation of species related to minimum light intensity necessary to reach compensation point. (Modified from Hosokawa and Odani, 1957).

Habitat	Species	Minimum light intensity (lux)	Optimum light intensity (lux)
Uppermost branches	*Ulota crispula*	6000	± 20 000
Boughs	*Bouleya mittenii*	1200	10 000–20 000
Trunk	*Anomodon giraldii* *Pterobryum arbuscula*		
Tree base	*Thuidium cymbifolium* *Hylocomium cavifolium* *Thuidium sandei* *Homalodendron scalpellifolium*	400	± 10 000

and Hosokawa, 1961) suggest that plants on the lower parts of trunks are shade species and those on the crowns light-demanding species.

In a recent study of the physiology of four epiphytic mosses growing on *Acer saccharum* in the Adirondack Mountains, USA, Tobiessen *et al.* (1978) found that *Plagiomnium cuspidatum,* a facultative epiphyte, required a higher water content for optimum photosynthesis but had a more rapid assimilation rate than did the three obligate epiphytes, *Anomodon rugelii, Neckera pennata* and *Ulota crispa.* The *Ulota* was found to be more tolerant of desiccation than the *Neckera* with which its vertical range overlaps.

7.3.6 Nutrition

Species on twigs and outer branches probably obtain their nutrients from bark, rain, dust and perhaps small quantities of leaf leachate. Those species on the lower parts of the trunk probably obtain nutrients from leaf leachates and leachates from bryophytes higher up. Species of the upper trunk and lower branches are in an intermediate position (Grubb *et al.*, 1969). After comparative measurements of nutrients in *Hypnum cupressiforme,* the bark of the host species (*Fraxinus excelsior* and *Fagus sylvatica*) and bulk precipitation in north Denmark (Rasmussen and Johnsen, 1976), it would appear that the moss accumulates nutrients mainly from atmospheric precipitation and possibly also from stem-flow.

Evidence that epiphytes are, to some extent, dependent upon stem-flow for nutrients is provided by Pitkin (1973b). He showed that growth of *Hypnum cupressiforme, Cryphaca heteromalla* and *Lophocolea heterophylla* was generally greater on an extract of bark made from stem-flow water than on a bark extract made with pure water. The topic is dealt with further in Chapter 11, pp. 404–14.

It would appear that species, such as *Ulota crispa,* on ultimate branches are able to grow with a lower internal content of certain nutrients such as nitrogen, potassium and sodium than species of larger branches and trunks (e.g. *Hypnum cupressiforme* and *Dicranum scoparium*) which have nutrient contents similar to those of terrestrial bryophytes (Grubb *et al.*, 1969). It is not known if the low supply of nutrients to the species of outer branches is a limiting factor. Material of *Ulota crispa* from the Lake District and North Wales has a higher nutrient content than specimens from north-west Scotland but this is possibly due to growth restriction by atmospheric pollutants (Grubb *et al.*, 1969).

In considering the nutrient content of epiphytes it is necessary that lichens are also taken into account. Pike *et al.* (1972) estimated the total nitrogen content of the epiphytes of an old growth *Pseudotsuga menziesii* in a forest, in which there are about 10 trees per acre, (i.e. 25 trees per hectare) about 75 km east of Eugene, Oregon, USA. They found that of the 225 gm total, bryophytes contained 82 gm and lichens 143 gm (including 127 gm held by the single species *Lobaria oregana*). They consider that this could represent a significant imput into the forest ecosystem and that nitrogen may enter the soil from stem-flow leachates from the epiphytes.

7.3.7 Growth, productivity and biomass

Measurements of the growth rate of epiphytic mosses have only been made on pleurocarpous species which, because of their 'two-dimensional' habit, are more easily dealt with than 'three-dimensional' acrocarpous species. The only example of growth rate of an acrocarpous species is of *Ulota crispa* with an annual diameter increment of up to 2 mm (T.J. Bines in Pitkin, 1975). Growth of corticolous bryophytes is slow and depends upon environmental conditions. In studies on the effects of moss gathering by florists in the montane deciduous forests of Transvaal, South Africa, it was estimated that regeneration of bark species took seven to ten years after harvesting (Jacobsen, 1978). Table 7.6 shows the growth rate of four pleurocarpous mosses from Wychwood, Oxfordshire, England.

The position on the tree, the tree species and the geographical location affect growth rate (Pitkin, 1975). In Oxfordshire, growth of *Isothecium myosuroides* is more rapid at a height on the trunk of 30–100 cm than at 150–200 cm and the growth of *Hypnum cupressiforme* is more rapid on the upper side than the under side of inclined trunks. These two examples suggest that the reason for the greater luxuriance of bryophytes near tree bases than higher up the trunk and on the upper side of inclined trunks is more rapid growth. *Hypnum cupressiforme* grows faster on *Fraxinus excelsior* than on *Quercus* spp. but the growth rates of both *H. cupressi-*

Table 7.6 Annual growth in mm in 1971 of four corticolous mosses from Wychwood, Oxfordshire. (Modified from Pitkin, 1975).

Hypnum cupressiforme	14.5 mm
Isothecium myosuroides	16.0 mm
Homalothecium sericeum	15.9 mm
Neckera pumila	10.6 mm

forme and *Platygyrium repens* differ from tree to tree of the same species under apparently identical environmental conditions. The effect of different climatic conditions is illustrated in Table 7.7.

There does not appear to be any inherent seasonal growth pattern, at least in Britain, growth being directly related to prevailing weather conditions (Pitkin, 1975). There is a close correlation between monthly rainfall minus evapotranspiration and shoot growth rate.

The only example of productivity of corticolous bryophytes is from the montane deciduous woods in Transvaal (Jacobsen, 1978). Up to a height of 2 m (the height to which Bantu gatherers collect bryophytes for florists) on tree trunks, the annual productivity is estimated to be 4700 g ha^{-1} (equivalent to a growth rate of 7.64%). This is not very informative as no data are provided to relate this to tree number, climate, etc.

Table 7.7 Growth rate of two corticolous pleurocarpous mosses over a six month period (May–October, 1971) at four localities in England. From data kindly provided by Dr P.H. Pitkin.

Locality	Rainfall	Altitude	*Hypnum cupressiforme*	*Isothecium myosuroides*
Wytham	224.2 mm	70 m	1.38 mm	±0
Wychwood	301.9 mm	150 m	1.56 mm	1.36 mm
Roborough	437.0 mm	110–115 m	4.80 mm	4.99 mm
Black Tor	696.0 mm	345–420 m	5.47 mm	6.00 mm

There are two examples of biomass, one from Hungary, the other from the USA. Simon (1974) estimated the biomass of bryophytes (the predominant species being *Hypnum cupressiforme* accompanied by *Platygyrium repens, Bryum flaccidum, Brachythecium salebrosum, Plagiomnium cuspidatum, Anomodon attenuatus, Frullania dilatata, Ulota crispa* and *Neckera besseri*) from oak forest (*Quercetum petraeae-cerris*) from the

Bükk Mountains, Hungary, as 43.375 kg ha^{-1}. Simon considers that in this type of Hungarian forest the epiphyte flora is less rich than in several other types of Hungarian forest. If one takes the annual productivity of Transvaal montane forests (Jacobsen, 1978) of 7.64% as 4700 g ha^{-1}, then bryophyte biomass up to a height of 2 m would be in the region of 61 kg ha^{-1}.

Pike *et al.* (1977) estimated the biomass of epiphytes on a 450-year-old *Pseudotsuga menziesii* near Eugene, Oregon and their results are given in Table 7.8. In a second example (Pike *et al.*, 1972), the weight of bryophytes on a 65 m tall tree was estimated to be 8.9 kg with 3.9 and 4.9 kg from the trunk on main branches respectively. When wet the bryophytes will be three to four times heavier and it is considered that this may be a significant factor affecting branch fall.

Table 7.8 Biomass in kilograms of epiphytic bryophytes on a 450 year old tree of *Pseudotsuga menziesii* (1.46 m dbh, 77 m height). Modified from Pike *et al.*, 1977.

	Mosses	Liverworts
Trunk	2.49 ± 0.13	0.00
Branches	2.06 ± 1.39	0.07 ± 0.07
Branchlets	0.03 ± 0.04	0.03 ± 0.04
Total	4.57 ± 1.58	0.10 ± 0.11

7.3.8 Epiphytes and pollution

Generally speaking, epiphytes are more susceptible to atmospheric pollution than epiliths and the lower part of a tree provides a more sheltered habitat than does the upper part. In Britain, obligate epiphytes such as *Orthotricum, Ulota* and *Zygodon* species and *Lejeunes ulicina* are usually more susceptible to pollution than are facultative epiphytes such as *Dicranoweisia cirrata, Hypnum cupressiforme* and *Homalothecium sericeum.* It has been shown (Syratt and Wanstall, 1968) that *Dicranoweisia cirrata* has a very high content of chlorophyll *a* and is capable of converting sulphite to sulphate with a high degree of efficiency compared with other species. This may account for its wide distribution as an epiphyte in Britain even in areas of high atmospheric pollution. Figs 7.1 and 7.2 show the distributions of *Dicranoweisia* and *Lejeunea ulicina* in Great Britain and Ireland. Whilst the former species is widespread, the latter is absent from most of lowland Britain downwind from areas producing pollutants despite the occurrence of suitable habitats.

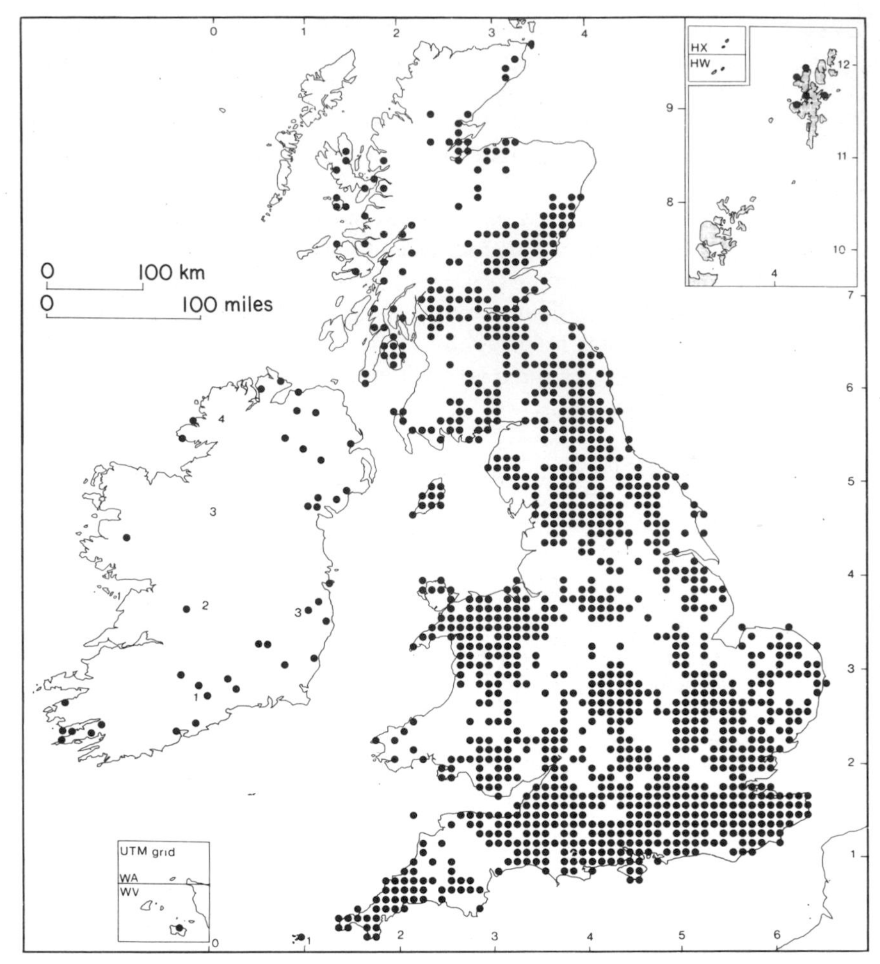

Fig. 7.1 Distribution map of *Dicranoweisia cirrata* in Britain. This species is tolerant of atmospheric pollution. Reproduced from Smith (1978).

7.4 EPILITHS

Studies on epilithic bryophytes date back to the early years of this century, commencing with the work of Grebe (1911) and Schade (1912) although there had been earlier attempts at classifying bryophytes on the basis of the substrate upon which they grew (e.g. Molendo, 1865; Pfeffer, 1871). By the end of the first 30 years of this century a considerable amount of phytosociological work had been done leading to the recognition of various types of saxicolous communities or associations (Gams, 1932). There

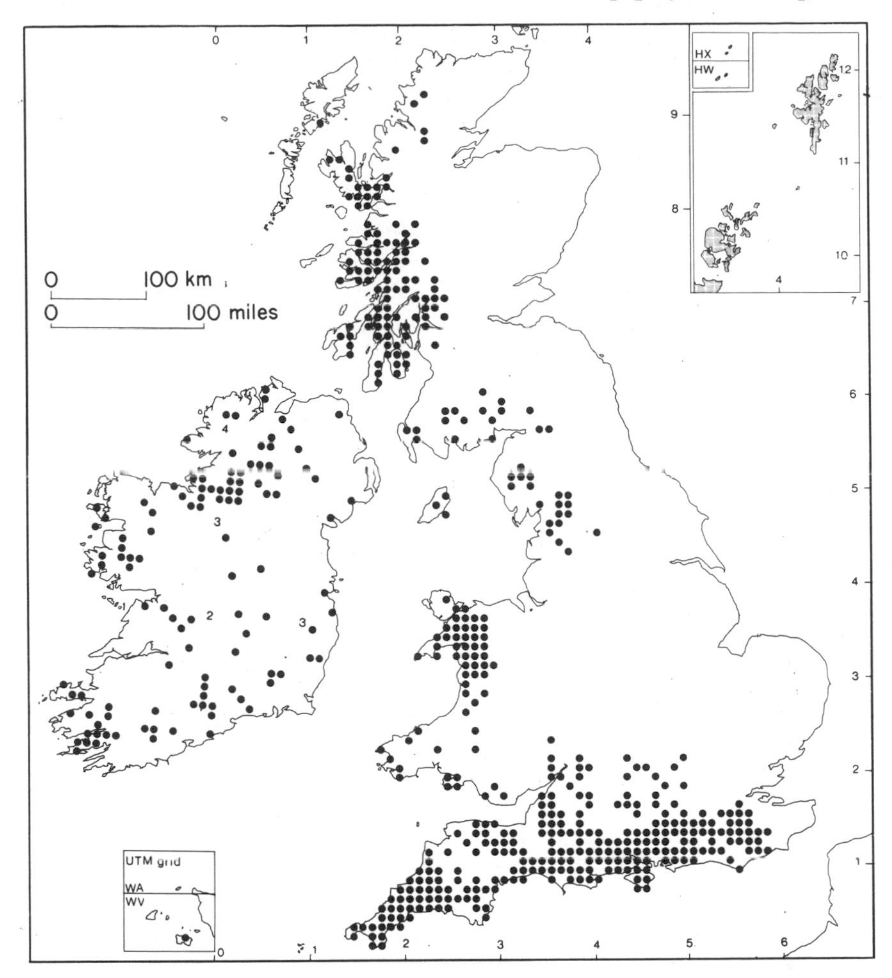

Fig. 7.2 Distribution map of *Lejeunea ulicina* in Britain. This species is intolerant of atmospheric pollution (prevailing winds are from the south-west). Reproduced from Smith (1978).

have been few studies on succession or the dynamics of epilithic communities but in recent years there have been some detailed investigations into the factors affecting the distribution and frequency of rock species, for example by Bates (1975, 1978), Nagano (1969, 1972) and Yarranton (1967a,b,c,d), although the efforts of this last author mainly centre round statistical methodology.

As pointed out by Gams (1932) species growing on rock may be obligate or facultative epiliths. Examples of the former are *Tortella* and

Gymnostomum spp. (including *Hymenostylium*), almost all members of the *Grimmiaceae, Seligeria* and *Blindia* spp., *Orthotrichum anomalum, O. cupulatum, O. rupestre, Ulota hutchinsiae, Gymnomitrion crenulatum, Porella thuja, Marchesinia mackaii* and *Frullania microphylla.* These species are only very rarely found on any other substrates and when this does occur may be of sufficient interest to merit note (e.g. Duncan, 1965). There are numerous examples of facultative epiliths, such as *Dicarnum* spp., *Tortella tortuosa, Bryum capillare, Thamnobryum alopecurum, Heterocladium heteropterum,* species of *Isothecium, Brachythecium, Eurhynchium,* and *Rhynchostegium, Homalothecium sericeum, Hypnum cupressiforme* s.l., *Conocephalum conicum, Metzgeria* and *Plagiochila* spp., *Diplophyllum albicans, Scapania nemorosa, S. gracilis, Porella platyphylla, Frullania* spp. and some Lejeuneaceae.

7.4.1 Epilithic communities

Workers have attempted to describe and classify communities on the basis of species composition and to elucidate the importance of environmental factors in the distribution of epiliths. Rock communities have been defined on the basis of the most common or constant species. Thus, several saxicolous communities are dominated by members of the Grimmiaceae (Gams, 1932). European examples of these are the *Dryptodontetum hartmanii* (= *Grimmietum hartmanii*) of dry shady siliceous rocks in forest regions, the *Grimmietum elatioris* from similar but moister habitats, the *Grimmietum ovalis* (= *G. commutatae*) and *Grimmietum laevigatae* (= *G. campestris*) from dry sunny siliceous rocks, the *Schistidietum apocarpi* from less dry and less acid rocks and the *Grimmietum orbicularis* from dry sunny limestone rocks. Other examples of associations are the *Diplophylletum albicantis,* containing many leafy liverworts and sometimes *Hymenophyllum* in oceanic parts of Europe and the *Conocephaletum* (= *Fagatelletum*).

Whilst some of these communities such as the *Grimmietum laevigatae* (e.g. from south-eastern USA), the *G. orbicularis* (e.g. from limestone outcrops in north-west Wales) and the *Diplophylletum albicantis* (e.g. from western Britain and Ireland) are readily recognizable, these associations have been based on subjective assessments of species frequencies and are only applicable where there are clear-cut discontinuities between associations. On a wider geographical basis, there may be sufficient similarities between certain communities to arouse comment. Thus, apline communities dominated by *Andreaea rupestris* and *Racomitrium* and *Grimmia* spp. in Japan (Horikawa *et al.*, 1961) correspond very closely with those described from montane parts of Germany (Hübschmann, 1955).

In many areas, objective assessments of species associations using random sampling techniques and statistical analyses indicate that there are continua rather than distinct species groupings and that these can be related to environmental gradients, factors that may be overlooked in subjective studies. Also, it may be misleading to compare communities from different geographical areas, at least without details of the micro-environmental data, as species habitat requirements may vary from region to region. Two extreme examples illustrate this. *Polytrichum commune* is apparently a xeric epilith in the USA but a plant of wet peaty habitats in Europe; *Scapania undulata* occurs on dry sandstone in Illinois, USA (Stotler, 1976), but is a plant of submerged or flushed rocks in Europe.

There are a number of examples which indicate the differences in the structure of the communities and of the gradation between one association and another within communities. On different sandstone exposures in two canyons in southern Illinois (Stotler, 1976; West and Stotler, 1977) aspect and moisture are important micro-environmental factors affecting species distribution. The importance value (Greig-Smith, 1964) of *Conocephalum conicum* at the lowest levels is 17.12 and it is the most important element (i.e. the community is a *Conocephaletum* to use Gam's (1932) terminology), whilst at the highest levels on the sandstone it is twenty-seventh in importance with a value of 1.44. Presumably, with increasing height the importance value of *Conocephalum* drops from 17.12 to 1.44 making difficult any attempt to define any part of the community as a *Conocephaletum* unless arbitrary divisions of the habitat are used.

Other areas that provide examples of the continuum between associations or degree of frequency of species are the bryophyte-macrolichen communities on Old Red Sandstone outcrops on Cape Clear Island, southern Ireland (Bates, 1975) and on limestone outcrops in the driftless area of Wisconsin, USA (Foote, 1966). On Cape Clear Island the most maritime communities, dominated by *Schistidium maritimum* and *Ulota phyllantha,* grade into the least maritime communities where the predominant species are *Grimmia trichophylla, Hypnum cupressiforme* var. *resupinatum* and *Isothecium myosuroides.* There are no clear-cut species associations.

On the limestone in Wisconsin, 103 bryophyte and macrolichen species were recorded in random quadrats (Foote, 1966), mostly infrequent, but 12, including *Schistidium apocarpum,* occurred in most habitats so that the cryptogamic vegetation could only be considered a single community. Many of the other species were limited to particular ecological niches and earlier workers in the area had recognized distinct associations. Because the frequency and cover of species varies in different stands, the naming of associations varied with each previous worker. With the continuous nature of the variation when the vegetation is analysed on a larger random scale it is not possible to delimit associations. This is in contrast with the limestone associations delimited in Germany (Klement, 1950, 1959), Poland (Motyka, 1925) and Gotland, Sweden (Du Rietz, 1925). Some of the European associations correspond with some of the stands on the Wisconsin limestone, but it seems likely that with larger scale and less subjective techniques of sampling and data analysis continuous variation or at least some degree of intergradation between associations would be found.

7.4.2 Establishment and succession

There is no definite information on the mode of establishment of saxicolous species but it is evident from field observation and experiment that colonization of virgin rock may be by spores or by vegetative fragments.

The Icelandic island of Surtsey first appeared off the coast of Iceland as the result of a volcanic eruption in November 1963 and the first bryophytes, *Funaria hygrometrica* and *Bryum argenteum,* were discovered on a sandbank in August 1967 (Fridriksson, 1975). The first saxicolous species, *Pogonatum urnigerum* and *Racomitrium canescens* were found in 1969 and by 1972 there were at least 17 species, several of them obligate epiliths, on the island. These plants could only have arrived on Surtsey as wind-blown spores. The spread of some of these, especially members of the *Grimmiaceae,* was extremely rapid. In 1971, *Schistidium strictum* was known from only one site, in 1972 it was the fifth commonest bryophyte; *Schistidium apocarpum* in 1971 was known from only two quadrats, in 1972 it inhabited 31 quadrats. Both these species fruit freely and their rapid dispersal could well be by spores.

On exposed granite rock in North Carolina, USA, *Grimmia laevigata* is very common, forming coalescing patches, but fruits very rarely. In the laboratory (Keever, 1957) shoots and portions of shoots produced protonemata which adhered firmly to both hard and soft granite fragments and gave rise to new leafy shoots. Once developed, drying enhanced the adherence of the protonemata to the rock. From experiments it was evident that most favourable conditions for establishment of fragments are under a regime of wetting every four of five days (by rain or dew) for three weeks during cool weather followed by a period of dry sunny weather. It is likely that in nature fragments of *Grimmia laevigata* may be caught in irregularies of the surface of the granite and, assuming the conditions approximate to those above, give rise to new plants. It is also possible, on the basis of laboratory observations, that *Orthotrichum anomalum* becomes established in a similar fashion (Johnsen, 1969). It is unlikely in the case of the *Grimmia* that establishment is from spores in view of the rarity of fruiting.

It is usually assumed that lichens are the first colonizers of bare rock. On Surtsey (Fridriksson, 1975) the first saxicolous mosses appeared in 1969, the first lichens did not appear until 1970 although lichens are the first colonizers of lava flows on Iceland itself, especially at high altitudes. Even where they are the first colonizers it is unlikely that such lichens play any further part in the subsequent succession. This is partly because of the extremely slow growth of crustose lichens and partly due to their life-form. Over a period of 17 years, there was no detectable growth in crustose lichens on rock at Isle Royale, Lake Superior, USA (Cooper, 1928). In North Carolina, *Grimmia laevigata* at low altitudes and *Andreaea rothii* and *Racomitrium geterostichum* at high altitudes are thought to be the first effective colonizers (Oosting and Anderson, 1937, 1939; Keever *et al.*, 1951). The reason for this is that the early lichens accumulate negligible quantities of soil particles, unlike mosses, *Grimmia laevigata*, for example, accumulating debris to the extent that only the tips of the tufts protrude. The *Grimmia* mats are successively invaded by stages dominated by (1) *Cladonia-Selaginella*, (2) *Polytrichum ohioense*, (3) *Andropogon* and (4) conifers (Oosting and Anderson, 1939). The *Grimmia* forms more or less circular mats (Keever *et al.*, 1951) on which later stages in the succession develop in concentric rings of the different species, successive rings being made of plants requiring more soil and moisture, with most recent invaders nearest the centre.

Although *Grimmia laevigata* is the most important pioneer species on the granite in the south-eastern United States, *Grimmia pilifera* (Keever *et al.*, 1951), *G. olneyi* and *Hedwigia ciliata* (McVaugh, 1943) may also play a role. In Texas, where the climate is hotter and drier, *Grimmia laevigata* appears to be the only moss species that is a primary colonizer (Whitehouse, 1933). At altitudes of 1100–1170 m in North Carolina (Oosting and Anderson, 1937) the first colonizers are *Andreaea rupestris* and *Racomitrium microcarpon* and these are invaded by *Cladonia, Selaginella* and other mosses, then higher plants and, when mats reach sufficient size, woody plants. By contrast with both the low and the high altitude situation in North America, on Hawaii there is no evidence of cushion formation on lava flows (Jackson, 1971), angiosperms growing in rock crevices and mosses (*Racomitrium lanuginosum* and *Campylopus densifolius*) occurring on intervening rock surfaces.

On some rock habitats, especially where the surface is steeply sloping or vertical, bryophytes may form the climax community. On superficial examination these communities appear stable but if studied over a period of time it is evident that cyclic changes take place; further the behaviour of particular species with regard to recolonization varies with environmental conditions.

On shaded sandstone rocks in Sussex, England, (Paton, 1956) a cycle may be started by the falling away of thalloid mats of *Pellia epiphylla*, the loss of moss cushions or the death of *Tetraphis pellucida*. Primary colonizers invading the spaces so left, such as small hepatics forming compact mats, are displaced by mosses and foliose lichens. On wet rocks there appears to be a cycle from bare rock to *Isopterygium elegans* to the filmy fern *Hymebophyllum tunbrigense* or bare rock to *Dicranella heteromalla* and *Tetraphis pellucida* to *Pellia epiphylla*. Loss of the mats so formed will start the cycle all over again. On drier rocks, the situation is more complex and, on dry rock for example, there seem to be at least four possible cycles. These are bare rock to (1) the lichen *Crocynia* to *Tetraphis*, (2) *Isothecium myosuroides*, with or without an intermediate hepatic mat, to *Hymenophyllum*, (3) hepatic mat (*Cephalozia media*, *Lepidozia reptans*, *Lophocolea heterophylla*, *Odontoschisma denudatum*) to large hepatics (e.g. *Lophozia attenuata*, *Bazzania trilobata*) and/or moss cushions (e.g. *Mnium hornum*, *Dicranum scottianum*), (4) *Scapania gracilis*. The duration of the various cycles is unknown.

Table 7.9 shows the rate of colonization of 5×5 cm (approximately) quadrats scraped clear of vegetation within pure stands on rocks with varying degrees of moisture. That most species have their maximum re-colonization rate on a particular rock type indicates the importance of micro-environmental factors, in these cases particularly, water. The rock

Table 7.9 Rate of recolonization of denuded 5 × 5 cm quadrats on (a) wet rocks, (b) moist rocks and (c) dry boulders. Modified from Paton (1956).

Species	Habitat	6 months	18 months	4½ years
Cephalozia nedia	(b)	Almost closed	—	—
	(c)	Present	Less than half-closed	—
Isopterygium elegans	(a)	Almost closed	Closed	—
	(b)	Present	Half-closed	Closed
	(c)	Sometimes present	Present	Half-closed
Lepidozia reptans	(a)	Present	Very scattered	—
	(b)	Present	Almost closed	—
	(c)	Present	Less than half-closed	—
Pellia epiphylla	(a)	Almost closed	Closed	—
	(b)	Present	Half-closed	
Tetraphis pellucida	(a)	Present	Very scattered	Scattered
	(b)	Young plants	Closed	—
	(c)	Present	Closed	—

types upon which a species shows the maximum rate of recolonization is also the rock type upon which the species is most abundant.

Brief observations on succession on boulders in a Killarney oakwood in Ireland (Richards, 1938) are of interest. There are two associules, the open boulder associule and the closed boulder associule. The former is characterized by the presence of *Diplophyllum albicans, Heterocladium heteropterum* var. *flaccidum* and *Sematophyllum micans,* the latter by *Hylocomium brevirostre, Rhytidiadelphus loreus, Thuidium tamariscinum, Dicranum majus, Plagiothecium undulatum, Polytrichum formosum* and *Hymenophyllum tunbrigense.* It appears that the open boulder associule is a stage in the succession to the closed boulder associule and there is an intermediate stage where there is a high frequency of *Plagiothecium undulatum.* The species of the open boulder associule tend to be small and adhere to the rock surface by means of rhizoids, hence are adapted as primary colonizers. The species of the closed associule form coarse wefts which are easily detached, so starting a new cycle of associules.

7.4.3 Life-forms

There is little published information on life-forms of saxicolous bryophytes. In exposed habitats, the predominant life-forms are small cushions and smooth and rough mats; in sheltered, humid or damp habitats, short and tall turfs, small and large cushions and various mat forms are important constituents. It is suggested by Gimingham and Robertson (1950) that the life-form spectra within any one habitat type are essentially similar. They give as an illustration communities on montane siliceous and limestone boulders. Of the 31 and 28 species respectively recorded from the two types of boulder, there are only eight in common, although the life-form compositions are more or less similar. This lends support to the argument on pp. 194 and 198 that it is the life-form that is ecologically important.

Observations on siliceous and limestone boulders in the vicinity of Bangor (see Table 7.10) (Bartholomew, 1980), from habitats with essentially similar macroclimates, show that the life-form spectra are very different. It is likely that the differences in the life-forms are explicable in terms of the water-holding capacity of the rock, the limestone holding more water than the acidic rock. Other data in Table 7.10 indicate that the life-form spectra of rocks are very much influenced by environmental factors, especially light intensity and moisture or humidity. This is in keeping with the findings of Yarranton and Beasleigh (1968) who concluded that the distribution of species in limestone grikes was determined by microclimate as affected by the microtopography of the grikes. Gimingham and Birse (1957) found that the zonation of life-forms on a partly shaded wall could be related to humidity and light intensity.

Table 7.10 Proportions (% dry weight) of various life-forms from boulders in various ypes of habitat in the vicinity of Bangor, North Wales. Modified from Bartholomew (1980).

Life-form	Siliceous boulders in humid wood	Limestone boulders in humid wood	Siliceous boulders in woodland stream	Exposed, dry limestone boulders
Tall turfs	15	20	4	—
Rough mats	73	16	80	—
Smooth mats	4	—	11	46
Thread forms	2	18	—	—
Wefts	6	—	—	—
Short turf	—	4	2	38
Dendroid	—	42	3	—
Small cushions	—	—	—	16
No. of species	16	10	12	12

7.4.4 Influence of environmental factors

It is often extremely difficult to ascertain the precise effect of particular environmental factors either on the occurrence of individual species or on the structure of communities. Early attempts at determining the effects of environmental factors began with the work of Schade (1912) on saxicolous communities in German mountains but it is evident from recent work that there is no straightforward answer to problems of frequency and distribution. With some species it is evident that one ecological feature is of great significance in determining distribution, although whether it is that particular feature of the habitat that is significant or whether, in that habitat, competition is less severe for the species concerned is not clear. In the British Isles, for example, *Schistidium maritimum* is found only on acidic rocks within about 400 m of the shoreline (unless the substrate is a relict of a former shoreline); *Seligeria paucifolia,* except for one location in Northern Ireland is restricted to shaded chalk. Distribution maps of these species (Figs. 7.3 and 7.4) reflect very closely the distribution in Britain and Ireland of chalk and acidic maritime rocks.

There are several examples of the complex interactions between microenvironmental and microclimatic factors from recent studies of epilithic communities. In limestone grikes (vertical fissures) near Kemble, Kettle township, Ontario, Canada, (Yarranton and Beasleigh, 1968) *Anomodon attenuatus, Brachythecium salebrosum, Schistidium apocarpum, Gymnostomum aeruginosum, Leskea nervosa, Thamnobryum alleghaniense,* and *Seligeria doniana* are all highly sensitive to width, depth and aspect of

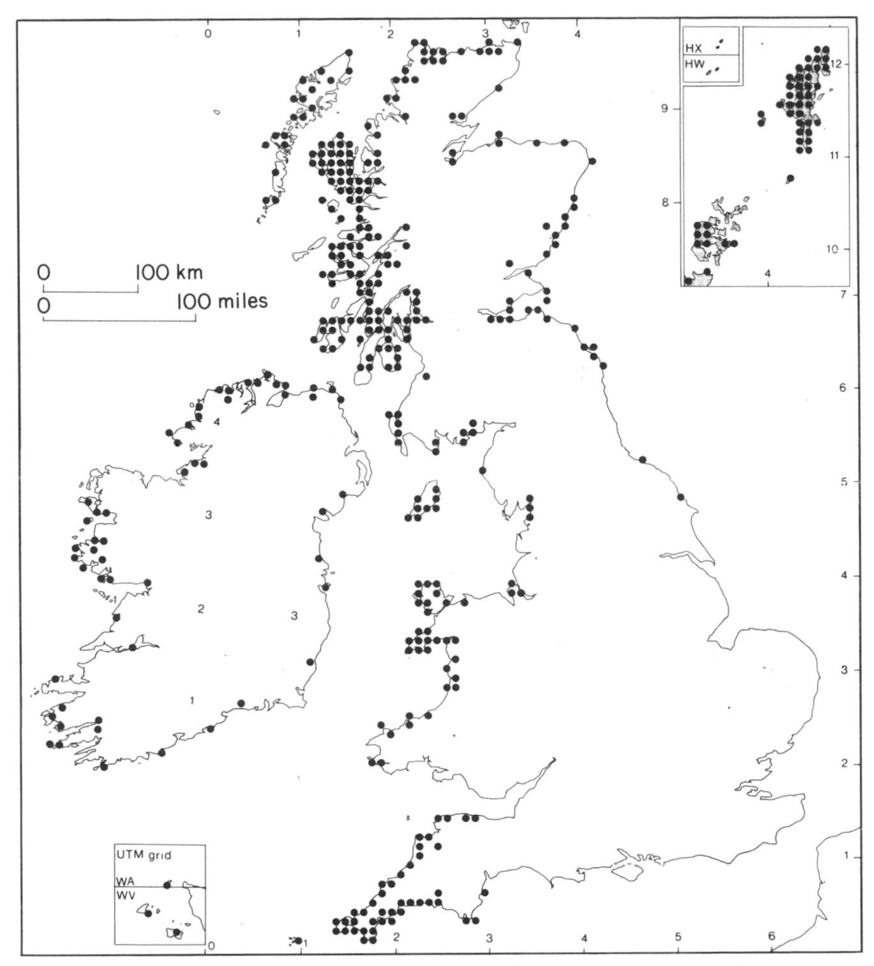

Fig. 7.3 Distribution map of *Schistidium maritimum,* a species of acidic maritime rocks in the British Isles. Reproduced from Smith (1978).

grikes. It seems, however, that microtopography does not affect the distribution of bryophtes directly, but that the chemical nature of the rock and the micrometerorology as affected by the microtopography are likely to be the controlling factors.

In the limestone bryophyte–macrolichen communities of southern Illinois, USA (Stotler, 1976; West and Stotler, 1977), about 50% of the species have restricted distributions, suggesting that micro-environmental conditions are important relative to species occurrence. On the limestone outcrops of the driftless area of Wisconsin (Foote, 1966) there is a moisture continuum; the frequency of each species of the bryophyte–macrolichen

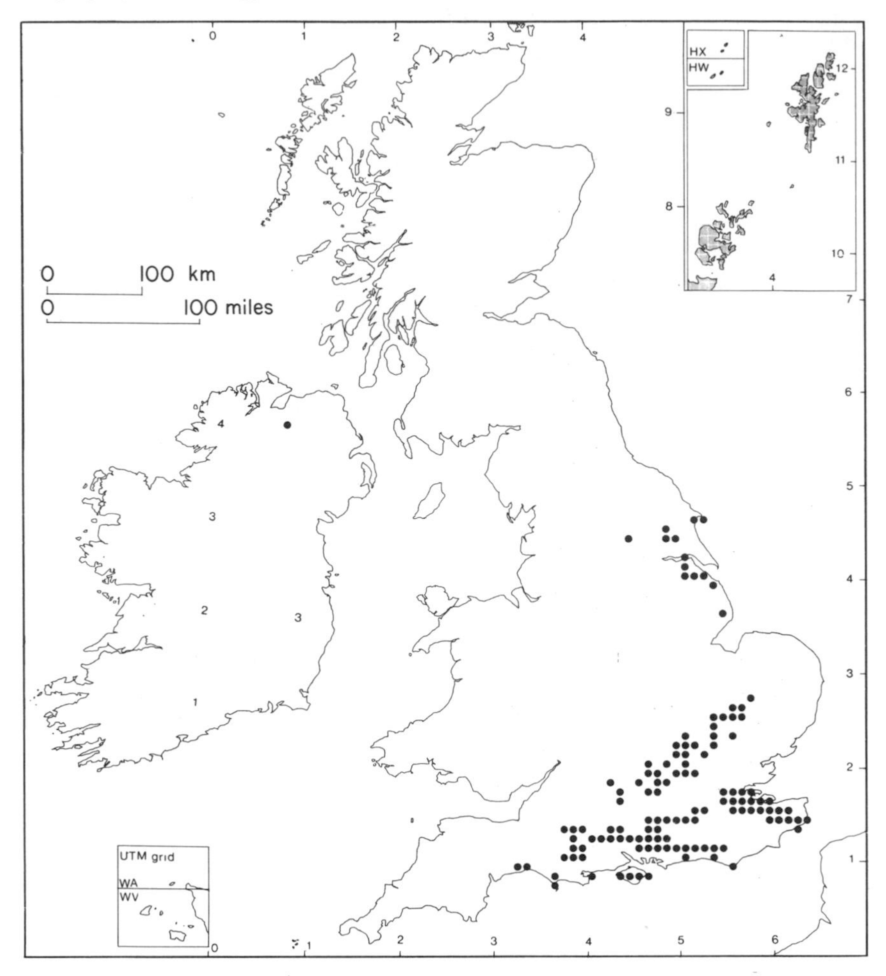

7.4 Distribution map of *Seligeria paucifolia,* a species mainly of shaded chalk in the British Isles. Reproduced from Smith (1978).

community varies with moisture and frequencies of different species peak at different points along the moisture gradient.

On Cape Clear Island, southern Ireland, there is a continuum in the bryophyte–macrolichen community composition related to environmental conditions (Bates, 1975). The continuum is from the most maritime species, *Schistidium maritimum* and *Ulota phyllantha,* to the least maritime, *Grimmia trichophylla, Hypnum cupressiforme* var. *resupinatum* and *Isothecium myosuroides.* Associated with damp rocks and salt spray are *Tortella flavovirens, Trichostomum brachydontium* and *Schistidium maritimum* whilst *Ulota phyllantha* occurs on drier rocks. Inland species fall into two groups: associated with shelter and shaded sites are *Isothecium myosuroides, Frullania germana, F. tamarisci, Plagiochila spinulosa* and *Scapania gracilis*; species of exposed sites are *Frullania dilatata, Grimmia trichophylla, Hedwigia ciliata, Hypnum cupressiforme* var. *resupinatum* and *Polytrichum piliferum*. Neither of these, however, form discreet associations.

It has long been known that some bryophytes are calcicole, some calcifuge and some neutral and that the composition of saxicolous communities depends upon the chemical nature of the rocks but the precise reasons for the calcicole–calcifuge situation is obscure.

In the Chichibu Mountain area of Japan (Nagnao, 1969, 1972) there are a number of different rock types including limestone, calcareous conglomerate, calcareous graywacke, blackslate and chert. At a number of sites some of these rock types abut and since, at any one locality, macro-environmental conditions are uniform, differences in the structure and composition of the bryophyte communities can be attributed to differences in the rock substrate upon which they grow. Such differences exist and are most marked between the limestone and the chert communities. The communities on calcareous conglomerate, calcareous graywacke and blackslate each have their own characteristics but show much greater affinities with the limestone associations than with the chert. Clearly, rock type has a marked effect on species compositon of communities, although not necessarily on the life-form spectra. It is considered (Nagano, 1969) that the differences may be attributed to pH controlled by the quantities of calcium, silicon and sometimes aluminium in the substrate.

A similar example is provided by the different rock types on the Islands of Rhum and Skye off the west coast of Scotland (Bates, 1978). Rock types include Durness limestone, (calcareous sedimentary), Torridonian sandstone (acid sedimentary), basalt (basic igneous) and peridotite and allivalite (ultrabasic igneous). Table 7.11 shows the number of species common to the different rock types. The flora of the limestone is clearly

Table 7.11 Bryophyte and macrolichen taxa common to the four rock types on Skye and Rhum (the number of bryophyte species is in brackets). Only taxa for which there were more than five records in the 121 sample grids are included. From Bates, 1978.

	Durness limestone	Torridonian sandstone	Basalt	Ultrabasic rocks
Durness limestone	23(14)	1(1)	0	1(1)
Torridonian sandstone	1(1)	26(11)	19(10)	13(9)
Basalt	0	19(10)	24(14)	13(9)
Ultrabasic rocks	1(1)	13(9)	13(9)	22(12)

markedly different from those of the other rock types whereas the floras of the non-calcareous rocks have many species in common. Analyses of data suggest that it is the high availability of calcium that plays the predominant role in determining the species differences between the limestone and other rocks. There are some differences between the acid Torridonian sandstone and the basic non-calcareous igneous rocks but it is not considered likely that these are due to chemical differences in the rocks and it is also unlikely that the physical nature of the substrate is important.

In the bryophytes from Rhum and Skye, calcicole species had a calcium content 17 times that of species from non-calcareous rocks (Bates, 1978). In the Chichibu Mountains (Nagano, 1972) although the situation is not so extreme, calcicole species contained far more calcium than calcifuge ones. Thus, the calcicole *Tortella tortuosa* contained 9 times as much calcium as the calcifuge *Bartramiopsis lescurii*. In the Hebridean species, differences in iron and aluminium content are not related to amounts in the substrate rocks and similarly with quantities of nitrogen, magnesium, potassium and phosphorous in the Japanese species. It is evident that the distribution of calcicole and calcifuge species is determined by the high solubility of calcium in calcareous rocks. A possible reason for this is discussed in Chapter 11, pp. 414–17.

7.4.5 Nutrition

Little is known about the nutrition of saxicolous bryophytes. Bates (1978) points out that the high iron and aluminium contents of *Andreaea rothii* growing on Torridonian sandstone, which has low contents of these elements, suggests that they may come from external sources. Also the levels of potassium and phosphorus in bryophytes are remarkably constant and bear no relation to the levels in the substrate rock (Nagano, 1972; Shacklette, 1965). Bates (1978) suggests that the various elements may be derived from rain, dust, salt spray and animal excrement. He also observes that on ultrabasic rocks on Skye and Rhum bryophytes and lichens are often limited to the upper surface of boulders which may serve as bird perches and hence receive supplies from that source.

In the five commonest lichen and moss species that cover 80–90% of the bare granite rock surface of the Piedmont Plateau in Georgia, USA (Snyder and Wullstein, 1973), nitrogen fixation by *Azotobacter* is greater in *Grimmia laevigata* than in the four lichen species. This is probably because the bacteria are associated with soil particles trapped in the moss tufts. It is not possible to quantify the nitrogen fixation by soil bacteria in mosses but the activity of *Azotobacter* in such mosses may well contribute to the nitrogen pool of the communities concerned.

7.5 CONCLUSIONS

Most of the studies of epiphytic and epilithic bryophytes have been carried out in temperate regions, but even these present a picture more of what is not known than what is, especially with regard to epiliths. Most emphasis has been placed on phytosociological studies, relatively little on the life-form in relation to habitat and the significance of micro-environmental factors in determining occurrence and frequency of species. Many of the conclusions concerning the latter point are speculative and require experimental confirmation. Modern quantitative techniques have proved valuable in suggesting the possible significance of various environmental features (e.g. Bates, 1975, 1978; Yarranton, 1967a–d; Yarranton and Beasleigh, 1968) and these are likely to prove of similar value with respect to epiphytes.

REFERENCES

Barkman, J.J. (1958), *Phytosociology and Ecology of Cryptogamic Epiphytes*, Assen.

Bartholomew, F.P. (1980), *Some observations on the relationship between bryophyte community structure and environment in north-west Wales*. MSc. Thesis, Univ. Wales.

Bates, J.W. (1975), *J. Ecol.*, **63**, 143–62.

Bates, J.W. (1978), *J. Ecol.*, **66**, 457–82.

Beals, W. (1965), *Oikos*, **16**, 1–8.

Billings, W.D. and Drew, W.B. (1938), *Am. Midl. Nat.*, **20**, 302–30.

Coleman, B.B., Muehsher, W.C. and Charles, D.R. (1956), *Am. Midl. Nat.*, **56**, 54–87.

Cooper, W.S. (1928), *Ecology*, **9**, 1–5.

Culberson, W.L. (1955), *Ecol. Monogr.*, **25**, 215–31.

Duncan, U.K. (1965), *Trans. Br. Bryol. Soc.*, **4**, 828.

Du Rietz, G.E. (1925), *Svensk Växtsoc. Sallis Handl.*, **2**, 1–65.

Engle, M.J. (1960), *Bryologist*, **63**, 238–41.

Foote, K.G. (1966), *Bryologist*, **69**, 265–72.

Fridriksson, S. (1975), *Surtsey: Evolution of Life on a Volcanic Island*, Butterworth, London.

Gams, H. (1932), In: *Manual of Bryology* (Verdoon, F., ed.), pp. 323–66. The Hague.

Gimingham, C.H. and Birse, E.M. (1957), *J. Ecol.*, **45**, 533–45.

Gimingham, C.H. and Robertson, E.T. (1950), *Trans. Br. Bryol. Soc.*, **1**, 330–44.

Gough, L.P. (1975), *Bryologist*, **78**, 124–45.

Grebe, F.C. (1911), *Festschr. Vereins f. Naturkinde zu Cassel*, **1911**, 195–258, 259–83.

Greig-Smith, P. (1964), *Quantitative Plant Ecology*, Butterworth, London.

Grubb, P.J., Flint, O.P. and Gregory, S.C. (1969), *Trans. Br. Bryol. Soc.*, **5**, 802–17.

Hale, M.E. (1952), *Ecology*, **33**, 398–406.

Hale, M.E. (1955), *Ecology*, **36**, 45–63.
Hoffman, G.R. (1971), *Bryologist*, **74**, 413–27.
Hoffman, G.R. and Boe, A.A. (1977), *Bryologist*, **80**, 32–47.
Hoffman, G.R. and Kazmierski, R.G. (1969), *Bryologist*, **72**, 1–19.
Horikawa, Y., Ando, H. and Kawai, I. (1961), *Ecological Studies of Hakusan Quasi-National Park*, Tokyo.
Horikawa, Y. and Nakanishi, S. (1954), *Bull. Pl. Ecol. (Jap.)*, **3**, 203–10.
Hosokawa, T. and Kubota, H. (1957), *J. Ecol.*, **45**, 579–91.
Hosokawa, T. and Odani, N. (1957), *J. Ecol.*, **45**, 901–5.
Hosokawa, T., Odani, N. and Tagawa, H. (1964), *Bryologist*, **67**, 396–411.
Hosokawa, T. and Omura, M. (1959), *Mem. Fac. Sci. Kyushu Univ. Ser. E (Biol.)*, **3**, 43–50.
Hübschmann, A. von (1955), *Mitt. For.-soz. Arbeitsgem N.F.*, **5**, 50–7.
Iwatsuki, Z. (1960), *J. Hattori bot. Lab.*, **22**, 159–350.
Iwatsuki, Z. and Hattori, S. (1956), *J. Hattori bot. Lab.*, **16**, 106–116.
Iwatsuki, Z. and Hattori, S. (1966), *J. Hattori bot. Lab.*, **29**, 223–37.
Iwatsuki, Z. and Hattori, S. (1970), *Mem. Natn. Sci. Mus., Tokyo*, **3**, 365–74.
Jackson, T.A. (1971), *Pacif. Sci.*, **25**, 22–32.
Jacobsen, N.H.G. (1978), *Jl. S. Afr. Bot.*, **44**, 297–312.
Johnsen, A.S. (1969), *Bryologist*, **72**, 397–403.
Johnsen, I. and Søchting, W. (1976), *Bryologist*, **79**, 86–92.
Jones, E.W. (1959), *Trans. Br. Bryol. Soc.*, **4**, 611–12.
Keever, C. (1957), *Ecology*, **38**, 422–9.
Keever, C., Oosting, H.J. and Anderson, L.E. (1951), *Bull. Torrey Bot. Club*, **78**, 401–21.
Klement, O. (1950), *Ber. Bayer. Bot. Gesell.*, **28**, 1–26.
Klement, O. (1959), *Decheniana*, **7**, 5–56.
LeBlanc, F. (1963), *Can. J. Bot.*, **41**, 591–638.
Martin, N.M. (1938), *J. Ecol.*, **26**, 82–95.
McVaugh, R. (1943), *Ecol. Monogr.*, **13**, 121–66.
Miyata, I. and Hosokawa, T. (1961), *Ecology*, **42**, 766–75.
Molendo, L. (1865), *Jahresber Naturhistor. Verein Augsburg.*
Motyka, J. (1925), *Bull. Acad. Bot. Sci. Lett.*, **B 1924**, 835–50.
Nagano, I. (1969), *J. Hattori bot. Lab.*, **32**, 155–203.
Nagano, I. (1972), *J. Hattori bot. Lab.*, **35**, 391–8.
Nakanishi, S. (1966), *Jap. J. Bot.*, **19**, 231–54.
Olsen, C. (1917), *Bot. Tidsskr.*, **34**, 313–42.
Omura, M. and Hosokawa, T. (1959), *Mem. Fac. Sci. Kyushu Univ. Ser. E (Biol.)*, **3**, 51–63.
Omura, M., Nishikara, Y. and Hosokawa, T. (1955), *Rev. bryol. lichénol.*, **24**, 50–68.
Oosting, H.J. and Anderson, L.E. (1939), *Bot. Gaz.*, **100**, 750–68.
Paton, J.A. (1956), *Trans. Br. Bryol. Soc.*, **3**, 103–14.
Patterson, P.M. (1953), *Bryologist*, **56**, 157–9.
Pfeffer, W. (1871), *Neue Denkschr. der all gem. Schweizer Ges.*, **24**, 1–42.
Phillips, E.A. (1951), *Ecol. Monogr.*, **21**, 301–16.
Pike, L.H., Denison, W.C., Tracy, D.M., Sherwood, M.A. and Rhoades, F.M. (1975), *Bryologist*, **78**, 389–402.

Pike, L.H., Rydell, R.A. and Denison, W.C. (1977), *Can. J. For. Res.*, **7**, 680–99.
Pike, L.H., Tracy, D.M., Sherwood, M.A. and Nielson, D. (1972), In: *Research on Coniferous Forest Ecosystems* (Franklin, J.F., Dempster, L.J. and Waring, R.H., eds), pp. 177–187. Pacific Northwest Forest and Range Experiment Station, Portland, Oregon.
Pitkin, P.H. (1973a), *J. Bryol.*, **7**, 522–3.
Pitkin, P.H. (1973b), *Aspects of the Ecology and Distribution of some Widespread Corticolous Bryophytes*, D.Phil. thesis, Univ. Oxford.
Pitkin, P.H. (1975), *J. Bryol.*, **8**, 337–56.
Rasmussen, L. (1975), *Lindbergia*, **3**, 15–38.
Rasmussen, L. and Hertig, J. (1977), *Rev. bryol. lichénol.*, **43**, 207–17.
Rasmussen, L. and Johnsen, I. (1976), *Oikos*, **27**, 483–7.
Richards, P.W. (1938), *Ann. Bryol.*, **11**, 108–30.
Schade, F.A. (1912), *Bot. Jahrb.*, **48**, 19.
Shacklette, H.T. (1965), *U.S. Geog. Surv. Bull.*, **1189-D**, 1–21.
Simon, T. (1974), *Acta Bot. Acad. Sci. Hung.*, **20**, 241–8.
Sjögren, E. (1961), *Acta phytogeog. suec.*, **48**, 1–149.
Slack, N.G. (1975), *XII Int. Bot. Congr. Abstr.*, Leningrad, **I**, 87.
Slack, N.G. (1976), *J. Hattori bot. Lab.*, **41**, 107–32.
Slack, N.G. (1977), *Bull. N.Y. St. Mus. Sci. Surv.*, **428**, 1–70.
Smith, A.J.E., ed. (1978), *Provisional Atlas of the Bryophytes of the British Isles*, Natural Environmental Research Council, London.
Snyder, J.M. and Wullstein, L.H. (1973), *Bryologist*, **76**, 196–9.
Staxang, S. (1969), *Oikos*, **20**, 224–30.
Stotler, R.E. (1976), *Bryologist*, **79**, 1–15.
Syratt, W.J. and Wanstall, P.J. (1968), In: *Air Pollution: Proc. First European Congress on the Influence of Air Pollution on Plants and Animals*, pp. 79–85.
Tagawa, H. (1959), *Jap. J. Ecol.*, **9**, 178–84.
Tobiessen, P.L., Mott, K.A. and Slack, N.G. (1978), *Bryophytorum Biblioth.*, **13**, 253–77.
West, V. and Stotler, R.E. (1977), *Bryologist*, **80**, 612–18.
Whitehouse, E. (1933), *Ecology*, **14**, 391–405.
Yarranton, G.A. (1967a), *Lichenologist*, **3**, 392–408.
Yarranton, G.A. (1967b), *Can. J. Bot.*, **45**, 93–115.
Yarranton, G.A. (1967c), *Can. J. Bot.*, **45**, 229–47.
Yarranton, G.A. (1967d), *Can. J. Bot.*, **45**, 249–58.
Yarranton, G.A. and Beasleigh, W.J. (1968), *Can. J. Bot.*, **46**, 1591–9.

Chapter 8

The Ecology of *Sphagnum*

R.S. CLYMO AND P.M. HAYWARD

I held it truth with him who sings . . .,
That men may rise on stepping stones
Of their dead selves to higher things.

Tennyson In Memoriam A.H.H.

8.1 INTRODUCTION

It may be true of men; it is certainly true of *Sphagnum*. The plants grow at the apex, as do most other mosses. The apex produces initials which develop into branches of determinate growth, though in a few cases (*S. cuspidatum* var. *plumosum* for example), the branches may themselves branch. Whilst the branches are increasing in length the internodes of the main stem do not elongate. This results in the branches and attendant leaves forming a compact hemispherical head – the coma or capitulum. Later, after branch growth has finished, the internodes do elongate; it then becomes obvious that the branches are borne in groups (fascicles) on the central stem. Each branch bears 30 to 150 spirally arranged imbricate leaves. These leaves are one cell thick, with the unique and well known differentiation into porose hyaline cells with thickened hoops, and enclosed or chlorophyllose cells. The branches may be spreading or pendent and clasping the stem (Fig. 8.1). The individual leaves live for a year or two, but by that time the growth of branches above has put them into dense shade and they die. The only parts left alive at this distance below the apex seem to be the axillary buds. These usually remain inactive, and eventually die, but if the apex is destroyed, either artificially (for example, by cutting if off) or accidentally (for example, by drought) then one or more lateral buds may begin to grow again from as far as 10 cm below the apex. These buds can survive for at least 18 months in a refrigerator (dark, about 2°C). The shoots formed from such buds are very thin at first but reach full width

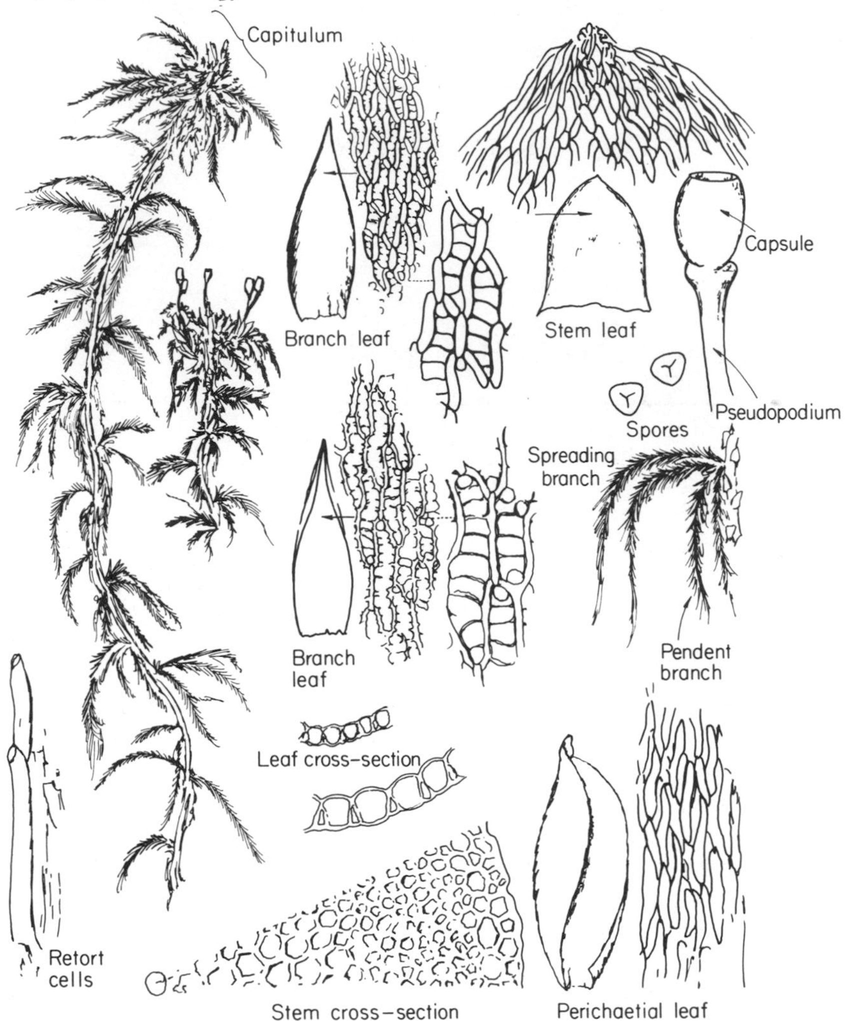

Fig. 8.1 The structure of a *Sphagnum* plant, *S. recurvum (S. fallax)*. Imbricate leaves on branches which are either spreading or pendent are shown. The hyaline cells of the leaf have thickened hoops, and the walls are perforated by pores. Branches are invested with a sheath containing retort cells, which have a single large apical pore (lower left). From Nyholm (1969).

within a year or two. This suppression of lateral buds is similar to that seen in vascular plants in which it is controlled by hormones from the apex itself, and suggests that there may be more vertical transport in *Sphagnum* plants than is commonly supposed. The apex of a *Sphagnum* plant may sometimes be replaced by two smaller ones. Whether this results from direct division of the original apex or from development of an axillary bud prematurely released from subjugation is not clear.

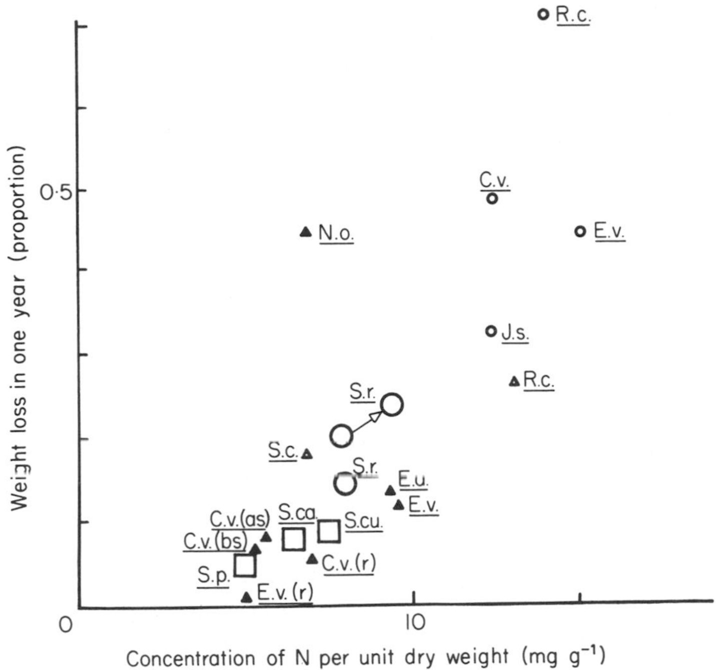

Fig. 8.2 Decay rate just below the surface of blanket bog at Moor House of *Sphagnum* and of other plant materials in relation to concentration of nitrogen. *Sphagnum: S.ca, S. capillifolium; S.cu, S. cuspidatum; S.p., S. papillosum; S.r., S. recurvum.* Other plants: *C.v., Calluna vulgaris* shoots; *C.v.* (as), *C. vulgaris* above-ground stems; *C.v.* (b.s.), *C. vulgaris* below-ground stems, *C.v.* (r), *C. vulgaris* roots; *E.a., Eriophorum angustifolium* leaves; *E.v.*(r), *E. vaginatum* roots; *J.s., Juncus squarrosus; N.o., Narthecium ossifragum* leaves; *R.c., Rubus chamaemorus* leaves; *S.c., Scirpus cespitosus* shoots. Symbols: □, Clymo (1965); △, Heal *et al.* (1978); ○, Coulson and Butterfield (1978). Larger size symbols are *Sphagnum*. The arrow shows the effect of experimental enrichment with nitrogen.

Much the same growth pattern is shown by many other species of moss. What makes *Sphagnum* peculiar is that the rate of decay of the dead material of the commonest species is unusually slow so that the dead plants accumulate as peat (Fig. 8.2). There are several reasons for this. One is the unusually low concentration of nitrogen in the plants – usually less that 1% of dry mass. Not only is there a positive correlation between decay rate and nitrogen concentration, but increasing the nitrogen concentration of the plants by fertilizing the live plants increases the rate of decay when the plants die (Coulson and Butterfield, 1978). This is a specific effect of nitrogen enrichment, and does not occur when the plants are enriched in phosphorus. A second reason may be the acid conditions which are produced by the *Sphagnum* itself (Skene, 1915; Clymo, 1963). The third reason for slow decay is associated with the generally wet environment which most species of *Sphagnum* require. Not far below the apices – perhaps 2 to 20 cm down – the peat is water-saturated. There is a continuum of water potential but it is convenient to recognize three points in the continuum. First, water may be present as free liquid: a hole dug in the surface may fill to give a free water surface. Secondly, the water may be held in capillary spaces, roughly 1 μm to 1 mm across. Thirdly, the water may be present in smaller spaces, as it is in a jelly. A hole in the jelly will not spontaneously fill with water but the jelly is water-saturated nevertheless.

Micro-organisms living on the plants just at and below the level of water-saturation use the molecular oxygen in solution. They are mainly aerobic or facultatively anaerobic fungi and bacteria. Oxygen does diffuse down from the air, but the rate is very slow, being only 1/10000th of the rate of diffusion in air. There is probably very little convective or other mass (non-diffusive) movement in the water. Anaerobic conditions prevail, therefore, and the rate of decay in such conditions is much slower than it is in aerobic conditions, though why this is so is not clear. The water and the associated anaerobic, low decay-rate conditions, are not static. They fluctuate over the course of minutes to days as rainfall, run-off, evaporation and temperature determine. But in the long term the level of water-saturation rises steadily too, because at some point below the living surface the stems of the dead *Sphagnum* have decayed to such an extent that they lose their mechanical strength and the plant its integrity, so that the whole open plant structure collapses. The bulk density increases from about 0.01 to about 0.1 g cm^{-3}. The leaves retain their structure but pack down more tightly; the lateral hydraulic conductance falls, and the water table rises when it rains. Thus, as the plants grow at the apex so the water-saturated level follows them upwards. This concept of an aerobic layer above and an anaerobic layer below is too simple though (Fig. 8.3). The upper layer (acrotelm of Ingram, 1978) contains some anaerobic spots. These become more frequent, and probably larger and more anaerobic, nearer to the water-saturated layer (catotelm). Even here the roots of vascular plants, such as *Eriophorum* spp., create local aerobic channels (Armstrong, 1964). Nevertheless, the general effect is that the *Sphagnum* branch dies and decays predominantly aerobically until its structure collapses and thereby raises the water-saturated layer to cover it. From that point onward decay is predominantly anaerobic. It is clear that the rate at which dead material enters the relative safety of the anaerobic zone depends partly on the rate at which new plant material is produced and partly on the proportion of that material which is lost whilst running the gauntlet of the dangerous aerobic zone. Calculations show how the initial advantage of a relatively slow rate of aerobic decomposition of *Sphagnum* can lead to its considerable over-representation in peat (Clymo, 1982). There are differences between species of *Sphagnum* too: the

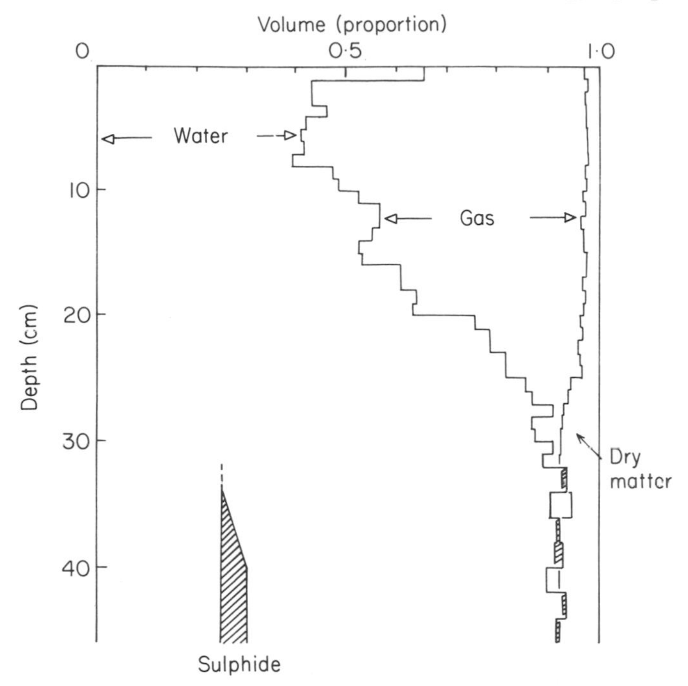

Fig. 8.3 The proportion of dry matter, water and gas at different depths in a carpet of *S. fuscum* at Nordmjele, Andøya, Norway. A 20-cm diameter core was collected in a tube which was then sealed to prevent loss of water. The core was stood vertical for 24 h and then ejected vertically and sliced in 1 cm steps. Clymo (unpublished).

proportional loss, in one year, of newly dead *S. cuspidatum, S. papillosum,* and *S. capillifolium* placed in the top layer of a valley bog in Southern England was 0.16, 0.06 and 0.11 respectively (Clymo, 1965).

The extensive carpet-like growth and the slow rate of decay are two of the reasons why *Sphagnum* is such an important bryophyte genus: there is an enormous amount of it, and more dead than alive. Peatlands cover about 150×10^6 ha, which is between 1 and 2% of the earth's land surface (Tibbetts, 1968). Many of them – perhaps most – have *Sphagnum* as a major component. The depth of peat can exceed 15 m, but 1–5 m is more usual. If one assumes a mean depth of 2 m and bulk density of 0.1 g cm^{-3}, then the mass of dead plant material is about 300×10^9 t. This may be compared with an estimate for terrestrial productivity of 72×10^9 t $year^{-1}$ (Woodwell *et al.*, 1978). If only half this peat is *Sphagnum* then there is more carbon locked up in *Sphagnum,* alive and dead, than is fixed by all terrestrial vegetation in one year. It is interesting to speculate that there may be more carbon in *Sphagnum* than in any other genus of plants, vascular or non-vascular. The comparison is rather artificial because the *Sphagnum* genus is so taxonomically isolated that the comparison ought perhaps to be at the level of the family, or higher. But it seems clear that no other bryophyte can approach the general success of *Sphagnum*, though some (*Polytrichum commune, Racomitrium lanuginosum*) may have higher productivity over small areas.

8.2 GENERAL ECOLOGY

The distribution and rate of growth of *Sphagnum* plants and the performance of one species relative to another are determined primarily by the supply of water and by the concentration of solutes, particularly of Ca^{2+} and H^+. The general requirements are an assured water supply with a relatively low concentration of Ca^{2+}.

In many places these conditions are provided by relatively high-rainfall, equitably distributed throughout the year, with no long periods when evaporation exceeds precipitation. Labrador and Ireland (both oceanic coastal islands) are examples. In such climates *Sphagnum*-dominated vegetation may blanket the whole countryside on slopes up to 20°. In regions with summer drought, *Sphagnum* may be more localized in basins where the accumulated peat is sufficient to insulate the plants from the ground-water and to provide a reservoir (which shrinks during summer) allowing *Sphagnum* to survive. In a few places, a carpet of *Sphagnum* or sedges (or both) grows out over a deep pool. Here again the surface may in time become insulated from the water below and float up and down on it (Green and Pearson, 1968). The same continuity of water supply may be found in some valleys but these are suitable for extensive *Sphagnum* growth only if the water has flowed through solute-poor rocks with soils such as the Bagshot sand of southern England or the erosion-resistant rocks of the Canadian Shield.

These four cases are recognized types of mire: blanket bog, raised-bog, schwingmoor and valley-bog respectively. There are many other types of mire, and seemingly no end to the schemes classifying them (Moore and Bellamy, 1974; Kivinen *et al.*, 1979) but they need not be considered further here.

The taxonomy of the genus *Sphagnum* is still in flux. The main sections are generally agreed and in most cases the taxa in these sections share a distinctive anatomy and ecology. Thus, species of the sections *Cuspidata* and *Subsecunda* are often found with the capitulum at or slightly above or below the free water surface. *Sphagnum cuspidatum* and *S. subsecundum* themselves may be found in pools and, when growing in slowly flowing water in ditches, may grow in length by 50 cm or more during the summer months of one year (Overbeck and Happach, 1956). Species of the section *Sphagnum*, including *S. papillosum* and *S. magellanicum*, are robust and often form carpets or low hummocks, whilst several species of the section

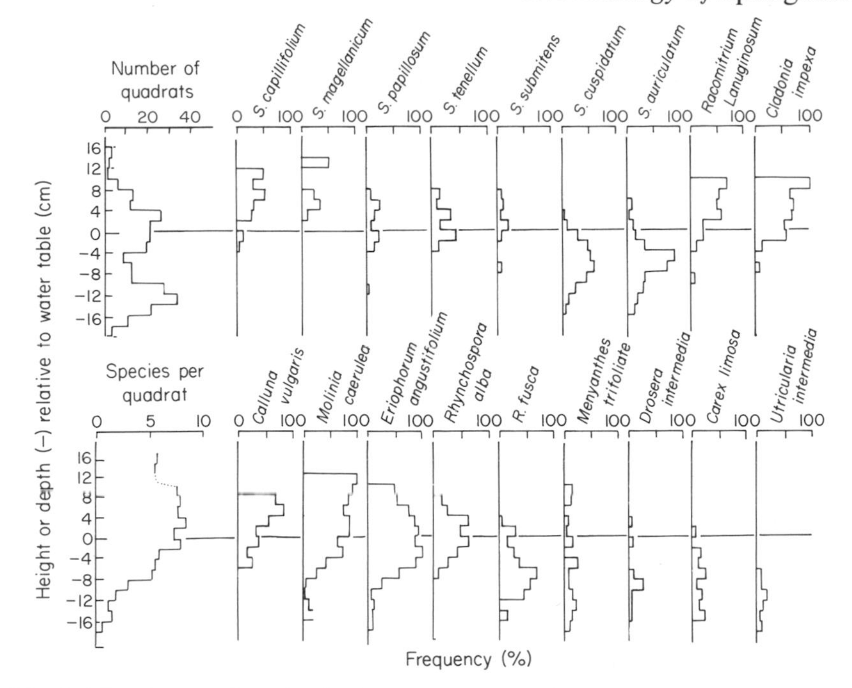

Fig. 8.4 Frequency of species (in 25×25 cm quadrats) in relation to the water-table on the patterned mire at Kentra, north-west Scotland. The observations were made during a short time in September 1979 during which the water-table may have been unusually high. The number of quadrats, and the total number of species per quadrat, are shown too. Unpublished results of the following members of the Mires Research Group: A.D.Q. Agnew, S. Agnew, O. Bragg, A. Coupar, H.A.P. Ingram, M.C.F. Proctor. Similar observations were made by Ratcliffe and Walker (1958).

Acutifolia, including *S. capillifolium* and *S. fuscum,* usually grow on hummocks some way above the water table. Some of these tendencies are shown in Fig. 8.4.

Within the sections, however, the taxonomy is in dispute. *S. acutifolium, S. rubellum, S. capillaceum* and *S. capillifolium* may be nearly synonymous: *S. recurvum* has been variously renamed and subdivided in the *S. flexuosum* aggregate. More important is the plasticity of form: *S. recurvum* grown submerged may be difficult to distinguish from *S. cuspidatum* grown in unusually dry conditions, though they are clearly distinct when grown together in the same conditions. Similarly, the 'species' of *S. subsecundum* of some authors may be interconverted by changing the growing conditions. Because of these and other problems the nomenclature of Hill (1978) is used in this account.

There are perhaps 300 species of *Sphagnum* world-wide. The greatest bulk of *Sphagnum* (probably of fewer than 30 species) grows in the North

Temperate and Boreal zones. At other latitudes *Sphagnum* does grow, but usually at high altitudes, for example in the Snowy Mountains of Australia, in the Chilean Andes, and near the top of the high equatorial African mountains. An example of a single species with a disjunct distribution of this kind is *S. junghuhnianum* ssp. *junghuhnianum* growing on mountains in Malaysia, China, Taiwan and Japan (Johnson, 1960).

Small amounts of *Sphagnum* of many species may sometimes be found in the most unlikely habitats. For example, Lange (1973) describes 15 species, with *S. teres* and *S. subnitens* the commonest, around hot springs in many parts of Iceland. The pH of water in the springs ranged from 2 to 9 (though that amongst the plants was not measured) and the temperature of the water amongst the plants was in some cases more than 40°C, the plants growing adjacent to water at 90°C or more.

8.3 INTER-RELATION OF *SPHAGNUM* AND SOLUTES

8.3.1 Effects of solutes on plant growth

Most species of *Sphagnum* cannot survive in water which has flowed through calcareous rocks or soil. There are a few exceptions: *S. squarrosum*, *S. teres* and *S. fimbriatum* are usually found in places where the water supply is moderately calcareous – perhaps reaching a concentration of

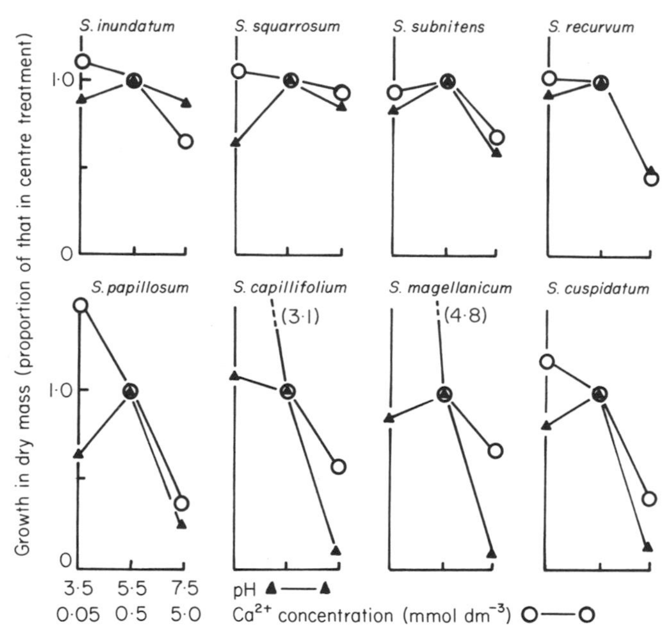

Fig. 8.5 The Relative growth in mass of eight species of *Sphagnum* in relation to pH (△–△) and concentration of Ca^{2+} (O–O) in the water. Results scaled to the value 1.0 for the central treatment: pH 5.5 and ½Ca^{2+} concentration 0.5 mmol l^{-1} respectively. Redrawn from Clymo (1973).

½Ca $^{2+}$ of 1 mmol dm^{-3}. In experiments, *S. squarrosum* grew almost as much when supplied with solutions containing a concentration of ½Ca^{2+} of 0.5 and 5.0 mmol dm^{-3} as they did in 0.05 mmol dm^{-3}, whilst seven other species grew much less well (Fig. 8.5). Water with a high calcium concentration usually has a high pH too, and this may be inimical to the growth of most species of *Sphagnum* (Olsen, 1923). It is not easy, however, to achieve and maintain a high pH in *dilute* solutions around *Sphagnum* plants. When this is done it becomes apparent that high concentrations of Ca^{2+} alone or high pH alone are not sufficient to reduce the growth rate of most species significantly (Fig. 8.6). It is the combination which is lethal. Amongst the major peat-forming species, hummock species such as *S. capillifolium* seem to be particularly sensitive to combined high Ca concentration and high pH whilst immersed species are less so.

The growth rate of *Sphagnum* plants is affected not only by concentration but also by the rate of supply of solutes – what might be called the rate of flushing (Fig. 8.7). Increasing the rate of flow of solution to eight species of *Sphagnum* from zero to about 20 cm day^{-1} caused an increase in

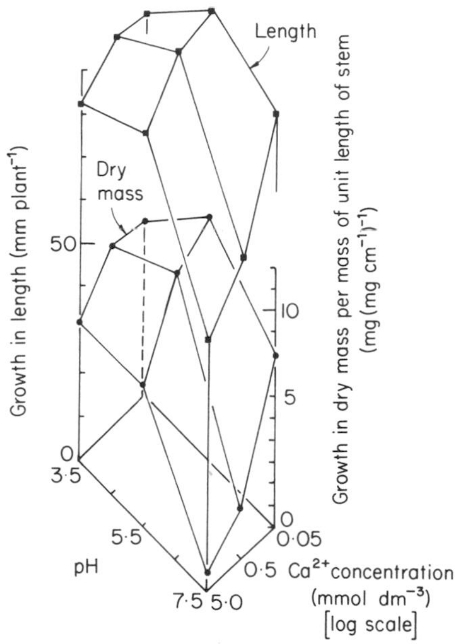

Fig. 8.6 Mean growth in mass (●) and in length (□) of eight Species of *Sphagnum* in an experiment with factorial combinations of pH and of Ca^{2+} concentration in the water. The growth in mass is expressed per mass of unit length of stem, so that the effect of specific plant size is minimized. Redrawn from Clymo (1973).

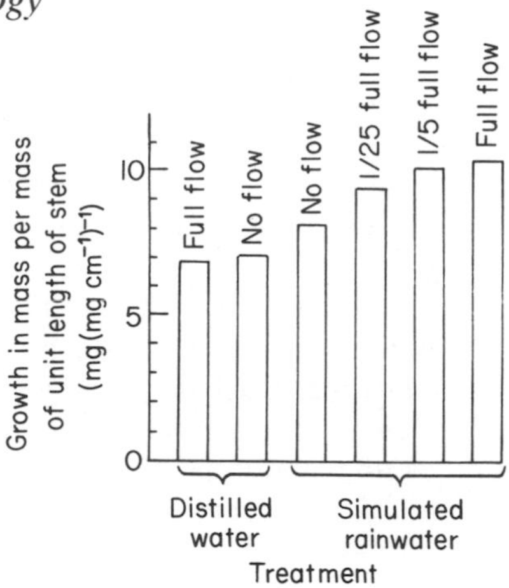

Fig. 8.7 Mean growth in mass of eight species of *Sphagnum* in relation to rate of flow of a solution simulating rainwater, and of distilled water. The full rate of flow was at a velocity of about 20 cm day^{-1}. Redrawn from Clymo (1973).

growth rate of about 30%. This experiment was made outdoors, but with artificial solutions. Much the same result was found by Sonesson *et al.* (1980) who grew *S. riparium* in open-topped cylinders placed in a wet, flushed area in which the species usually grows. Some cylinders were closed at the base; but the base of most were covered by a nylon net. The cylinder walls were perforated, so that water could move in and out. The closed cylinders contained eight times the volume of water that the perforated ones did. The concentration in the water of PO_4, Cl, SO_4, Fe, Zn, Mn, Mg, Ca, K, Na and NO_3 and NH_4 was measured at weekly intervals. In the perforated cylinders no concentration (except Fe) changed by more than two-fold during the 12-week growing season. In the closed cylinder, however, the concentration of PO_4 dropped from 0.6 to <0.04 μmol dm^{-3} within a week. The concentration of chloride fell from 47 to 2 μmol dm^{-3} over four weeks; that of Mg was halved within a week but then remained steady; Fe fell from 15 to 2 μmol dm^{-3} over a month (but rose steadily in the perforated containers to 126 μmol dm^{-3} after nine weeks). The concentration of total N in the closed container fell to about a fifth that in the perforated ones after six weeks. These differences were paralleled by the differences in the amount of growth: plants in the perforated containers made about twice the growth that those in the closed containers did.

Although some species – *S. squarrosum* for example – are almost always found in habitats with a relatively high concentration of solutes, or a high rate of flushing, the behaviour of other species is less consistent. At Cranesmoor (southern England) *S. papillosum* occupies the more oligotrophic lawns and *S. magellanicum* the flushed areas (Newbould, 1960), but at the Åkhult mire (southern Sweden) the reverse tends to be true (Malmer, 1962a).

The concentration of nitrate, ammonium, and of phosphate in water squeezed from around *Sphagnum* is very low; Gorham (1956), for example, records that at Moor House, northern England, the concentration of $\frac{1}{2}Ca^{2+}$, NO_3–N, and PO_4–P amongst *S. cuspidatum* and *S. auriculatum* was about 50, <3, and <0.1 μmol dm^{-3}.

The growth rate of adult plants of *Sphagnum* may be limited by the supply of phosphate: when ground rock-phosphate was applied to a bog surface then the growth rate of *S. compactum*, *S. auriculatum*, *S. cuspidatum*, and *S. tenellum* appeared to increase, and the plants became apple green (McVean, 1959). Grouse droppings and sheep dung usually kill the adjacent plants, but are later covered by bright green, robust, and apparently fast-growing plants which invade from the side. The same effect on grass-growth is seen around cow-pats. Certainly *S. papillosum* is able to

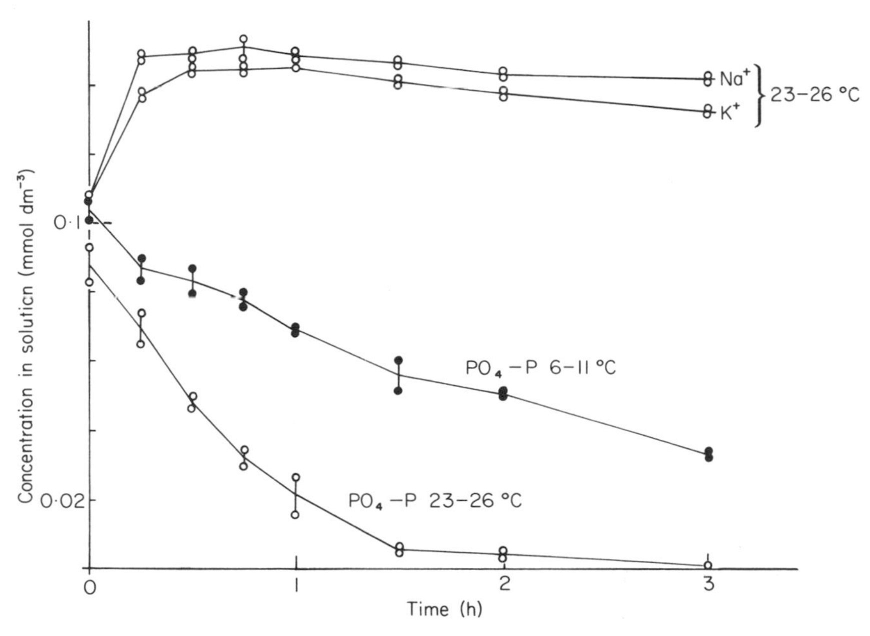

Fig. 8.8 Course of accumulation of phosphate from a carpet of *S. papillosum* at different temperatures. The carpet was 5 cm deep and 25 cm diameter. A total volume of 2.5 l of solution was sprayed on the plants at 78 ml min^{-1}, and then recycled. The initial increase in concentration of Na^+ and K^+ results from cation exchange, and change in concentration of cations in the solution by subsequent accumulation in the cells would be small because of the buffering action of the exchange sites. Clymo (unpublished).

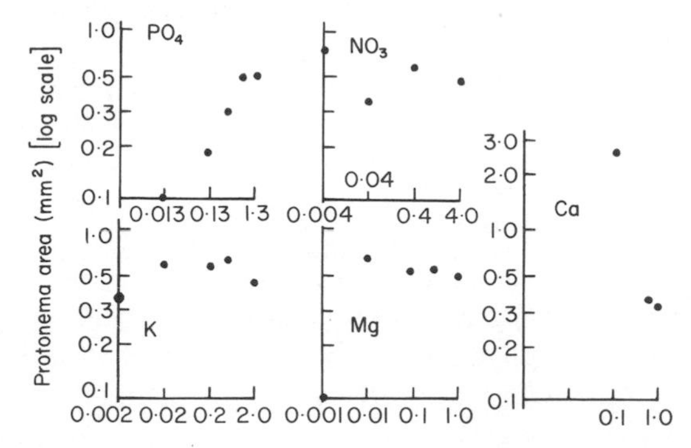

Fig. 8.9 Growth in area of the protonema of *S. papillosum* in relation to concentration of five ions. In these conditions phosphate is a nutrient, calcium a toxin, and the plants are indifferent to nitrate, potassium, and magnesium. Redrawn from Boatman and Lark (1971).

take up phosphate fairly rapidly from solution (Fig. 8.8) just as most other plants can.

It is surprising to find that the response of the protonema of *Sphagnum* to solutes is different from that of the adult haploid plants. The protonema seems to thrive on solutions of relatively high concentration of solutes; it grows better on agar made with full strength Moore's medium which contains concentrations of Ca^{2+}, NO_3–N, and PO_4–P of 0.7, 12.5 and 1.5 mmol dm^{-3}, than it does on any lower concentration (Clymo, unpublished). Of these solutes it seems that phosphate is the most important to *S. cuspidatum* and to *S. papillosum* (Fig. 8.9): a concentration of PO_4–P of about 1 mmol dm^{-3} seems to be necessary if the morphological transitions from filamentous to plate growth, and from plate growth to shoots are to be made. The concentration of PO_4 in bog water is about 10000 times smaller. How new plants establish from spores in the field is a mystery. Perhaps they do so only in very exceptional circumstances, though the rapidity with which *Sphagnum* re-invades suitably wet habitats argues against such an explanation.

8.3.2 Effects of plants on solute concentrations

That most *Sphagnum* plants grow in unusually acid conditions and that the *Sphagnum* plants contribute to the production of these very conditions has been known for a long time (Paul, 1908; Skene, 1915). The process of acidification was recognized as one of cation exchange by Williams with Thompson (1936) and Anschutz and Gessner (1954), and this approach has been extended to allow quantitative predictions (Clymo, 1963, 1967).

The cation exchange sites are probably the carboxyl (COO^-) groups on long-chain polymers containing uronic acids (Theander, 1954). A uronic acid may be visualized as a sugar in which the CH_2OH side chain at C6 has been replaced by COOH. There is a very close correlation of the concentration of uronic acid residues in polymers with the cation exchange capacity, and the regression of exchange capacity on uronic acid concentration is almost exactly the calculated one (Clymo, 1963; Spearing, 1972). Uronic acids constitute approximately 10% of the dry mass of immersed *S. cuspidatum* and up to about 30% of *S. fuscum* growing on hummocks. This is reflected in the exchange ability of the different species, and of plants of the same species growing in different habitats (Table 8.1). Any one species tends to have a higher exchange ability the further above the water table that it is growing, and at a given height above the water table those species which commonly occur in the higher habitats have a greater exchange ability than those usually found at lower levels. It is possible to suggest hypotheses to account for this striking pattern, but none is easily testable.

Table 8.1 Cation exchange ability (Ca^{2+} 7.4 mmol dm^{-3}, pH 6.0) of species of *Sphagnum* growing at various heights above the water table (-w-). For each of the ten sites the heights are ranked, but the vertical intervals are not the same either at one site or between sites. From Clymo (1963). *S. cus, S. cuspidatum; S. pul, S. pulchrum; S. sub, S. subnitens; S. pap, S. papillosum; S. pal, S. palustre; S. aur, S. auriculatum; S. mag, S. magellanicum; S. cap, S. capillifolium.*

S. cus	*S. pul*	*S. sub*	*S. pap* (1)	*S. pap* (2)	*S. pap* (3)	*S. pal*	*S. aur*	*S. mag*	*S. cap*
–	–	–	–	0.91	–	–	–	–	1.13
–	–	–	–	0.01	–	–	–	1.13	1.12
–	–	0.98	1.03	0.96	–	–	–	1.13	1.25
–	0.88	0.97	1.01	0.95	1.00	1.04	1.14	1.14	1.22
–	0.87	1.00	0.99	0.95	0.93	0.83	1.14	1.11	1.19
0.73	0.88	0.96	0.89	0.91	0.89	0.68	1.01	1.07	1.15
-w-	-w-	-w-	-w-	-w-	-w-	-w-	-w-	-w-	-w-
0.83	0.76	–	–	0.78	0.87	–	0.84	–	–
0.73	–	–	–	–	–	–	–	–	–

Many of the phenomena of cation exchange in dead plants can be accounted for by a simple model based on two compartments. One, probably the plant cell walls, contains indiffusible anions (the carboxyl groups) and the other compartment is the water outside the cell walls. The equilibrium conditions in such a system can be calculated (Donnan, 1911) and show that in the chemical environment that is usual for *Sphagnum*:

(a) the walls will contain a much higher concentration of cations than will the solution;

(b) that the higher the cation valence the greater the selective concentration in the walls; and

(c) both effects are more marked the more dilute the solution outside the walls.
The general equilibrium is described by:

$$(x_i^{+v})^{1/v}/(x_o^{+v})^{1/v} = K$$

where x^{+v} is the concentration of cation of valence v, i and o refer to the inside and outside (wall and solution) compartments, and K is a constant for the particular set of conditions. (The same relationship is the basis of Schofield's ratio law applied to soils). This equation is satisfied for all cations simultaneously and, given the high concentration of exchange sites, implies that the concentration of cations in solution is strongly buffered. There are complications however. The exchange capacity may be considered as measured by the total number of carboxyl groups per unit volume or mass but the special affinity of H^+ for the COO^- (describable by a dissociation coefficient) ensures that as pH falls below about 5 a significant and increasing number of the carboxyl groups is tied up as COOH. This means that the effective exchange capacity ('exchange ability') falls as pH falls. The other cations may also have individual specificities not included in the simple Donnan analysis. In particular, the larger ones such as Ni, Pb etc. show a special affinity which may be described as chelation.

None of these processes is unique to *Sphagnum* – they are found in most bryophytes (Brown, Chapter 11 of this volume). Nor is *Sphagnum* particularly extraordinary in its concentration of exchange sites; other mosses have concentrations that are half or more those of *Sphagnum* (Clymo, 1963) and exceed the concentration in *S. cuspidatum* growing in wet places.

Possession of high cation exchange capacity does not of itself confer the ability to maintain the surrounding flowing water acid: if a freshly killed *Sphagnum* plant in a tube is slowly flushed with a large volume of solution the pH of the effluent solution falls dramatically during the first few minutes, but then rises steadily until eventually the pH of the effluent is the same as that of the inflowing solution (Clymo, 1967). In fact, as Brehm (1971) points out, a whole *Sphagnum* hummock might behave as if it were a cation exchange column. A similar sequence of changes in pH is found if live plants are flushed at a velocity of about 100 cm day^{-1} with a dilute solution of salts at concentration similar to those in rainwater. It is only when the flushing velocity falls to say, 0.3 cm day^{-1} (corresponding to rainfall of about 100 cm year^{-1}) that the pH of the effluent remains consistently below that of the inflow. This could be because the plants are excreting whole organic acids, but the amount of such excreted acids seems to account for less than 10% of the observed acidity (Clymo, 1967). Another possibility is that the *Sphagnum* plants can in some way regenerate the exchange sites – that is to replace other cations by H^+. This possibility has not been tested experimentally and could not contribute much in the long term because other cations would accumulate. In any case, Brehm (1970) shows that, if the sites are regenerated artificially with dilute HCl, the plants grow more slowly, probably as a result of the removal of other cations. In normal conditions, the concentration of cations inside the cytoplasm seems to remain fairly constant (Brehm, 1968) even if the concentration in the cells and outside solution changes greatly. A peroxidase extracted from *S. magellanicum* (Tutschek, 1979), has maximum activity at pH 5, so it may be that the cytoplasm is isolated by the plasmalemma from the more acid conditions outside. It is clear, however, that newly produced *Sphagnum* tissue contains large amounts of exchangeable H^+, and it seems reasonable to suppose that the carboxyl

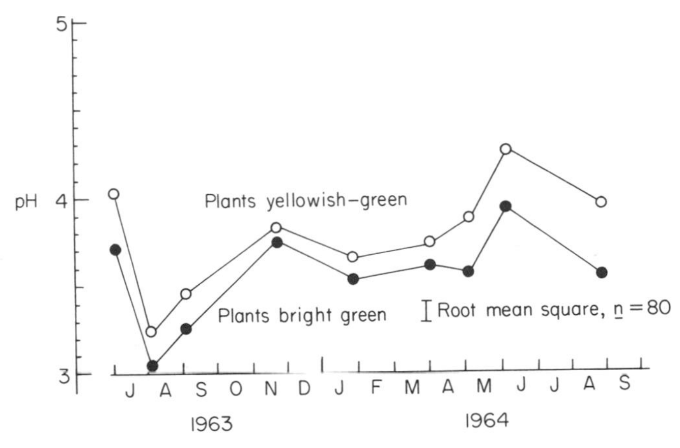

Fig. 8.10 Seasonal course of pH of water squeezed from *S. papillosum* plants growing at Thursley Common, southern England. On each occasion, five measurements were made on water from plants which were bright green and five on water from plants which were live but less bright green. The samples were paired, the individuals in each pair being no more than 2 m apart. The difference between bright green and less bright green plants is highly significant, as is the seasonal change. Clymo (unpublished).

groups in the cell are formed as COOH. The ability to *maintain* an acid environment whilst rain (a dilute solution of salts) is flushing the *Sphagnum* carpet must then depend on continued growth and production of new exchange sites. Green and apparently rapidly growing *S. papillosum* plants do maintain a lower pH around them than do less bright green plants (Fig. 8.10). Values for the annual production and net water supply (precipitation – evaporation) can be combined with the Donnan system exchange model to predict the equilibrium pH which would be produced (Fig. 8.11). From such calculations, it is apparent that, with productivity of 1 kg m^{-2} year^{-1} (10 t ha^{-1} year^{-1}) and net water supply of 50 cm containing about 0.1 mmol dm^{-3} of dissolved salts, then an average pH of about 4 could be maintained by the plants. Such an average conceals the effects of three factors which could cause much greater acidity at particular times in particular places. Firstly, the net supply of water is not uniformly distributed during the year and the times of least supply (lowest rainfall, highest evaporation) coincide with the times of most rapid plant growth, *D*. Both should lead to greater acidity in summer. Second, the hummock species have a greater exchange capacity and can cause correspondingly more acid conditions. Third, the amount of water around hummock plants

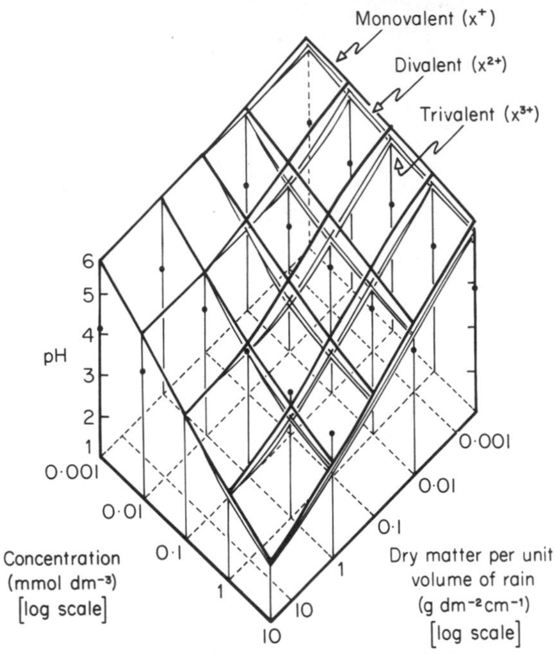

Fig. 8.11 Calculated equilibrium pH of the solution around a cation exchange phase (the *Sphagnum* cell walls) in relation to the initial concentration of cations of different valence (cation concentration in the rain). The cell walls are assumed to have 20% uronic acid of dissociation constant 10^{-4} and chemical concentration 1.0 mol l^{-1}. These values are typical of *Sphagnum* (Clymo, 1963). The third axis gives the quotient of new dry mass produced by the plant (growth, *D*) and rainfall (*V*). The units are g dm^{-2} year^{-1} (= t ha^{-1} year^{-1}) and cm year^{-1}. For growth of 5 t ha^{-1} and rainfall 50 cm year^{-1} then $D/V = 0.1$ g dm^{-2} cm^{-1}. Filled circles define a plane of pH 4. Redrawn from Clymo (1967).

is much smaller than it is around pool plants, so the water volume, V, with which the plants can equilibrate is much smaller. During periods without rain it may happen that there is very little water movement around the plants, and the addition of new plant material effectively increases D in the D/V expression, whilst V remains constant or even decreases. One might then expect that the hummocks in summer would become considerably more acid than pH 4. All these effects may be seen in the field measurements shown in Fig. 8.12: the high acidity during periods of rapid growth and low water supply in summer, most marked on hummocks, and the autumn 'washout' after which the pH in all habitats is close to that of rain.

It is perhaps surprising to find the rain to be so acid, though the general phenomenon of acid rain is now well known. The acidity is associated with sulphur oxides produced by burning coal and oil. This additional load of acid is probably relatively recent and one might expect it to have favoured the growth of *Sphagnum,* but this does not seem to be so. *Sphagnum* has almost disappeared from large areas of the Southern Pennines where it was abundant (as evidenced by its dominance in recent peat) even a hundred years ago (Tallis, 1964). The sulphur containing anions, particularly bisulphite, HSO_3^-, damage *Sphagnum* even in low concentration. For *S. recurvum,* HSO^-_3 concentration of 0.1 mmol dm^{-3} reduces both ^{14}C fixation and O_2 evolution rates by about 30–40% at pH about 4.5, and by even more if the pH is lower (Ferguson and Lee, 1979), though very low HSO_3^- concentration may stimulate ^{14}C fixation. Both HSO_3^- and SO_2, at concentrations commonly found in the Pennines, reduce the growth of *Sphagnum,* though the extent of impairment differs between species: the general order of sensitivity is *S. tenellum* > *S. imbricatum* = *S. papillosum* > *S. capillifolium* ≥ *S. magellanicum* > *S. recurvum.* The immersed species *S. cuspidatum* seems to be very sensitive too (Ferguson *et al.*, 1978). Most convincing of all though is the fact that the growth of *Sphagnum* in field conditions is substantially reduced by occasional treatments with HSO_3^- solutions, simulating the natural episodic nature of acid rain 'events' (Fig. 8.13).

The high cation exchange capacity of *Sphagnum* led to its selection as a material suitable for trapping, and hence sampling, aerial heavy metal pollutants. It proved highly successful, but probably mainly because of its ability to trap airborne particulate matter. It has been used for trapping water-borne particles and solutes (for example those containing uranium) too. Live plants of various species of *Sphagnum* have also been used to estimate deposition rates over large geographic areas (for examples, see Ruhling and Tyler 1971; Pakarinen and Mäkinen, 1976). The vertical distribution of the commoner cations, of heavy metals, and of radioactive isotopes such as ^{137}Cs have also been studied (for example Pakarinen and

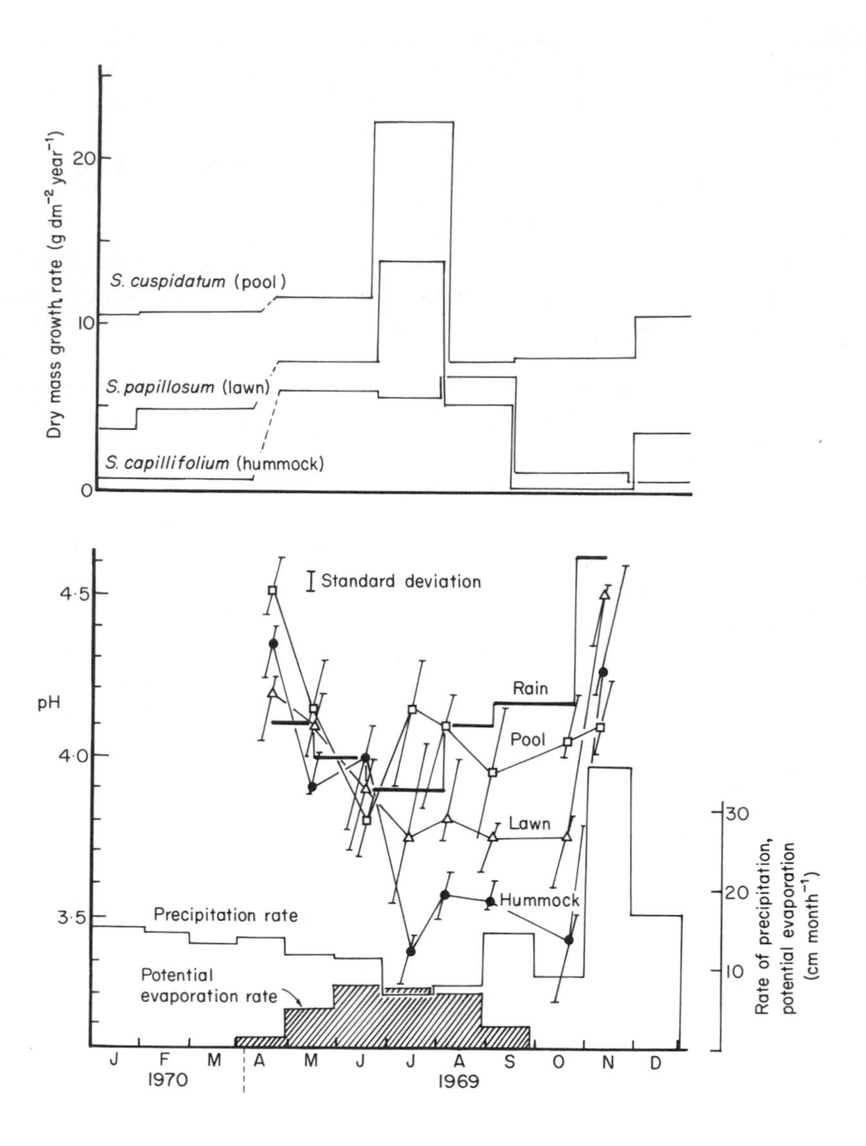

Fig. 8.12 Growth of *Sphagnum*, precipitation, potential evaporation, and pH of rain and water in three habitats at Moor House, northern England. Pools contained *S. cuspidatum*, lawns were of *S. papillosum* and hummocks of *S. capillifolium*, with less than 5% cover of other species. The same five sites in each habitat were sampled on each occasion. Values shown are the median; diagonal bars show range (in many cases the extremes are the same at sites on most occasions). The standard deviation of the difference between 15 samples on successive days is shown by a vertical bar. The first three months of 1970 are shown before the last nine of 1969. Clymo (unpublished).

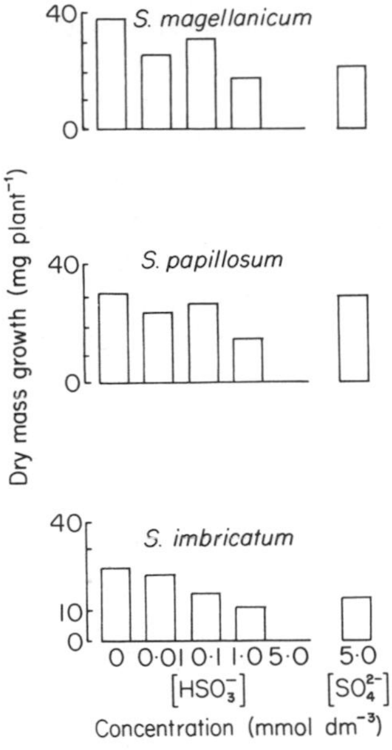

Fig. 8.13 Growth of three species of *Sphagnum* between February 1977 and July 1978 on unpolluted blanket bog in Snowdonia treated with various concentrations of bisulphite or with sulphate. The rainfall is 200 cm year^{-1} and solutions were sprayed on the experimental plots at about weekly intervals from February 1977 to March 1978 at a rate corresponding to about 50 cm year^{-1}. The mean concentration of HSO_3^- in rain in Manchester during the winter of 1975–76 was 0.02 mmol l^{-1}, with a range from undetectable to 0.15 mmol l^{-1}. Redrawn from Ferguson and Lee (1980).

Tolonen, 1977a, 1977b; Clymo, 1978, 1981; Damman, 1978; Pakarinen, 1978b). Most of these studies are more concerned with peat than with the live plants however.

8.4 INTER-RELATIONS OF *SPHAGNUM* AND WATER

It is convenient to consider how *Sphagnum* as an inert physical system affects the states of water surrounding and permeating it, first at equilibrium and then when the water is moving. The responses of the plants can then be considered.

8.4.1 Equilibrium states

The concepts of water potential (ψ) and water content (volume per volume, ϕ, or volume per unit dry mass, ϕ_m) give a convenient framework and allow recognition of the relationship with structural features of the plant (Fig. 8.14). The potential, ψ, may be given in the form of a pressure. The units of measurement are conveniently given as Pa (Pascal). The equivalence with older units is 100 kPa = 10^5 Pa = 10^5 N m^{-2} = 1 bar $\simeq$ 1 atmosphere. Dilks and Proctor (1979) give a schematic diagram more complex than Fig. 8.14, including cytoplasmic water and osmotic and plasmolysis effects, but for a moss plant (rather than a carpet of plants) and using a linear scale for water content. The proportion of the *Sphagnum* carpet occupied by the chlorophyllose cells is so small however that

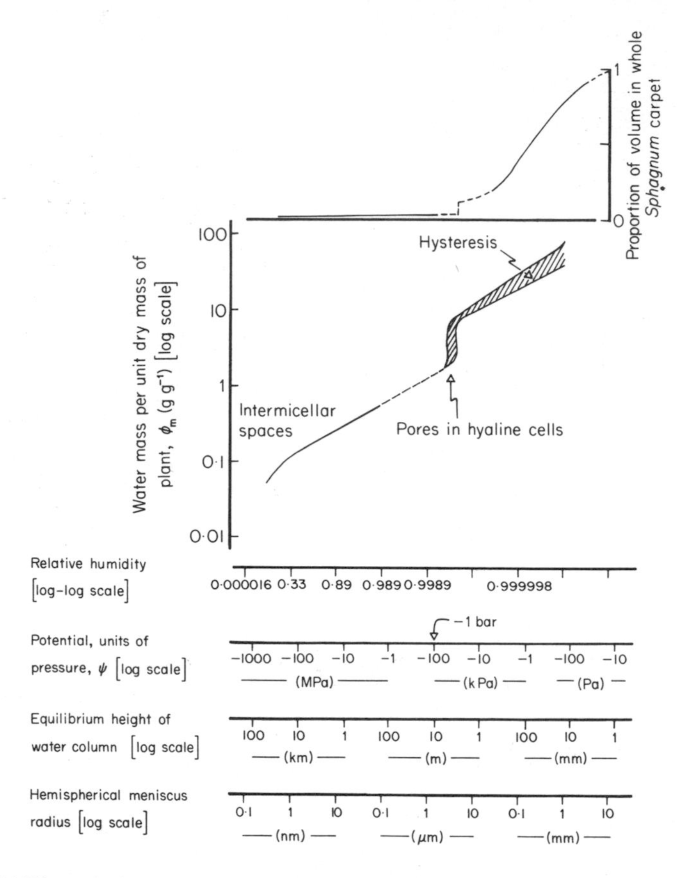

Fig. 8.14 The relationship between water potential (ψ) and water mass per unit dry mass of *Sphagnum* (ϕ_m), and with the proportion of space in a *Sphagnum* carpet occupied by water (ϕ, upper graph). The equivalence between water potential and relative humidity, the radius of a hemispherical meniscus, and the height to which water could rise in a circular capillary of the same radius and with contact angle 0° are also shown. Full lines are measured (see Figs 8.3, 8.15, 8.17, 8.18, 8.19); dashed lines are inferred. Hatched regions show hysteresis.

osmotic and plasmolysis phenomena are not visible in Fig. 8.14. Potentials spanning seven orders of magnitude are involved. At the lower end, the water content increases slowly as potential rises from −1000 MPa to about −2 MPa. In this range, it is water absorbed on surfaces and filling the spaces within the cell walls and differences in relative humidity of the air which are involved. At relative humidity of about 0.99, just distinguishably different from saturated, the water content has reached about 0.8 and is rising rapidly (Fig. 8.15). The water content is greater than that of cellulose, perhaps because of the high concentration of carboxyl groups and associated cations. The water content is almost the same if the plants gain water to reach equilibrium as it is if they lose it: there is little hysteresis. The dry matter of the cell wall occupies about 1–2% of the total space in a *Sphagnum* carpet (Fig. 8.3). The wall and the water which it contains and encloses in chlorophyllose cells occupies about 10% of the total space (Figs 8.14 and 8.17).

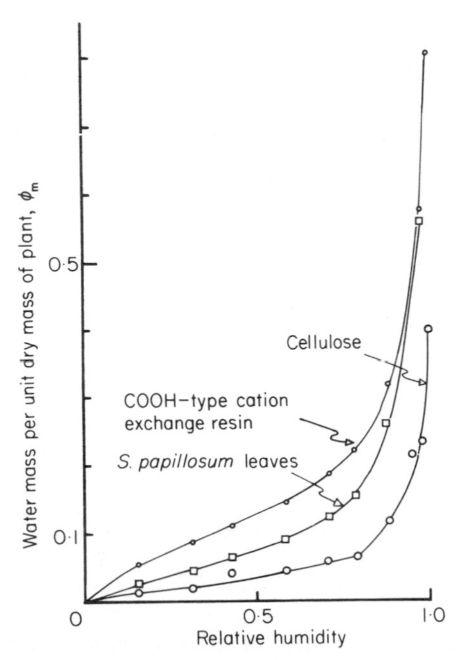

Fig. 8.15 Isopiestic water sorption by dry *S. papillosum*, by a cation exchange resin with carboxyl functional groups, and by cellulose. The samples were in closed tubes with humidity controlled by sulphuric acid solutions of different concentration (Hodgman *et al.*, 1961). Very similar results were obtained if the *S. papillosum* started wet: hysteresis was usually less than 2% of the value at any humidity. Clymo (unpublished).

The next important point is reached at $\psi \simeq -20$ kPa, corresponding to a circular wettable capillary radius of about 5 μm. The upward step in water content corresponds to the filling of hyaline cells. Water can move by diffusion through the wet cell wall but it does so much more rapidly by mass flow through the pores which occur in almost all hyaline cells. These pores are circular or nearly elliptical holes. Scanning electron micrographs (Mozingo *et al.*, 1969; Troughton and Sampson, 1973; Dickinson and Maggs, 1974) show that in the species examined the pores have a border, and that a hyaline cell may have several such areas which are covered by a thin wall, but it also has others which are perforated. It is not known whether the pores form in different ways initially, or form in the same way initially but differ later on in that some pores have the wall resorbed to different degrees, or are broken mechanically. The pores have a diameter characteristic for a particular species (Table 8.2). The diameter is less than

Table 8.2 Size and number of pores in hyaline cells, and number of branches, in some species of *Sphagnum*. From Hill (1978).

	Abaxial pores		Adaxial pores		Branches	
	Diameter (μm)	No. per cell	Diameter (μm)	No. per cell	Pendant	Spreading
S. imbricatum	1–18	3–8	usually none		1–2	2
S. papillosum	3–22	3–10	absent		1–2	2
S. magellanicum	1–19	3–6	absent or few		1–3	2
S. palustre	8–25	3–16	absent or few		1–4	2
S. capillifolium	8–25	4–7	absent or few		1–2	2
S. fuscum	2–30	3–7	absent or few		1–2	2
S. subnitens	6–30	4–10	absent		1–2	2
S. compactum	5–12	0–4	absent		1–5	1–2
S. auriculatum	3–8	0–30	3–12	0–25	0–4	2–3
S. subsecundum	2–5	20–40	absent or few		2–4	2–3
S. cuspidatum	2–6	0–1	4–8	4–10	3–5	3–5
S. tenellum	4–12	0–1	4–18	0–2	1	2
S. pulchrum	3–8	0–1	4–8	3–9	2	2
S. recurvum	4–9	0–1	5–9	3–7	3	2

that of the hyaline cell, and the constriction causes hysteresis (Fig. 8.16). Because there are many hyaline cells and pores of similar dimension the ψ *vs* ϕ graph shows a step transition from water content of about 1 to about 8. The potential, about −30 kPa, corresponds to a circular capillary able to support a water column of about 3 m, so the hyaline cells will never be emptied by any likely fall in the water table. As the water potential is raised further, however, the remaining 90% of space in the *Sphagnum* carpet gradually becomes filled. Here it is, at first, the capillary spaces between

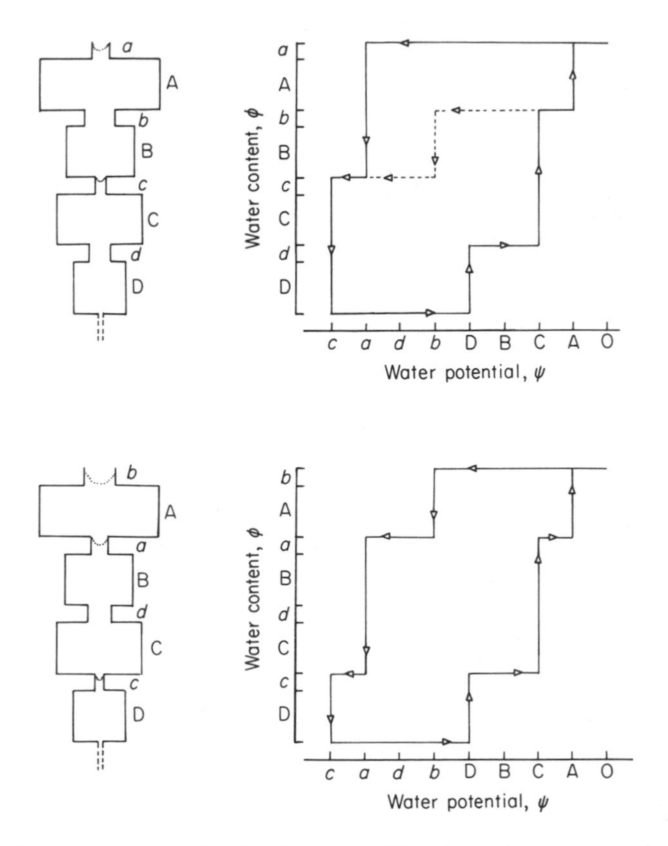

Fig. 8.16 Hysteresis in the ψ *vs* ϕ relation resulting from the presence of a multichambered system with constrictions (the 'cavern' or 'ink-bottle' effect). The water content axis identifies which part of the system is water-filled. The water potential axis identifies the rank order in which the meniscus in capillaries of the radius of the caverns and constrictions would just be maintainable. In the upper example suppose the system is water-filled at $\psi = 0$, and that the water column is put under tension (ψ being slowly reduced) and that the cavern D is in contact with capillary spaces smaller than any of those above. The system remains full until *a* can no longer maintain a meniscus, at which point the system empties to the next smaller constriction, *c*. This in turn will eventually break allowing C and D to empty. If the water potential is now increased it has to reach a point high enough for the cavern D to be bridged before filling begins. Next, C is bridged and both C and B fill (because B is of smaller radius than C). Finally A fills. These are the primary drying and wetting curves. If, at the point where B and *b* have just refilled, the potential is lowered, then the dashed line is followed. This is a secondary curve. The lower example shows the effect of rearranging the constrictions, but retaining the same order of caverns. The drying curve is changed but the wetting one is not.

For very small capillaries, and water potential changed by changing the relative humidity, the model should be one which is open at both ends.

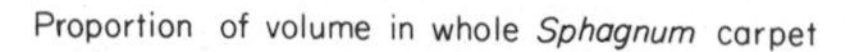

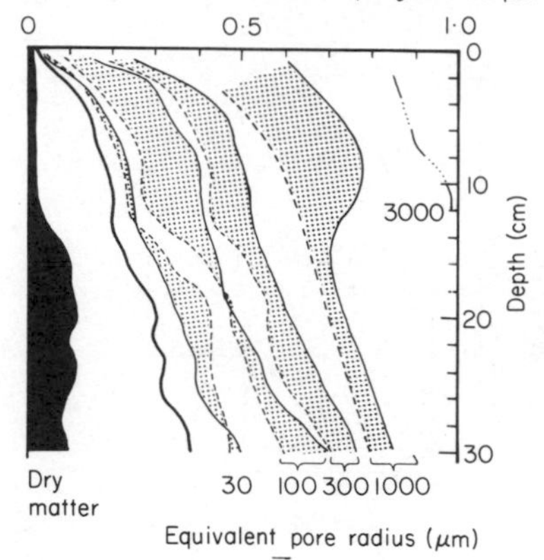

Fig. 8.17 Profiles of cumulative proportion of pore volume (circular capillary equivalents) of various sizes, calculated from profiles of water content (measured by absorbance of soft gamma radiation) with the water table at different depths below *S. capillifolium*. The black area at the left is dry matter; to the left of the bold line are pores which remain full under tensions equivalent to 120 cm of water ($\psi = -120$ kPa). The solid lines to the right of the bold line are those obtained with a falling water table; dashed lines are for water table rising from 120 cm deep. Shaded areas show hysteresis. Redrawn from Hayward and Clymo (1982).

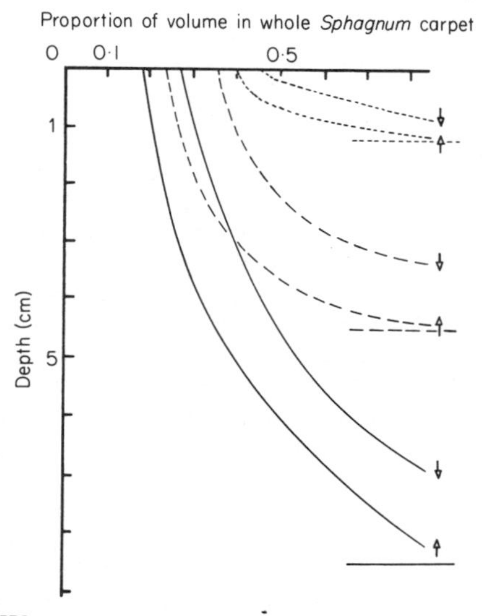

Fig. 8.18 Water content profiles for *S. papillosum* with the water table at three different depths. For each depth there are two curves: one when the water table rose to the specified level, the other when it fell to that level. The difference is a measure of the amount of hysteresis. Redrawn from Hayward and Clymo (1982).

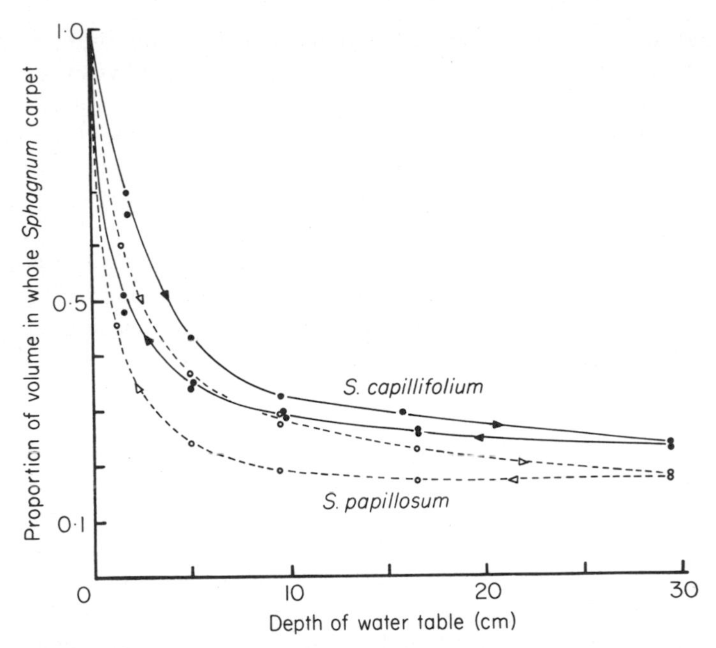

Fig. 8.19 Effect of raising and lowering the water table between the surface and 30 cm deep on the proportion of water in the capitulum (top 1 cm) of *S. papillosum* and *S. capillifolium*. Redrawn from Hayward and Clymo (1982).

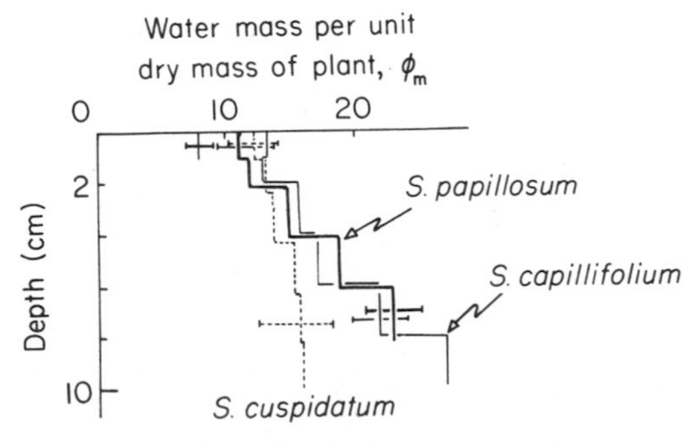

Fig. 8.20 Water content profiles of three species of *Sphagnum*. Plants were removed, with forceps, from carpets of plants in the field and cut into 2 cm-sections whilst held vertically. All plants were collected within two hours on a dry day. Bars show 95% confidence intervals. The left-most isolated 0–2 cm result is for capitula of *S. papillosum* growing as isolated plants in a carpet of *S. capillifolium*. Redrawn from Clymo (1973).

imbricate leaves, and between pendent branches and stem, that are involved. Again, there are constrictions which cause the hysteresis effects shown in Fig. 8.17. This is the range within which the water content depends very much upon the water table, upon whether the water table is rising or falling (Fig. 8.18), and upon the detailed distribution and proportion of spaces of different sizes. This last differs considerably between species (Fig. 8.19): *S. capillifolium* has a higher water content (ϕ_m) in its capitula than does *S. papillosum,* whatever the water potential (ψ). The same phenomenon may be seen in field measurements (Fig. 8.20) but these are far less accurate than the laboratory ones because the plants must be removed and this must necessarily disturb the capillary films. Similar values may be seen in Fig. 8.3.

8.4.2 Dynamic states

The details of water movement about the *Sphagnum* plant are still unclear but are of great importance. The height to which water might rise in a wettable circular capillary at *equilibrium* is of less importance as water potential or capillary size decrease, partly because below 100 kPa (equivalent to a meniscus of about 1 μm radius) the height (10 m) is vastly greater than is necessary, and mainly because the rate of movement in such small spaces is so slow. For a straight-sided tube and a constant potential drop the volume rate of flow in a *single* tube is proportional to the fourth power of the radius. But it is possible to fit more small capillaries than large ones into a given cross-sectional area, and the *total* volume rate of flow over the whole cross-section is then proportional to the square of the radius.

The rate of movement of water up the *Sphagnum* plant is rapid, and large fluxes may occur. This was shown by Overbeck and Happach (1956), and

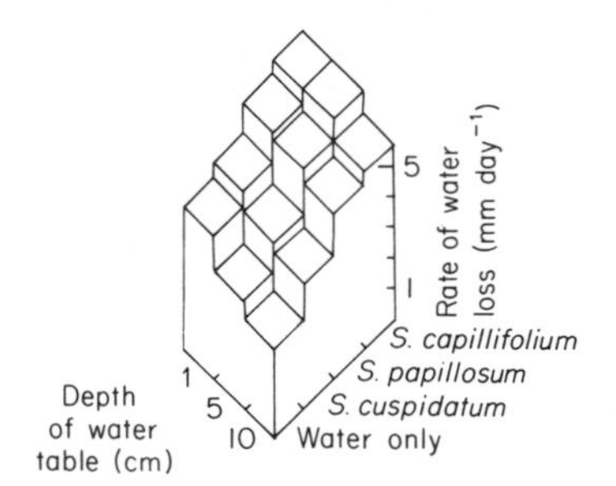

Fig. 8.21 Rate of loss of water from three species of *Sphagnum* in 0.6 l beakers with water table at one of three levels. The capitula were level with the top of the beaker. Guard rows of unmeasured *Sphagnum*-filled beakers surrounded the experimental ones, and the experimental ones were re-randomized each day. Measurements were made for nine consecutive days. Redrawn from Clymo (1973).

may be seen in more detail in Fig. 8.21. The greater rate of loss from *S. capillifolium,* and the increase in this difference as the water table was lowered, may be partly explained by the capillary structure already revealed by the equilibrium measurements. It seems unlikely that higher water content *per se* is the explanation. Rather, it is likely to be a smaller total resistance to transport in capillary spaces outside the cell walls. If one inspects a *Sphagnum* carpet during a drought it is common to find individuals or groups of *S. papillosum* plants which have dried out and become papery and white while adjacent (touching) plants of *S. capillifolium* are still wet. Cavers (1911) noted that if 'the tufts of branches be removed from the end of the stem, which is then dipped into water, the plant remains dry, hence the stem-tissue does not serve a conducting function'. If most of the water moves in the pendent branches then the linear velocity of water on a fine day with rapid evaporation may be well above 1 mm min^{-1}, a rate which, if Stokes' Law applied, could keep pollen grains and spores of 10 μm diameter in suspension (Clymo, 1973). Experiments have confirmed this (Mackay, unpublished).

That such mass transport is possible is easily shown. A few wet leaves of *S. papillosum* may be put on a microscope slide, without a coverglass, and observed as they dry. When the surface water has evaporated there is a short pause, for a minute or so, until, with explosive suddenness, a hyaline cell becomes part filled with air (Fig. 8.22). Within a few seconds many hyaline cells are 'popping' in this way and water, with entrained particles, is moved violently from one place to another. The series of events seems to be as follows. As water evaporates from the wet hyaline cell walls the menisci in the pores, which were flat, become increasingly concave though the major and minor radii of curvature are still greater than those of the pore. Beneath this curved meniscus there is a reduction in pressure appropriate to the head of water which a meniscus with these curvatures is capable of supporting. The net of chlorophyllose cells is relatively strong and not easily deformed. The hoops of thickening on the hyaline cells are not so strong, and the unthickened wall between is even weaker, so the abaxial and adaxial walls are sucked in. Under the microscope the walls can be seen to go slowly out of focus. That this reduction in pressure is real can be seen if a small air bubble is trapped in a hyaline cell. The bubble grows as the pressure falls – indeed the growth of the bubble can be used to measure the change in pressure. A pore, in *S. papillosum,* of 8 μm radius can support a pressure reduction of 19 kPa – about 0.2 bar. Eventually, so much water has evaporated that the meniscus reaches a radius of curvature smaller than that of one of the pores. The meniscus then breaks, the cell wall springs out again, and the remaining water, which is sufficient to fill perhaps half the cell may appear at one end of the cell. In *S. papillosum* the hyaline cell is lined with papillae which have capillary spaces less than 1 μm

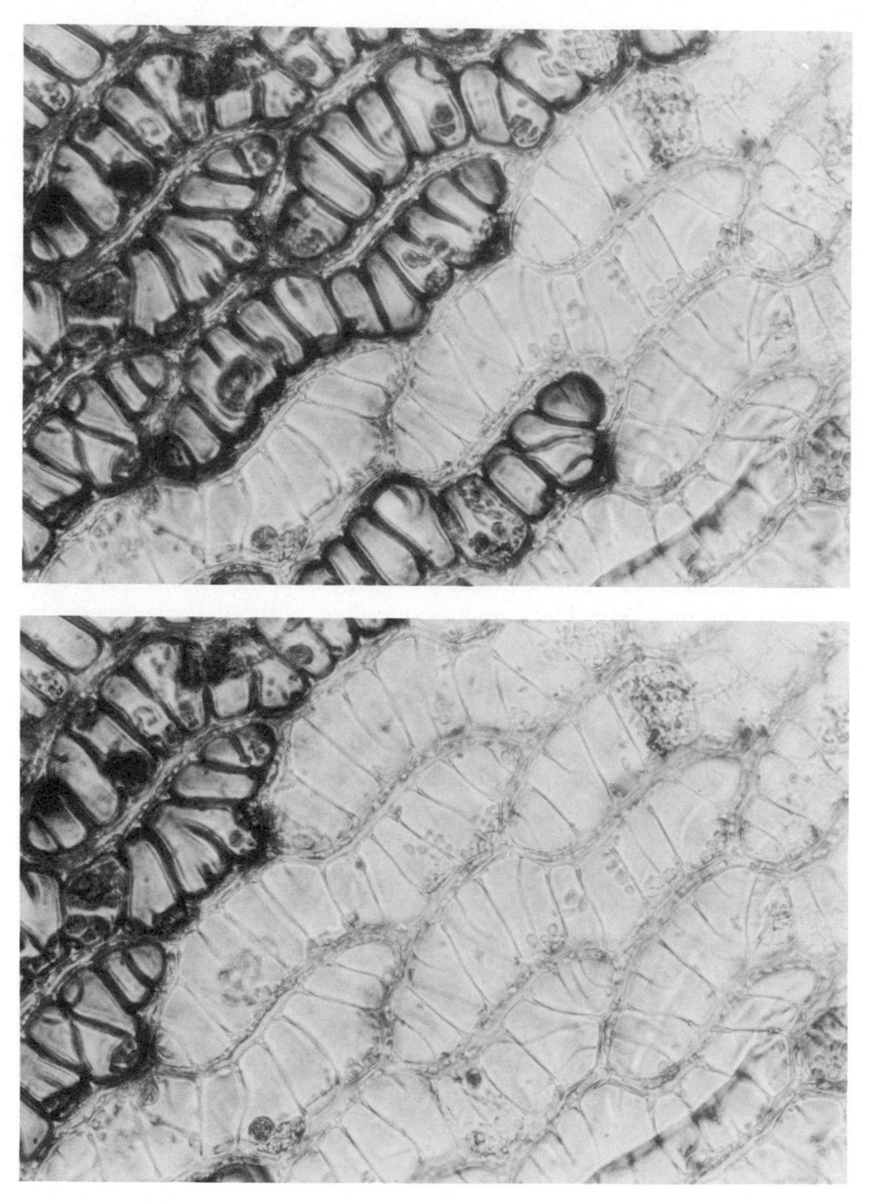

Fig. 8.22 Part of a leaf of *S. papillosum* drying out. The lower photograph was taken 5 s after the upper one. In this time three more hyaline cells have 'popped' and (apparently) become air filled. See text for further explanation. The width of the hyaline cells is about 25 μm. Photographs by R. S. Clymo.

across between them. When the meniscus breaks, the water therefore forms a thin layer lining the cell and with the centre filled by a large air bubble. The lining layers of water are not easily visible, so the cell appears to be magically completely empty of water. A little later the hyaline cell appears to flicker, as the mensci retreat between the papillae. The whole leaf may empty in this way in less than five minutes, and then appears white and papery.

Such simple observations reveal in a dramatic way the importance of surface tension as it changes the physical pressure and moves water bodily in the microscopic world.

It is not only the size and proportion of spaces of different sizes which affect the movement of water. The exact position of the spaces in relation to one another is crucial. In a system showing marked hysteresis of equilibrium states there must be interlinked 'caverns' (Fig. 8.16). The dynamic behaviour of this system will be determined largely by the bottlenecks, because the volume rate of flow through these is so much restricted. The primary 'drying' curve (the case where the water table starts at the surface and is slowly lowered) is determined largely by the size, number and position of bottlenecks, whilst the primary 'wetting' curve is determined mainly by the maximum width, number, and position of caverns. The difference between the two curves then gives some indication of the shape and arrangement of the spaces. It seems that the widest parts are no more than twice the size of the bottlenecks at any water potential, probably because, in this three-dimensional system, many of the spaces are open round much of their perimeter. In any case this gives only a crude indication because the *order* of caverns matters too: the same bottlenecks and maximum widths would, if re-ordered, give a different hysteresis result but might not affect the volume rate of flow. It is also possible to imagine reservoirs connected through a single opening to the main line of flow. These would contribute to hysteresis in the ψ *vs* ϕ curve, but would scarcely affect the volume rate of flow.

The general pattern is shown if an undisturbed *Sphagnum* carpet is arranged in a glass jar with the water table 10 cm below the capitula and the whole apparatus placed in a stream of dry air so that evaporation occurs. An anionic dye solution (for example, eosin) is injected on one plant just above the water table and its progress followed. This experiment is similar to, but much more complex than, the systems used in gas, gas–liquid, or liquid chromatography. The dye remains mostly around the stem and pendent branches to which it was applied but it moves upward. The concentration profile becomes attenuated. The profile is clearly not just that of laminar mass flow with diffusion broadening: there is mixing with reservoirs along the route too. There is a small amount of lateral spread to other plants, and much more occurs if the pendent branches are removed from one node of the treated plant. In this case the water is moving sideways along the spreading branches to adjacent plants. It is important to recognize that it is not only the potential difference but also the resistance to flow which determines the volume rate of flow: an isolated plant with pendent branches removed may lose water by evaporation and develop a lower water potential, but the water can move directly upward only in the very small intermicellar spaces and at a very slow rate. These effects may be seen in the results (Fig. 8.23) of an experiment in which different parts of the plant were removed 8–10 cm below the apex and the water content of higher parts measured. One set of plants was placed in still, damp air, and another was placed in moving drier air so that evaporation was rapid. In still air, the water content of capitula was only slightly reduced by removing either or both the spreading and pendent branches, though it was much reduced if the stem too was replaced by a water filled 1 mm diameter glass tube – exactly why is not obvious. When evaporation was rapid however the water content of capitula was markedly reduced if both pendent and spreading branches were removed but less so if only one or the other was

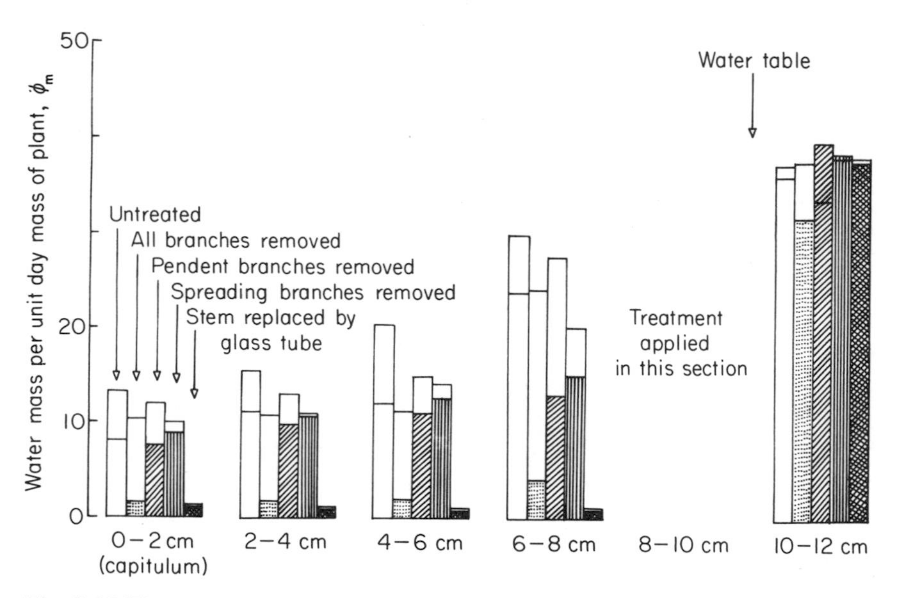

Fig. 8.23 Water content of successive 2 cm-sections of *S. papillosum* treated in various ways. The treated plant was surrounded at natural density by untreated ones. All were in 1 l beakers. The upper bars are for plants kept in still, moist air; the lower ones are for plants in moving unsaturated air. The water table was at 10 cm depth, and treatments were applied to the section between 8 and 10 cm. The treatments were: all branches removed; pendent branches removed; spreading branches removed; all branches removed and stem replaced by a water-filled glass capillary of 1 mm diameter into which the cut stems were pushed. Clymo (unpublished).

removed. On the segment immediately above the one from which branches were removed, however, the water content was much reduced if either sort of branch had been removed below. This strongly suggests that, when pendent branches had been removed, water then moved sideways to adjacent plants and returned again higher up: the further from the site of removal the more touching branches there were through which lateral movement would occur.

Change in water content has other effects on *Sphagnum* too. If the water table starts at the surface and is then lowered the water potential at a fixed point decreases and the water content decreases. The reduction in pressure below the menisci is sufficient to suck the pendent branches in around the stem. At a later stage, imbricate leaves may be similarly sucked down onto one another. The distribution of spaces of different size is not constant therefore: as the water table is lowered the proportion of smaller spaces increases. From the point of view of water transport, a small but water-filled space is infinitely better than a larger empty one.

Another rather unexpected effect is that the whole carpet shrinks as the water table is lowered. For example, a cylindrical carpet of *S. papillosum* (30 cm diameter, 17 cm deep) was put on top of a membrane filter covering a water-filled space, so that water tensions of more than 17 cm could be applied (Hayward, 1980). As the water table was lowered from the surface (0 cm) to the equivalent of 30 cm deep so the 17 cm carpet shrunk by 2.5 cm to 14.5 cm. This effect was reversible and probably results from the replacement of water by air, so that the weight of water-filled *Sphagnum* is no longer supported by the water (just as one's weight increases on getting out of a bath). Capillary forces may contribute too. The movement is distinct from the bodily upward and downward floating movement of some peat systems with rainfall and evaporation (Green and Pearson, 1968).

In a *Sphagnum* carpet, the height of the water table at a particular time is determined by many processes: precipitation, evaporation, hydraulic conductivity in both vertical and horizontal directions. The flow of water through a mat of recently dead *Sphagnum* usually obeys Darcy's law (Fig. 8.24) though this is not true of more highly humified peat (Ingram *et al.*, 1974; Rycroft *et al.*, 1975). The consequent local variations in water level below hummocks, in lawns and in pools are complex (Goode, 1970; Boatman and Tomlinson, 1973; Hulme, 1976; Ingram, 1981; Bragg, unpublished). The vertical range of the water table level may be about 20 cm in an average year, but the water table may rise 5–10 cm within a few hours in some cases (see, for example, Tallis, 1973). The vertical range of water table amplitude below some hummocks is more than it is in pools, and the water table in hummocks may, at some times, be above that in adjacent pools and at other times below. Such temporal and spatial variations have obvious implications for the ecology of *Sphagnum* but cannot be pursued here.

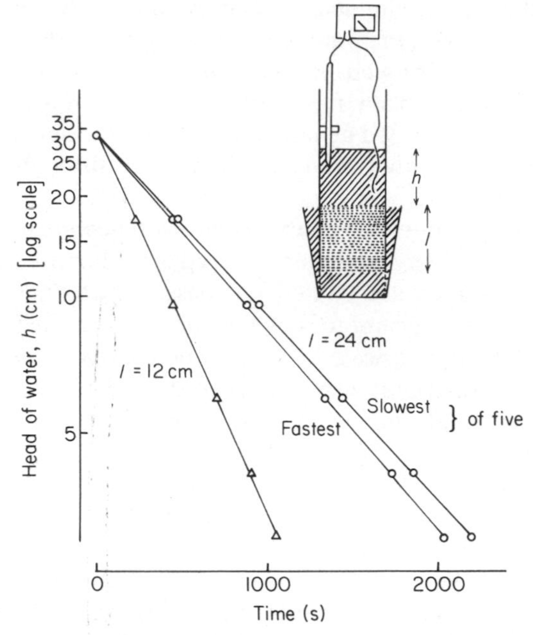

Fig. 8.24 Vertical hydraulic conductivity, *K,* of *S. fuscum* at Stordalen, Abisko, in Artic Sweden. A core, 20 cm diameter and 12 or 24 cm deep (from 3 cm below the capitula) was a tight fit in a plastic cylinder. The velocity with which water flowed through was recorded. The straight line on a plot of log (h) *vs* t shows that Darcy's law is obeyed. It can be shown that $1n(h/h_0) = -K/lt$, and that $K = 0.693/t_{1/2}$ where $t_{1/2}$ is the time needed for the head of water to fall to half what it was. This gives $K = 0.035$ cm s^{-1}. Clymo (unpublished).

8.4.3 Effects of water supply on *Sphagnum*

It is convenient to recognize two ranges in the continuum of states; 'normal' conditions and 'exceptional' (or 'catastrophic') conditions. To be successful a plant must survive both. The commonest catastrophe for *Sphagnum* is desiccation. The effects of this under fairly natural conditions were studied in the desiccation experiments shown in Fig. 8.25, upper half. The results are shown as survival of 5 cm-long plants of different species drained of water and allowed to dry for different periods. Two interacting effects can be seen: the larger plants, such as *S. imbricatum* and *S. papillosum* survived relatively well, presumably because they held larger amounts of water. In *S. imbricatum* these water stores were not exhausted by the end of the experiment after 16 days but once the stores had gone from *S. papillosum* the plants died rapidly. Plants of the smaller species dried out early during the experiment. Of these, *S. fimbriatum* showed poor survival, but some plants of *S. capillifolium* survived even after 16 days, and plants with dead capitula produced many new lateral shoots from lower down the stem. The results of another experiment examining the inherent ability to resist low water potential are shown in Fig. 8.25 (lower half). This shows the survival of capitula of seven species which had reached equilibrium after 21 days with atmospheres of relative humidity corresponding to menisci of radius less than 1 μm (less than the hyaline pore sizes). A larger proportion of capitula remained green than were able to resume growth but the two measures gave parallel results. The ability to survive these treatments seems to be uncorrelated with capitulum size, pore size or ecological habitat: *S. auriculatum* survived desiccation best but normally grows immersed whilst *S. capillifolium,* which grows on hummocks, survived rather poorly. The most sensitive species was *S. papillosum.* But the conditions in this second experiment were very artificial: no air movement, continuous dim light and a temperature of 25°C. Desiccation resistance *per se* seems to have rather little part in determining which species can survive in which habitat. The rate of transport of water up from the water table is probably of greater importance.

Particularly puzzling is the ability of species such as *S. compactum* and, especially, the very delicate *S. tenellum* to survive and grow on wet heath. Both species are minor though widespread components of wet *Sphagnum* lawns, usually growing as isolated individuals amongst other species. On wet heath, however, they form tussocks or single-species carpets which may become dried out and even burned over during droughts.

The responses of *Sphagnum* to rare but extreme drought conditions (say once in 20 to 100 years) may be of importance but are poorly documented. Catastrophes of this kind are not usually predictable and long-term recording of detailed changes, though it would almost certainly be valuable, appears at any one time to be a time-consuming and risky venture.

The responses of *Sphagnum* to water supply in 'normal' conditions are equally important and experiments to measure the growth of *Sphagnum* in such conditions are less risky than long-term recording. Results of such work show that light flux is important too, and that water supply and light flux have interactive effects on *Sphagnum* growth. The experiments will therefore be considered in the next section.

It is worth commenting here, however, on the peculiar case of *S. imbricatum.* Records and studies of peat deposits show that this species was once widespread and, at least locally, abundant in the southern Pennines (Tallis, 1964). It seems to have declined in abundance from perhaps the fourteenth to early nineteenth century and is now absent from the southern Pennines altogether. Various possible causes have been suggested amongst which burning, grazing, draining and poisoning (by aerial pollutants) are the most often given. The species grows today in two sorts of habitats: in a compressed growth form on hummocks and (for example in western Ireland) as a lax growth form in wet *Sphagnum* carpets. This lax growth form is the one which is abundant in Pennine peat. The

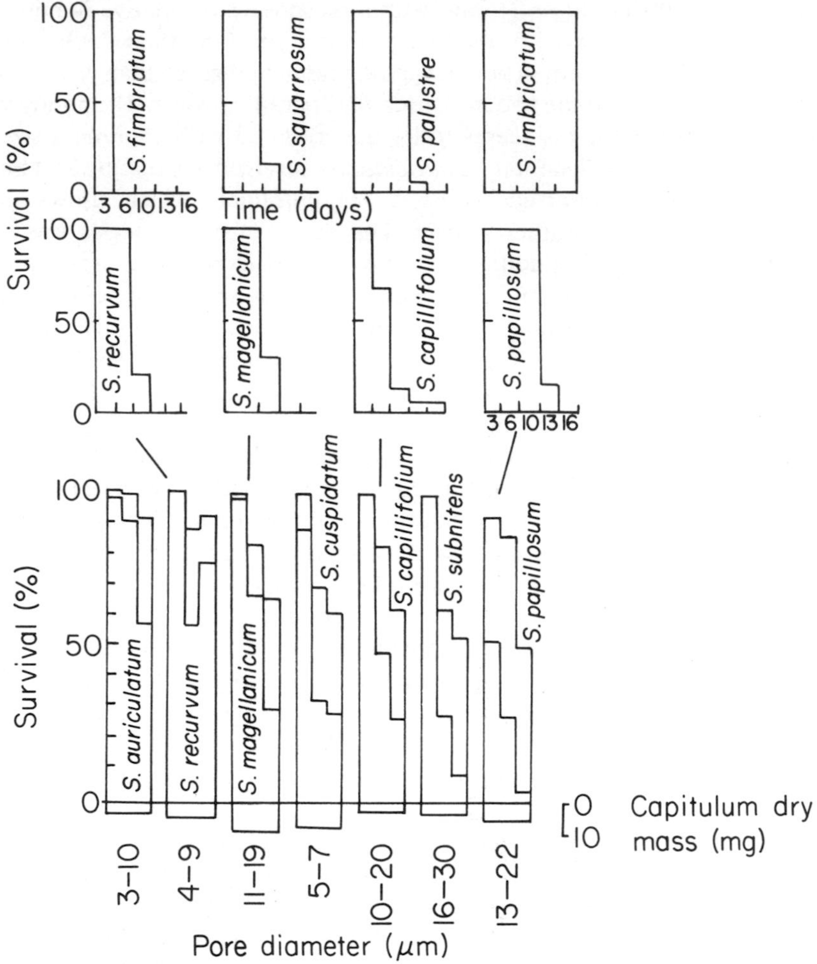

Fig. 8.25 Survival of *Sphagnum* spp. in desiccation experiments. Upper half: plants in beakers drained of water and allowed to dry for up to 16 days. Some plants were re-watered after 3, 6, 10, 13 or 16 days and the proportion resuming growth at the apex was recorded. Lower half: capitula put in closed containers, with humidity kept constant by sulphuric acid solutions (Hodgman *et al.*, 1961), and kept in dim light for 21 days. After this the capitula were kept on damp filter paper in the light for a futher 21 days. Humidity (left to right): 0.998, 0.991, 0.981. Upper histograms are proportion of capitula with branches remaining green; lower are proportion showing new growth. Below the axis are histograms of capitulum size, and the size of pores in leaves. Lines join species common to the two experiments. Redrawn from Green (1968) and Clymo (1973), and from Hill (1978).

difference is not genetically determined but is a response to water supply conditions (Green, 1968). The rate of elongation of shoots is particularly sensitive to change in water potential, but there is no evidence that the species is particularly drought sensitive (Fig. 8.25); its occurrence on hummocks makes this unlikely anyway (though *S. capillifolium* appears to be desiccation-intolerant, yet grows on hummocks). Further, it is recorded that *S. imbricatum* has disappeared from undissected (and still very wet) peat bogs in the middle Pennines, just as it has from the southern Pennines (Tallis, 1964). The work of Ferguson and Lee (1979), 1980) already described, shows that *S. imbricatum* is usually sensitive to low concentration of HSO_3^-, and the reduction of its growth rate from this cause seems to be the most likely explanation for the disappearance of *S. imbricatum* from the industry-girt southern Pennines.

8.5 THE GROWTH OF *SPHAGNUM*

8.5.1 The growth of individuals and populations

In some species there is a marked annual variation in the length of branches and in their density on the stem (Malmer, 1962b). In *S. papillosum,* the branches formed during winter are relatively short and the internodes are also short. In early summer, there is a sharp change to longer branches and longer internodes followed by a gradual reduction in shoot- and internode-length in late summer and winter. Such a cyclic pattern of growth may be seen in other species too, but it is more often absent than present and it is not obvious in Fig. 8.1 for example. Both branch and internode length can be markedly affected by water supply in experimental conditions (Fig. 8.26) but, in the field, the variation during a year is not necessarily, or even probably, related primarily to water supply. In vascular plants, the length of internodes is linked with factors such as plant hormones, temperature, and the balance between red and far-red irradiation.

The same factors are often involved in the development of red pigments, so it is of interest to note that *S. magellanicum*, for example, is often apple green in early summer (and remains so all the year round in shade) but produces a wine-red pigment later in the year. Production of red pigment can be induced by low temperature (Rudolph, 1978). The same is true of *S. capillifolium.* In this species the pigment forms not only in the leaves but also in the stem. The stem can be seen to have annual bands consisting of a green section, formed in early summer, through a transition zone to a deep red section formed in late summer. The boundary between dark red and apple green is often very abrupt, just as the winter–summer transition in branch pattern is sharp. The red pigment is related to the anthocyanins but is unusual in being firmly attached to the cell wall (Goodman and Paton,

1954; Rudolph, 1964, 1968; Paton and Goodman, 1955; Bendz *et al.*, 1966). Different species produce a variety of other pigments: *S. fuscum* and some forms of *S. recurvum* are well known for their brown colouration. *Sphagnum subnitens* has an irridescent blue sheen when dry, but not when wet. This has been shown to result from thin-film interference (Morris, 1977) – an unusual phenomenon in plants. In all cases the pigment-produced colours are at their best in late summer, when, as Braithwaite (1880) observes: 'Few persons can have traversed our moorlands without having had their attention attracted to the great masses of *Sphagnum* which adorn their surface – now in dense cushions of lively red – now covering some shallow pool with a vast sheet of light green, inviting it may be by its bright colour, but woe betide the inexperienced collector who sets foot thereon, for the spongy mass may be many feet in depth, and he may run the chance of never reaching *terra firma* again.'

Growth hormones are also probably involved in the peculiar growth of *S. recurvum* (Overbeck and Happach, 1956). This species has many growth forms. One is commonly found where the water is flowing past the plants in spring mires, in the channels in eroded peat, on flushed hillsides and so on. The plants in such habitats have stems which are mostly horizontal, and may be traced back for 30 to 60 cm or more to the point where they have

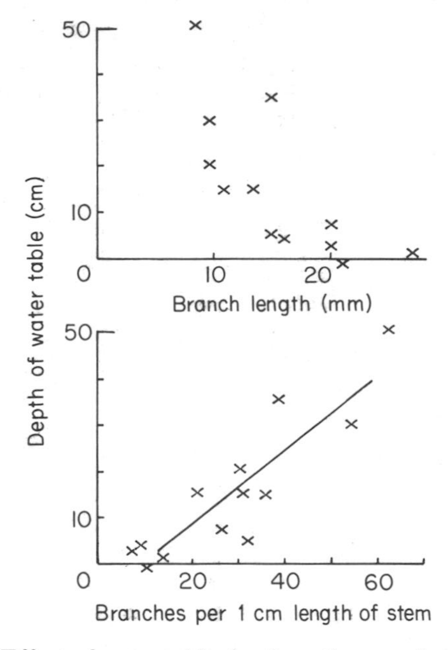

Fig. 8.26 Effect of water table depth on the morphology of *S. imbricatum*. The plants were grown in beakers with controlled water table. Redrawn from Green (1968).

decayed. The stems lie in a thick mat on top of one another but the capitula, which are often a bright and inviting green, are densely packed in the usual upright position with the stem immediately below them standing vertically. About a centimetre below the capitulum the stem turns through 90° to run horizontally. As the capitulum grows on, extruding more stem and branch behind it, the position of the bend in the stem moves too. This series of growth movements appears to be controlled by starch statoliths (Bismarck, 1959) which form in the tissue just behind the stem apex. A similar but less obvious pattern of growth is seen in *S. cuspidatum*. Both this species and *S. recurvum* have juicy brittle stems, which crack across very easily and seem to have high turgor. Most other species have tougher stems.

Again, by analogy with vascular plants, it seems likely that fruiting is controlled by hormones, day length, and perhaps by the balance of red and far-red light. *S. subnitens* is reported to be a short-day species (Benson-Evans, 1964), but capsules appear on different species at different times. For example, *S. tenellum* has abundant capsules in early May almost every year around the margins of southern English valley bogs, though *S. papillosum* a few metres away more often produces capsules in June or July.

The pattern of branch formation varies, and so does the frequency with which the capitulum divides (forks) to form two separate stems. The

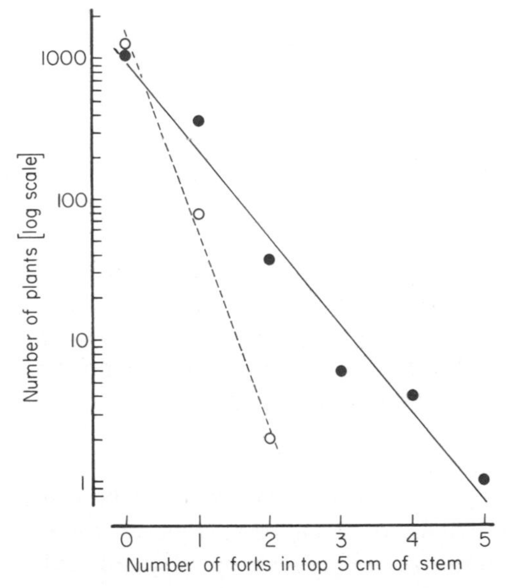

Fig. 8.27 Frequency of forking of the main stem in the top 5 cm of *S. papillosum* in relation to spatial density. Symbols: O, at Moor House, northern England; ●, at Thursley Common, southern England. Redrawn from Hayward (1980).

number of plants with 0, 1, 2 etc. forks for two populations of *S. papillosum* is shown in Fig. 8.27. The relationship is

$$\log (n/n_0) = -kf$$

where n is the number of plants, n_0 the number of unforked plants, and f the number of forks. The slope, k, is a measure of the probability that a plan will fork. Casual observation suggests that large capitula are the most likely to fork and that k is largest for populations that appear to be colonizing or invading a new area. The straight line implies that the probability that a plant will fork is independent of whether it has forked already or not.

The daughter apices produced by forking are each smaller than the parent. By the time the leaves on their branches have died the daughter plants are, for most functions, independent. A carpet of *Sphagnum* may thus contain several genets each represented by several ramets. It would be interesting to know how many of each there are in a *Sphagnum* carpet. Detailed recording over many years, or isoenzyme studies might provide the answer. A genet with recognizably distinct characters would give a partial answer. No such observations have been reported however.

Individuals of the same species growing together differ considerably in size, as do the different parts of the plants. Some of these differences may be related to local water supply and to spatial density of plants. Fig. 8.28 shows, for *S. papillosum,* the mass per plant per 1 cm-depth as an indication of size. The samples chosen were taken from sites over the full range of water table depth, and were collected on a single day during a dry period, so the water table depth is probably a useful indicator of the relative water supply to the different populations. The spatial density of plants shows a maximum (about 180 dm^{-2}) falling to about 130 dm^{-2} on the drier side and to about 80 dm^{-2} for an immersed population. In all but one population the average mass (per cm) of capitulum is greater than that of the product (stem plus branches) which it produces. This is simply a result of stem internode elongation. The exception (population D) had very densely packed branches, but population E, with a similar water table and spatial density, had much laxer growth. Most populations showed rather skewed distributions of individual plant size, generally with a tail of larger sizes, but the degree of skewness differed considerably. This phenomenon is now well known in vascular plant populations (for example, Bazzaz and Harper, 1976). There is clearly a lot more to be discovered about the factors which control size of *Sphagnum* plants, and the spatial arrangement of plants of different size has not been investigated at all.

The bulk density of capitula and of their product (stem plus branches) seems to be rather uniform over a range of densities and water tables, though the product of population D and both capitula and product of

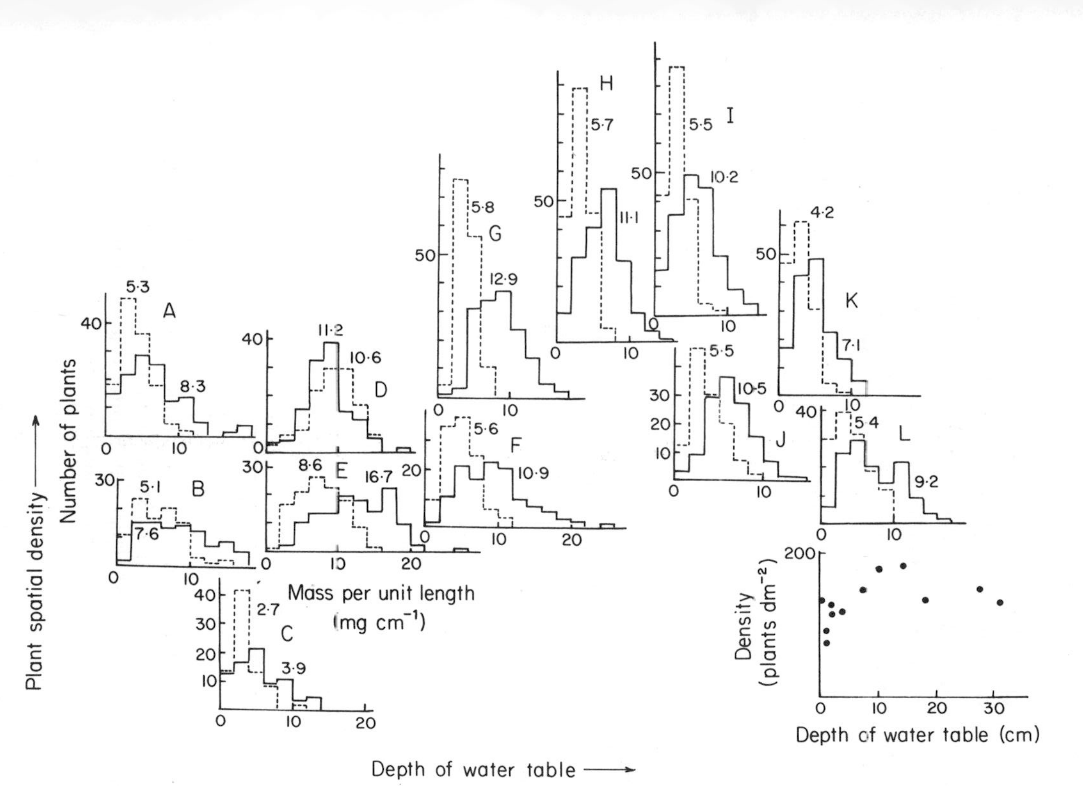

Fig. 8.28 Distribution of size of capitulum (0–1 cm, full lines) and of the top 5 cm of plant (broken lines) in twelve populations of *S. papillosum* at Moor House, northern England. Numerical values are bulk density (mg cm^{-3}). The histograms are arranged approximately in relation to spatial density and the depth of the water table on the day of collection (which was in a dry period). The detailed relationship is at the lower right. Hayward (unpublished).

population E seem to be exceptional. A wider range of results is shown in Fig. 8.29. The populations sampled were all from the centre of their range of normal occurrence. It appears that the bulk density of capitula of plants growing anywhere other than in immersed habitats is nearly constant whatever the species: high density compensates for small individual size. (The estimates for capitula in Fig. 8.29 are rather larger than the central ones in Fig. 8.28, probably because the definition of 1 cm of capitulum is an operational one and depends on how hard the capitulum is pressed against the end of the cutting board). There is much greater variation, for a single species, in the bulk density of product, however, which may be related to the species' response to environmental forces. The species differ too. Clearly, the structure of the individual plants can itself modify the environment to which the branches are exposed, and this effect is now considered.

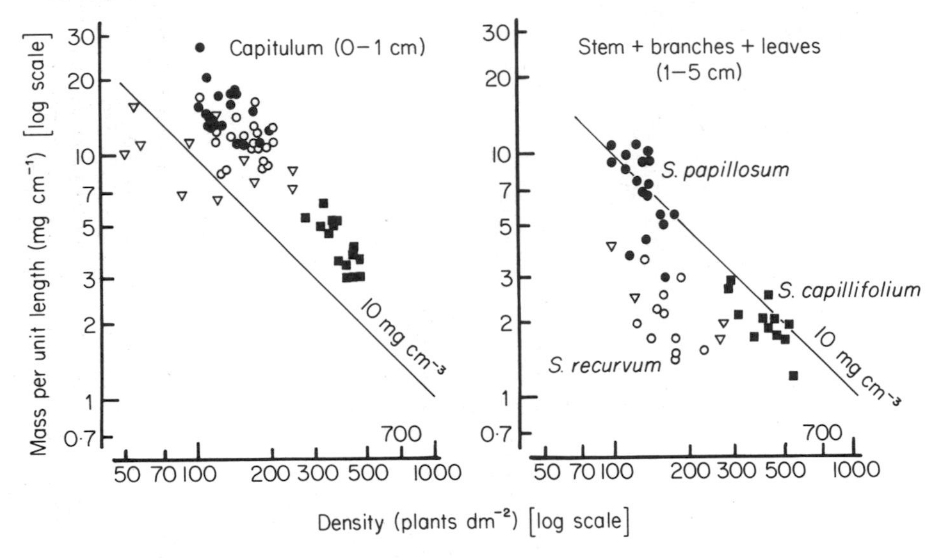

Fig. 8.29 Relationship between spatial density and individual size for capitula (0–1 cm) and stem plus branches (1–4 cm) of *Sphagnum* spp. from various sites in all of which the plants seemed to be growing well. Symbols: ■, *S. capillifolium;* ●, *S. papillosum;* △, *S. cuspidatum;* ○, *S. recurvum.* Redrawn from Clymo (1970).

8.5.2 Physical environment in a *Sphagnum* carpet

The way in which *Sphagnum* affects water supply and potential has already been considered. The plants themselves also affect the light and temperature microclimates.

The light climate in a *Sphagnum* carpet is shown in Fig. 8.30. The semi-logarithmic plot should give a straight line if Beer's law is obeyed. On the whole the results are quite close to a straight line, but they pass through 40–60% transmission at the surface, rather than the expected 100%. There are several possible explanations for this. Definition of the surface is rather arbitrary because the capitula are not flat either individually or *en masse*. An error of only 3 mm would account for most of the difference. Where light has passed through only a few mm of *Sphagnum* canopy then there may be 'light fleck' effects which are less common lower down. The depth would be better measured in units of cumulative mass below the surface because it is the matter which is absorbing and reflecting light. The effects of this are also shown in Fig. 8.30. Some of the surface anomaly is thus accounted for, but not all. The depth of the euphotic zone (within which all but 1% of the incident light is absorbed) is about 1–2 cm in unshaded stands of both *S. papillosum* and *S. capillifolium* and 4–5 cm in the lax growth of shaded stands. Much the same is true in both natural conditions and in

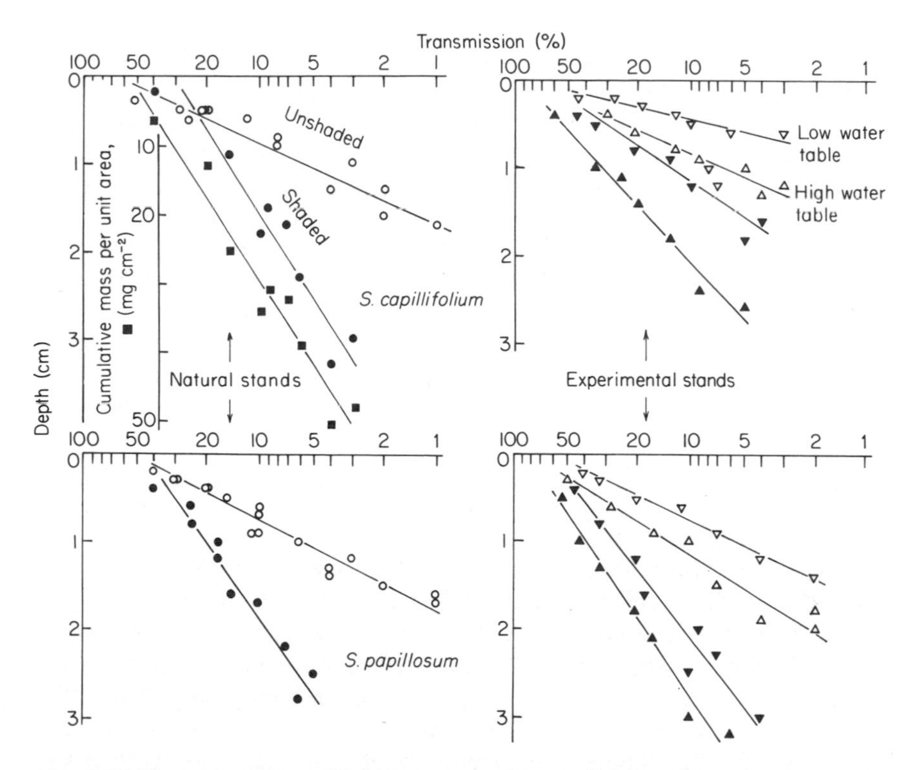

Fig. 8.30 The attenuation of light in carpets of *S. capillifolium* (upper) and *S. papillosum* (lower) both in natural stands (left), and in stands grown with controlled water table and shade in 50 cm-diameter experimental containers (right). Open symbols, unshaded; filled symbols, shaded. The results for the shaded natural stand of *S. capillifolium* are plotted again with depth measured by cumulative mass (squares, top left). Redrawn from Hayward (1980).

experiments. The latter show that the euphotic zone is deeper when the water table is high than when it is low. Again, the growth form of the plants is probably the main cause.

The temperature climate of *Sphagnum* carpets varies in a complex way. Some of these variations are shown in Fig. 8.31, which shows measurements made over five fairly fine days at the end of May and beginning of June. The second, fourth and fifth days had intermittent cloud, and about 6 mm of rain fell on the fifth day. The daily range in air temperature (at 152 cm height) was about 15°C. The temperature about 5 mm below the capitula of a *S. papillosum* carpet was consistently higher than the air temperature except for 2–3 hours just after sunrise; water has a large thermal capacity. By noon the *Sphagnum* temperature was 20–25°C which was about 5°C warmer than the air temperature on sunny days, but little different on the cloudy fourth day. The daily amplitude of *Sphagnum* temperature was about 20°C. Deeper down the temperature fluctuated less. At 5 cm below the surface the temperature oscillations were damped to about 6°C amplitude, lagging 2–3 hours, and with a slower fall than rise; at 29 cm deep the daily fluctuation was less than 0.5°C, but the temperature did rise steadily by 2°C during five days. The reduced amplitude and lag are well known, and from them the thermal diffusivity of the surface layers can be estimated (very approximately). There are distinct differences in the thermal regime in different bog habitats. For example, *S. cuspidatum* in pools warms surprisingly rapidly and on sunny days is about 2°C warmer by noon than is *S. papillosum* and remains so until the early hours of the morning. The north face of a *S. capillifolium* hummock warmed more quickly on a sunny morning than did *S. papillosum* in the early morning and the south-west face of the hummock warmed more slowly. By mid-morning, however, the north face of the hummock was about 5°C cooler than was *S. papillosum*, whilst by noon the hottest place was the south-west face of the hummock. Rain on the fifth day – a heavy shower – reduced the temperature of all habitats by about 7°C.

The temperatures shown in Fig. 8.31 were measured with thermistors shaded from direct radiation if in the air, and not visible from above if just below the capitulum of the plants. The true surface temperature however is probably considerably higher. An infra-red thermometer is necessary for accurate measurements. Values in *S. capillifolium* consistently above 40°C during the greater part of the day have been recorded, during drought, at a site close to that where the measurements for Fig. 8.31 were made (Tattersfield, 1976). The plants were not permanently damaged.

The following points seem clear: *Sphagnum* must tolerate quite high temperatures during summer; it must cope with a fairly large daily fluctuation of temperature; hummock species are exposed to a greater range of temperature than are other species; and, rather surprisingly, lawn

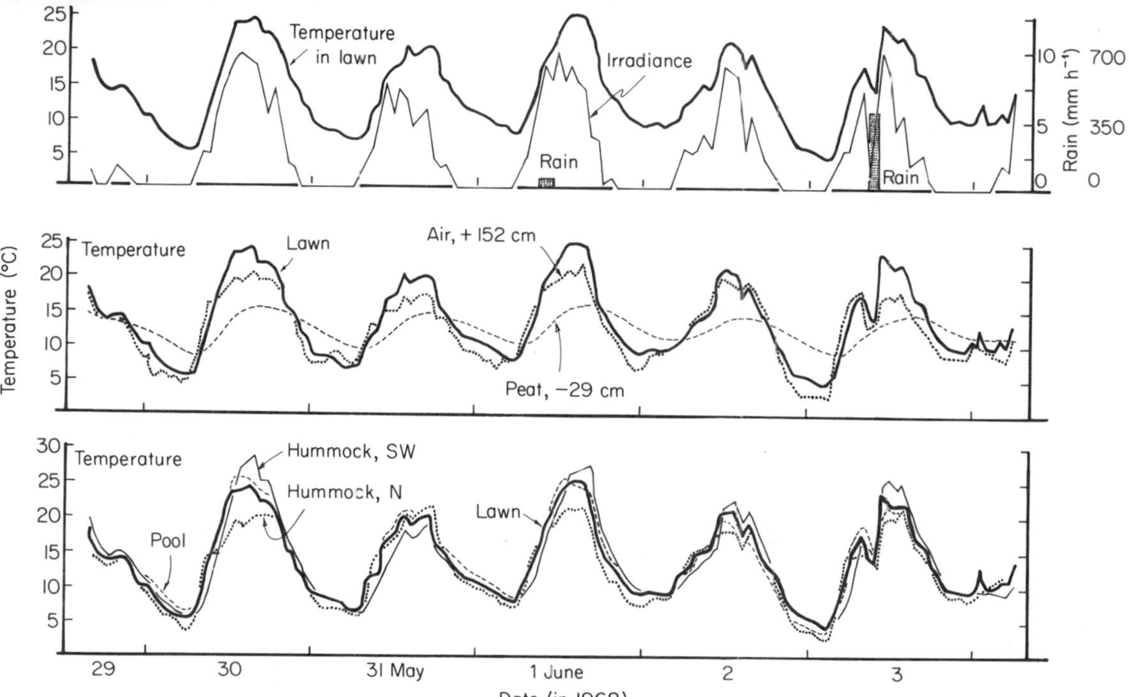

Fig. 8.31 Course, on blanket bog at Moor House, northern England, during 5 days in summer 1968, of irradiance, of rainfall, and of temperature just below the surface of a *S. papillosum* lawn (upper graph), of temperature 15 cm above the surface, just below the surface of a *S. papillosum* lawn, 5 cm below and 29 cm below (centre graph), and of temperature just below the surface of *S. cuspidatum* in a pool, of *S. papillosum* in a lawn, and of *S. capillifolium* on the north face of a hummock and on the southwest face of the same hummock (lower graph). (The *S. papillosum* lawn temperature is repeated to provide a reference trace). Records were made once an hour for 5 s: the irradiance record is therefore rather noisy. Above-ground thermistors were shaded. Those at the surface were horizontal and just sufficiently far below the capitula to be not visible – usually about 3 mm deep. Clymo (unpublished).

species such as *S. papillosum* appear to be exposed to the smallest range of temperature (in sunny weather at least).

There are important features of the regimes in other seasons too (Popp, 1962). In the spring and autumn the bog surface temperature may be high enough to permit active growth for a few hours each day when air or peat temperature, or both, are too low to allow growth of vascular plants.

8.5.3 Growth in relation to shade and water supply

The most detailed experiments are those of Hayward (1980). For *S. capillifolium, S. papillosum,* and *S. recurvum* grown in a factorial design with the water table at 0, 3, 6, 10 or 14 cm below the capitula, and with 0,

Table 8.3 Main effects of shading and of water-table depth on the growth of three species of *Sphagnum*. The experimental design was factorial (5×5×3) and a log transform was used to remove the dependence of error on treatment effect. There were no significant interactions between treatments for growth in mass, but, for growth in length, there were very highly significant interations ($P<0.001$) between species and shade, and between species and water-table level. These are shown in Fig. 8.32. Water-table depth had little effect on growth in mass. From Hayward (1980).

	Growth in mass		Growth in length (cm)
	per plant (mg)	per unit of stem (mg(mg cm^{-1})$^{-1}$)*	
Species	$P < 0.001$	$P < 0.001$	$P < 0.001$
S. capillifolium	3.0	6.2	2.7
S. papillosum	8.5	7.3	2.6
S. recurvum	6.2	13.1	4.9
Shade (proportion of full light absorbed)	$P < 0.001$	$P < 0.001$	$P < 0.001$
0.0	9.2	14.4	2.2
0.54	7.4	12.0	4.0
0.80	5.8	8.1	4.4
0.91	4.0	6.5	3.9
0.96	2.9	4.7	2.5
Water table depth (cm)	$P < 0.05$	$P < 0.05$	$P < 0.001$
0	4.8	7.6	4.2
3	5.4	8.7	4.4
6	5.4	8.4	3.3
10	5.9	9.1	2.7
14	5.6	8.4	2.3

*This measure attempts to make allowance for the difference in size between species by using the mass of a unit length of stem for comparison.

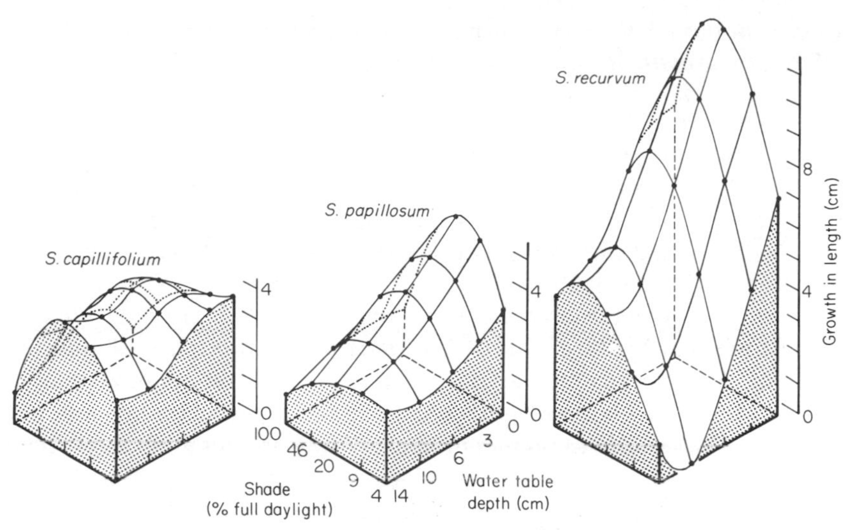

Fig. 8.32 Growth in length of three species of *Sphagnum* in experiments with constant water table and shade. The shapes shown are the least squares fit of the cubic:

$$L = a + bW + cA + dW^2 + eA^2 + fWA + gW^3 + hA^3 + iW^2A + jWA^2$$

where L = growth in length, W = depth to the water table, A = absorbance of the nylon net used for shading, and a to j are parameters. The underlying shapes are not so easy to see in the original results, in which 'noise' is added to the undulating surface but analysis of variance shows highly significant effects of all three main treatments and first order interactions. Redrawn from Hayward (1980).

0.54, 0.80, 0.91, and 0.96 shading (using 0–5 layers of black nylon gauze), there were marked effects associated with species, water table level and shade. For dry matter increase there were no significant interactions, so the effects are summarized in Table 8.3. Water level had little effect, shading reduced growth and there were specific differences associated with plant size. None of these effects is surprising, and they agree in general with the results of other less detailed experiments (Clymo, 1973). For growth in length, however, there were highly significant interactions, individual species behaving differently in response to shade, and to a lesser extent, in response to water level (Fig. 8.32). In unshaded conditions, water level had a large effect on elongation of *S. capillifolium* and *S. recurvum* but not of *S. papillosum*. All three species have an 'optimum' shade, and for *S. papillosum* (but not *S. capillifolium*) a fall in the water table reduces the elongation at the shade optimum. These results are again consistent with those of Clymo (1973). The results of experiments in arctic Sweden are also consistent. *Sphagnum riparium* was grown in differing degrees of shade in natural conditions (Sonesson *et al.*, 1980) and the elongation of the plants measured, though the growth in mass was, unavoidably, not measured. The control of growth in length of *Sphagnum* is obviously not simple, and different species differ in their behaviour. The association of red pigments with unshaded, unetiolated, plants has been mentioned.

A prediction can be made on the basis of the experimental results (Clymo, 1973; Hayward, 1980): the roughness or irregularity of a carpet of *Sphagnum* will be smaller the lower the water table is. Figure 8.33 shows that this is indeed observed, as it is too in experiments with controlled water table (Hayward, 1980).

The effects of water table depth in natural conditions are shown in Fig. 8.34. Batches of similar plants of four species were transplanted to pools, lawns or hummocks. Of particular interest is the observation that *S. capillifolium* which normally grows on hummocks, grew best in pools, but it out-grew *other* species on hummocks. All four species grew better than other species in that habitat in which they are usually found. This emphasizes that relative performance and interspecific competition are probably important in field conditions – a point which is often assumed but less often demonstrated – and that the results of laboratory experiments which have not been confirmed by field experiments must be treated with caution.

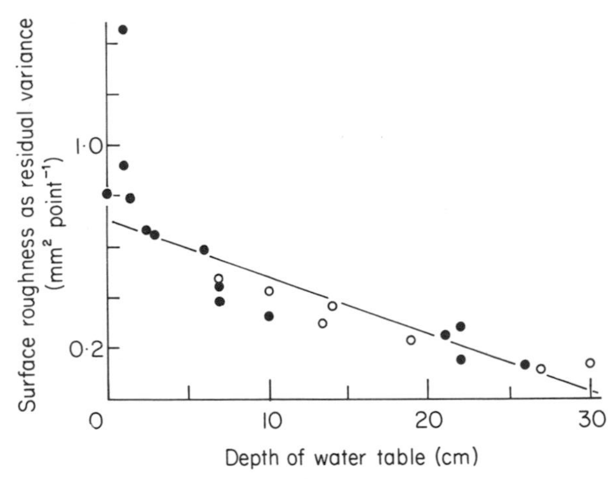

Fig. 8.33 Roughness of the surface of *S. papillosum* lawns in relation to water table depth measured after several days without rain. Open symbols, at the Silver Flowe, southern Scotland; filled symbols, at Moor House, northern England. Roughness is the residual variance in height of a grid of 4×8 light pointers (10 cm lengths of culms of *Deschampsia caespitosa*), 0.8 cm apart, after allowing for larger scale curvature of the surface by fitting.

$$H = a + bX + cY + dX^2 + eY^2 + fXY$$

where H is the height of the pointer, X and Y the grid co-ordinates, and a to f are parameters. The point at (1, 1.45) was excluded from the linear regression shown. Redrawn from Hayward (1980).

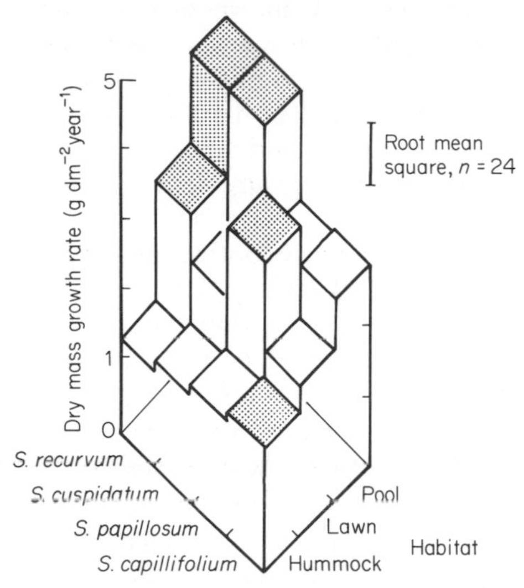

Fig. 8.34 Growth of four species of *Sphagnum* in three habitats at Moor House, northern England. The plants, cut to 5 cm long, were transplanted in groups and allowed to grow for 12 months. Stippling shows natural habitat. Redrawn from Clymo and Reddaway (1971).

It seems that specific structural differences are sufficient, acting through control of the water supply, to affect the growth of the individual plants, and that the habitat preferences of different species may be determined mainly by competitive growth during 'normal' conditions, sharpened by selective survival of some plants during drought. Nothing in the detailed ecology of the species encourages the view that there is likely to be a hummock and hollow cyclic succession (Tansley, 1939; Watt, 1947), but nor can widespread synchronous changes (Walker and Walker, 1961) involving the extension or contraction of carpets of a particular species be explained either. Climatic change still seems the most likely cause of such changes.

8.5.4 Productivity

Accurate measurement of productivity is not easy. The methods causing least disturbance to the intact *Sphagnum* make use of the carpet plant's innate pattern of branch density (Malmer, 1962b) or utilize some sort of external marker such as vertical wires. More disturbance is caused by removing plants, cutting them to known length, and replacing them (Clymo, 1970) or by marking them with thread (Overbeck and Happach, 1956) though these methods do allow transplant experiments to be made. It may be necessary to use a method which allows for that part of the capitulum carried up passively by internode elongation; for the different size of individual plants; and, in the measurements of growth in length, for the average bulk density of the stem-plus-branch product produced by the capitulum factory (Clymo, 1970). The method using vertical wires has been refined by Sonesson (personal communication) to the point where the standard deviation of replicate measurements of length in the field is less than 0.3 mm.

Estimates of productivity of fairly continuous carpets of *Sphagnum* seem generally to be in the range 100–600 g m^{-2} year^{-1}. On British blanket bog, productivity (g m^{-2} year^{-1}) of 150 (hummocks), 500 (lawns) and 800 (pools) has been recorded (Clymo, 1970; Clymo and Reddaway, 1974; Smith and Forrest, 1978). The high value for pools is largely because *S. cuspidatum* seems to continue to grow throughout most of the year, whilst growth in the other habitats stops in winter. Similar values (100–600 g m^{-2} year^{-1}) are reported for the drier hummocks and wet depressions of Stordalen mire near Abisko in arctic Sweden (Sonesson, 1973) and in Finland (Pakarinen, 1978a). The arctic values are the more remarkable because the summer there is short, and the average temperature over the whole year is little above 0°C. The rate of growth in wet habitats during the few weeks of continuous light in midsummer is remarkably high: the linear growth rate of *S. riparium* is about 1 cm week^{-1} (Sonesson *et al.*, 1980), corresponding to an instantaneous rate of about 2500 g m^{-2} year^{-1}. Of course, this high rate is sustained for a brief period only, and presumably, there is no growth at all during the long arctic winter.

8.5.5 Rate of photosynthesis and respiration

It is likely that the rate of photosynthesis and respiration by a given branch of a particular species of *Sphagnum* will depend on the light flux, temperature, water content and history of the branch at least. The earliest measurements for *Sphagnum* are those of Stålfelt (1938). He showed that *S. girgensohnii* was light-saturated at about 20% of full summer midday light flux, and that in these conditions assimilation was maximal at 18°C. Nowadays these measurements would be considered to be technically crude, and Stålfelt gives no indication of the plant's water content or history. More recently, Grace (1970) examined rates of photosynthesis and respiration in *S. capillifolium* in relation to light flux, temperature and water content in laboratory experiments in which he measured the change in CO_2 concentration of a gas stream passing over the plants. He found that the rates varied greatly from one experiment to the next. If plants were allowed to dry out slowly then the rate of photosynthesis was greatest at a water content of about 10 g g^{-1}. At water content above that value it declined slowly, presumably as a result of longer diffusion paths for CO_2, as found for other bryophytes (Tallis, 1959; Willis , 1964). In drier conditions,

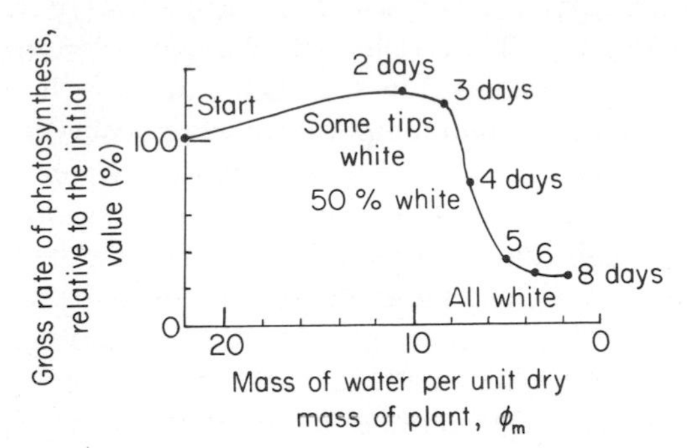

Fig. 8.35 Rate of gross photosynthesis (relative units) of *S. capillifolium* in relation to water content. The plants were allowed to dry during the experiment, so water content is confounded with history, as it is in the field. Redrawn from Grace (1970).

the rate declined abruptly as the hyaline cells emptied (Fig. 8.35). It is of interest that this threshold value of water content is just that most often found in natural conditions (see Fig. 8.20). It is not clear how far the rapid decline as the water content dropped below 10 g g^{-1} was a result of reduced water content *per se* or of a lowered water potential. Grace (1970) showed that neither the temperature of storage, for up to one month, nor the time of year at which measurements were made, had much effect on the relationship of rate of photosynthesis with light flux or temperature. One set of results is shown in Fig. 8.36. The water content in these experiments was greater than 10 g g^{-1}. There is some indication of light saturation at the highest light flux used (about half that of full summer midday values), but it is much less clear than in Stålfelt's experiments with *S. girgensohnii.* The photosynthesis rate, at moderate to high light fluxes, increases with temperature up to 27°C at least. This may be important in the field, because the temperature on *Sphagnum* hummocks is often above this, as already described. The compensation point increases as temperature does: dark respiration rate increases approximately exponentially with temperature (Grace, 1970) and it seems likely that the temperature optimum for photosynthesis is lower than that for respiration, as it is in so many plants, though there is no published direct experimental evidence of this in *Sphagnum.*

Complementing these laboratory experiments are two sets of field measurements. Johansson and Linder (1980), working in arctic Sweden, used the rate of incorporation of ^{14}C to measure the net rate of photosynthesis of *S. fuscum* (a hummock species) and *S. balticum* (a species of wetter habitats, similar to *S. recurvum*). Their experimental plants were in unnatural conditions to the extent that the plants were removed from the carpet and enclosed at near-natural spatial density in an experimental chamber. They relied on changing climatic conditions to create natural experiments. In practice, only a limited range of *combinations* of light flux and temperature were observed and the effects of these are confounded. Nor did they record the water content, though Grace's results suggest that this may be of less importance in natural conditions than might have been expected. The response to light flux is shown in Fig. 8.37, which includes measurements made with temperature between 10 and 20°C. The rate of photosynthesis is much the same as that of *S. capillifolium* at the same light flux, and with the same variability. The rate for *S. balticum* is perhaps rather higher than that for *S. fuscum*. Both

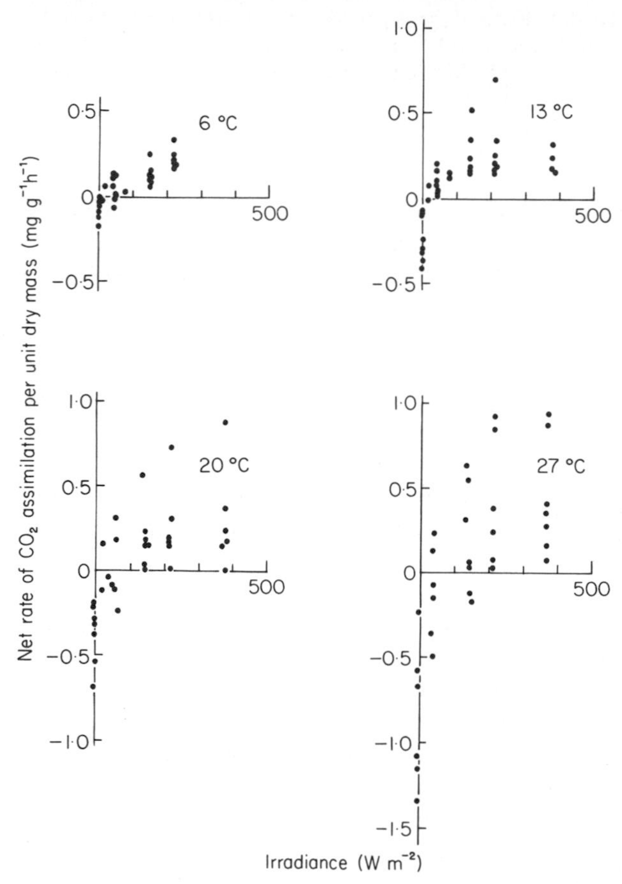

Fig. 8.36 Rate of net photosynthesis of *S. capillifolium* in relation to irradiant flux and temperature. The plants were in a container through which a stream of air passed. The rate of photosynthesis was calculated using the change in CO_2 concentration of the gas stream. The light source was artificial. Redrawn from Grace (1970).

show signs of becoming light-saturated at the highest irradiance, but the hyperbola of best fit for *S. balticum* gives a saturation asymptote for photosynthetic rate (as CO_2 per unit time per unit dry mass of plant) of 1.2 mg g^{-1}, h^{-1}, and half-saturation at a light flux of 300 W m^{-2}. The daily

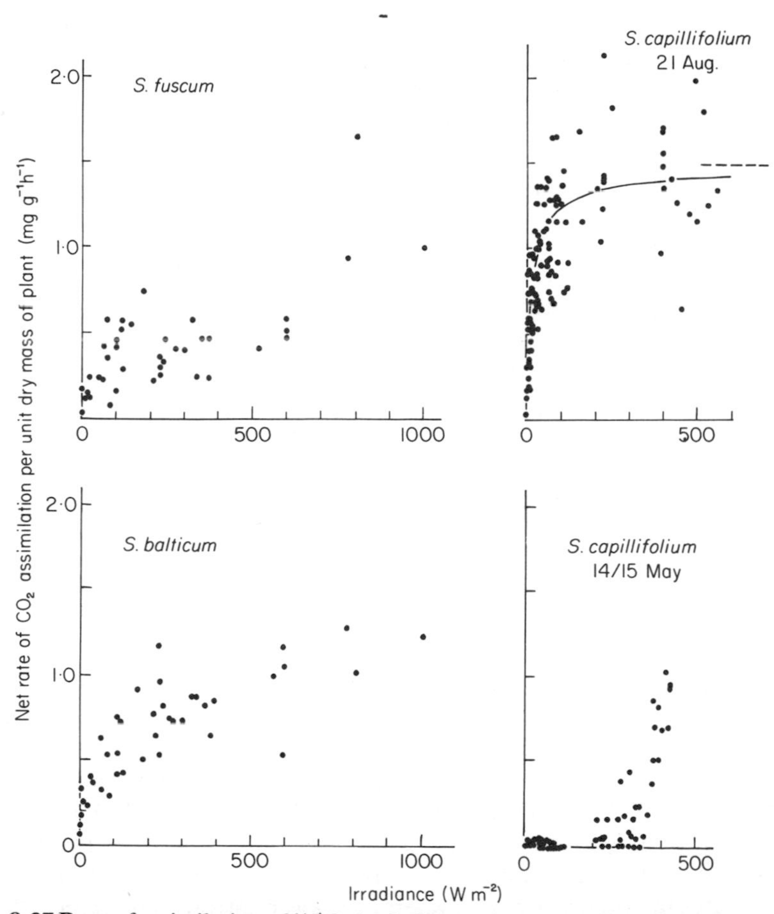

Fig. 8.37 Rate of assimilation of $^{14}CO_2$ by *S. fuscum* and *S. balticum* throughout the growing season at Stordalen, Abisko, arctic Sweden and of *S. capillifolium* at Moor House, northern England on two single days. The hyperbola of best fit is shown for one day; broken line is asymptote. Capitula of the plants (top 1 cm for *S. capillifolium*) were put in experimental chambers in the field and measurements made at the prevailing temperature and water content (10–20°C in the Swedish plants, values shown in Fig. 8.39 for the British plants). The original values for *S. capillifolium* are per unit area of carpet, and light flux was measured as quantum flux. To allow comparisons to be made it has been assumed that the capitula have a bulk density of 20 mg cm^{-3} (Fig. 8.29) and that on average 1 W = 4.15 μmol s^{-1}. The results for 14–15 May 1980 are exceptional, being taken when the plants were unusually dry (Fig. 8.39). Redrawn from Johansson and Linder (1980) and from S. Daggitt (unpublished).

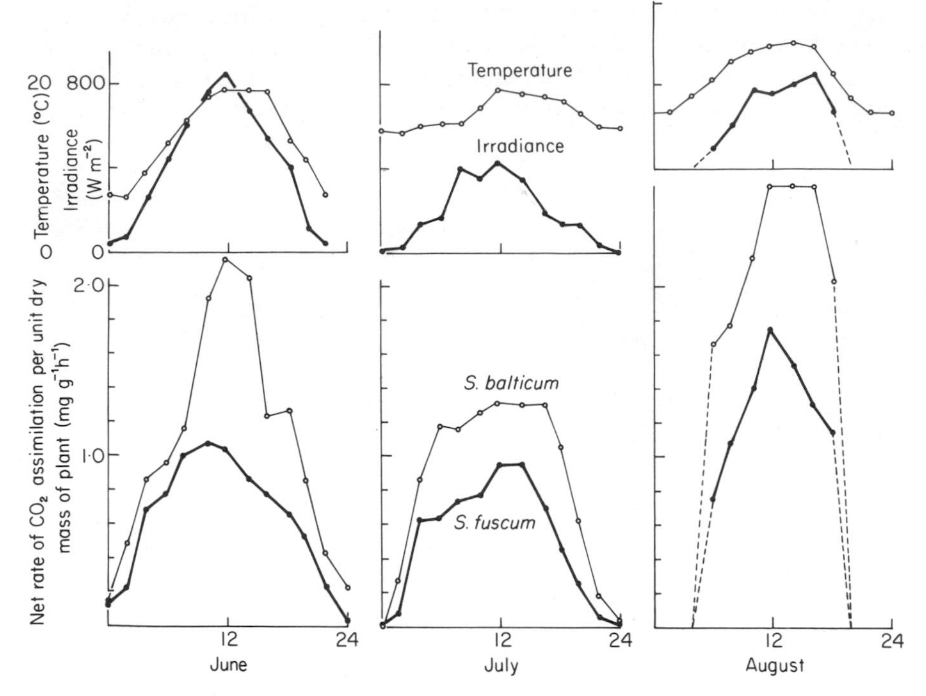

Fig. 8.38 Course of irradiance, temperature, and the rate of CO_2 assimilation by *S. fuscum* and *S. balticum* on three days during the growing season at Stordalen, Abisko, arctic Sweden. Redrawn from Johansson and Linder (1980).

course of photosynthetic rate is shown in Fig. 8.38. Differences between species are not easy to interpret, because the sampling method was necessarily destructive, and because the proportion of photosynthetic machinery is not necessarily closely related to dry mass: one species may, for a given cell volume, have thicker walls and this would make it appear to be less efficient. This may be one cause of variability between plants too. There is need for an investigation of the best basis for comparisons. The higher rates in August, when temperature was relatively low, than in June or July may be because the rate of respiration is proportionately lower than is the rate of gross photosynthesis.

The other set of field measurements were made by Daggitt (unpublished) working on *S. capillifolium* at Moor House, northern England. He too removed capitula (the top 1 cm) of plants, packed them at natural density in an experimental chamber, and measured the incorporation of ^{14}C using a method which gives a result somewhere between gross and net photosynthesis. Some of his results are shown in Fig. 8.37. Those for 21 August, typical of most days, again show considerable scatter but a tendency to saturate. The hyperbola (maximum likelihood fit)

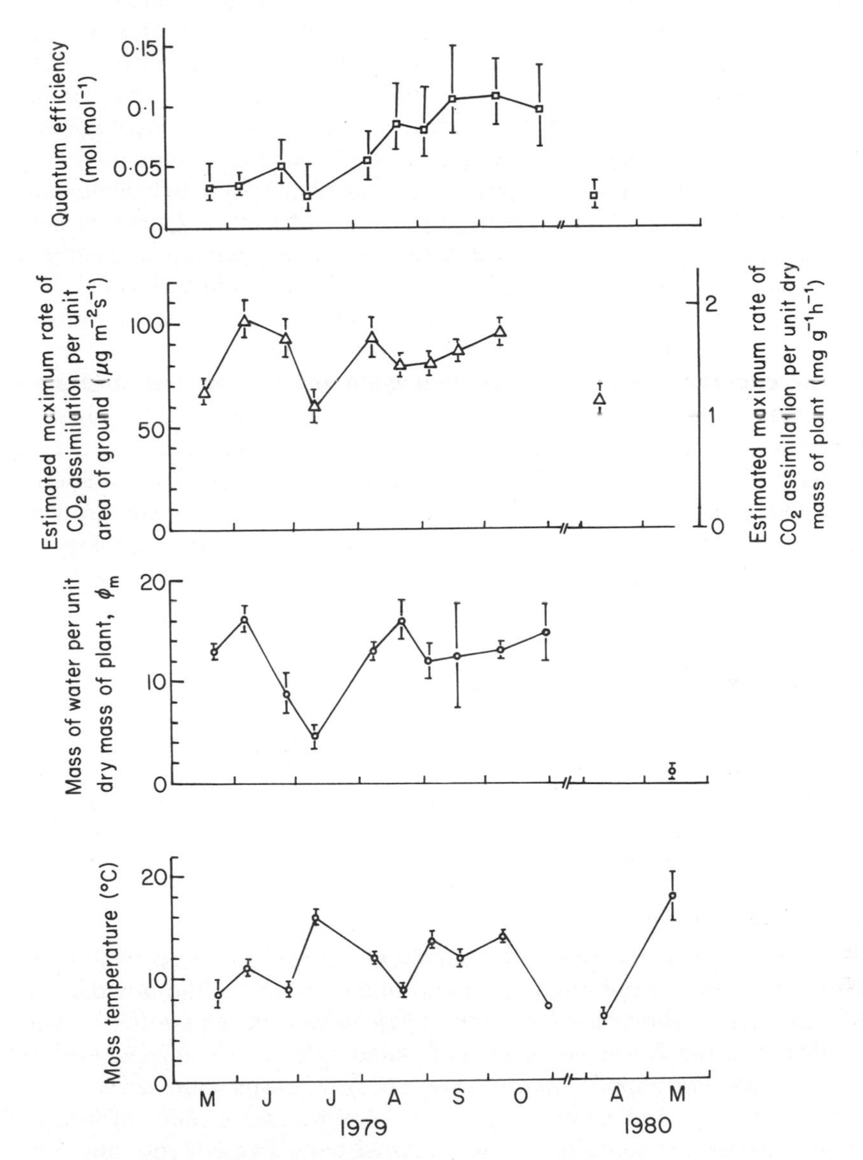

Fig. 8.39 Seasonal course for *S. capillifolium* at Moor House, northern England, of quantum efficiency (the slope of the maximum likelihood hyperbola, Fig. 8.37, at zero light flux), of maximum rate of photosynthesis (the asymptote of the maximum likelihood hyperbola, Fig. 8.37), of capitulum water content, and of mean temperature in the *Sphagnum* capitula during the day of measurements. (For much of the time the temperature was above the mean). From S. Daggitt (unpublished).

has an asymptote, of about 1.4 mg g^{-1} h^{-1}, similar to that of the arctic measurements. Daggitt found that this estimated maximum rate was fairly constant throughout the growing season, except for lower values in May and July (Fig. 8.39). The low early summer value may be attributed to damage caused to capitula during the winter. The lower midsummer value is correlated with extremely dry conditions when the water content fell well below the 10 g g^{-1} level. An extreme case of this kind was found in May 1980. The water content was about 2 g g^{-1} and the response to light flux was so atypical (Fig. 8.37) that no hyperbola could be fitted. Quantum light fluxes were measured so it is possible to estimate the quantum efficiency as the slope of the hyperbola at zero light flux. The units are moles of carbon assimilated per mole (6.02×10^{23} quanta) of light. The efficiency starts at about 0.03 in May and rises to about 0.1 by September.

The experiments of Grace, of Johansson and Linder, and of Daggitt agree in showing that nominally replicate samples differ considerably, and that the light saturated rate of photosynthesis is about 1–2 mg g^{-1} h^{-1}. Daggitt's results show that this rate is much lower in very dry conditions, and that the quantum efficiency increases steadily during the growing season. But there is still much to be learned about the physiology of photosynthesis in *Sphagnum*.

8.6 ASSOCIATED PLANTS AND ANIMALS

Sphagnum carpets and the climate in which they flourish combine to create a habitat characterized by a physically open interdigitating network of gas and liquid filled spaces over a waterlogged peat, and which is chemically unusually acid and poor in inorganic solutes. Relatively few other species seem to be able to flourish in these conditions.

8.6.1 Plant associates

Amongst the vascular plants the commonest associates are probably the cotton grasses (*Eriophorum* spp.) and other 'sedges' in the broad sense (*Carex* spp., *Rhynchospora* spp., *Trichophorum cespitosum*), and members of the Ericaceae (such as *Calluna vulgaris, Erica* spp., *Kalmia* spp., *Vaccinum* spp., *Andromeda* spp.). Conspicuous shrubs are *Empetrum* spp. and *Ledum* spp., and there are also a range of insectivorous plants including the closely related pairs *Drosera* spp. and *Sarracenia* spp., and *Pinguicula* spp. and *Utricularia* spp. It is remarkable that the insect-trapping mechanisms in these closely related pairs are so different: sticky leaf and pitcher, sticky leaf and bladder respectively. The nutritional advantages of the insectivorous habit to a plant growing amongst *Sphagnum* are obvious and have been demonstrated by experiment. The name of the beautiful yellow bog asphodel, *Narthecium ossifragum,* conceals an instructive confusion of correlation and regression concerned with the low supply of calcium. Plants with broad soft leaves are rare on bogs, the notable (and delicious) exception being *Rubus chamaemorus*. Some of these species (*Calluna vulgaris* for example) grow some way above the water table (Fig. 8.4) and have few or no roots in the waterlogged peat. Others, such as *Eriophorum* spp. and *Rubus chamaemorus* have at least some roots with large aerenchyma which provides an effective conduit for the downward movement of O_2 to the root tip (Armstrong, 1964). The roots may grow down 180 cm or more (Stavset, 1973). Most of the species are rhizomatous perennials; annuals are rare on bogs. *Calluna vulgaris* is particularly interesting. On dry, sandy soils in southern and eastern England it has a determinate pattern of growth and life-span of about 30 years (Watt, 1947); in the Cairngorm mountains it grows in arcs, driven before the prevailing wind, and appears to be potentially immortal. It seems to be potentially immortal too in association with *Sphagnum capillifolium* on blanket bog: the *Sphagnum* plants grow up around the stem and create conditions which stimulate the development of roots (Smith and Forrest, 1978). The *Calluna* plant behaves rather like the *Sphagnum* itself, growing at the shoot tips and dying where completely buried, but persisting as an individual and moving steadily upwards, as *Ammophila arenaria* does in sandy habitats. The equilibrium between *S. capillifolium* and *Calluna* seems as though it might easily be disturbed, but this has not been demonstrated. A similar rejuvenation with potential immortality has been shown for *Calluna* growing on Swedish sand dunes with much accumulation of organic matter (Wallén, 1980). This last example provides an interesting comparison with the behaviour of *Calluna* on southern and eastern English sandy soils.

The physical and chemical conditions created by *Sphagnum* seem to favour the growth of a number of other bryophytes, particularly small leafy liverworts. These may sometimes be abundant forming a fibrous network just below the *Sphagnum* capitula and binding the whole mass of plants together. Some of these leafy liverworts (such as *Mylia,* spp., *Lepidozia setacea* and *Odontoschisma sphagni* are not only common amongst *Sphagnum* but also rare elsewhere. Others (such as *Gymnocolea inflata, Cephalozia biscuspidata* and *C. connivens*) are found in almost any acid habitat. Similar patterns are seen in the associated mosses. The grouping of *Sphagnum recurvum* and *Polytrichum commune* (often with the grass *Molinia caerulea*) is characteristic of rather flushed habitats, but *P. commune* is widespread in other habitats too. So are *Aulacomnium palustre, Racomitrium lanuginosum,* and, in oceanic Britain, *Campylopus atrovirens*. A few mosses are almost entirely restricted to *Sphagnum* bogs however: *Dicranum undulatum* (= *D. bergeri*) is an example.

Amongst the algae there is a large number of species associated with *Sphagnum*. Almost any gathering of plants when examined with a microscope, is seen to contain abundant algae, many of them motile and passing in and out of the hyaline cells of the leaf. The most extensive studies of algal abundance and diversity amongst *Sphagnum* are those of Flensburg (1965, 1967), Flensburg and Malmer (1970), and Hooper (in press). Desmids and diatoms predominate, but several species of heterocystous blue-green algae also occur. In the arctic mire studied by Basilier *et al.* (1978), these algae were especially associated with the immersed species *Sphagnum riparium* growing in flushed depressions. The commonest algae were of the genera *Calothrix, Chlorogloea, Fischerella, Hapalosiphon, Nostoc, Scytonema, Tolypothrix,* and free-living *Anabaena*. This association of *S. riparium* and algae was found to reduce acetylene to ethylene at appreciable rates – an indication of potential nitrogen-fixing ability. Reasonable assumptions about the relationship between N_2 fixation and acetylene reduction gave a rate for nitrogen fixation of up to 6 g m^{-2} $year^{-1}$, which is a significant contribution to the nitrogen budget of these oligotrophic areas. The rate was independent of pH in the range 4.5 to 7, but reduced to 10% at pH 3. The rate was positively correlated with light flux and with temperature, and was greatest at mid-day and in mid-summer. The greatest rate occurred a few cm below the *Sphagnum* capitulum.

Associations of this kind may be fairly common in flushed habitats, but seem to be absent, or at least less active, in the majority of *Sphagnum* carpets with less lateral movement of water.

8.6.2 Animal associates

A remarkable feature of *Sphagnum* is that almost nothing eats it. In spring, frogs may spawn in bog pools amongst *S. cuspidatum*. They often die there and their bodies may perhaps be significant as concentrators of plant nutrients in the pools. Owls often perch, and defaecate, on *Sphagnum* hummocks. Again, they leave a local concentration of nutrients. Many invertebrates burrow in *Sphagnum,* particularly tipulids and enchytraeid worms, and one tipulid *Phalacrocera replicata,* ingests *Sphagnum*. Adult midges (chironomids) can be so numerous as to make work impossible; the larvae live amongst *Sphagnum*. Some invertebrates are restricted almost entirely to *Sphagnum*: the ant *Formica transkaucasica,* for instance, often nests in *Sphagnum* hummocks. The hunting spiders *Pirata piraticus* and *Lycosa pullata* both live in the top few cm of 'forests' of *Sphagnum recurvum,* but *P. piraticus* spends most of the time a few cm below the surface (out of sight to casual inspection) whilst *L. pullata* runs on the surface of the *Sphagnum* capitula (Nørgaard, 1951). *Pirata piraticus* will bring cocoons to the surface, but the two species do not usually meet or compete. When given the choice *L. pullata* prefers a temperature in the range 28–36°C, whilst *P. piraticus* generally prefers the range 18–24°C (and prefers higher humidity too). These preferences, combined with the *Sphagnum* microclimate already described, are sufficient to account for the separation of the species. Clear associations between oribatid mites and different degrees of wetness have been shown by Tarras-Wahlberg (1952). Here again, it seems that *Sphagnum* simply provides suitable physical and chemical conditions. Rotifers are common amongst *Sphagnum*. Again, the degree of restriction varies: *Keratella serrulata* is free-swimming and common in other habitats too, but *Habrotrocha angusticollis* builds a 'house' in a hyaline cell of a *Sphagnum* leaf and is not common elsewhere.

Perhaps the best known invertebrates associated with *Sphagnum* are the various species of testate rhizopod (Paulson, 1952; Heal, 1962; Corbet with Harding, 1973; Meisterfeld, 1977; and many others). These beautiful protozoa cover themselves with a house (test) made of plates. The houses, whose structure is characteristic of the species, persist after the occupant dies and become incorporated in peat. Different species prefer habitats of different wetness, so the discovery of tests in peat may allow the conditions

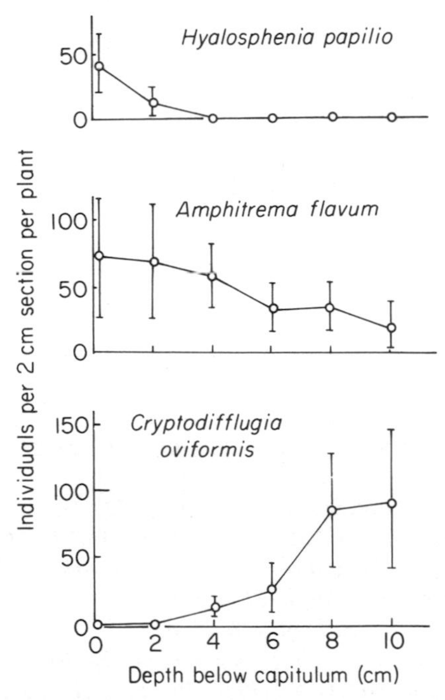

Fig. 8.40 Abundance of three species of testate amoeboa (rhizopods) at different depths below the capitulum of *S. recurvum*. The vertical bars are 90% confidence limits. Redrawn from Heal (1962).

at the time the peat formed to be deduced (e.g. Aaby and Tauber, 1975). For example, species such as *Difflugia bacillifera* are associated with the wettest conditions, *Amphitrema* spp. and *Nebela* spp. with drier conditions, and *Trigonopyxis arcula* and *Bullinularia indica* with the driest conditions on hummocks. Within these broad divisions there are clear differences associated with vertical position on the *Sphagnum* too (Fig. 8.40). Information on other invertebrates is given in Chapter 9.

8.7 CONCLUSION

The bulk of *Sphagnum* is greater than that of any other bryophyte genus (or perhaps of any plant genus) and the plants have the ability to produce and maintain an unusually acid environment given a suitable water supply. They grow upwards, decay slowly beneath, and accumulate as waterlogged peat. This combination of acidity, upward blanket-like growth, and waterlogging determines what other species, both plant and animal, can grow with them. The individual species of *Sphagnum* differ in their structure and water relations in ways consistent with their ecology.

All in all, *Sphagnum* is a beautiful and remarkable genus of plants.

ACKNOWLEDGMENTS

We thank Mrs P. Ratnesar for technical help over many years, S. Daggitt and the members of the Mires Research Group named in the caption to Fig. 8.4 for permission to use as yet unpublished results, Dr K.E. Clymo for comments on the manuscript and Dr B. Bolton, Dr J.C. Coulson, and Professor J.E. Green for information about animal associates of *Sphagnum*.

REFERENCES

Aaby, B. and Tauber, H. (1975), *Boreas,* **4,** 1–18.
Anschutz, I. and Gessner, F. (1954), *Flora, Jena,* **141,** 178–236.
Armstrong, W. (1964), *Nature,* **204,** 801–2.
Basilier, K., Granhall, U. and Stenstrom, T-A. (1978), *Oikos,* **31,** 236–46.
Bazzaz, F.A. and Harper, J.L. (1976), *J. appl. Ecol.,* **13,** 211–16.
Bendz, G.O., Mårtensson, O. and Nilsson, E. (1966), *Arkiv für Kemi,* **25,** 215–21.
Benson-Evans, K. (1964), *Bryologist,* **67,** 431–5.
Bismarck, R. von (1959), *Flora, Jena,* **148,** 23–83.
Boatman, D.J. and Lark, P.M. (1971), *New Phytol.,* **70,** 1053–9.
Boatman, D.J. and Tomlinson, R.W. (1973), *J. Ecol.,* **61,** 653–66.
Braithwaite, R. (1880), *The Sphagnaceae or Peat Mosses of Europe and North America,* Bogue, London.
Brehm, K. (1968), *Planta, Berlin,* **79,** 324–45.

Brehm, K. (1970), *Beitr. Biol. Pflanzen,* **47,** 91–116.
Brehm, K. (1971), *Beitr. Biol. Pflanzen,* **47,** 287–312.
Cavers, F. (1911), *New Phytol.,* **10,** 1–21.
Clymo, R.S. (1963), *Ann. Bot. N. S.,* **27,** 309–24.
Clymo, R.S. (1965), *J. Ecol.,* **53,** 747–58.
Clymo, R.S. (1967), Control of cation concentrations, and in particular of pH, in *Sphagnum*-dominated communities. In: *Chemical Environment in the Aquatic Habitat* (Golterman, H.L. and Clymo, R.S., eds), pp. 273–84. North Holland, Amsterdam.
Clymo, R.S. (1970), *J. Ecol.,* **58,** 13–49.
Clymo, R.S. (1973), *J. Ecol.,* **61,** 849–69.
Clymo, R.S. (1978), A model of peat bog growth. In: *Production Ecology of British Moors and Montane Grasslands* (Heals, O.W. and Perkins, D.F., eds), pp. 187–223. Springer, Berlin.
Clymo, R.S. (1982), Peat. In: *Ecosystems of the World,* (Gore, A.J.P., ed.), Vol. 4, Mires: Swamp, Fen, Bog and Moor, pp. 159–224. Elsevier, Amsterdam.
Clymo, R.S. and Reddaway, E.J.F. (1971), *Hidrobiologia,* **12,** 181–92. Reproduced without arbitrary cuts as: *Moor House Occasional Papers,* No. **3.**
Clymo, R.S. and Reddaway, E.J.F. (1974), *J. Ecol.,* **62,** 191–6.
Corbet, S.A. (1973), *Field Studies,* **3,** 801–38.
Coulson, J.C. and Butterfield, J. (1978), *J. Ecol.,* **66,** 631–50.
Damman, A.W.H. (1978), *Oikos,* **30,** 480–95.
Dickinson, C.H. and Maggs, G.H. (1974), *New Phytol.,* **73,** 1249–57.
Dilks, T.J.K. and Proctor, M.C.F. (1979), *New Phytol.,* **82,** 97–114.
Donnan, F.G. (1911), *Zeit. Elektrochem.,* **17,** 572–81.
Ferguson, P., Lee, J.A. and Bell, J.N.B. (1978), *Environ. Pollut.,* **16,** 151–62.
Ferguson, P. and Lee, J.A. (1979), *New Phytol.,* **82,** 703–12.
Ferguson, P. and Lee, J.A. (1980), *Environ. Pollut.,* Ser. A, **21,** 59–71.
Flensburg, T. (1965), *Acta phytogeog. suec.,* **50,** 159–60.
Flensburg, T. (1967), *Acta phytogeog. suec.,* **51,** 1–132.
Flensburg, T. and Malmer, N. (1970), *Bot. Notiser,* **123,** 269–99.
Goode, D.A. (1970), *Ecological Studies on the Silver Flowe Nature Reserve.* PhD thesis, Univ. Hull.
Goodman, P.J. and Paton, J.A. (1954), *Trans. Br. bryol. Soc.,* **2,** 470.
Gorham, E. (1956), *J. Ecol.,* **44,** 375–82.
Grace, J. (1970), *The Growth-Physiology of Moorland Plants in relation to their Aerial Environment.* Ph.D. thesis, Univ. Sheffield.
Green, B.H. (1968), *J. Ecol.,* **56,** 47–58.
Green, B.H. and Pearson, M.C. (1968), *J. Ecol.,* **56,** 245–67.
Hayward, P.M. (1980), *Effects of Environment on the Growth of Sphagnum.* Ph.D. thesis, Univ. London.
Hayward, P.M. and Clymo, R.S. (1982), *Proc. Roy. Soc. B.* (in press).
Heal, O.W. (1962), *Oikos,* **13,** 35–47.
Heal, O.W., Latter, P.M. and Howson, G. (1978), A study of the rates of decomposition of organic matter, In: *Production Ecology of British Moors and Montane Grasslands,* (Heal, O.W. and Perkins, G.F., eds), pp. 136–59. Springer, Berlin.

Hill, M.O. (1978), Sphagnopsida. In: *The Moss Flora of Britain and Ireland* (Smith, A.J.E., ed.). Cambridge University Press, London.
Hodgman, C.D., Weast, R.C. and Selby, S.M. (1961), *Handbook of Chemistry and Physics*, 43rd edn., Chemical Rubber Co., Cleveland, Ohio.
Hooper, C.A. (in press), *Oikos*.
Hulme, P.D. (1976), *Developmental and ecological studies on Craigeazle Bog, the Silver Flowe National Nature Reserve, Kirkcudbrightshire*. Ph.D. thesis, Univ. Hull.
Ingram, H.A.P. (1978), *J. Soil Sci.*, **29**, 224–7.
Ingram, H.A.P. (1982), Water relations of mires. In: *Ecosystems of the World* (Gore, A.J.P., ed.), Vol. 4, Mires: Swamp, Bog, Fen and Moore. Elsevier, Amsterdam.
Ingram, H.A.P., Rycroft, D.W. and Williams, D.J.A. (1974), *J. Hydrol.*, **22**, 213–18.
Johansson, L-G. and Linder, S. (1980), Photosynthesis of *Sphagnum* in different microhabitats on a subarctic mire. In: *Ecology of a Subarctic Mire*, (Sonnesson, M., ed.), pp. 181–90. Swedish Natural Science Research Council, Stockholm.
Johnson, A. (1960), *Trans. Br. bryol. Soc.*, **3**, 725–8.
Kivinen, E., Heikurainen, L. and Pakarinen, P. (1979), *Classification of Peat and Peatlands*, International Peat Society. Obtainable from I.P.S., V.V.T. Fuel Research Lab., Biologinkiya 3, SF 02150, Espo 15, Finland.
Lange, B. (1973), *Lindbergia*, **2**, 81–93.
Malmer, N. (1962a), *Op. Bot.*, **7:1**, 1–322.
Malmer, N. (1962b), *Op. Bot.*, **7:2**, 1–67.
McVean, D.N. (1959), *Ecol.*, **47**, 615–18.
Meisterfeld, R. (1977), *Arch. Hydrobiol.*, **79**, 319–356.
Moore, P.D. and Bellamy, D.J. (1974), *Peatlands*, Elek Science, London.
Morris, R.B. (1977), *J. Bryol.*, **9**, 387–92.
Mozingo, H.N., Klein, P., Zeevi, Y. and Lewis, E.R. (1969), *Bryologist*, **72**, 484–8.
Newbould, P.J. (1960), *J. Ecol.*, **48**, 361–83.
Nørgaard, E. (1951), *Oikos*, **3**, 1–21.
Nyholm, E. (1969), *Illustrated Moss Flora of Fennoscandia*. II. *Musci*. Fasc. 6., pp. 647–799. Natural Science Research Council, Stockholm.
Olsen, C. (1923), *C. R. Trav. Lab. Carlsberg*, **15 (1)**, 1–166.
Overbeck, F. and Happach, H. (1956), *Flora, Jena*, **144**, 335–402.
Pakarinen, P. (1978a), *Ann. Bot. Fenn.*, **15**, 15–26.
Pakarinen, P. (1978b), *Ann. Bot. Fenn.*, **15**, 287–92.
Pakarinen, P. and Mäkinen, A. (1976), *Suo*, **27**, 77–83.
Pakarinen, P. and Tolonen, K. (1977a), *Oikos*, **28**, 69–73.
Pakarinen, P. and Tolonen, K. (1977b), *Suo*, **28**, 95–102.
Paton, J.A. and Goodman, P.J. (1955), *Trans. Br. bryol. Soc.*, **2**, 561–7.
Paul, H. (1908), *Mitt, K. Bayer. Mookulturanstalt*, **2**, 63–117.
Paulson, B. (1952), *Oikos*, **4**, 151–65.
Popp, E. (1962), *Int. Rev. ges. Hydrobiol. Hydrograph.*, **47**, 431–64.
Ratcliffe, D.A. and Walker, D. (1958), *J. Ecol.*, **46**, 407–45.
Rudolph, H. (1964), *Flora, Jena*, **155**, 250–93.

Rudolph, H. (1968), *Planta,* **79,** 35–43.
Rudolph, H. (1978), *Bryophytorum Bibliotheca,* **13,** 279–309.
Ruhling, A. and Tyler, G. (1971), *J. Appl. Ecol.,* **8,** 497–507.
Rycroft, D.W., Williams, D.J.A. and Ingram, H.A.P. (1975), *J. Ecol.,* **63,** 535–56, 557–68.
Skene, M. (1915), *Ann. Bot.,* **29,** 65–87.
Smith, R.A.H. and Forrest, G.I. (1978), Field estimates of primary production. In: *Production Ecology of British Moors and Montane Grasslands* (Heal, O.W. and Perkins, G.F., eds), pp. 17–37. Springer, Berlin.
Sonesson, M. (1973), Studies in production and turnover of bryophytes at Stordalen, 1972. In: *International Biological Programme – Swedish Tundre Biome Project Progress Report 1972* (Sonesson, M., ed.), pp. 66–75. Swedish Natural Science Research Council, Stockholm.
Sonesson, M., Persson, S., Basilier, K. and Stenström, T.A. (1980), Growth of *Sphagnum riparium* Ångstr. in relation to some environmental factors in the Stordalen mire. In: *Ecology of a Subarctic Mire* (Sonesson, M., ed.), pp. 191–207. Swedish Natural Science Research Council, Stockholm.
Spearing, A.M. (1972), *Bryologist,* **75,** 154–8.
Stavset, K. (1973), *Medd. Norsk. Myrselskap,* **4,** 153–156.
Stålfelt, M.G. (1938), *Planta,* **27,** 30–60.
Tallis, J.H. (1959), *J. Ecol.,* **47,** 325–50.
Tallis, J.H. (1964), *J. Ecol.,* **52,** 345–53.
Tallis, J.H. (1973), *J. Ecol.,* **61,** 1–22.
Tansley, A.G. (1939), *The British Islands and their Vegetation.* Cambridge University Press, Cambridge.
Tarras-Wahlberg, N. (1952), *Oikos,* **4,** 166–171.
Tattersfield, D.M. (1976), A physiological study of *Sphagnum rubellum* in relation to microclimate. In: *Moor House 17th Annual Progress Report* (Rawes, M. ed.), p. 23.
Theander, O. (1954), *Acta Chem. Scand.,* **8,** 989–1000.
Tibbets, T.E. (1968), Peat resources of the world. In: *Proceedings of the Third International Peat Congress, Quebec 1968* (Lafleur, C. and Butler, J. eds), pp. 8–22. No publisher. Sponsored by Department of Energy, Mines and Resources, Ottawa, Canada & National Research Council of Canada.
Troughton, J.H. and Sampson, F.B. (1973), *Plants. A Scanning Electron Microscope Survey,* Wiley, Sydney.
Tutschek, R. (1979), *Phytochemistry,* **18,** 1437–9.
Walker, D.A. and Walker, P.M. (1961), *J. Ecol.,* **49,** 169–85.
Wallén, B. (1980), *Oikos,* **35,** 20—30.
Watt, A.S. (1947), *J. Ecol.,* **35,** 1–22.
Williams, K.T. with Thompson, T.G. (1936), *Int. Rev. Hydrobiology,* **33,** 271–5.
Willis, A.J. (1964), *Trans. Br. bryol. Soc.,* **4,** 668–83.
Woodwell, G.M., Whittaker, R.H., Reiners, W.A., Likens, G.E., Delwiche, C.C. and Botkin, D.B. (1978), *Science,* **199,** 141–6.

Chapter 9

Bryophytes and Invertebrates

URI GERSON

Mosses are useful to the insect tribe, countless numbers of which find homes among their branches, and roam about in their shades as in mighty forests, looking with their thousand eyes upon the wonders of their leaves, and sunning their wings of purple and of gold, and burnishing their shining armour upon the polished columns of their urns

Frances Tripp British Mosses, *1888*

9.1 INTRODUCTION

These thousand eyes have also been looking upon naturalists for quite a while, but only few have looked back. Usually they were zoologists interested in specific groups which live on, in or under bryophytes; in the role these animals play during initial land colonization by cryptogams; in freshwater associations and in diverse other aspects. Botanists have published far fewer observations, these dealing mainly with fertilization or spore dispersal by invertebrates. Although scattered and uneven, taken as a whole the compiled data offer suggestive insights into the relationships between bryophytes and invertebrates, especially in regard to their coevolution.

Ramazzotti (1958) and Gadea (1964a), following Heinis, distinguished between bryophilous and bryoxenous moss animals, including those which spend their entire lives in bryophytes in the former group, those that stay for only part of their cycle in the latter. These authors addressed themselves mostly to the microscopic water fauna (Protozoa, Rotifera, Nematoda and Tardigrada). Also considering larger invertebrates, it becomes more useful to divide the fauna of bryophytes – the bryofauna – into four categories:

(i) Bryobionts: animals which occur exclusively in association with bryophytes.

(ii) Bryophiles: animals which are usually found on bryophytes, but may survive elsewhere.

(iii) Bryoxenes: animals which regularly spend part of their life cycle on bryophytes.

(iv) Occasionals: animals which may at times be found in bryophytes, but do not depend on these plants for their survival.

The components of the bryofauna may also be categorized according to the humidity of the plants, thereby separating an aquatic from a terrestrial fauna (Travé, 1963). However, as some of the invertebrates occur in aquatic as well as terrestrial habitats, the faunas are not separately treated here. They are introduced in taxonomic order, Richards and Davies (1977) being the authority on the largest group, the insects. Concerning the names of organisms, it should be noted that the presented data were gleaned from a variety of botanical, zoological and ecological publications. Some discretion must therefore be exercised in applying scientific names.

9.2 INVERTEBRATES

9.2.1 Protozoa

Three of the four major divisions of the protozoa occur in bryophytes. These include the Sarcodina (rhizopods), characterized by moving with protoplasmatic flow or by pseudopodia; the Ciliophora (ciliates), which swim about with cilia, and the Mastigophora (flagellates), which use one or more flagella for motion.

The Sarcodina, and especially the testate or shell-bearing rhizopods, appear to be particularly abundant in moss and sphagnum, and were extensively studied there (Bonnet 1973, and many citations therein). The genera *Nebela* and *Hyalosphenia* and the species *Difflugia pyriformis* and *D. globularis* are considered as sphagnicolous (Bovee, 1979 and citations therein). Others, like *Euglypha ciliata, Trinema lineare* and *Difflugiella oviformis* were obtained from European as well as North American mosses (Bovee, 1979). The ciliates were recently studied by Grolière (1978 and citations therein). He concluded that *Cyclidium sphagnetorum* and

Bryometopus sphagni (among others) are sphagnophilic; as the former is only known from sphagnum, it may be considered a bryobiont. Data on protozoan groups in bryophytes were presented by Fantham and Porter (1945) and by Bovee (1979).

Sphagnum appears to be richer in species (and in their numbers) than other bryophytes; Bovee (1979) obtained 145 species of Protozoa from the former habitat, only 65 from forest mosses. Fantham and Porter (1945) recorded a maximum of 220 000 protozoans (mostly flagellates) from 1 g of *Sphagnum girgensohnii,* as compared to a maximum of 150 000 individuals from *Campylium chrysophyllum.* Calculating differently, Heal (1962) estimated that there were about 16 million testaceans per m^2 of sphagnum sward. Populations of sphagnum-inhabiting protozoans also differ between alkaline fens and acid bogs, the former apparently being richer (Stout and Heal, 1967).

Protozoans living on *Sphagnum* have a vertical distribution which does not appear to be seasonal (Heal, 1962). This distribution is affected by tolerance to desiccation and to extreme temperatures, by animal size and food, as well as by various other factors (Bovee, 1979; Corbet, 1973). The effect of cryptogamic vegetation on the composition of ciliate populations in a sphagnum peat bog was explored by Grolière (1977). Ciliates in bryophyte-containing trenches resembled those from other *Sphagnum* areas, whereas those from algae-rich trenches were similar to populations inhabiting freshwater ponds. Much remains to be studied regarding the distribution of Protozoa in bryophytes; 'Every moss [including *Sphagnum*] in our collection had a different fauna of Protozoa' (Fantham and Porter, 1945).

Stout and Heal (1967) pointed out that in acid *Sphagnum* peat bacterial and fungal growth and metabolism are rather limited. The main growth activity in this habitat is due to rhizopods, which may play a direct role in the organic cycle.

The Protozoa probably invaded bryophytes via three major pathways: freshwater, by species colonizing land (Stout, 1963); air, by wind-borne, desiccation-resistant forms (Kühnelt, 1976), and soil, by protozoans moving into corticolous mosses (Bonnet, 1973).

The shells of Testacea enable their recognition even after many years. Štěpánek (1963) identified rhizopods in bryophyte samples collected even 80 years earlier. Grospietsch (cited by Corbet, 1973) showed that analysis of subfossil rhizopod tests in peat may supplement pollen studies.

Like some other microscopic components of the bryofauna, several Protozoa endure dry periods by forming cysts or other forms which resist desiccation. Cysts are wind-born and thus widely dispersed. Alternatively, they may remain on the mosses until rewetted, when both groups of organisms resume their activities.

9.2.2 Turbellaria

These mostly non-parasitic flat worms usually – but not always – live within soil-water films. Little has been published about their specific relationships with bryophytes. Kühnelt (1976) mentions that *Acrochordopostia* spp. live in very dry saxicolous mosses. Berg and Petersen (cited by Macan, 1963) found a few turbellarians in *Fontinalis dalecarlica* beds, and Lindegaard *et al.* (1975) reported 278 *Dugesia gonocephala* m^{-2} of *Cratoneuron commutatum* and *C. filicinum* in a small Danish lake. Turbellarians may survive desiccation or saturation conditions by forming cysts (Kühnelt, 1976).

9.2.3 Rotifera

The rotifera or rotatorians, also called 'wheel animalcules', are small (c. 0.5 mm) animals whose anterior end is formed into a ciliary feeding apparatus (thus 'wheel) which crawl or settle on various substrates.

They appear to be associated with bryophytes mostly in alpine and polar regions (Heinis, 1959; Hyman, 1951), being rarer in the tropics (Donner, 1966). One group, the Bdelloidea, abound on bryophytes, especially those which are submerged, emergent or terrestrial with intermittent wetting

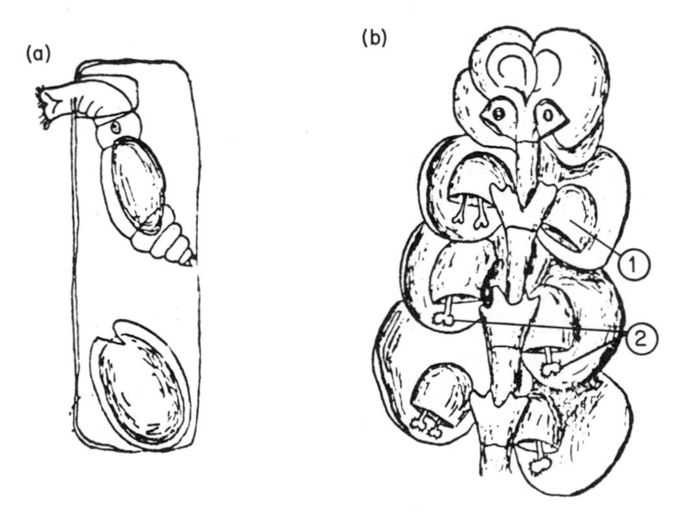

Fig. 9.1 (a) The rotifer *Habrotrocha reclusa* inside a cell of *Sphagnum*. (b), The rotifer *Mniobia symbiotica* (marked 1) inside postical lobes (marked 2) of leaves of *Frullania*. From *The Invertebrates* Vol. III by L.H. Hyman (1951). By permission of McGraw-Hill, New York.

(Pennak, 1978). Certain rotifers inhabit very specific positions on these plants. *Habrotrocha reclusa* and *H. roeperi* live inside hyaline *Sphagnum* cells (Fig. 9.1a), and *Mniobia symbiotica* dwells in the pitcher-like leaves of *Frullania* (Fig. 9.1b) (Hyman, 1951). *Ptygura velata* settles only, but in large numbers, near the leaf tips of *Sphagnum erythrocalyx* (Edmonson, 1944). The rotatorians occur exclusively on the leaves' concave side, in the tubular part, so that they are almost entirely enclosed; only their anterior ends project above the edges. But *P. velata* was also collected from *Fontinalis* and angiosperms (Edmonson, 1944), therefore, despite its habitat specificity, it is merely a bryophilous species.

Terrestrial bryophyte-inhabiting rotifera are active only while their moss cushions are permeated with water; thus they actually lead an aquatic existence in the water films between plant parts. The ubiquity and survival of rotifera on bryophytes have led Hyman (1951) to write: 'Almost any bit of dried sphagnum will be found to contain desiccated rotifers'. Their ability to withstand desiccation is due to contracting, drying and shrinking their body contents and cuticle to the smallest possible volume, usually without forming cysts (Hyman, 1951; Pennak, 1978). During this period of anabiosis they are very resistant to extreme low or high temperatures, which explains rotatorian survival – for long periods – under harsh climatic conditions. The animals, along with their eggs, may then be blown about by winds over wide areas. Consequently, many species are cosmopolitan in distribution (Hyman, 1951). Recovery from anabiosis depends on immersion in fresh water, and can be quite rapid, requiring from minutes to one day.

Alpine moss-dwelling rotifers are often red, a colour imparted to them by carotenoids (Hyman, 1951). They may thus feed on particles of bryophyte origin. Rotifers may build up large populations in bryophytes. Fantham and Porter (1945) reported the following (numbers/g bryophyte): *Ceratodon purpureus:* 700; *Schistidium apocarpum:* 600; *Polytrichum juniperum:* 1108; *Sphagnum fuscum:* 920; *S. girgensohnii:* 1160; *S. wulfianum:* 872.

9.2.4. Nematoda

The nematodes or roundworms are elongated, cylindrical animals which are parasitic in animals and plants, besides being very abundant in diverse aquatic and terrestrial situations. In the latter habitat, they may be associated with bryophytes, constituting an unspecialized, soil-derived fauna (Nicholas, 1975). Nematodes of the bryofauna occur all over the world; *Plectus cirratus* inhabits moss in the Galapagos Islands (Gadea, 1977) and high up in the Pamirs (Nicholas, 1975) as well as being abundant in European soils (Kühnelt, 1976). Several Dorylaimoidea, like *Dorylaimus macrodorus, D. stagnalis,* and *Eudorylaimus carteri* are often found in bryophytes and some may feed on these plants (Gadea, 1964b). Another Dorylaimoid recorded from mosses is *Funaria thornei* (Goodey, 1963).

Nematodes survive dry spells by forming resistant resting stages and by encystment. *Plectus rhizophilus*, abundant in dry moss, may at times remain as the sole nematode in shallow bryophyte cushions exposed to extreme climatic conditions. It is resistant to dehydration during its entire development (Nielsen, 1967). Vertical movements along the moss – up at night or during rain, down when it gets dry – also facilitate nematode survival. Their numbers in bryophytes approximate those of rotifers. The following values were reported by Fantham and Porter (1945) (animals/g bryophytes): *Brachythecium acutum:* 510; *Ceratodon purpureus:* 440; *Funaria hygrometrica:* 336; *Sphagnum girgensohnii:* 4680; *S. fuscum:* 740; *S. wulfianum:* 720. Nielsen (1967) counted 200 *P. rhizophilus* cm^{-2} of *Tortula ruralis*, 330 cm^{-2} of *C. purpureus*, both collected on a roof.

9.2.5 Annelida

This phylum, which comprises the segmented worms, includes such well-known animals as earthworms and leeches. The former belong to the class Oligochaeta and the latter to the class Hirudinea, and members of both are found in bryophytes. Lindegaard *et al.* (1975) counted 7520 *Nais elinguis*

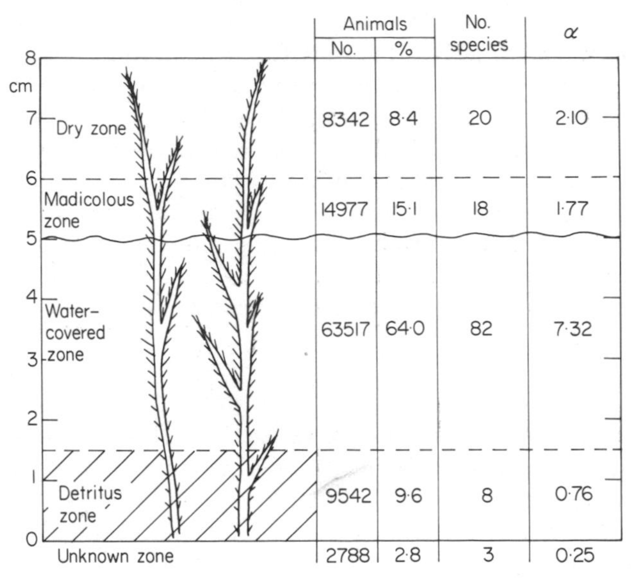

Fig. 9.2 Vertical zonation of a Danish *Cratoneuron* carpet. Number and percent of animals, number of species and Margalef's index of diversity ($\alpha = (S-1)/\ln N$, where S is the number of species and N the number of individuals) for each zone. Data from Lindegaard *et al.*, 1975.

and 615 *Eiseniella tetraedra* (both Oligochaeta) in 1 m^2 of the detritious zone (Fig. 9.2) of *Cratoneuron* spp. in a Danish spring. Terrestrial oligochaetes were found under mosses on tree trunks (Larsson, 1978). Berg and Petersen (in Macan, 1963) reported 67 oligochaetes and five leeches per 1 m^2 of Scandinavian *Fontinalis*; members of both groups also occur in moss balls derived from *Fontinalis* and *Drepanocladus* (Luther, 1979). Finally, commercial leeches were packed and shipped while embedded in the moisture-retaining *Hypnum* (Thieret, 1956, citing Welch).

9.2.6 Mollusca

Representatives of two of the major molluscan classes, Gastropoda (snails and slugs) and Bivalvia (mussels and clams) associate with bryophytes. Some snails graze heavily upon *Octodiceras fontanus,* preferring it to *Fontinalis* and *Drepanocladus* (Lohammar, 1954). This author consequently believed that the scarcity of *O. fontanus* in certain North European

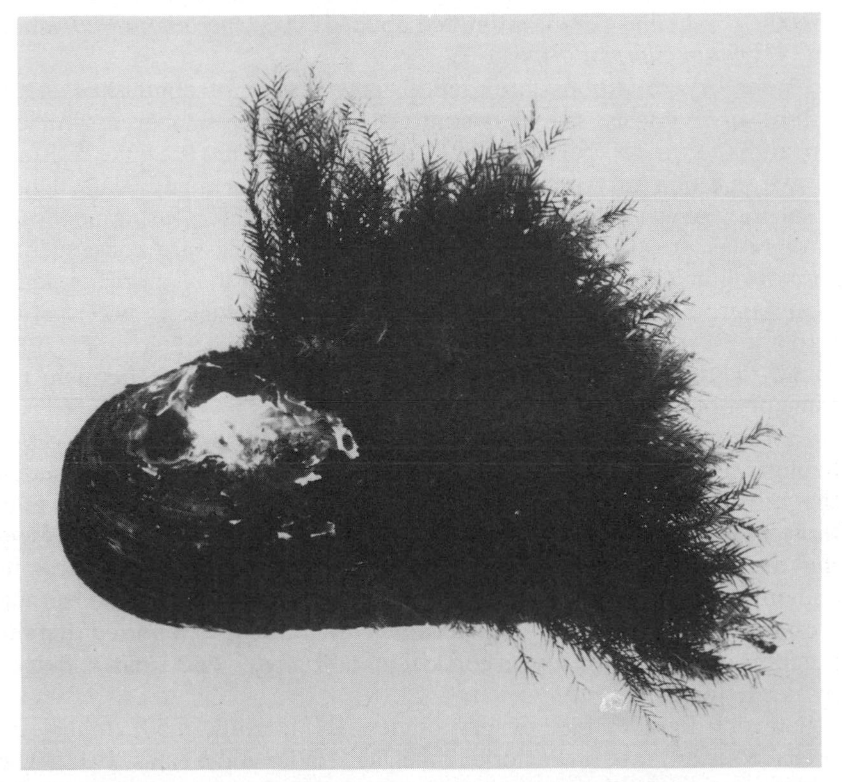

Fig. 9.3 *Octodiceras fontanum* growing on the live mussel *Anodonta cygnea.* From Lohammar (1954) used with the permission of *Svensk Botanisk Tidskrift.*

water bodies results from the snails feeding. On the other hand, *Octodiceras* uses molluscs as a substrate. It grows on dead as well as live mussels, like *Anodonta cygnea* (Fig. 9.3), these sometimes moving about and thus possibly dispersing the bryophyte (Neumann and Vidrine, 1978). *Leptodictyum riparium* was also reported by these authors to grow on mussels.

Lindegaard *et al.* (1975) collected 1247 gastropods (*Galba truncatula*) m^{-2} *Cratoneuron* from the plant-water interface (madicolous zone in Fig. 9.2).

9.2.7 Tardigrada

The tardigrades or 'water bears' are small, eight-legged invertebrates closely related to the arthropods. They are very common in moss, less so in *Sphagnum*. Moss is believed to be the most favourable habitat for tardigrades, besides being well suited for population studies (Morgan, 1977). The largest numbers of tardigrades were reported by this author: as many as 2287000 m^{-2} *Bryum argenteum* and *Ceratodon purpureus*, or 132000 g^{-1}. Hallas (1975) estimated about 15000 *Macrobiotus hufelandii* m^{-2} *Hypnum cupressiforme*.

Under dry conditions tardigrades enter a state of diminished metabolism, or anabiosis. The body contracts in a regular manner, assuming a barrel shape which is called a 'tun' (Cuénot, 1949). Tuns are quite resistant to extreme climatic conditions and may survive for many years. Upon absorption of water there is a rapid return to normal life. During anabiosis, tardigrades are disseminated by winds, most species are consequently cosmopolitan (Morgan and King, 1976), but appear to be more abundant in temperate as compared to tropical regions (Mehlen, 1972). Moss tardigrades may also be disseminated by arthropods: Bertrand (1975) found them on carabid beetles and on 'Myriapoda' which frequent the same bryophytes (see also Ramazzotti, 1958).

Tardigrades on the whole do not appear to be host-specific but certain habitats contain more animals than others. Morgan and King (1976) stated that button and erect mosses (acrocarpus) harbour more tardigrades than those of creeping growth habits (pleurocarpous). Bertrand (1975) found that the relative abundance of several species differed between saxicolous and epiphytic bryophytes, and also between cushion-forming and 'carpet' mosses (i.e. *Polytrichum*, Hypnaceae). This author also reported that the numbers of *Diphascon* spp. in corticolous bryophytes were clearly affected by tree height.

Some tardigrades feed on bryophytes (Ramazzotti, 1958), others are predaceous or graze on bacteria and algae (Hallas and Yeates, 1972). The bryophagous tardigrades pierce host cells with their stylets and suck out the contents. *Racomitrium* and *Polytrichum* colonies contain very few specimens, probably because of their more heavily protected cell walls (Morgan and King, 1976; Ramazzotti, 1958). Moss growth and architecture affect tardigrade populations. Morgan (1977) obtained correlations between prevailing daylight hours, temperatures and tardigrade-numbers, indicating some links between the growth patterns of both groups of organisms. The arrangement of *Hypnum cupressiforme* leaves in a mat determines the amount of humidity available there after rain. When water bridges are formed, tardigrades may move vertically among the various layers (Hallas, 1975) and thus explore the entire mat.

9.2.8 Arthropoda

Arthropods, animals with segmented legs, are the largest phylum in the animal kingdom, and proportionately represented in the bryofauna. For the sake of convenience, they are discussed under the headings of Crustacea, 'Myriapoda', Insecta and Arachnida.

(a) Crustacea

Crustaceans, like crabs, shrimps and barnacles, are mostly aquatic animals, which either have a head, thorax and abdomen, or else the former two parts fuse to form a cephalothorax. They breathe through the body walls or by gills. Members of two subclasses, Copepoda and Malacostraca, often associate with aquatic bryophytes.

Copepods occur in fairly large numbers in moss-overgrown waters; Rylov (1963) recorded *Acanthocyclops nana* in populations up to 200 m^{-2} from this habitat. Fantham and Porter (1945), on the other hand, counted only 50 *Cyclops* g^{-1} of *Climacium dendroides*. A total of 18 copepod species was collected by Frost (1942) from an Irish river, of which eight were considered as normal components of the bryofauna. The tropical *Muscocyclops* and *Bryocyclops* live in minute aquaria formed by epiphytic bromeliaceous leaves as well as in mosses (Lindberg, 1954; Vandel, 1965). The bryophylous *Epactophanes* is resistant to desiccation, which facilitates its passive dissemination and explains its cosmopolitan occurrence in bryophytes (Menzel, 1921).

Few malacostracans are found in bryophytes. Lindegaard *et al.* (1975) obtained 6819 *Gammarus* m^{-2} *Cratoneuron* in a Danish lake. Berg and Petersen (cited by Macan, 1963) reported 198 *Asellus* m^{-2} *Fontinalis* in Lake Gribso. This isopod was also observed browsing on moss balls by Luther (1979).

Members of two other subclasses, the Branchiopoda and the Ostracoda may occur, in rather small numbers, in aquatic mosses (Frost, 1942).

(b) 'Myriapoda'

Members of three disparate groups, the Chilopoda, Diplopoda and

Symphyla, used to be combined within the Myriapoda. Although outmoded, this arrangement is retained here for convenience as well as due to the relative scarcity of the animals in byrophytes. 'Myriapoda' – elongated arthropods whose bodies consist of numerous similar segments each bearing one or two pairs of legs – may be collected from corticolous mosses (Larsson, 1978; Pschorn-Walcher and Gunhold, 1957) or in sphagnum peat (Pedroli-Christen, 1977). Little is known about their activities there. Stebaev (1963) believed that predaceous Lithobiidae (Chilopoda) were characteristic of the moss stage of soil colonization by plants in the Urals. Bryophyte-inhabiting tardigrades were collected from symphylans found in the same habitat, indicating that the latter were disseminating the former from one bryophyte habitat to another (Bertrand, 1975).

(c) Insecta

Richards and Davies (1977) recognized 29 orders within this class; members of 21 associate with bryophytes. Examples of the wingless insects – subclass Apterygota – come from the Thysanura (bristletails), Protura and Collembola (springtails). Great swarms of thysanurans, the most primitive of all insects, feed on dry moss, lichen and other plant material on high mountains (Mani, 1962); others associate with moist forest bryophytes (Larsson, 1978). The Protura, antenna-less soil dwellers, are often found in bryophytes (Nosek, 1973), but little is known about them. The moss springtails, on the other hand, were studied in regard to colonization, succession, habitat-specificity and feeding. Although they occur in fairly large numbers on temperate bryophytes (Lindegaard *et al.,* 1975, recorded 4182 *Isotomurus palustris* m^{-2}*Cratoneuron* from Denmark), collembolans really abound under harsh climatic conditions. Bengtson *et al.* (1974) working in Spitsbergen, compared springtail populations from lichens with those collected from mosses; there were 20000–40000 m^{-2} in the former habitat, 243500 m^{-2} in the latter. Rather similar data were obtained by Chernov *et al.* (1977) from Cape Cheluskin, the northernmost mainland of Europe. Collembola also abound in high altitudes, where they live in and feed on the available colonizing mosses (Mani, 1962). Antarctic springtails like *Isotoma klovstadi* and *Gomphiocephalus hodgsoni* feed extensively on bryophytes available there, the latter preferring mosses to fungi (Pryor, 1962; Wise *et al.,* 1964). Changes in collembolan populations consequent upon the succession of sphagna in a British bog were followed by Murphy (1955). Bonnet *et al.* (1975 and former papers) sampled bryophytes growing in various habitats and demonstrated contiguous changes in collembolan populations from soil to soil mosses to aerial mosses. Humidity preferences, as in Murphy's series, were believed to determine which springtail occurred in the various plants. Some species (i.e. *Xenylla tullbergi*) appeared to be quite bryophylous.

The Pterygota, winged or secondary wingless insects, are divided between those undergoing simple metamorphosis, the Exopterygota or Hemimetabola, and those with complete metamorphosis, the Endopterygota or Holometabola. Eleven Exopterygote orders are represented in bryophytes. The mayflies (Ephemeroptera) *Ephemerella* and *Baetis* abound in aquatic mosses (Frost, 1942), both feeding mainly on algae, only occasionally on bryophytes (Chapman and Demory, 1963; Hynes, 1961). About 530 mayfly nymphs 200 g^{-1} mosses were recorded by Frost (1942) from Ireland. Some Odonata (dragonflies), an order of predators, deposit their eggs into bryophytes (Macan, 1963), and their nymphs live there. Examples are the Himalayan *Calicnemia miles* (Kumar and Prasad, 1977) and the European *Leucorrhinia dubia* (Matthey, 1971). Plecoptera (stoneflies) may live in and feed on aquatic bryophytes; *Protonemura meyeri* is commonly found in *Fontinalis* (Hynes, 1941). Stonefly populations may reach substantial numbers: Lindegaard *et al.* (1975) counted 16 500 *Nemurella picteti* m^{-2} *Cratoneuron*.

A member of the small, obscure order Grylloblattodea deposits its eggs in mosses (Richards and Davies, 1977). Orthoptera (crickets and grasshoppers) feed on *Sphagnum* (Vickery, 1969), *Hypnum* (Verdcourt, 1947) and other bryophytes; Uvarov (1977) believed that some grasshoppers may eat mosses because of their water content. The establishment of electric light in caves in New Zealand has stimulated moss growth there, on which local orthopterans browse (Richards, 1962). Among these is *Pallidoplecteron turneri,* which feeds on thick growths of *Marchantia.* Some Phasmida (stick insects), which are predaceous, closely resemble, by colour and in their foliate body and limbs, the epiphytic mosses on which they live and hunt (Robinson, 1969). A webspinner (order Embioptera) was collected from a moss cushion in Israel (unpublished data). The termite (Isoptera) *Hospitalitermes umbrinus,* which is unusual in feeding on live plants, prefers lichens but will also take bryophytes (Collins, 1979). Booklice (psocoptera), common on plants, may also feed on moss (Larsson, 1978).

Bugs (Hemiptera) are divided into the Homoptera and the Heteroptera, and members of both suborders occur in bryophytes. Peloridiidae, considered as the most primitive bugs and even 'living fossils', live and feed on southern-hemisphere mosses (China, 1962; Helmsing and China, 1937). Several aphids live on bryophytes, those of the subtribe Melaphidina forming galls on *Rhus* and migrating to mosses (Eastop, 1977). *Aspidaphium cuspidati* lives in water-logged *Calliergonella cuspidata* and *Drepanocladus aduncus* (Stroyan, 1955). Upon being placed on stems of *C. cuspidata,* nymphs of this aphid immediately moved below the water level (Müller, 1973). The latter author suggested that feeding on bryophytes was a secondary feature in aphid evolution, the transfer to these plants having taken place several times in aphid history. Southwood and Leston (1959) listed bryophyte-associated bugs which attach their eggs to moss stalks (*Megalonotus chiraga*), overwinter there (*Sehirus biguttatus*) or feed on the plants (*Drymus sylvaticus*). Lacebugs of the genus *Acalyptus* primarily feed on bryophytes, but under adverse conditions some move to phanerogams which thus serve as secondary hosts. This was interpreted as a case of incomplete transference to mosses (Horn *et al.*, 1979). A few thrips (Thysanoptera) occur in bryophytes, like *Lissothrips muscurum* (Rhode, 1955) and *Bournierothrips,* only found in mosses (Bournier, 1979).

Representatives of seven out of the nine holometabolous orders associate with bryophytes. Lacewing (Neuroptera) larvae lurk about mosses in search of prey (Richards and Davies, 1977). Many Coleoptera (beetles), the largest of all animal orders, occur in bryophytes. Many find refuge in this habitat. Bordoni (1972) obtained 179 species belonging to 25 families from mosses in a Tuscan fen, but only a few were bryophilous. The Staphylinidae, known to feed on moss (Mani, 1962) were best represented there. Frost (1942) found many Elmidae in Irish bryophytes and stated that this environment was most favourable for the beetles. Other Coleoptera obtained from bryophytes are the Limnebiidae, called 'minute moss beetles', and the Sphaeriidae ('minute bog beetles') (Arnett, 1971). The hydrophilid *Cretinis punctatostriata* is a true bryobiont, spending its entire life in *Sphagnum* (Matthey, 1977). Its eggs are deposited therein, larvae live there and pupation takes place in a cell formed from these bryophytes. The chrysomelid *Mniophila muscorum* is another bryobiont, living in ground litter and feeding on moss (Kühnelt, 1976). Many other bryophilous beetles subsist on epiphytic algae (LeSage and Harper, 1976). New Guinean weevils whose dorsum was extensively overgrown with bryophytes were discussed by Gressitt *et al.* (1968). Mosses (*Daltonia*) as well as liverworts (*Lejeunea, Metzgeria*) occurred on these beetles.

One family of scorpion flies (Mecoptera), namely the Boreidae, commonly occur in and feed on mosses. They are rather catholic in their tastes; *Boreus notoperates* was found on Grimmiales as well as Isobryales (Cooper, 1974). Bryophyte texture may be a more important factor in determining boreid preferences. Mosses growing in low, compact cushions, whose rhizoids are tightly matted, usually contain more Boreidae than those growing as loose cushions (Penny, 1977). One reason may be greater predation pressure from carabid beetles in the latter bryophytes. Larvae and adults of *B. brumalis* eat mosses throughout winter in Ontario, presumably feeding on bryophytes even below snow cover (Shorthouse, 1979). *B. notoperates* survives dry Californian summers by hiding inside small insect-made earthen chambers which are hidden within bryophyte rhizoids (Cooper, 1974).

Chironomid (Diptera) larvae made up about five-sixths of the nearly 600 000 organisms Frost (1942) collected from Irish submerged mosses. Lindegaard *et al.* (1975), who obtained about 100 000 invertebrates m^{-2} *Cratoneuron,* noted that flies made up more than 40% of the total. These data will emphasize that the Diptera (flies) is the insect order most intimately associated with bryophytes. The crane flies (Tipulidae) *Dolichopeza, Liogma* and *Triogma* live in and feed on mosses. Larvae of the latter two species remarkably resemble their host plants (Alexander, 1920), that of *D. americana* being green and irregularly marked with dark lines, so that it blends into its moss background (Byers, 1961). Other tipulid adaptations to life in bryophytes include the special anal papillae found in larvae confined to these plants (Brindle, 1957). Choice of specific moss species by *Dolichopeza* appears to depend mostly on bryophyte growth habit and leaf texture. Very compact-growing species, like *Bryum argenteum* or *Ceratodon purpureus*, hinder the larvae's tunnelling, while loose-growing mosses like *Climacium, Polytrichum* or *Plagiomnium cuspidatum* are too diffuse for constructing suitable tunnels. Furthermore, coarse-leafed *Polytrichum* or thallose liverwort are also rejected in favour of softer leaves (Byers, 1961).

Larvae of many fly families of medical and veterinary importance occur in bryophytes. These include Psychodidae (sand flies) (Quate, 1955), Culicidae (mosquitoes) (Fantham and Porter, 1945), Simuliidae (Snow *et al.*, 1958), Ceratopogonidae (Séguy, 1950) and Tabanidae (horse flies) (Teskey, 1969). Thomas (1971) developed a quick method to extract tabanid larvae from moss. Adults of coprophilous and dung loving flies, like *Sarcophaga* and *Scatophaga,* are attracted to the capsules of certain Splachnaceae and disperse their spores (Garjeanne, 1932; Ingold, 1965; Koponen and Koponen, 1978).

The caterpillars of a few Lepidoptera (butterflies and moths) live in bryophytes. Prominent among them are the primitive Micropterygidae. *Neomicropteryx nipponensis* deposits its eggs on the liverwort *Conocephalum conicum* and the larvae feed on its tissues (Yasuda, 1962). *Sabatinca,* which also occurs in hepatics, resembles them in its greenish colour and large setae (Tillyard, 1922). Larvae of another primitive group, the Meessiinae, feed on lichens and mosses and incorporate fragments thereof in their cases, which thus resemble the substrate (Zagulyayev, 1970). The North European *Nudaria mundana,* which feeds on saxicolous lichens and liverworts (Forster and Wohlfahrt, 1960) and the Australian *Eudonia* (Anonymous, 1970), are examples of more advanced bryophyte-associated Lepidoptera. The hemlock looper, a forestry pest in North America, deposits its eggs on mosses, and special means had to be devised to separate them from this substrate (Shepherd and Gray, 1972). Caddis flies (Trichoptera), which live in bryophytes construct their cases from these plants, some showing a clear predilection for certain liverworts (Glime, 1978). Bryophytes constituted most of the diet of *Rhyacophila verrula,* apparently because they were the dominant aquatic plants in a stream (Thut, 1969). Berg and Petersen (in Macan, 1963) collected 260 Trichoptera m^{-2} *Fontinalis* in Lake Gribso, and Frost (1942) obtained 492 200 g^{-1} of submerged moss in Ireland.

The Hymenoptera (ants and wasps) are the last insect order to be considered. *Myrmica ruginosis* and *Formica picea* form nests in sphagnum, being the principal predators there when bogs dry up (Matthey, 1971). Plitt (1907) observed ants which were gnawing the capsules of *Diphyscium foliosum* (called *Webera sessilis* in Plitt's paper). Loria and Herrnstadt (1981) observed harvester ants (*Messor* spp.) climbing on the setae of *Aloina aloides* and *Bryum bicolor* in Israel and cutting off their capsules (Fig. 9.4). These were then gathered to the nest, where they

Fig. 9.4 The ant *Messor* harvesting capsule of *Bryum bicolor.* Note that ant is standing upside-down, its mouthparts (right arrow) about to cut the seta. To the left (arrows), several harvested setae (M. Loria, unpublished).

probably served as substitute food until phanerogam seeds became available, later in the season. Spruce sawfly larvae were reported to pupate within European mosses (Nägeli, 1936).

(d) Arachnida

Four orders of arachnids (Savory, 1977) are so far known from bryophytes; the first three are strictly predaceous. Spiders (Araneae) abound in moss and sphagnum; their abundance there has prompted Müller (1973) to warn collectors of moss aphids against them. Spiders were also established in small *Fissidens* colonies growing around lamps in caves (Dobat, 1972). The distribution of two spiders within a sphagnum carpet was affected by microclimates prevailing within different layers of that carpet (Nørgaard, 1951). Pseudoscorpiones (false scorpions or book-scorpions) may be common enough in bryophytes to be called 'moss scorpions' (Larsson, 1978). Some members of that order, like *Microbisium femoratum,* are bryobionts, whereas *Neobisium muscorum,* despite its name, is merely bryoxenous (Cloudsley-Thompson, 1968). Opiliones (harvestmen) were collected from moss mats in North America by Rhode (1955).

The Acari (mites) are the fourth order, and they have diverse associations with bryophytes. One large group, the Oribatei or Cryptostigmata, is so abundant there that they have been called 'moss mites' ('Moosmilben'). They are common in sphagna (Tarras-Wahlberg, 1961); some, like *Hydrozetes lacustris* and *Limnozetes ciliatus,* living on floating *Sphagnum* mats (Karppinen and Koponen, 1973). Gerson (1969) observed oribatids feeding on bryophyte capsules in Canada. A total of 4600 *Mucronothrus* m^{-2} *Cratoneuron* were recorded by Lindegaard *et al.* (1975). Many Prostigmata – usually reddish, predaceous or phytophagous mites – are present in mosses; of 217 species collected by Schweizer and Bader (1963) in Switzerland, 57 (c. 26%) were obtained from bryophytes. *Eustigmaeus,* a moss-feeding mite, is commonly found in North American mosses (Gerson, 1972). Some species of the usually pestiferous *Bryobia* live in bryophytes (unpublished). Adult water mites, whose larvae parasitize mosquitoes, aggregate, mate and oviposit in submerged mosses. Fantham and Porter (1945) obtained 90 Hydrachnida g^{-1} *Climacium dendroides,* and Frost (1942) 147 200 g^{-1} submerged mosses. The mostly-predaceous Mesostigmata also occur in bryophytes. Lindegaard *et al.* (1975) counted 1604 and 545 individuals of *Pergamasus* and *Platyseius,* respectively in 1 m^2 of *Cratoneuron.* Some of these mites feed on bryophagus species, as Gerson (1972) collected Mesostigmata which had been devouring the moss-feeding *Eustigmaeus.* Ticks, obligate parasites of vertebrates, were reported from Canadian bryophytes by Fantham and Porter (1945).

9.3 BRYOPHYTES AS A HABITAT FOR INVERTEBRATES

9.3.1 General considerations

Bryophytes possess several attributes which affect the distribution and abundance of dependent invertebrates. They are among the earliest of soil colonizers (Kühnelt, 1976; Stebaev, 1963) and important components of high-mountain ecosystems (Mani, 1962). Bryophytes rapidly absorb large quantities of water, retain them, and thereby also retard the drying-out of their underlying substrate. They serve as insulation against heat, cold and wind (Corbet, 1973; Gressitt, 1967; Strong, 1967) thereby cushioning invertebrates which live within them against climatic changes. Bryophytes also muffle sounds in the 30–50 KHz frequency range (Kolb, 1976). Last but not least, due to the perennial life-forms of many bryophytes, they are there when invertebrates need them. On the other hand, bryophytes are usually inconspicuous, being in a subordinate position in most ecosystems (During, 1979). In this context, the concepts of apparency and unapparency (Feeny, 1976) will now be introduced. An 'apparent' plant is one which, due to its size, form, persistence and relative abundance in the community is 'bound to be found', or 'susceptible to discovery' by its enemies. An unapparent plant, then, is 'hard to find' or 'unsusceptible to discovery' by its adversaries. Bryophytes, as major components of the Forests of Lilliput (Bland, 1971), are rather unapparent. It is only under extreme climatic conditions, above the timber line or while colonizing new land that bryophytes dominate entire ecosystems, thus becoming apparent.

9.3.2 Physical effects

Different bryophytes have different life-forms and grow in different colonies. Strong (1967) and Tilbrook (1967), working in Antarctica, wrote that mats of *Polytrichum* and *Dicranum* are less wet and cold than those of *Pohlia,* and consequently harbour more arthropods. *Polytrichum* supports these animals to a greater depth probably due to its more developed rhizoid system. Mat compactness is important for larvae of the mecopteran *Boreus* and the crane fly *Dolichopeza. B. notoperates* avoids mosses with coarse and open clumps, which do not have a fine rhizoid mat (Cooper, 1974). *Dolichopeza* shuns *Bryum argenteum* and *Ceratodon purpureus* because their very compact mats are too resistant to its tunnelling. On the other hand, the framework formed by the loosely-growing *Climacium* or *Plagiomnium cuspidatum* is too diffuse for constructing suitable tunnels by these larvae (Byers, 1961).

Shape, texture and thickness of bryophyte leaves affect the associated invertebrates. The predilection of the rotatorian *Ptygura relata* for the curved leaves of *Sphagnum erythrocalyx* (Edmonson, 1944) has been noted in Section 9.2.3. The liverwort *Frullania dilatata* bears under its leaves minute pockets in which numerous rotifers occur (Donner, 1966). Given a choice, larvae of *Dolichopeza* prefer thin, soft-textured leaves like those of *Plagiomnium* to the coarse leaves of *Polytrichum* (Byers, 1961). Gerson (1972) correlated the inability of *Eustigmaeus frigida* to feed on several thick-leafed Polytrichaceae with the mite's short mouth parts. Barr (1973) stated that more water mites may be obtained from fine-structured aquatic bryophytes than from those with larger leaves.

The tendency of bryophytes to accumulate plant debris provides the first stable microhabitat for many invertebrates in newly colonized soils (Kühnelt, 1976). Consequently, the bryofauna becomes quite varied (Stout, 1974). This matter will be further discussed in Section 9.6.1.

9.3.3 Shelter

Strong (1967) believed that bryophytes, widely spread in Antarctica, serve primarily as shelter for the arthropods found therein. This is supported by the data of Chernov *et al.* (1977), who reported that bare soils at Cape Cheluskin, sparsely populated with algae and lichens, harboured only 5–10% of the arthropod populations found in mosses there. Moving from cold to hot deserts, Hammer (1966) collected hygrophilous mites from small moss colonies in an oasis near the Red Sea. The mites are rather sensitive to low air humidities; their long-term survival was therefore believed to be wholly dependent on these moss enclaves.

In the milder regions, bryophytes still serve as refugia. Bordoni (1972) found 179 species of beetles in Tuscan moss cushions, from bryophilous to occasional species, most probably overwintering there. Bryophilous but predaceous Pselaphid beetles overwinter, as adults, in the interstices of frozen moss mats (Reichle, 1966). The water beetle *Hydroporus morio* lies in English sphagnum pools which dry out during summer. The beetle bores small round holes in the damp bryophytè 'floor' and aestivates in this moist microhabitat until rains come (Jackson, 1956).

9.3.4 Oviposition and pupation

Invertebrates which feed on bryophytes usually also oviposit their eggs there. Examples are the crane fly *Dolichopeza* and the mite *Eustigmaeus* (Byers, 1961; Gerson, 1972). *E. frigida* places only a single egg on each leaf, but *E. schusteri* deposits 3–4, these eggs imparting a reddish hue to moss shoots. Another bryophyte feeder, the ground cricket *Pteronemobius,* punctures *Sphagnum* leaves with its ovipositor and places its eggs in the resultant cavity (Vickery, 1969). Several predators which hunt in bryophytes also deposit their eggs therein. These include the bug *Myrmedobia tenella,* which preys upon aphids (Southwood and Leston, 1959) and the fly *Eurynogaster* (Williams, 1939). The water mite *Thyas barbigera*, a parasite of mosquitoes, begins its life as an egg deposited on water mosses (Mullen, 1977).

Insects associated with bryophytes may pupate therein, at times utilizing the plants for constructing special cells. Pupation of the water beetle *Crenitis punctatostriata* takes place within a small cell which the larva had formed from decomposed *Sphagnum* (Matthey, 1977). Glime (1978) reported on blackflies which live among *Fontinalis* leaves and use them to support their pupal cases. Nägeli (1936) noted that larvae of a spruce-feeding sawfly preferred to pupate under *Hylocomium* and *Thuidium* as compared to *Polytrichum*.

9.3.5 Camouflage

Invertebrates which live in bryophytes gain an obvious advantage by camouflaging themselves with these plants. As such disguise favours them only while they remain within, or against the background of bryophytes, the use of this particular concealment may mark the stage at which some invertebrates become bryophiles (and mimicry, below, will denote bryobionts). Camouflage may be used by larvae which construct cases, building them from bryophyte pieces. Glime (1978) noted that the caddis *Palaegapetus celsus* always composed its cases from leafy liverworts. These were cut into nearly circular pieces and cemented together along their margins to form flattened cases, shaped like eyeglass cases. Larvae of another trichopteran, *Adicrophleps hitchcocki* construct their cases from blades of *Fontinalis dalecarlica, Hygrohypnum* sp., *Chiloscyphus polyanthus* var. *rivularis* and *Plagiochila asplenioides* (Glime, 1978). Larvae of the crane fly *Dicranomyia badia* spin about themselves a web onto which they fasten bits of *Bryum* and *Amblystegium* (Alexander, 1920). The spider *Spilasma tubulofaciens* incorporates moss fragments into its retreat and web, but this may well be fortuitous (Quintero, 1974).

Gressitt *et al.* (1968) reported that bryophytes (and other cryptogams) were growing on the backs of New Guinean weevils (Curculionidae). The beetles, mostly of *Gymnopholus* subgenus *Symbiopholus*, are often structurally modified to promote plant growth; they have dorsal pits surrounded by ridges, stiff setae and produce a sticky secretion which may encourage cryptogam establishment there (Gressitt, 1966). *Daltonia angustifolia* was found on five species of *Symbiopholus*. Liverworts, apparently rare on these beetles, grew extensively on weevils of the genus *Poropterus* living in the same habitat. *Odontolejeunea, Microlejeunea* and *Metzgeria* were reported from the latter animals. Most of these bryophytes grew on nearby plants, indicating their origin. Finding moss protonema on the weevils suggested that they originated from spores (Gressitt *et al.*, 1968). The beetle-cryptogam association apparently benefits both partners (hence 'epizoic symbiosis'), the animals providing a suitable environment for the plants, the latter camouflaging the insects. Successful disguise was inferred from the scarcity of predation on the beetles, which feed on the upper shoots of various trees, thus being rather exposed. Another possibility was that the cryptogams made the beetles distasteful, a form of chemical protection.

As noted in Section 9.2.6, certain water bryophytes grow on live mussels. The molluscs' shells are colonized like any other submerged calcareous substrate. However, as predation on Mollusca in aquatic environments may be visual (Heller, 1975), bryophyte overgrowth could be of advantage.

9.3.6 Mimicry

The moss-overgrown beetles and mussels may also thrive without their masks of living plants, but invertebrates which mimic bryophytes will not survive elsewhere. Mimicry implies special coloration and/or special morphological modifications based on a model; without the latter, mimicry makes no 'evolutionary sense' and the imitating animals will disappear. Mimicry thus provides examples of bryobionts, but not necessarily also of bryophagy. Examples will be drawn from various insect orders. The predaceous phasmid *Trychopeplus laciniatus* has greenish foliations and lamellations on its body, which make it 'extraordinarily mossy' in appearance (Robinson, 1969). The peloridiid bug *Hemiodoecus veitchi* feeds on bryophytes, and with its cryptically shaped and coloured body it greatly resembles the plants' growing tips, being consequently rather difficult to see on the mosses (Helmsing and China, 1937). Larvae of the primitive *Sabatinca* (Lepidoptera) resemble hepatic leaves, carrying large setae similar to spines found on the host liverworts (Tillyard, 1922). Finally, crane flies which feed on bryophytes exhibit behavioural and morphological modifications which help them to blend into the plants' background. Larvae of *Phalacrocera replicata,* carrying specialized outgrowths, are greenish and sluggish in their movements. Clinging to mosses and remaining almost motionless for hours, they closely resemble bryophytes. Larvae of *Triogma trisulcata* have rows of leaf-like appendages on their dorsal, pleural and ventral surfaces, making them quite similar to *Fontinalis* on which they live (Alexander, 1920).

9.4 BRYOPHYTES AS FOOD FOR INVERTEBRATES

Diverse invertebrates feed on bryophytes. Orthopterans, beetles and moth caterpillars bite and chew, whereas bugs, aphids and mites suck out the contents of moss cells. Ants feed on the spores within the capsule (Plitt, 1907) or harvest the entire capsule (Loria and Herrnstadt, 1981). The scorpion-fly *Boreus brumalis* thrusts its rostrum into young moss shoots, chews off the apex, and eats its way down into the core (Shorthouse, 1979). Larvae of the fly *Dolichopeza* tunnel in bryophyte rhizoids (Byers, 1961), and those of *Lycoria* in the thallus of *Marchantia*. Gerson (1972) presented photographs of *Didymodon* cells sucked out by the mite *Eustigmaeus*; the tardigrade *Echiniscus testudo* probably feeds in a similar way (Morgan. 1977).

These feeders, while restricted to bryophytes, are not particularly host-specific. Even aphids, usually quite narrow in their feeding choices (Eastop, 1973), have many moss species as hosts (Müller, 1973). Of eleven species of *Eustigmaeus* collected by Gerson (1972) from Canadian mosses, those which were collected more than once were found in different bryophytes. Similar data were presented for the fly *Dolichopeza* (Byers, 1961) and for the scorpion fly *Boreus* by Cooper (1974) and Penny (1977). These and other authors concluded that habitat is probably more important than specific bryophyte host-plant in determining the distribution of invertebrates which feed on these plants.

Some bryophyte-consuming animals are not restricted to these plants. Fortuitous alternate feeding takes place when bryophytes and other plants – algae, lichens or even phanerogams – occur together, as in freshwater habitats (Chapman and Demory, 1963) or on mountain tops (Mani, 1962). An Antarctic springtail preferred moss to fungi (Wise *et al.,* 1964), while the termite *Hospitalitermes* selected lichens rather than bryophytes (Collins, 1979). Alternate feeding also occurs when the bryophyte host dries up, as happened when the lacebug *Acalypta barberi* shifted from moss to hops (Drake and Lattin, 1963). This case shows that, under certain conditions, bryophagous invertebrates may attain economic importance. Regular movement between moss and phanerogam constitutes the third mode of alternate feeding; aphids of the subtride Melaphidina, which shift between bryophytes and *Rhus* (Eastop, 1977), are an example.

The digestion and metabolism of bryophytes in invertebrates, and the energetics of the association, are poorly known. Numbers of various animals occurring in bryophytes were presented above, and most of the cited sources contain additional data. Standing crop and/or annual production of some bryophytes were compiled by During (1979) and by Rieley *et al.* (1979). Data on bryophyte amounts ingested are available for an orthopteran and a beetle. Duke and Crossley (1975) tagged *Grimmia laevigata* with ^{134}CsCl and showed that the rock grasshopper *Trimerotropis saxalitis* ingested 27.25 mg moss day^{-1}. They calculated that the entire population in the sample area consumed 391.2 mg moss m^{-2} year^{-1}. Using a simple feeding technique, Smith (1977) offered various plants to the beetle *Ectomnorrhinus similus.* It consumed 1.67 mg *Brachythecium rutabulum* per beetle day^{-1}, or 3–13% of the bryophyte's total production.

Frankland (1974) stated that 'consumption of bryophytes by animals appears to be insignificant'. Many invertebrates observed on these plants actually subsist on algae or detritus occurring on aquatic and terrestrial bryophytes (Glime and Clemons, 1972; LeSage and Harper, 1976; Matthey, 1971; Stout, 1974), and many, as noted, only shelter, oviposit, pupate or hunt there. And yet the data presented above show that members of a dozen insect orders, as well as some mites, gastropods and tardigrades, feed on bryophytes. On balance, it seems that more invertebrates feed on these plants than meets the eye; many others, however, do not. Crum's (1976) observations that 'In nature animals rarely eat bryophytes possibly because of their taste', and that *Preissia quadrata* when tasted, was hot enough to tingle the tip of the tongue and burn the back of the mouth, provide the clues. Taste, as well as smell, are imparted to plants by secondary chemical compounds, and it is this group of substances which are believed to protect plants against herbivores (Fraenkel, 1959; Swain, 1974), making the former unpalatable or even poisonous for the latter.

Bryophyte chemicals (including secondary substances) were reviewed by Markham and Porter (1978) and by Suire (1975). Some representative compounds known to affect arthropods are listed in Table 9.2, p. 324, but this is barely the tip of the iceberg. As bryophyte chemicals become better known, many additional substances will be found which repel, deter, inhibit or poison invertebrates. Certain animals, however, have managed to breach the chemical (and other) defenses of bryophytes by various ways – of which we are quite ignorant – and thrive on them. The archaic and the recent feeding associations of insects with bryophytes, as expounded in Section 9.2.8 (c), suggest that this breaching is of ancient as well as relatively modern origin. When Crum (1976) complained about the taste of *P. quadrata* he was probably savouring the bitter flavour of glycosides present in that liverwort (Markham and Porter, 1978).

Some of the secondary substance of bryophytes (Suire, 1975), like the terpenoids, pinene, limonene, camphene and cardinene, and the sterols, sitosterol and stigmasterol, induced the formation of mixed-function oxidases (MFO) in a moth caterpillar (Brattsten *et al.*, 1977). It was argued that MFO play a major part in protecting herbivores against chemical stress from secondary plant chemicals. These substances, then, while protecting bryophytes from some herbivores, may enable others to devour them.

A distinction should here be made between 'generalist' and 'specialist' herbivores (Feeney, 1976). The former are animals which attack plants of many chemically unrelated families, subsisting on most plants unless these are specifically repellent to them. Specialists feed on plant species only within one family, or group of families which are chemically related. Their choice of food is strongly influenced by chemical signals from the host plant. Invertebrates devouring bryophytes and other plants (with the exception of aphids) are generalists. Peloridiid bugs, *Dolichopeza, Boreus* and *Eustigmaeus* are examples of arthropods (drawn from diverse groups) specializing on bryophytes. Special adaptations to locate these plants, on the one hand, and to overcome their array of chemical (and other) defences, on the other, might most profitably be sought among these arthropods. Strict specialists, which exclusively feed on one diet (i.e., plants, wood, blood) require internal symbionts to supplement their fare with essential sterols and vitamins. Bryophyte specialists can therefore be expected to have them, and H.J. Müller (cited by Buchner, 1965) described the single symbiont of an archaic Peloridiid bug. Nothing is known about its specific role or how other bryophyte specialists obtain essential rare nutrients.

Certain bryophytes may absorb and store in their tissues heavy metals like copper, lead and nickel (Rühling and Tyler, 1970), elements toxic to all animals. If ingested by herbivores, the heavy metals will move along the food chain, continuously harming all participants. Much the same might happen when invertebrates consume mosses which had accumulated radionuclides (Lowe, 1978; Svensson, 1967).

9.5 INVERTEBRATE UTILIZATION BY BRYOPHYTES

9.5.1 Dispersal

Symbiopholus weevils which participate in epizoic symbiosis (Section 9.3.5) carry their cryptogam gardens about. During their long lives (c. 5 years)the growths become cumbersome and may fall off as a result of fighting and mating (Gressitt and Sedlacek, 1970). Dispersal occurs when the gardens drop off away from their initial sites, and bryophytes then continue to grow. A similar mechanism may operate with freshwater mussels overgrown by *Octodiceras.* The animals migrate with fluctuations in water levels, carrying their moss growth along (Lohammar, 1954; Neumann and Vidrine, 1978). Bryophyte dispersal in both cases is quite fortuitous, not dependent on animal presence, as the plants grow on other substrates in the same habitat.

The association between certain Splachnaceae and flies which disseminate their spores is another matter altogether. Its apparent long duration (Koponen, 1978), and the role insects appear to have played in the bryophytes' evolution, make it of special interest for those fascinated by plant–insect co-evolution.

Upon ripening the capsule of *Splachnum* (and some other Splachnaceae) remains fleshy, the hypophysis being large and coloured.

The sticky clumps of small, smooth spores, which are not air-borne (Crum, 1976), then become exposed. A 'musty' smell which attracts flies is emitted at that time (Erlanson, 1930, writing about *Tetraplodon mnioides,* described it as a 'strong and rank valerianic odour'). The odour apparently has two components, one resembling decaying protein (Faegri and van der Pijl, 1979), the other consisting of alcohols and acids (Pyysalo *et al.,* 1978). The latter authors identified the fetid butyric acid as well as rather specific 8-carbon compounds in some Splachnaceae; amount and distribution of these substances differed among the bryophytes assayed (Table 9.1). *Splachnum luteum* which has the largest hypophysis, contained more volatiles than the other species. The differences between the two *Splachnum* species suggest that others of this genus might also vary in their attractant substances. The entomophilous Splachnaceae are further characterized by enlarged, brightly coloured hypophyses, which, standing out against the generally green-mossy background, will likewise allure insects. Comparative observations on flower pollination by insects (Faegri and van der Pijl, 1979) suggest that odours preceded visual stimuli as attractants, and it is likely that the same sequence was followed in the Splachnaceae.

Table 9.1 Volatile compounds identified from several Splachnaceae (from Pyysalo *et al.*, 1978).

Compound	*Splachnum luteum*	*Splachnum vasculosum*	*Tayloria tenuis*
1-octen-3-ol	+	+	+
3-octanone	+	−	−
3-octanol	+	−	−
2-octen-1-ol	+	−	−
trans-2-octenal	+	−	−
octanal	+	−	−
octanol	+	−	−
acetic acid	−	+	+
butyric acid	−	+	+
propionic acid	−	+	+

Koponen and Koponen (1978), who reviewed reports on insect attraction to Splachnaceae, baited traps with these mosses and caught several flies. Representatives of the families Muscidae, Sepsidae and Calliphoridae (which include the Sarcophaginae) were obtained. All belong to the more advanced order Cyclorrhapha and are attracted to dung and cadavers, in which they oviposit and where their larvae (maggots) live. Dung and cadavers are the specific habitats of the entomophilous Splachnaceae. A representative list (Crum, 1976) includes cow, deer, horse, human and moose faeces, cattle tracks, a site at which whales were slaughtered, wolf bones and owl pellets. Entomophilous Splachnaceae and the flies they allure are thus restricted to the same habitats, and the bryophytes appear to be making the most of it. They entice the insects by a two-component (odour and colour) system, and, as these arrive, their bodies and appendages become covered with sticky spores. Flies which later alight on dung or on cadavers and move around will cause the spores to fall off, thereby dispersing Splachnaceous propagules to their preferred sites.

Splachnum and its relatives, upon becoming adapted and then restricted to animal droppings and remains, were faced with the need to transport their spores to other similar habitats. Unlike a possibly analogous group, the copper mosses, *Splachnum* had adapted to growing on substrates which were inhabited by, and attractive to, highly mobile organisms like insects. Initially, the latter probably only rested on these bryophytes, but some spores could have adhered to them. Insofar as the insects then flew to animal dung or cadavers, and the spores dropped off there, dispersal was actually effected. The advantages accruing to Splachnaceae whose spores had thus been transported from one suitable habitat to another would then have exerted strong selective pressures to preserve and enhance this mode of dispersal. Consequently, *Splachnum* and its relatives evolved means to attract and hold flies on the hypophysis long enough for the spores to stick to them. In other words, steps in the evolution of the Splachnaceae towards entomophily were determined by insect response. The association between the entomophilous Splachnaceae and flies is therefore claimed as a case of bryophyte–insect co-evolution.

Many questions arise. A selection includes the specific role visual stimuli play in attracting flies; the variety of chemical attractants and the significance of 8-carbon compounds; a possible 'reward' for moss-visiting flies (such as high-energy and/or fecundity-promoting substances from the hypophysis) and whether there are any special behavioral or morphological insect modifications for spore dispersal. Whatever the answers to those (and many other) questions, the presence of specific fly allurements, the distinct growth sites of the mosses and the relatively large numbers of entomophilous species (Koponen, 1978) attest to the diversification and thus evolutionary success of the Splachnaceae. It is tempting to speculate that their unique relationship with insects has contributed to their success.

The remarkable morphological uniformity of entomophilous Splachnaceae obtained from various parts of the world, their similar habitats, and their associations with insects led Koponen (1978) to postulate a monophyletic origin for these mosses. Their strict entomophily precludes long-range dissemination, which suggests that the spread of the Splachnaceae to their present disjunct distribution could only have been along land connections in the geological past. This in turn indicated an approximate era, about 200 million years ago, for the rise of these mosses (Koponen, 1978), a date not inconsistent with fossil evidence (Lacey, 1969). Two problems intrude here. The first concerns the original dung and cadavers producers, because cattle, moose and humans are rather recent newcomers to the geological scene. Koponen (1978) recognized this difficulty and suggested that reptiles were the early source of animal remains. The other problem concerns the disseminating insects. As noted, all flies baited or reported from Splachnaceae to date belong to the advanced suborder Cyclorrhapha. These are believed to have originated in the Cretaceous (Oldroyd, 1964), approximately 100 million years later than the entomophilous Splachnaceae. The difficulty may be overcome by assuming that these mosses originally attracted archaic flies, whose place was then taken by modern relatives. To sum up, as reptiles gave way to mammals, *Splachnum* and its relatives 'shifted' to growing on the latter's remains (Kühnelt, 1976, mentions a case of preferential moss growth on snail droppings – a possible early link). And as modern flies replaced ancient ones (or other insects), the bryophytes attracted them also. Should this model turn out to be generally correct, it would show that the entomophilous strategy of these Splachnaceae must have been quite successful, as well as constituting a rather long-running, extant case of plant-insect co-evolution.

9.5.2 Fertilization and fertilizers

Harvey-Gibson and Miller-Brown (1927) observed mites, springtails and other arthropods on the antheridia of *Polytrichum vulgare*, and noted that their bodies became smeared with male gamete-containing mucilage. Fertilization was believed to result from such activities. A more restrained point of view was expressed by Garjeanne (1932), who noted that the conspicuous 'inflorescences' of the Polytrichaceae are not known to attract insects. More recent data on this potential bryophyte–invertebrate interaction are not available.

As the colonization of land passes from the initial lichen stage to the moss phase, a layer of humus is formed below the latter plants which enables many more invertebrates to live there (Kühnelt, 1976). This additional mass of dead animals and their excretions may well provide bryophytes with general and specific growth factors, thereby in turn enhancing their development. The case of a moss growing preferentially on snail droppings (Kühnelt, 1976) will serve to illustrate this point.

9.6 BRYOPHYTE–INVERTEBRATE COMMUNITIES

9.6.1 Colonization

While bryophytes participate in the early stages of plant succession, their associated invertebrates form similar stages in faunal succession (Gerson,

1969). The hardiest and most mobile animals arrive first, being later replaced by species of narrower ecological requirements but better competitive ability. Moss cushions developing on rock faces were first colonized by rhizopods, rotifers and tardigrades, nematodes and ciliates also occurring there (Kühnelt, 1976). As dead material formed under the cushions, rotifers and tardigrades became very abundant, nematodes also increased, and arthropods, like springtails and oribatid mites, began to appear. As a thicker decomposition layer formed below, rhizopods, rotifers and tardigrades declined, but nematodes, oribatids and springtails increased. 'Myriapods' appeared, the composition of the fauna becoming similar to that of the soil fauna. Other succession sequences involving bryophytes and invertebrates, but affected by other physical factors, were also described by Kühnelt (1976) and by Stebaev (1963).

The first bryophyte and invertebrate colonizers usually reach new sites by wind. The animals must therefore be able to withstand extreme temperatures, solar radiation and infrequent, widely fluctuating humidities. The ability of certain bryophyte-associated Protozoa, Turbellaria, Rotifera, Nematoda and Tardigrada to survive such conditions has already been mentioned (Section 9.2). To this may be added Kühnelt's (1976) remark that colonizing Collembola and Acari are usually drought-resistant. The end result of such strong selection is that disparate components of the bryofauna show remarkable convergence in their adaptation to harsh, unpredictable environments. Wind dispersal of the colonizing bryofauna partially explains its 'surprising' cosmopolitanism (Hesse *et al.,* 1951): seven of 12 Antarctic rotiferans, and most of the tardigrades, were already known from Europe and the Arctic. Nicholas (1975) remarked that common moss nematodes, like *Plectus cirratus,* are found in very different parts of the world. Menzel (1921) noted the same in regard to bryophilus copepods.

Bryophytes play only a transitory part in plant colonization in most habitats, but retain a dominant role on high mountains or under very cold conditions, where succession cannot continue. Bryophytes may be the dominant producers in such situations and strongly affect their faunas. The dependence of Antarctic arthropods on mosses was emphasized by Janetschek (1967), who defined the bryosystem as one of the dominant Antarctic ecosystems. When this ecosystem matures, bryophytes become increasingly important and most of the fauna occurs in moss tufts and mats.

Under favourable conditions bryophytes form perennial colonies in many environments, thereby becoming apparent to feeders and predictable as habitats. This in turn causes a shift from r-selected to K-selected moss invertebrates. As noted above, the change from a colonizing moss tuft to a perennial cushion brings about a change from the small, mobile, rapidly reproducing r-selected Protozoa and Rotifera to larger, less mobile, slower reproducing, K-selected arthropods. Year-round availability of bryophytes promotes monophagy, itself conductive to K-selection. Examining *Eustigmaeus* mites, Gerson (1972) found species like *E. gersoni* and *E. rhodomela* to be associated with colonizing mosses, while *E. frigida* occurred only on perennial pleurocarpous species in forests and other humid situations. Life strategies of the host bryophytes thus impose themselves on their dependent faunas.

9.6.2 Succession, annual activity and exclusion

Changes in bryophyte composition affect the associated animals. This aspect of bryophyte–invertebrate inter-relationships was studied by Murphy (1955) in sphagnum bogs. Accumulations of acid peat are built up, at certain sites, by a complex of sphagnum hummocks and water-filled depressions which are constantly replacing and succeeding each other, with a concomitant change in *Sphagnum* species. Each of the latter tends to occupy a characteristic position relative to the water table and the age pattern of the hummock. The springtail *Sminthurides malmgreni* was associated with the most humid *S. cuspidatum* and *S. subsecundum, Folsomia brevicauda* and *Isotoma sensibilis* with the mesophiles *S. papillosum* and *S. magellanicum,* and other Collembolans with the drier *Calluna* and *Cladonia* habitat. Similar series were described by Tarras-Wahlberg (1952–53) for oribatid mites, and by Reichle (1966) for pselaphid beetles.

An interesting feature of some members of the bryofauna is their year-round activity, even under extreme cold conditions. Immatures and adults of the snow scorpionfly *Boreus brumalis* feed on mosses the year around, even below winter snows (Shorthouse, 1979). Active *Eustigmaeus* mites were dug out of their 3-ft snow cover in eastern Canada by Gerson (1972). Upon being offered mosses in the laboratory, they immediately began to eat and oviposit. Morgan (1977) stated that tardigrades living in a moss on a roof in South Wales reproduced throughout the year.

Active cases of exclusion between bryophytes and invertebrates are rare. When *Fontinalis antipyretica* was introduced into South Africa, it quickly became established but caused a decline in the populations of the indigenous water-insect larvae. The latter were adapted for living on slime algae, and the blanketing of these microhabitats by *Fontinalis* did not enable the larvae to survive there (Richards, 1947). Another water moss, *Octodiceras fontanus,* is grazed by snails to the extent that they are mutually exclusive on some North European rock surfaces (Lohammar, 1954).

9.6.3 Specific communities

Four bryophyte–invertebrate communities will be discussed below briefly, emphasizing the special environmental features of each. Readers interested in arthropods living in saxicolous mosses are referred to Bonnet *et al.* (1975) and Simon (1974).

Water bryophytes support large and varied, specialized and unspecialized invertebrate populations. The former include the primitive *Fontinalis*—simulating crane fly *Triogma* (Alexander, 1920), the aphid *Aspidaphium cuspidati,* which feeds on *Calliergonella cuspidata* only below water level (Müller, 1973), and the tardigrade *Macrobiotus macronyx,* exclusively collected from submerged mosses (Morgan and King, 1976). Many of the unspecified, bryoxenous or occasional components of the aquatic bryofauna were found by Lindegaard *et al.* (1975) during their rather complete study of the animals on *Cratoneuron.* A special feature of their work was the comparison between invertebrates found in the various zones. These were defined as the underlying detritus zone, above it the one holding water-covered mosses, then the 'madicolous' zone ('just above the water surface where the moss is constantly wetted by capillary water') and uppermost, the dry zone (Fig. 9.2). One may thus see how the fauna changes from springtails, beetles, spiders and predaceous mites in the dry region to mainly flies, caddis flies and molluscs in the madicolous and water zones, to worms and more flies in the detritus. The water-covered zone harboured the largest numbers of both species and individuals as well as having the biggest diversity. A special aquatic bryofauna, developing around moss balls (formed from *Fontinalis antipyretica* and *Drepanocladus tenuinervis*), includes the isopod *Asellus aquaticus,* oligochaetes and leeches (Luther, 1979).

Glime and Clemons (1972) compared number and diversity of insects found on *Fontinalis* with those collected from similar but artificial substrates. They concluded that this moss probably serves only as a physical substrate for insects (which browse on accumulating detritous and algal growths). Studying submerged Antarctic mosses, Priddle and Dartnall (1978) reached a somewhat similar conclusion. The six rotifers which they investigated had clear settling preferences on various sites along the submerged stem of *Calliergon sarmentosum,* but none of the invertebrates found appeared to feed on the moss.

Sphagna in bogs have their own faunas, whose composition is determined by the special characteristics of these plants. Their acidity, for instance, limits the number of animal groups which may live there. Protozoa therefore abound in number as well as species (Bovee, 1979; Groliére, 1978; Heal, 1962), while tardigrades are rare (Morgan and King, 1976). *Sphagnum* bogs undergo periods of drying out, which result in various invertebrate succession patterns (Section 9.6.2). Some of the plants actually float, thereby serving as habitats for oribatid mites (Karppinen and Koponen, 1973) and beetles (Dybas, 1978). Sphagnum 'carpets' can be divided into two layers, the upper one consisting of the heads which, as they grow close together, form a smooth surface. Small animals may run there, and it effectively insulates the lower layer from temperature and humidity fluctuations (Nørgaard, 1951). The lower or stalk layer, on the other hand, is much thinner, being filled with small spaces and cavities in which animals live, rather protected from the surface. *Sphagnum* has hollow cells, in which some rotifers live (Fig. 9.1). And finally, these plants appear to be rather unedible to invertebrates (Smirnov, 1961), although odd feeders, like the ground cricket *Pteronemobius,* feed on sphagna (Vickery, 1969). *Sphagnum* peat is usually sterile *in situ.* However, as peat particles enter a bog, they are colonized by micro-organisms, which enhances their nutritional value and makes them edible to chironomid fly larvae (McLachlan *et al.,* 1979).

Arboreal bryophytes occur as continuous growth from the underlying soil, or as disconnected patches. In the former case, the plants serve as 'bridges' for the fauna between soil and trees; many animals are known to have followed bryophytes upwards (Bonnet, 1973; Larsson, 1978). The upper limit for such invertebrates is determined by the prevailing humidity, the animals arranging themselves according to their drought resistance at various tree heights (Pschorn-Walcher and Gunhold, 1957). Invertebrates in strictly arboreal bryophytes live in a biotype characterized by spatial and temporal discontinuity (Bonnet, 1973). Their fauna consists mainly of wind-borne species, capable of surviving intermittent humidities.

The only invertebrates living in bryophytes on roofs are very resistant species, which, having this environment all to themselves, may produce considerable populations. The tardigrades *Macrobiotus hufelandii, Hypsibius oberhaeuseri* and *Milnesium tardigradum* abound in British as well as Danish roof mosses (Morgan, 1977; Nielson, 1967). Another species very common in this habitat is the nematode *Plectus rhizophilus.* Corbet and Lan (1974) recorded the protozoan *Arcella arenaria* and the rotifer *Mniobia* from bryophytes growing on a west Norfolk roof.

9.7 BRYOPHYTES AS 'STEPPING STONES'

Bryophytes have been called the amphibians among plants (Richards, 1959), as they straddle, so to speak, the aquatic and the terrestrial worlds. This unique position has rendered them suitable to serve as 'stepping stones' or 'halfway houses' for invertebrates as these emerged from archaic water bodies onto land. The ability of bryophytes to live under intermittent wetting and drying was probably of considerable importance in this respect.

The transition of arthropods from aquatic to terrestrial habitats probably began in the Silurian and the Lower Devonian (Kevan *et al.*, 1975), and was dramatically described by Størmer (1976): 'plants and animals of the seas, lakes and rivers were on their way to invade and conquer the land. The swamps, the early "mangroves" and the beaches with an abundance of plant litter were the "land of promise" for the animals struggling for life in the less nutritive waters along the shores'. Bryophytes (apparently present from the Devonian; Lacey, 1969) which grew at the water–land interface could also serve as a 'land of promise'. The ability of bryophytes to maintain high humidities might have been of special importance, as the dampness would protect newcoming invertebrates against desiccation, while enabling aquatic ones to breathe without drying out.

It is upon confronting change that organisms evolve. The intermittently-wetted bryophyte cushions constitute such changing, uneven habitats, their variability by itself exerting strong selection pressure. Such contingencies would favour animals capable of coping with them. The co-existence of invertebrates employing various respiratory mechanisms is likewise encouraged by moisture gradations available within the same cushion.

Emigration of animals from an aquatic to a semi-aquatic habitat will increase the weight placed upon the legs. Some aquatic arthropods walk on the bottoms of water bodies, the displaced volume of liquid taking some of the weight off their appendages. As these animals wander into bryophyte cushions, the humidity prevailing there will help them to 'take the weight off their feet'. Variable moisture encountered within such cushions would encourage stronger legs while affording the animals areas of 'rest', even of survival, between bouts of walking. The newly colonized bryophyte environment, being partially dry, affords some animals sanctuary from wholly aquatic predators. The plants also offer an abundant, diverse food supply. For carnivores, many protozoans, nematodes and tardigrades are available, while the bryophytes themselves, their algal periphyton and the detritus accumulating underneath provide opportunities for phytophages and omnivores.

If bryophytes had indeed played a role in the colonization of land by invertebrates, we would expect to find primitive, relict animals still living in moss, sphagna and hepatics. The extant presence of some rather archaic insects in bryophytes thus lends support to the above hypothesis. The Peloridiidae, for example, are a family of archaic moss-feeding bugs, regarded as 'living fossils' which have changed little since Triassic times (China, 1962). The two most ancient fly families, Tipulidae and Nymphomyiidae, are represented in bryophytes. Oldroyd's (1964) words are appropriate about the former family: 'We believe that flies, as a group, have arisen from ancestors whose larvae, though terrestrial, lived in wet moss and so had, as it were, a foot in both worlds . . . Perhaps nearest to the ancestral way of life are those Tipulid larvae that live in moss'. The larvae of *Palaedipteron walkeri,* a member of the Nymphomyiidae, live among mosses in rapid streams (Cutten and Kevan, 1970). Among Lepidoptera, the Micropterygidae are the most primitive, and species of the genera *Micropteryx, Sabatinca* and *Neomicropteryx* are found in bryophytes, on which they may feed (Tillyard, 1922; Yasuda, 1962).

Bryophytes serve as 'stepping stones' for arthropods moving in other directions also. The alpine water mite *Partnunia steinmanni* is a recent newcomer to water, and can still be found in moist as well as partially dry moss (Bader, 1969). Bryophytes were used in this case as a 'halfway house' on the mite's way into water. Cavernicolous copepods are believed to be relicts of a tropical fauna which flourished in Europe during the early Tertiary (Vandel, 1965, citing papers by Menzel). As the climate changed most tropical forms perished, except those which managed to adapt to subterranean waters. Mosses appeared to have a role as refugia during that adjustment period, because tropical copepods, like *Bryocyclops* and *Muscocyclops,* are closest to present-day cavernicolous species. Transition, through mosses, from an epigean to a cavernicolous mode of life may be observed in amphibious copepods. The genus *Morania,* for example, contains a series of species which live in dead leaves, in decaying wood, in bryophytes and finally in caves (Vandel, 1965). Another series begins with the beetle *Bathysciola,* which lives in mosses. Quite similar genera may be found in caves, these beetles having highly specialized, strictly cavernicolous, relatives (Vandel, 1965).

Passing to a totally different group, Stout (1963) believes that terrestrial Protozoa are freshwater species which had invaded land, probably via the moss-sphagnum of stream beds, and hence into forest litter and soil.

9.8 EFFECTS OF HUMAN ACTIVITIES

Modern technology is altering bryophyte distribution, abundance and composition, with resultant changes in the accompanying faunas. New habitats become available for cryptogams (bryophytes and algae) by installing electric lights in caves. Insects, mites and spiders – representing two additional trophic levels – rapidly colonize these plants (Dobat, 1972).

Bryophytes are sometimes considered to be weeds, necessitating chemical control. The effect of bryocides on the fauna would be mostly indirect, by reducing plant cover and biomass. Direct effects on animals can be expected upon using compounds like calomel (mercury chloride) (Pycraft, 1975) or chloroxuron and diuron (both of which contain urea) (Hackemesser and Lichte, 1978). Calomel is a recognized insecticide, having been used against pestiferous maggots (Martin and Worthing,

1977), while urea is known to repel soil arthropods (Marshall, 1974). Residues of another pesticide, methoxychlor, were found in aquatic mosses several weeks after treating the water against simuliid flies (Wallace *et al.*, 1976). A broad-spectrum pesticide like this can be expected to have a severe influence on the non-target bryofauna.

More pernicious effects may be anticipated from accumulations of heavy metals and radionuclides. Grodzinska (1978) reported considerable concentrations of heavy metals (including the dangerous cadmium, nickel and lead) in *Pleurozium schreberi* and *Hylocomium splendens*. Increasing radionuclide levels were found in temperate as well as subtropical bryophytes (Svensson, 1967; Lowe, 1978). The effects of these types of pollution on the bryofauna are not known. Increased SO_2 levels are currently causing major changes in bryophyte cover and diversity. While most species decrease, some, like *Hylocomium splendens* and *Ptilium crista-castrensis,* increase their coverage (Winner and Bewley, 1978). Such changes are bound to alter the associated invertebrate fauna, decreasing the populations of many species while increasing those of few.

9.9 MISCELLANEOUS

Bryophyte–invertebrate associations have a few minor economic consequences, cited by Thieret (1956). Thus sphagnol was recommended for relieving itching caused by insect bites and even preventing them, and *Hypnum* was used for packing leeches for shipment as well as by anglers as a medium for scouring worms. Commercial use implies shipment; the lacebug *Acalypta sauteri* was intercepted by quarantine authorities at New York port, on moss used for packing nursery stock imported from Japan (Drake and Lattin, 1963). The ubiquity of bryophytes suggested to Sayre and Brunson (1971) and to Corbet and Lan (1974) that these plants and their microfauna might be suitable material for classrooms.

Finally, an etymological note. Invertebrates were named after bryophytes (or musci) although often they were only occasionals (i.e. the snail *Pupilla muscorum,* some Protozoa and other animals mentioned in Section 9.2). An entire phylum, Bryozoa ('moss animalcules') was given its name because their (mainly marine) colonies look like 'a mat of moss' (Pennak, 1978); they do not appear to be associated with bryophytes. The prefices 'bryo' and 'musci', when applied to the names of invertebrates, therefore do not necessarily indicate a close relationship with bryophytes. On the other side of the fence, neither does the name *Ctenidium molluscum,* a moss sometimes growing on calcareous soil with molluscan remains (Marstaller, 1979). Notwithstanding these reservations, one is struck by the congruence (in Latin) between *musca* (fly) and *muscus* (moss).

9.10 DISCUSSION

9.10.1 The origin of the bryofauna

The data presented indicate at least five sources of the bryofauna. Desiccation-resistant species came by winds. Others moved up from the underlying soil. Several invertebrates arrived from freshwater bodies. The phytophagous species originated from other host plants. The fifth group are relict species which apparently survived in bryophytes from archaic eons, their current distribution being quite disjunct (Hammer, 1965). The present diversity of the bryofauna probably reflects the disparate origins of its components.

9.10.2 The preponderance of flies in the bryofauna

Reviewing the bryofauna, one is struck by the variety of flies (Diptera) which occur on and in these plants. They predominate in numbers (Lindegaard *et al.*, 1975) and in diversity, being the only invertebrates to which mosses (i.e., Splachnaceae) have clearly adapted. And yet, the Diptera is neither the largest insect order (which is the Coleoptera) nor, among the major orders, the one most consistently associated with plants (which is the Lepidoptera). Three factors have probably contributed to the success of the bryophyte–fly relationship. The great versatility in the life style of the flies may have been one factor. 'No other order of insects present so great a diversity of larval habits as the Diptera' (Richards and Davies, 1977). To this should be added that in the context of another plant–insect association, namely pollination (mostly conducted by winged adults), the Diptera show more variation of habits than any other insect group (Faegri and van der Pijl, 1979). The need of most fly larvae for a more or less humid medium (Oldroyd, 1964), could have been another factor. Bryophytes have similar requirements, thus they co-exist with flies more often, and in a greater variety of situations, than members of other insect orders. The third factor may well have been the highly evolved position of the Diptera which have attained the highest degree of structural specialization (Richards and Davies, 1977).

9.10.3 Bryophytes as a factor in invertebrate evolution

Insects simulating bryophytes in their behaviour, coloration and even appendages were noted in Section 9.3.6. 'Epizoic symbiosis' (Section 9.3.5) is another instance of insects adapting to bryophytes (and other cryptogams) and the same might be said about case-building insects which use only bryophytes, even in the presence of other plant material (Glime, 1978). More subtle adaptations must have occurred in regard to feeding. As noted in Table 9.2, bryophytes have their array of secondary chemical substances, and invertebrates feeding on them would have to develop

Table 9.2 Selected bryophyte chemicals (gleaned from Markham and Porter, 1978, and Suire, 1975) and their adverse effect on arthropods.

Chemical	Effect	Source
Oxalic acid	Repels moth caterpillars; repels a phytophagous mite	Dethier, 1947 Dubitzky and Gerson, unpublished
Lauric acid	May inhibit mite growth	Rodriguez, 1972
Limonene	Kills pine-feeding beetles	R.H. Smith (in Levin, 1976)
Pinguisone	Inhibits feeding by a polyphagous moth caterpillar	Wada and Munakata, 1971
Sesquiterpene lactones	Inhibits feeding by moth caterpillars	Burnett *et al.*, 1974
Benzyl benzoate	Kills insects and mites	Brown, 1956

Other groups of bryophyte chemicals known to affect arthropods (Levin, 1976) include alkaloids, cinnamic acid derivates, saponins and tannins.

mechanisms to overcome these nutritional defences. In this context, it will be repeated that bryophagous insects and mites usually feed on several, often unrelated bryophytes (Section 9.4); they do not appear to be specialists on any one moss species. This suggests that most bryophytes are protected by similar chemicals; breaching one thus means breaching all. Or else, coping with the defensive substances of one bryophyte enables the bryophages to cope with others also. The mixed-function oxidases (Brattsten *et al.*, 1977) noted in Section 9.4 may play an important role here. It is also noteworthy that some arthropods appear to feed only on bryophytes and lichens. The latter have their own rich and unique arsenal of secondary chemicals (Culberson, 1969) which feeders would also have to overcome. One could speculate that invertebrates which live in these cryptogams have evolved the means to overcome both systems of defensive chemicals. This, however, goes against the tenet that changes in the host range of a particular feeder are most likely to occur among plant taxa sharing similar secondary substances (Feeny, 1976). More data are clearly required.

9.10.4 Invertebrates as a factor in bryophyte evolution

The presence of secondary substances of possible defensive function in bryophytes is also meaningful when seen from the other side. Their very occurrence suggests that they were evolved (or selectively favoured once synthesized) in response to invertebrate feeding pressure. Several other attributes of bryophytes may also be hypothesized to have arisen (or to have been encouraged) in answer to animal grazing. Bryophytes appear to be deficient in such essential amino acids as methionine, tryosine and tryptophane (Suire, 1975, Table 9.2). This could result from selection for inedibility, as postulated by Gordon (1959) for higher plants. Another conjectured protective mechanism could be the external leaf texture of several bryophytes, especially Polytrichaceae (Gerson, 1972; Ramazzotti, 1958). The role of plant surface texture in defending plants from their enemies was discussed by Levin (1976). The ability of some bryophytes to grow on and accumulate heavy metals in their tissue could well be another means of warding off feeders. Sphagnum growing in very acid surroundings may thereby be protected from herbivores. And finally, the rugose, sculptured, papillose spores of many bryophytes (Horton and Murray, 1976; Sorsa and Koponen, 1973) could protect them from attack by animals, or from the latters' alimentary system, when ingested (Kevan *et al.*, 1975). The oft-repeated observation that bryophytes are relatively free from arthropod attack attests to the success of these or other defence mechanisms, and their very presence implies that invertebrates, although by no means the only adverse factor, influenced bryophyte evolution.

9.10.5 Bryophyte–invertebrate coevolution

The last two sections strongly suggested that bryophytes affected the evolution of invertebrates, as well as the other way around. The relationship between some Splachnaceae and flies (Section 9.5.1.) provides firm evidence for actual coevolution.

The apparent antiquity of these mosses (Koponen, 1978) suggests that their mechanisms of attracting insects by odour and sight may have antedated those of the flowering plants. Another relevant question is whether the relatively rare known cases of bryophyte–invertebrate coevolution are due to the plants' slow rate of evolution (Crum, 1972) or to lack of observations. Answers to this question could be found in the tropics or subtropics, regions of continuous, year-round growth and invertebrate development.

9.10.6. Tropical considerations

Bryophytes abound in the humid tropics and additional data on the matter at hand could be sought there, but little information is actually available. Donner (1966) and Mehlen (1972) referring to rotifers and tardigrades, respectively, stated that these animals are rarer in tropical than in temperate mosses, an opinion with which Hesse *et al.* (1951) concurred. Lack of data could therefore also result from lack of some invertebrates in tropical bryophytes. On the other hand, this scarcity may reflect fewer collections. Unique associations were recorded from the tropics; the singular 'epizoic symbiosis' was already noted in Section 9.3.5. Another case is the rare fungous gnat (Diptera) which lives in and feeds on central

African bryophytes (Matile, 1972). Additional novel relationships between bryophytes and invertebrates, which could contribute to our understanding of the co-evolution between these two groups of organisms may therefore be expected to be found in the tropics. It is appropriate to recall here that evolution in the tropics operates in fundamentally different ways than in temperate regions (Dobzhansky, 1950).

9.10.7 Suggested topics for further research

This overview of what must sadly be admitted as mostly gaps in our knowledge has indicated some research areas in which additional information may have special interest.

(i) Invertebrate feeding on bryophytes, encompassing specificity to plants and to their various parts, and the occurrence of secondary chemicals in these parts. Also: effect and fate of these substances in herbivores; digestion and energetics; influence of heavy metals, pesticides and radionuclides, which had accumulated in bryophytes, on their various feeders.

(ii) The effect of urban and industrial pollution on bryophyte–invertebrate communities.

(iii) Associations in the tropics.

(iv) Dispersal of Splachnaceae by flies.

(v) Aspects of economic entomology. A few pests, especially Diptera of medical importance, live in bryophytes. A better understanding of this specific association could lead to improved sampling and control measures.

This assay has hopped along the bryophyte–invertebrate interface, alighting here and there for specific topics but mainly skipping lacunae. One hopes that it will encourage others to look back at the thousand eyes which have been looking at us – askance – for so long.

ACKNOWLEDGMENTS

Dr A.J.E. Smith, of the University College of North Wales, suggested this topic, was always generous with his advice and sent me otherwise unavailable literature. Discussions with Mrs and Dr T.A. Koponen (University of Helsinki), Professor C. Heyn, Dr I. Herrnstadt and Mr M. Loria (all at the Hebrew University of Jerusalem) were always most interesting, stimulating and fruitful. The editor of *Svensk Botanisk Tidskrift* kindly provided the original of Fig. 9.3. Mr Loria kindly also allowed me to use some unpublished data and a photograph (Fig. 9.4). To all of them I wish to extend my most sincere thanks.

Preliminary work on this review was begun during a sabbatical at the University of Bradford. It is a pleasure to thank my host there, Dr M.R.D. Seaward, for his hospitality.

REFERENCES

Alexander, C.P. (1920), *Cornell Agric. Exp. Stat., Mem.* No. 38.

Anonymous (1970), *The Insects of Australia,* Melbourne University Press.

Arnett, R.H. Jr. (1971), *The Beetles of the United States,* American Entomological Institute, Ann Arbor, Michigan.

Bader, C. (1969), *Proc. 2nd Int. Cong. Acarol.,* pp. 89–92.

Barr, D. (1973), *Life Sci. Misc. Publ., R. Ont. Mus.*

Bengtson, S.-A., Fjellberg, A. and Solhöy, T. (1974), *Ent. Scand.,* **5,** 137–42.

Bertrand, M. (1975), *Vie Milieu,* **25,** 299–314.

Bland, J.H. (1971), *Forests of Lilliput. The Realm of Mosses and Lichens,* Prentice-Hall, New Jersey.

Bonnet, L. (1973), *Protistologia,* **9,** 319–38.

Bonnet, L., Cassagnau, P. and Travé, J. (1975), *Oecologia,* **21,** 359–73.

Bordoni, A. (1972), *Boll. Ass. romana. Ent.,* **27,** 9–25.

Bournier, A. (1979), *Bull. Soc. ent. Fr.,* **84,** 147–53.

Bovee, E.C. (1979), *Kansas Univ. Sci. Bull.,* **51,** 615–29.

Brattsten, L.B., Wilkinson, C.F. and Eisner, T. (1977), *Science,* **196,** 1349–52:

Brindle, A. (1957), *Entomologist's mon. Mag.,* **93,** 202–4.

Brown, A.W.A. (1956), *Insect Control by Chemicals,* John Wiley, New York.

Buchner, P. (1965), *Endosymbiosis of Animals with Plant Micro-organisms,* Interscience Publishers, New York.

Burnett, W.C., Jones, S.B., Mabry, T.J. and Padolina, W.G. (1974), *Biochem. syst. ecol.,* **2,** 25–9.

Byers, G.W. (1961), *Kansas Univ. Sci. Bull.,* **42,** 665–924.

Chapman, D.W. and Demory, R.L. (1963), *Ecology,* **44,** 140–6.

Chernov, J.I., Striganova, B.R. and Ananjeva, S.I. (1977), *Oikos,* **29,** 175–9.

China, W.E. (1962), *Trans R. Ent. Soc. Lond.,* **114,** 131–61.

Cloudsley-Thompson, J.L. (1968), *Spiders, Scorpions, Centipedes and Mites,* Pergamon Press, London.

Collins, N.M. (1979), *Ecol. ent.,* **4,** 231–8.

Cooper, K.W. (1974), *Psyche,* **81,** 84–120.

Corbet, S.A. (1973), *Field studies,* **3,** 801–38.

Corbet, S.A. and Lan, O.B. (1974), *J. biol, Educ.,* **8,** 153–60.

Crum, H. (1972), *J. Hattori bot. Lab.,* **35,** 269–98.

Crum, H. (1976), *Mosses of the Great Lakes Forests.,* Univ. of Michigan.

Cuénot, L. (1949), Les Tardigrades. In: *Traité de Zoologie, Anatomie, Systematique, Biologie* (Grassé, P.-P., ed.), Vol. VI, pp. 39–59. Masson, Paris.

Culberson, C.F. (1969), *Chemical and Botanical Guide to Lichen Products,* University of North Carolina Press, Chapel Hill.

Cutten, F.E.A. and McE. Kevan, D.K. (1970), *Can, J. Zool.,* **48,** 1–24.

Dethier, V.G. (1947), *Chemical Insect Attractants and Repellents,* Blakiston, Philadelphia.

Dobat, K. (1972), *Umschau in Wissenschaft und Technik,* **72,** 493–4.

Dobzhansky, T. (1950), *Am. Sci.,* **38,** 209–21.

Donner, J. (1966), *Rotifers,* Warne, London.

Drake, C.J. and Lattin, J.D. (1963), *Proc. U.S. Nat. Mus.,* **115,** 331–45.

Duke, K.M. and Crossley, D.A. Jr. (1975), *Ecology,* **56,** 1106–17.
During, H.J. (1979), *Lindbergia,* **5,** 2–18.
Dybas, H.S. (1978), *Am. Mid. nat.,* **99,** 83–100.
Eastop, V.F. (1973), Deductions from the present day host plants of aphids and related insects. In: *Insect/Plant Relationships* (van Emden, H.F., ed.), pp. 157–78, Blackwells, Oxford.
Eastop, V.F. (1977), Worldwide importance of aphids as virus vectors. In: *Aphids as Virus Vectors* (Harris, K.F. and Maramorosch, K., eds), pp. 3–62, Academic Press, London.
Edmonson, W.T. (1944), *Ecol. monog.,* **14,** 31–66.
Erlanson, C.O. (1930), *Bryologist,* **33,** 13–4.
Faegri, K. and van der Pijl, L. (1979), *The Principles of Pollination Ecology,* Pergamon Press, Oxford.
Fantham, M.A. and Porter, A. (1945), *Proc. Zool. Soc. Lond.,* **115,** 97–174.
Feeny, P. (1976), Plant apparency and chemical defense. In: *Biochemical Interaction Between Plants and Insects* (Wallace, J.M. and Mansell, R.L., eds), pp. 1–40, Plenum, New York and London.
Forster, W. and Wohlfahrt, T.A. (1960), *Die Schmetterlinge Mitteleuropas. Band III. Spinner und Schwärmer (Bombyces und Sphinges),* Franckische Verlag, Stuttgart.
Fraenkel, G.S. (1959), *Science,* **129,** 1466–70.
Frankland, J.C. (1974), Decomposition of lower plants. In: *Biology of Plant Litter Decomposition* (Dickinson, C.H. and Pugh, G.J.F., eds), Vol. 1, pp. 3–36, Academic Press, London.
Frost, W.E. (1942), *Proc. R. Irish Acad.,* **47B,** 293–369.
Gadea, E. (1964a), *Pub. Inst. Biol. apl. Barcelona,* **36,** 113–20.
Gadea, E. (1964b), *Pub. Inst. Biol. apl. Barcelona,* **37,** 73–93.
Gadea, E. (1977), *Inst. Munic. Cienc. Nat. Misc. Zool.,* **4,** 17–22.
Garjeanne, A.J.M. (1932), Physiology. In: *Manual of Bryology* (Verdoorn, F., ed.), pp. 207–32, Nijhoff, The Hague.
Gerson, U. (1969), *Bryologist,* **72,** 495–500.
Gerson, U. (1972), *Acarologia,* **13,** 319–43.
Glime, J.M. (1978), *Bryologist,* **81,** 186–7.
Glime, J.M. and Clemons, R.M. (1972), *Ecology,* **53,** 458–64.
Goodey, T. (1963), *Soil and Freshwater Nematodes,* Methuen, London.
Gordon, H.T. (1959), *Ann. N.Y. Acad. Sci.,* **77,** 290–351.
Gressitt, J.L. (1966), *Pacific Insects,* **8,** 221–80.
Gressitt, J.L. (1967), Notes on arthropod populations in the Antarctic Peninsula – South Shetland Island – South Orkney Islands area. In: *Entomology of Antarctica* (Gressitt, J.L., ed.), pp. 373–91, American Geophysical Union, Washington, D.C.
Gressitt, J.L., Samuelson, G.A. and Vitt, D.H. (1968), *Nature,* **217,** 765–7.
Gressitt, J.L. and Sedlacek, J. (1970), *Pacific Insects,* **12,** 753–62.
Grodzinska, K. (1978), *Water Air Soil Pollution,* **9,** 83–97.
Grolière, C.-A. (1977), *Protistologia,* **13,** 335–12.
Grolière, C.-A. (1978), *Protistologia,* **14,** 295–311.
Hackemesser, H. and Lichte, H.-F. (1978), *Nachrichtenb. Deut. Pflanzenschutz.,* **30,** 129–33.

Hallas, T.E. (1975), *Ann. Zool. Fennici,* **12,** 255–9.
Hallas, T.E. and Yeates, G.W. (1972), *Pedobiologia,* **12,** 287–304.
Hammer, M. (1965), *Acta univ. Lund.,* **2,** 1–10.
Hammer, M. (1966), *Zool. Anz.,* **177,** 272–6.
Harvey-Gibson, R.J. and Miller-Brown, D. (1927), *Ann. Bot.,* **41,** 190–1.
Heal, O.W. (1962), *Oikos,* **13,** 35–47.
Heinis, F. (1959), *Ber. Geobot. Forschlnst. Rübel,* **1958,** 110–23.
Heller, J. (1975), *Zool. J. Linn. Soc.,* **56,** 153–70.
Helmsing, I.W. and China, E.W. (1937), *Ann. Mag. nat. Hist.,* **19,** 473–89.
Hesse, R., Allee, W.C. and Schmidt, K.P. (1951), *Ecological Animal Geography,* Wiley, New York.
Horn, K.F., Wright, C.G. and Farrier, M.H. (1979), *North Carolina Agric. Exp. Stn., Tech. Bull.* No. **257.**
Horton, D.G. and Murray, B.M. (1976), *Bryologist,* **79,** 321–31.
Hyman, L.H. (1951), *The Invertebrates: Acanthocephala, Aschelminthes and Entoprocta; the Pseudocoelomate Bilateria.* Vol. III, McGraw-Hill, New York.
Hynes, H.B.N. (1941), *Trans. R. Ent. Soc. Lond.,* **91,** 459–557.
Hynes, H.B.N. (1961), *Arch. Hydrobiol.,* **57,** 344–88.
Ingold, C.T. (1965), *Spore Liberation,* Clarendon, Oxford.
Jackson, D.J. (1956), *Entomologist's mon. Mag.,* **92,** 154–5.
Janetschek, H. (1967), Arthropod ecology of South Victoria Land. In: *Entomology of Antarctica* (Gressitt, J.L., ed.), pp. 205–93, American Geophysical Union, Washington, D.C.
Karppinen, E. and Koponen, M. (1973), *Ann. Ent. Fennici,* **39,** 29–39.
Kevan, P.G., Chaloner, W.G. and Savile, D.B.O. (1975), *Palaeontology,* **18,** 391–417.
Kolb, A. (1976), *Z. Saugetierkunde,* **41,** 226–36.
Koponen, A. (1978), *Bryophytorum Bibliotheca,* **13,** 535–68.
Koponen, A. and Koponen, T. (1978), *Bryophytorum Bibliotheca,* **13,** 569–77.
Kühnelt, W. (1976), *Soil Biology, With Special Reference to the Animal Kingdom,* Faber & Faber, London.
Kumar, A. and Prasad, M. (1977), *Entomon,* **2,** 225–30.
Lacey, W.S. (1969), *Biol. Rev.,* **44,** 189–205.
Larsson, S.G. (1978), *Baltic Amber – A Palaebiological Study,* Scandinavian Science Press, Klampenborg.
LeSage, L. and Harper, P.P. (1976), *Ann. Limnol.,* **12,** 139–74.
Levin, D.A. (1976), *Ann. Rev. Ecol. Syst.,* **7,** 121–59.
Lindberg, K. (1954), *Hydrobiologia,* **6,** 97–119.
Lindegaard, C., Thorup, J. and Bahn, M. (1975), *Arch. Hydrobiol.,* **75,** 109–39.
Lohammar, G. (1954), *Svensk Botanisk Tidskrift,* **48,** 162–73.
Loria, M. and Herrnstadt, I. (1981), *Bryologist,* **83,** 524–5.
Lowe, B.G. (1978), *Health Physics,* **34,** 439–44.
Luther, H. (1979), *Ann. Bot. Fennici,* **16,** 163–72.
Macan, T.T. (1963), *Freshwater Ecology,* Longmans, London.
Mani, M.S. (1962), *Introduction to High Altitude Entomology,* Methuen, London.
Markham, K.R. and Porter, L.J. (1978), *Prog. Phytochem.,* **5,** 181–272.
Marshall, V.G. (1974), *Can. J. Soil Sci.,* **54,** 491–500.
Marstaller, R. (1979), *Feddes Repertorium,* **89,** 629–61.

Martin, H. and Worthing, C. (1977), *Pesticide Mannual,* 5th Edition. British Crop Protection Council.
Matile, L. (1972), *C.R. Acad. Sci. Paris,* **274,** 1927–30.
Matthey, W. (1971), *Rev. Suisse Zool.,* **78,** 367–536.
Matthey, W. (1977), *Bull. Soc. Ent. Suisse,* **50,** 299–306.
McLachlan, A.J., Pearch, L.J. and Smith, J.A. (1979), *J. Anim. Ecol.,* **48,** 851–61.
Mehlen, R.H. (1972), *Pac. Sci.,* **26,** 223–5.
Menzel, R. (1921), *Treubia,* **2,** 137–45.
Morgan, C.I. (1977), *J. Anim. Ecol.,* **46,** 263–79.
Morgan, C.I. and King, P.E. (1976), *British Tardigrades,* Academic Press, London.
Mullen, G.R. (1977), *J. med. Ent.,* **13,** 475–85.
Müller, F.P. (1973), *Ent. Abhand.,* **39,** 205–42.
Murphy, D.H. (1955), Long-term changes in collembolan populations with special reference to moorland soils. In: *Soil Zoology* (McE. Kevan, D.K., ed.), pp. 157–66, Butterworths, London.
Nägeli, W. (1936), *Mitt. Schweiz. Anst. forstl. VersWes.,* **19,** 211–381.
Neumann, A.J. and Vidrine, M.F. (1978), *Bryologist,* **81,** 584–5.
Nicholas, W.L. (1975), *The Biology of Free-Living Nematodes,* Clarendon, Oxford.
Nielsen, C.O. (1967), Nematoda. In: *Soil Biology* (Burges, A. and Raw, F., eds), pp. 197–211, Academic Press, London.
Nørgaard, E. (1951), *Oikos,* **3,** 1–21.
Nosek, J. (1973), *The European Protura, Their Taxonomy, Ecology and Distribution, with Keys for Determination,* Muséum d'Histoire Naturelle, Genève.
Oldroyd, H. (1964), *The Natural History of Flies,* Weidenfeld & Nicolson, London.
Pedroli-Christen, A. (1977), *Bull. Soc. Neuchatel Sci. Nat.,* **100,** 21–34.
Pennak, R.W. (1978), *Fresh Water Invertebrates of the United States,* Wiley, New York.
Penny, N.D. (1977), *Univ. Kansas Sci. Bull.,* **51,** 141–217.
Plitt, C.C. (1907), *Bryologist,* **10,** 54–5.
Priddle, J. and Dartnall, H.J.G. (1978), *Freshwater Biology,* **8,** 469–80.
Pryor, M.E. (1962), *Pac. Insects,* **4,** 681–728.
Pschorn-Walcher, H. and Gunhold, P. (1957), *Z. Morph. Ökol. Tiere,* **46,** 342–54.
Pycraft, D. (1975), *Garden,* **100,** 396.
Pyysalo, H., Koponen, A. and Koponen, T. (1978), *Ann. Bot. Fennici,* **15,** 293–6.
Quate, L.W. (1955), *Univ. California Pub. Ent.,* **10,** 103–273.
Quintero, D., Jr. (1974), *Psyche,* **81,** 307–14.
Ramazzotti, G. (1958), *Memorie Ist. Ital. Idrobiol.,* **10,** 153–206.
Reichle, D.E. (1966), *Syst. Zool.,* **15,** 330–44.
Richards, A.M. (1962), *Trans. R. Soc. N.Z.,* **2,** 121–9.
Richards, O.W. and Davies, R.G. (1977), *Imm's General Textbook of Entomology,* 10th edn, Chapman and Hall, London.
Richards, P.W. (1947), *Trans. Br. Bryol. Soc.,* **1,** 16.
Richards, P.W. (1959), Bryophyta. In: *Vistas in Botany* (Turrill, W.B., ed.), pp. 387–420, Pergamon Press, London.
Rieley, J.O., Richards, P.W. and Bebbington, A.D.L. (1979), *J. Ecol.,* **67,** 497–527.

Robinson, M.H. (1969), *Trans. R. Ent. Soc. Lon.*, **121**, 281–303.
Rodriguez, J.G. (1972), Inhibition of acarid mite development by fatty acids. In: *Insect and Mite Nutrition* (Rodriguez, J.G., ed.), pp. 637–50. North Holland, Amsterdam.
Rhode, C.J. Jr. (1955), *Studies on Arthropods From a Moss Habitat with Special Emphasis on the Life History of Three Oribatid Mites,* University Microfilms, Ann Arbor.
Rühling, A. and Tyler, G. (1970), *Oikos,* **21**, 92–7.
Rylov, V.M. (1963), *Fauna of the USSR, Crustacea, Freshwater Cyclopoidea,* Israel Program for Scientific Translations, Jerusalem.
Savory, T. (1977), *Arachnida,* 2nd edn., Academic Press, London.
Sayre, R.M. and Brunson, L.K. (1971), *Am. Biol. Teach.*, **33**, 100–5.
Schweizer, J. and Bader, C. (1963), *Denkschr. schweiz. Naturf. Ges.*, **84**, 209–378.
Séguy, E. (1950), *La Biologie des Diptères,* Lechevalier, Paris.
Shepherd, R.F. and Gray, T.G. (1972), *Can. Ent.*, **104**, 751–4.
Shorthouse, J.D. (1979), *Questiones Entomologicae,* **15**, 341–4.
Simon, J.C. (1974), *Graellsia,* **27**, 103–32.
Smirnov, N.N. (1961), *Hydrobiologia,* **27**, 175–82.
Smith, V.R. (1977), *Oecologia,* **29**, 269–73.
Snow, W.E., Pickard, E. and Moore, J.B. (1958), *J. Tenn. Acad. Sci.*, **33**, 5–23.
Sorsa, P. and Koponen, T. (1973), *Ann. Bot. Fennici,* **10**, 187–200.
Southwood, T.R.E. and Leston, D. (1959), *Land and Water Bugs of the British Isles,* Warne, London.
Stebaev, I.V. (1963), *Pedobiologia,* **2**, 265–309.
Štěpánek, M. (1963), *Hydrobiologia,* **22**, 304–27.
Størmer, L. (1976), *Senckenbergiana Lethaea,* **57**, 87–183.
Stout, J.D. (1963), *Tuatara,* **11**, 57–64.
Stout, J.D. (1974), Protozoa. In: *Biology of Plant Litter Decomposition* (Dickinson, C.H. and Pugh, G.J.F., eds), pp. 385–420, Academic Press, London.
Stout, J.D. and Heal, O.W. (1967), Protozoa. In: *Soil Biology* (Burges, A. and Raw, F., eds), pp. 149–95, Academic Press, London.
Strong, J. (1967). Ecology of terrestrial arthropods at Palmer Station, Antarctic Peninsula. In: *Entomology of Antarctica* (Gressitt, J.L., ed.), pp. 357–71, American Geophysical Union, Washington, D.C.
Stroyan, H.L.G. (1955), *Trans. R. Ent. Soc. Lond.*, **106**, 283–340.
Suire, C. (1975), *Rev. bryol. lichénol.*, **41**, 105–256.
Svensson, G.K. (1967), The increasing ^{137}Cs level in forest moss in relation to the total ^{137}Cs fallout from 1961 to 1965. In: *Radioecological Concentration Processes* (Åberg, B. and Hungate, F.P., eds), pp. 539–46, Pergamon Press, Oxford.
Swain, T. (1974), Biochemical evolution in plants. In: *Comprehensive Biochemistry* (Florkin, M. and Stotz, E.H., eds), Vol. 29, Part A., pp. 125–302, Elsevier, Amsterdam.
Tarras-Wahlberg, N. (1952–1953), *Oikos,* **4**, 166–71.
Tarras-Wahlberg, N. (1961), *Oikos,* Supp. 4, 1–56.
Teskey, H.J. (1969), *Mem. Ent. Soc. Can.*, **63**, 1–147.

Thieret, J.W. (1956), *Econ. Bot.*, **10,** 75–91.
Thomas, A.W. (1971), *Quaestiones Entomologicae*, **7,** 407–8.
Thut, R.N. (1969), *Ann. Ent. Soc. Am.*, **62,** 894–8.
Tilbrook, P.J. (1967), Arthropod ecology in the Maritime Antarctic. In: *Entomology Antarctica* (Gressitt, J.L., ed.), pp. 331–56, American Geophysical Union, Washington, D.C.
Tillyard, R.J. (1922), *Trans. ent. Soc. Lond.*, **1922,** 437–53.
Travé, J. (1963), *Vie Milieu,* Supp. 14, 1–267.
Uvarov, B. (1977), *Grasshoppers and Locusts, a Handbook of General Acridology,* Vol. 2. Centre for Overseas Pest Research, London.
Vandel, A. (1965), *Biospeleology, the Biology of Cavernicolous Animals,* Pergamon Press, Oxford.
Verdcourt, B. (1947), *Entomologist's mon. Mag.*, **83,** 190.
Vickery, V.R. (1969), *Ann. Soc. Ent. Quebec,* **14,** 22–4.
Wada, K. and Munakata, K. (1971), *Agric. Biol. Chem.*, **35,** 115–18.
Wallace, R.N., Hynes, H.B.N. and Merritt, W.F. (1976), *Environ. Pollut.*, **10,** 251–69.
Williams, F.X. (1939), *Proc. Hawaiian ent. Soc.*, **10,** 281–315.
Winner, W.E. and Bewley, J.D. (1978), *Oecologia,* **35,** 221–30.
Wise, K.A.J., Fearon, C.E. and Wilkes, O.R. (1964), *Pac. Insects,* **6,** 541–70.
Yasuda, T. (1962), *Kontyu,* **30,** 130–6.
Zagulyayev, A.K. (1970), *Ent. Rev.*, **49,** 657–64.

Chapter 10

Physiological Ecology: Water Relations, Light and Temperature Responses, Carbon Balance

M.C.F. PROCTOR

10.1 INTRODUCTION

Bryophytes face the same basic problems of life on land as flowering plants. Water is essential for metabolism, but is in limited and erratic supply in the above-ground environment where the leaves must be exposed in order to photosynthesize. Bryophytes and vascular plants exemplify two alternative patterns of adaptation to these conditions. Vascular plants have evolved roots and an efficient conducting system, bringing water from the soil where it is in relatively plentiful and constant supply. Bryophytes have evolved to utilize water where and when it is available above ground; some are confined to moist habitats, but many can tolerate drying out, and some are extremely desiccation-tolerant and highly adapted to a poikilohydric existence. But this contrast between vascular plants and bryophytes is by no means sharp. *Polytrichum* and *Dawsonia* species and various other large bryophytes have efficient water-conducting systems and maintain turgor for much of the year in many habitats. Conversely, many small vascular plants evade the problems of drought in moist habitats, or by passing the summer as desiccation-tolerant seeds, and there are a number of scattered genera of poikilohydric vascular plants.

Scale is also an important factor in the physiological and ecological adaptation of bryophytes. Surface/volume ratios are very different for vascular plants and bryophytes. Conduction of water up the many metres of xylem is paramount in a tree, and the foliage of a tree is borne high above the steep temperature and humidity gradients close to the surface of the ground. For bryophytes, the conditions in the micro-environment become of primary importance, and the water-conduction problems of a bryophyte are much closer in scale to those of a single leaf of an angiosperm. Superficial analogies between bryophytes and vascular plants are often invalidated by differences in scale or adaptive strategy, while, paradoxically, apparently far-fetched analogies can sometimes be valid and useful.

In this chapter I have not attempted to review the relevant literature exhaustively. More detailed accounts of some of the topics dealt with are given by Dilks and Proctor (1979), Longton (1980) and Proctor (1979a, 1981).

10.2 UPTAKE AND CONDUCTION OF WATER

A vascular plant takes in water through the roots, which is then conducted up the stem and evaporated from the leaves; opening and closing of the stomata allows some control over evaporation, and a more-or-less waterproof cuticle prevents water-loss from the whole surface of the plant. The electrical circuit of Fig. 10.1a represents a simple analogue of the whole process (Milburn, 1979). The voltage at various points in the circuit will be determined by the values of the various resistances and the current flowing through them. Similarly, the water potential (p. 338) at points in the plant and its surroundings will reflect the rate of water movement through it, and the magnitudes of the various hydraulic and diffusion resistances that the water encounters between the soil and the atmosphere. Usually, by far the greatest drop in water potential is between the mesophyll of the plant and the atmosphere, so the transpiration rate is determined mainly by the saturation deficit of the air and the diffusion

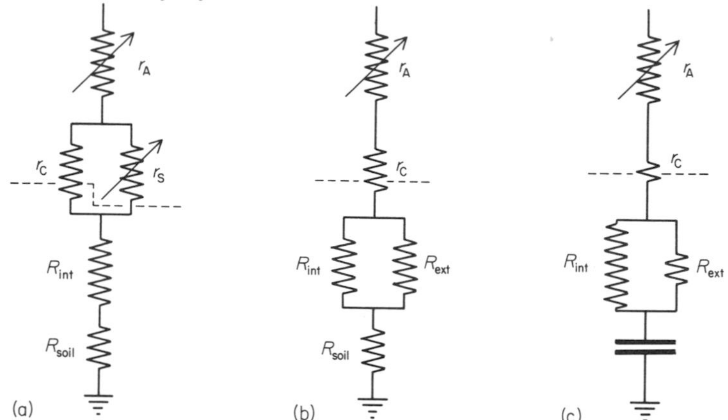

Fig. 10.1 Resistance to water flow in (a) a vascular plant, (b) an endohydric bryophyte, and (c) an ectohydric bryophyte, represented schematically as analogous electrical circuits. Capital letters represent hydraulic resistances (R_{int}, internal; R_{ext}, external), small letters (r) gaseous diffusion resistances. Cuticular resistance (r_C) may be regarded as substantially constant, but stomatal resistance (r_S) varies with stomatal opening, and boundary-layer resistance (r_A) with windspeed; these are the most important resistances in controlling water loss. The external hydraulic resistance varies with water content (Bayfield, 1973), but will generally be low.

resistances of the stomata and the boundary-layer of the leaf (p. 343). The hydraulic resistances within the plant and the soil then determine whether the water potential of the leaves drops sufficiently for the plant to wilt.

Bryophytes are more diverse in their adaptations to water uptake and movement than the great majority of vascular plants. In *endohydric* species (Buch, 1945; 1947), water is taken up from the substratum and conducted internally to the leaves or other evaporating surface; the surface of the plant is typically water-repellent and not readily stained by an aqueous solution of a basic dye such as toluidine blue. Such bryophytes (e.g. Polytrichaceae, Mniaceae, *Funaria hygrometrica,* Marchantiaceae) have a transpiration stream in much the same sense as a vascular plant (Fig. 10.1b). In general, they occur on moist, porous and often nutrient-rich substrata, and are well represented in base-rich woods, fens, and in weedy habitats on bare soil. In *ectohydric* bryophytes, water is readily absorbed (and lost) over the entire surface, which is quickly stained by basic dyes, and water movement is much more diffuse (e.g. Grimmiaceae, Orthotrichaceae, and many Hypnaceae and leafy liverworts). These plants often occur on impermeable substrata such as rock and bark, or are only loosely connected with the soil; they function on stored water after remoistening by rain or dew (Fig. 10.1c). In fact, these types represent extremes between which a range of intermediates can be found. Buch called species which combine a wettable surface with predominantly internal conduction *mixohydric,* but many bryophytes show various combinations of endohydric and ectohydric characteristics.

Some endohydric bryophytes have underground rhizomes (e.g. *Dawsonia, Polytrichum, Climacium*) or root-like structures (*Haplomitrium, Takakia;* Grubb, 1970). In the majority, the rhizoids provide intimate contact between the stem base and the soil surface. In mosses at least, the main function of the rhizoids in water uptake is probably to form an inert felt, like a layer of blotting paper between plant and substrate, and the same may be substantially true of liverworts. In ectohydric bryophytes, a connection to the ground is less important for water uptake, but the rhizoids are of course important for attachment in corticolous and saxicolous species (Odu, 1978).

There are four pathways by which water may move from one part of a bryophyte to another:

1. Specialized empty elongated conducting cells *(hydroids)* forming a central strand in the stems of many mosses; similar strands are present in a few liverworts (e.g. *Haplomitrium, Pallavicinia*).
2. The 'free space' of the cell walls.
3. Cell to cell, through the intervening walls and cell membranes.
4. External capillary spaces.

Well-developed central strands are characteristic of endohydric bryophytes, especially the larger mosses, and are responsible for the most rapid internal conduction (Hébant, 1977). External capillary conduction is especially important in many ectohydric species, and can bring about very rapid movement of water. However, these pathways account for only a part of the water movement in either case. In the stem cortex, the laminae of bryophyte leaves, and in liverwort thalli, much water movement must take place either along cell walls or from cell to cell. In vascular plants it is now generally accepted that a large part of the movement of water in the cortex of a root or the mesophyll of a leaf takes place through the free space of the cell walls, and the same is likely to be true of bryophytes. If we take the hydraulic resistance of a cell membrane to be 100 times that of a cell wall 1 μm thick (Laüchli, 1976), in cells 10 μm across the conductivity of the pathway via the cell walls will be c. 11 times that from cell to cell; this, for various reasons is likely to be a conservative estimate (see also Raven, 1977). Calculation on these lines indicates that for bryophytes with small, thick-walled cells, the cell–wall pathway should be overwhelmingly the more important, while cell-to-cell water movement may make a major contribution in bryophytes with large, thin-walled cells (as, for instance, in the parenchymatous tissue of some thalloid liverworts), but the whole subject needs experimental investigation.

Published measurements of rates of water movement in bryophytes reflect principally the rapid pathways through the central strand, or through external capillary channels. Many observations have been made of the movement of dyes or other solutes. These can demonstrate pathways of water movement qualitatively, but in general are likely to move more slowly than water, and may move either more or less freely than water through membranes. Cations (e.g. basic dyes, Li^+) will tend to be adsorbed onto cell wall materials and hence to move relatively slowly. Anions (e.g. acid dyes, NO_3^-) or non-ionic tracers (e.g. lead-EDTA; Trachtenberg and Zamski, 1979) are preferable. Measurements of water uptake and loss are more fundamental, but provide less detailed evidence of paths of movement.

Haberlandt (1886) demonstrated internal conduction of dyes in *Polytrichum juniperinum* and *Plagiomnium undulatum,* and his results have been confirmed by a number of subsequent dye-uptake and transpiration experiments (Bowen, 1931, 1933; Blaikley, 1932; Mägdefrau, 1935; Zacherl, 1956) on species of Polytrichaceae and Mniaceae. In *Funaria hygrometrica,* movement of fluorescent dyes indicates that water rises through the felt of rhizoids at the base of the stem, passes across the cortex into the central strand, and hence rapidly to the stem apex and the foot of the sporophyte; within the sporophyte conduction takes place mainly within the hydroids of the central strands (Bopp and Stehle, 1957; Bopp and Weniger, 1971). Grubb (1970) showed substantial and long-continued transpiration of water conducted internally up the stems of *Haplomitrium mnioides, Plagiomnium undulatum* and *Rhizomnium punctatum,* which possess central strands, contrasting with little water movement in *Plagiochila asplenioides* which does not. Rapid conduction through external capillary spaces has been demonstrated in a wide range of species, e.g. by Bowen (1931, 1933) using dyes and other tracers, and Mägdefrau (1935) in transpiration experiments which provided a direct comparison between internal and total conduction (Table 10.1). Of the species Mägdegrau studied, only *Polytrichum commune, P. formosum, Plagiomnium undulatum, P. elatum, Dicranum scoparium* and *Leucobryum glaucum* show rapid water movement both internally and externally, and only in *Polytrichum commune* and *Plagiomnium undulatum* at the highest humidity is internal conduction faster than movement in the external capillary spaces. The majority of species show much faster external than internal water movement. Of the species in Table 10.1, only *Thamnobryum alopecurum* shows little capacity for water movement either internally or externally (confirming the earlier findings of Bowen); it is noteworthy that it is a plant

Table 10.1 Mean daily rates of water loss from moss shoots (g water/g dry weight) with and without external capillary water conduction. Recalculated from the data of Mägdefrau (1935). Relative humidity 70% unless otherwise stated.

Species	Total conduction	Internal conduction	Internal as fraction of total
Sphagnum recurvum	32.7	0.35	0.01
Polytrichum formosum	14.3	4.3	0.30
P. commune	16.6	11.2	0.67
Dicranum scoparium	9.45	3.4	0.36
Leucobryum glaucum	16.9	5.0	0.30
Plagiomnium elatum	30.5	7.6	0.25
Plagiomnium undulatum (90% r.h.)	8.5	5.6	0.65
P. undulatum (70% r.h.)	11.1	5.7	0.51
P. undulatum (65% r.h.)	18.6	6.0	0.32
Aulacomnium palustre	64.6	5.3	0.08
Climacium dendroides	20.4	2.0	0.10
Neckera crispa	40.1	0.15	0.004
Thamnobryum alopecurum	0.095	0.035	0.37
Anomodon viticulosus	26.0	0.25	0.01
Drepanocladus vernicosus	113.7	4.0	0.04
Homalothecium lutescens	41.0	1.8	0.04
H. nitens	65.6	4.3	0.07
Pseudoscleropodium purum	18.8	0.65	0.03
Pleurozium schreberi	11.9	0.25	0.02
Rhytidiadelphus triquetrus	8.4	0.55	0.07
Hylocomium splendens	27.4	0.05	0.002

of shady or waterside habitats. Bowen also found little external water movement in *Mnium hornum, Plagiomnium undulatum* and *Rhizomnium punctatum* (which have effective internal conduction), and in *Brachythecium rutabulum* (which does not). External capillary movement of water has been demonstrated in *Plagiochila asplenioides* var. *major* and other leafy liverworts (Clee, 1937), *Pellia epiphylla* (Clee, 1939) and *Anthoceros laevis* (Isaac, 1941); the under surfaces of Marchantiales with their rhizoids and ventral scales are particularly richly supplied with capillary conducting spaces (McConaha, 1939, 1941; Clee, 1943). However a good deal of internal water movement must take place within the thick parenchymatous thalli of Marchantiales (whose upper surfaces are often water-repellent, see p. 343) and other thalloid liverworts, illustrated by the rapid movement of ^{14}C-acetate in solution in *Conocephalum conicum* (Rifot and Barrière, 1974).

Capillary conducting systems are diverse and often complex. The spaces between overlapping leaves, between stems and sheathing leaf bases, and amongst rhizoid tomentum and paraphyllia, are rather large, often in the range from 10–100 μm across. The interstices between the papillae that cover the leaf surfaces of many Pottiaceae (Fig. 1a, 2a), Encalyptaceae and other mosses, and the furrows between plicae and ridges on leaves and stems typically provide capillary spaces up to a few μm across. A few genera have capillary systems formed by cells which lose their contents and develop pores from one to another and to the exterior. The hyalocysts of *Sphagnum* are the best known example. The leaves of Leucobryaceae provide a parallel though differently organized instance (Fig. 10.2 e–f, and see Castaldo *et al.*, 1979), and the enlarged cells of the sheathing leaf bases of species of *Tortula* (Fig. 10.2c), *Encalypta* and Calymperaceae (Fig. 10.2d) characteristically develop into porous interconnecting hyalocysts which form a part of the capillary system surrounding the stem.

The tension (water potential) that can be developed by a meniscus in a capillary is given by:

$$\psi = -\sigma \left(\frac{1}{r_1} + \frac{1}{r_2}\right) \cos\theta, \qquad (10.1)$$

where ψ = water potential (Pa),
σ = surface tension ($N\,m^{-1}$),
θ = angle of contact,
r_2, r_2 = principal radii of capillary.

For a capillary of circular cross-section, $r_1 = r_2$, and if the cell walls are completely wettable $\theta = 0$, so that $\cos\theta = 1$. Then:

$$\psi = -\frac{2\sigma}{r}. \qquad (10.2)$$

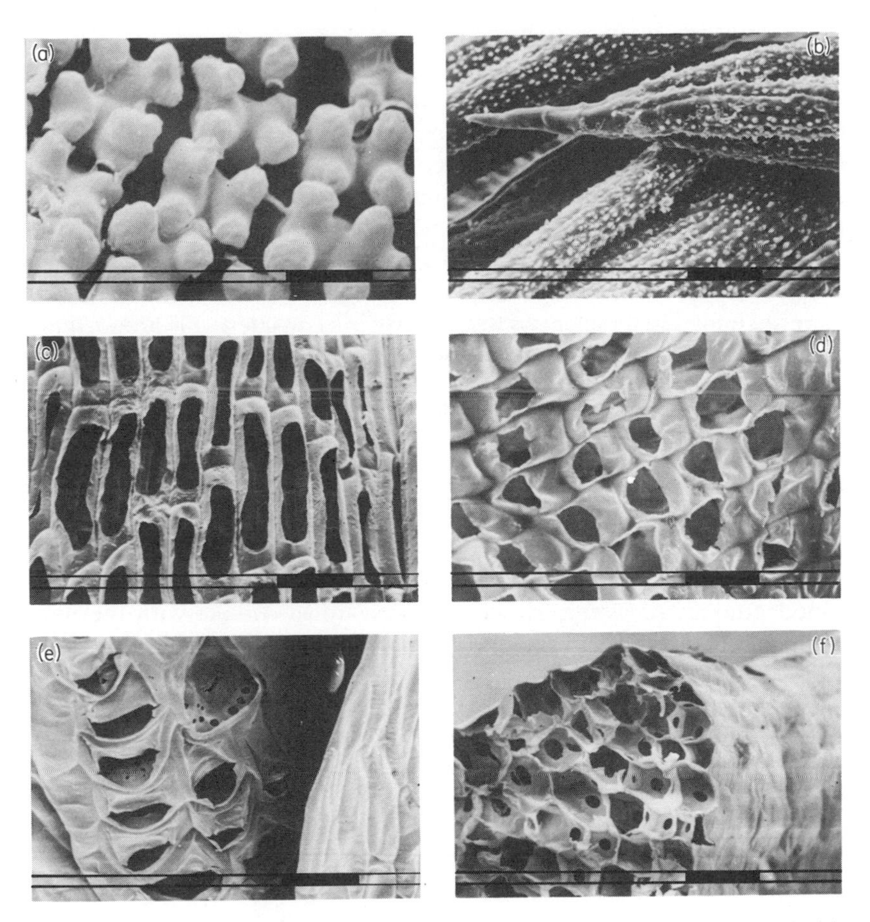

Fig. 10.2 Capillary structure in mosses. Scale bar at lower edge of photograph in (a) is 4 μm, in (d) 20 μm, and in all others 40 μm. (a) Leaf-surface papillae of *Tortula muralis* (Exeter, Devon). (b) Leaves of *Barbula hornschuchiana* (Malham, Yorks). (c) Hyalocysts in sheathing leaf-base of *Tortula ruralis* (Inchnadamph, Sutherland); these enlarged smooth-walled empty cells interconnect and open to both surfaces by elongated pores. (d) Hyalocysts in sheathing leaf-base of *Syrrhopodon japonicus* (Hiroshima prefecture, Japan, H. Ando); the cells interconnect and open by pores to both surfaces. (e) *Octoblepharum albidum* (Florida, H.S. Conard); pores in the dorsal leaf surface near the base. As is general in Leucobryaceae, the leaf consists largely of hyalocysts interconnecting by pores, with relatively few openings to the exterior, in this genus forming a tissue several cells thick. (f) *Octoblepharum albidum*; broken leaf showing hyalocysts with pores. The layer of chlorocysts which form a delicate network in the middle of the leaf cannot be seen clearly here; note the pires along the leaf margin at the top of the picture.

In an open half-cylindrical groove of radius r_1, $r_2 = \infty$, so the maximum tension that can be developed is half that given by Equation 10.2. If $\theta > 0$ in Equation 10.1, ψ will be correspondingly less negative, until in the extreme case of a water-repellent surface with $\theta > 90°$, ψ becomes positive and water can only be forced into the capillary under pressure.

From these relationships and Poiseuille's equation it is possible to calculate approximately the maximum rates of water movement that could be sustained by capillaries of different dimensions; in general, these turn out to be more than sufficient to balance fast evaporation rates in the field, as long as the channels remain full. With a constant pressure difference, the flow through a tube is proportional to r^4. Because the maximum capillary tension is inversely proportional to radius, the maximum rate of flow that can be sustained in a system of the kind we are considering is proportional to r^3 if the whole system is scaled up or down, or to r^2 if the spacing of the channels is kept constant. Other things being equal, larger channels will also store more water, but they will be more vulnerable to changes in the water potential of the surroundings. Table 10.2 lists water potentials and capillary rises for a range of radii, with some corresponding bryophyte structures.

The capillary structures of living bryophyte shoots are typically highly organized and of regular dimensions, so that they hold a rather well-defined amount of water. However, they are in contact with the more diffuse capillary spaces of the underlying soil, litter, or cushion or mat of

Table 10.2 Relationship of radius of curvature, capillary rise, water potential and relative humidity, with some bryophyte structures of corresponding dimensions. (Based on Slatyer, 1967.)

Radius of meniscus	Height of capillary rise	Water potential	Relative humidity at 15°C	Bryophyte structures in similar size-range
1 mm	1.5 cm	− 150 Pa	100.00	Large concave leaves; spaces amongst shoots
100 μm	15 cm	− 1.5 kPa	100.00	Spaces between leaves, paraphyllia etc.
10 μm	1.5 m	− 15 kPa	99.99	Spaces within sheathing leaf bases, tomentum etc.; hyalocysts of *Sphagnum* and *Leucobryum*
1 μm	15 m	− 150 kPa (1.5 bar)	99.89	Interstices between leaf-surface papillae
100 nm	150 m	− 1.5 MPa	98.90	? spaces within cell walls.
10 nm	1.5 km	− 15 MPa	89.54	Spaces between cell-wall microfibrils
1 nm	15 km	− 150 MPa	33.05	Glucose molecule

old shoots. The latter function as a reservoir, to which surplus water can drain, and then in part be drawn upon again by the shoots to maintain their characteristic water content until water potential drops sufficiently to break capillary contact (Dilks and Proctor, 1979; Klepper, 1963; Morton, 1977; Romose, 1940). As Buch points out, the papilla systems of individual leaves are often separated by regions where capillary continuity is broken at rather higher water potentials. The effect is that a leaf will have an abundant water supply or none, a feature which may be adaptive, minimizing the time leaves spend in a wilted state at which respiration may exceed photosynthesis.

Many ectohydric bryophytes have smooth leaves, or only sparse papillae which do not provide a continuous system of capillary channels over the surface (Fig. 10.2b). For these, as for endohydric species, water conduction must be internal, and as already pointed out probably takes place mainly in the cell walls. Calculation based on the measured or estimated hydraulic conductivities of *Nitella* or vascular plant cell walls suggests no serious difficulty in accounting for the rates of flow required by normal rates of evaporation, but extreme conditions may approach the limit of what can be sustained by physiologically tolerable water-potential gradients. Possibly bryophyte cell walls have higher hydraulic conductivities than those of vascular plants; scanning electron micrographs suggest that this may be so for some species. The very thick cell walls seen in many species of exposed habitats are probably primarily an adaptation to water conduction rather than mechanical support; in species of exposed habitats with papillose leaves and efficient external conduction (e.g. *Tortula* spp., *Encalypta* spp., *Anomodon viticulosus*) the walls are quite thin. The very long overlapping cells of the leaves of many Hypnaceae greatly reduce the number of cell membranes to be crossed between the leaf base and apex, and it is possible that cell-to-cell water movement is significant in these; there is as yet no experimental evidence one way or the other.

Ectohydric species face a particular problem in combining superficial conduction of water and gas exchange, because gaseous diffusion is slower in water than in air by a factor of about 10^4. In species with strongly concave leaves, water is held in the concavity while the outer surface remains dry: this is seen very clearly when a moss such as *Pseudoscleropodium purum* or *Pleurozium schreberi* is examined with a lens in the field. The physical reason for this follows from Equation 10.1; a convex surface (with r negative) will have a positive water potential. For the same reason, on a papillose leaf surface the tops of the papillae tend to remain dry while water is held in the interstices between them. Thus, *Thuidium tamariscinum* and other papillose mosses quickly take on a characteristic opaque bright green 'dry' appearance (their normal appearance when moist and turgid) as their waterlogged shoots drain after heavy rain. The effect may be accentuated in some species (e.g. *Encalypta streptocarpa*) by the tops of the papillae having water-repellent surfaces.

The cuticle of endohydric species provides a water-repellent surface as well as increasing resistance to water-loss. This is often obvious in the field in damp weather from the way water lies in discrete droplets on the leaves even of delicate moisture-loving species. In some mosses the leaves are covered with granules, platelets or ribbons of chloroform-soluble material

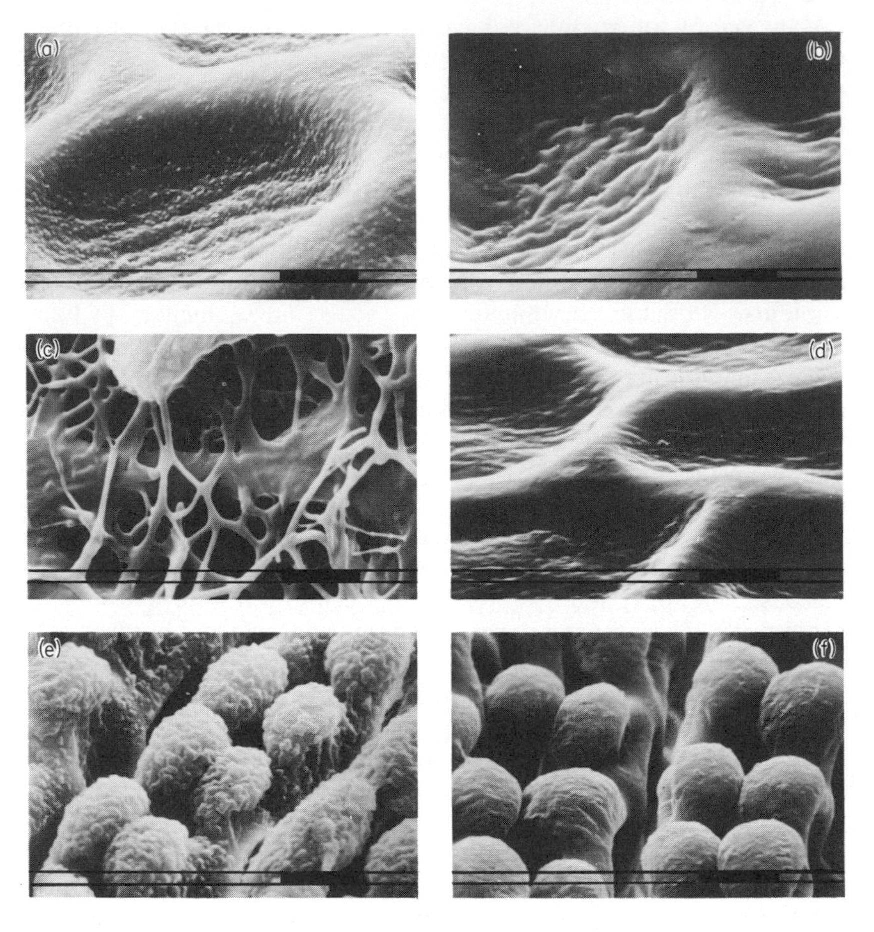

Fig. 10.3 Surface wax on moss leaves; the left-hand micrographs are of untreated leaf surfaces, those on the right of surfaces washed with chloroform. Scale bars, 4 μm. (a, b) *Mnium hornum*, Drewsteignton, Devon, April 1979. (c, d) *Conostomum tetragonum*, Lochnagar, Aberdeenshire, June 1953. (e, f) *Polytrichum commune*, near Chagford, Devon, Oct. 1978.

comparable with the epicuticular wax of vascular plants (Fig. 10.3 and Proctor, 1979b). When well developed this gives the leaves a glaucous bloom, as in *Pohlia cruda* and *P. albicans, Saelania glaucescens, Schistostega pennata, Pogonatum urnigerum* and many Bartramiaceae. The maximum attainable angle of contact over a smooth water-repellent surface is about 110°; a covering of angular platelets or needles or water-repellent papillae can raise the effective contact angle to 150° or more (as with water on a duck's back). The power to shed external water is likely to be particularly important to plants growing in shady crevices (*Bartramia pomiformis, Distichium capillaceum, Pohlia cruda, Schistostega pennata*) or in waterside habitats (*Pohlia wahlenbergii, Philonotis* spp.).

Exclusion of water from the air spaces of the complex photosynthetic systems of Polytrichaceae and Marchantiaceae is undoubtedly important to these plants, as is exclusion of water from the mesophyll spaces of vascular plants. *Polytrichum commune* and *Dawsonia superba* have a well-developed covering of wax platelets on the top edges of the photosynthetic lamellae (Fig. 10.3 e–f; Troughton and Sampson, 1973; Green and Clayton-Greene, 1977), and in Marchantiales the ledges surrounding the pores in the upper epidermis are strongly water-repellent (Schönherr and Ziegler, 1976).

10.3 EVAPORATION AND WATER LOSS

The rate of evaporation from a surface depends on two factors, (a) the difference between the water vapour density at the surface and that in the surrounding air, and (b) the diffusion resistance. The latter is made up of the 'leaf' resistance (r_L) and the air resistance (r_A).

$$E = \frac{\chi_S - \chi_A}{r_L + r_A}, \qquad (10.3)$$

where E = evaporation rate ($kg\ m^{-2}\ s^{-1}$),
χ_S = water vapour density at surface ($kg\ m^{-3}$),
χ_A = water vapour density in ambient air ($kg\ m^{-3}$),
r_L, r_A = leaf resistance, air resistance ($s\ m^{-1}$).

For a fully turgid leaf it is often accurate enough to take χ_S as the saturated vapour density at the temperature of the surface (measured with a thermocouple or small thermistor), but if the tissue has a substantial negative water potential, χ_S will be correspondingly reduced (Equation 10.8, p. 351). Saturated water vapour densities can be obtained direct from tables (Smithsonian Institution, 1951; Leyton, 1975), or calculated from the saturation vapour pressure e_s. χ_A is most easily obtained by looking up the partial pressure of water vapour, e, from wet and dry bulb thermometer readings in psychrometric tables (Meteorological Office, 1964). Then:

$$\chi = \frac{0.217e}{T}, \quad (10.4)$$

where χ = water vapour density (kg m^{-3}),
e = partial pressure of water vapour (millibars),
T = absolute temperature (K).

In vascular plants, r_L comprises the stomatal resistance and cuticular resistance in parallel; similar relations hold for moss sporophytes, the upper surfaces of *Polytrichum* leaves and thalli of Marchantiales. Some estimates of thallus resistances of Marchantiales are given in Table 10.3. Moss capsules have high resistances to water loss; I have found resistances (calculated over the surface area of capsule and seta) of about 5000 s m^{-1} for *Mnium hornum* and about 2000 s m^{-1} for *Bryum capillare*. For most bryophytes, r_L is low and probably often negligible, though a small cuticular resistance may be common; r_A is the major resistance to water loss.

Table 10.3 Internal diffusion resistances to evaporation for some liverworts (Proctor, 1980a, and unpublished data). Ranges given are for sets of five measurements. The increased resistance of water-stressed thalli of *Preissia quadrata* and *Marchantia paleacea* is at least in part due to closure of the compound epidermal pores, but the change is continuous and some increase is found in species with simple pores, so other factors are probably involved as well.

Species	Locality and date of measurement	Diffusion resistance (s m^{-1})
Reboulia	Trow Gill, Clapham, Yorks., 7 Sept. 1979	56–82
hemispherica	Missouri (Hoffman and Gates, 1970)	(min.) 25
Conocephalum	Bank of R. Teign, Drewsteignton, Devon,	
conicum	12 Aug. 1978	176–288
	Weed in glasshouse, Exeter, Devon.	
	14 Aug. 1978	447–740
	9 March 1979	306–608
Lunularia	Bank of R. Teign, Drewsteignton, Devon,	
cruciata	11 Aug. 1978	42–130
Preissia	Penyghent Ghyll, Yorks., 10 Sept. 1979	
quadrata	Fresh	79–250
	After 1 h (relative water content 74%)	200–335
Marchantia	Wicken Fen, Cambs., 18 June 1979	48–120
polymorpha		
Marchantia	Botanic Garden, Cambridge, 9 March 1979	
paleacea	Fresh	125–173
	Wilted	404–611
Pellia	Lydford, Devon, 11 Aug. 1978	125–173
epiphylla	Stoke Woods, Exeter, Devon, 20 March 1979	8–52

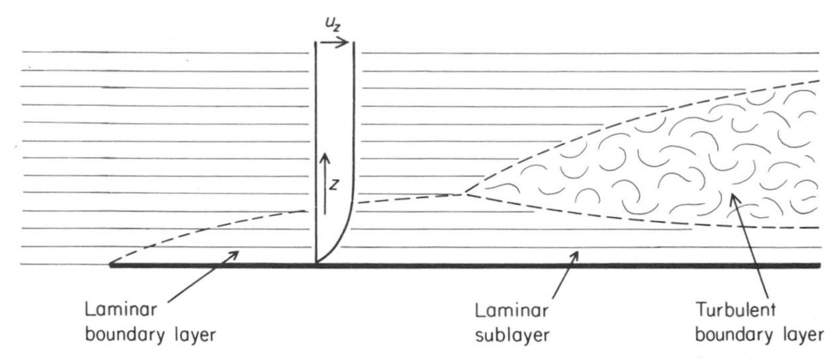

Fig. 10.4 Schematic diagram of the boundary-layer over a flat plate (mainly after Leyton, 1975). The velocity profile shows how the velocity u_z varies with height z above the surface. Note that all the transitions marked by broken lines will be gradual.

As air flows past a solid object a gradient of velocity develops, rising from zero in contact with the surface to the velocity of the surrounding air some distance above it. This layer of air influenced by the adjacent surface is the *boundary layer* (Fig. 10.4). If the windspeed is low or the solid object small (as will invariably be true of individual bryophyte parts), the streamlines of flow will be smooth and parallel with the surface; the boundary layer will be *laminar*. The effective thickness of a laminar boundary layer may be taken as:

$$\delta = 1.72\sqrt{(l\nu/V)}, \tag{10.5}$$

or

$$\delta = 1.72\, l/\sqrt{(\mathrm{Re})}, \tag{10.6}$$

where δ = displacement boundary layer thickness (m),
l = length parallel to flow,
V = wind velocity,
ν = kinematic viscosity of air (c. $1.5\times10^{-5}\ \mathrm{m^2\ s^{-1}}$ at 20°C),
Re = Reynolds number = Vl/ν.

At higher windspeeds, or with larger objects (Re > c. 2×10^4) the upper part of the boundary layer begins to develop eddies; it becomes *turbulent*, though a laminar sub-layer persists close to the surface. Fuller accounts of boundary layers are given by Monteith (1973) and Leyton (1975). The point that concerns us here is that water vapour must move through a laminar boundary layer (or the laminar sub-layer) by molecular diffusion, a very slow process compared with the turbulent mixing that takes place in the surrounding air. Consequently, r_A can be thought of as a boundary-layer resistance, and the air outside the boundary layer as being completely mixed by turbulence.

In many stands of vascular plants this distinction is a realistic

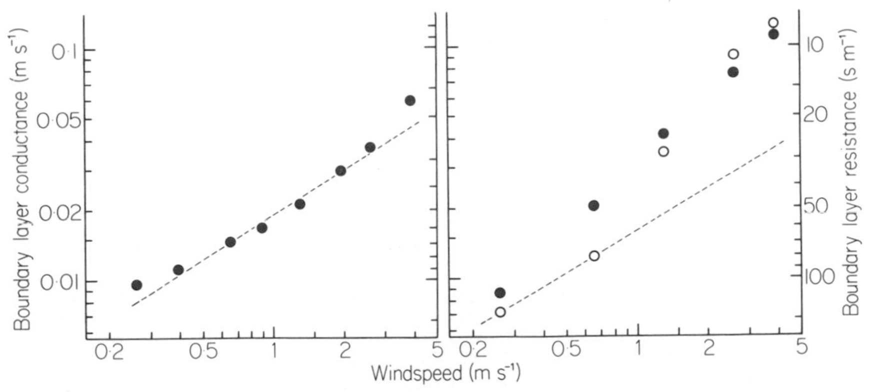

Fig. 10.5 The relation of boundary layer conductance and resistance to windspeed for (a) a *Leucobryum glaucum* cushion 5 cm diameter, and (b) cushions of *Mnium hornum* and *Dicranum majus* 9 cm diameter. The broken lines show theoretical relationships calculated from standard correlations for smooth spheres of the same diameters. Re-calculated from the data of Proctor (1980a).

approximation, but in bryophytes the leaves and shoots are often close enough together for their individual boundary layers to merge, so that molecular diffusion is also important in transfer processes between them. Surface roughness has no effect on the rate of water loss as long as it is submerged within a laminar boundary layer. Hence a dense moss cushion behaves as a large single object at low windspeeds (Fig. 10.5; Proctor, 1980a). Above a critical windspeed (approximately the speed at which the scale of surface roughness becomes comparable with boundary-layer thickness) the rate of water-loss rises steeply with increasing windspeed. This reflects the increasing effective surface area and decreasing diffusion path as turbulence increasingly penetrates among the leaves. It is easy to appreciate intuitively why rough, open moss cushions are characteristic of sheltered habitats, while mosses of exposed situations typically grow as tight cushions or smooth mats. Hair points can have the effect of trapping an additional thickness of stagnant air above the evaporating surfaces of the leaves, and may reduce the rate of water loss by 20–30% in *Grimmia pulvinata* or *Tortula* spp. (Fig. 10.6; Proctor, 1980a). Hair points and life-form characteristics cannot prevent water-loss, but they can retard it and so prolong the time available for photosynthesis.

Much more empirical study of water loss from bryophytes is needed, but it is important to record the relevant variables of Equation 10.3 if the results are to be fitted into a coherent theoretical framework to relate evaporation rates under different conditions. Many bryophytes form open cushions or wefts. Analysis of the water loss from these is a challenging problem because the diffusion resistance and hence the rate of evaporation is likely to be very dependent on the arrangement of the shoots and of

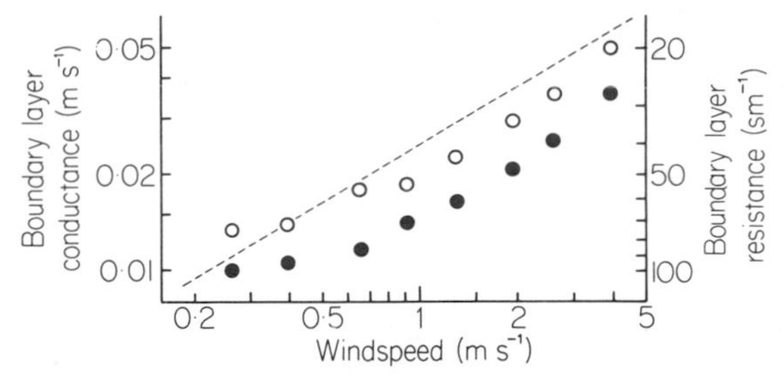

Fig. 10.6 The relation of boundary-layer conductance and resistance to windspeed for a cushion of *Grimmia pulvinata* 3 cm diameter, intact (solid dots) and with the hair-points clipped off (open circles). The broken line shows the theoretical relationship for a smooth sphere of the same diameter. Complete removal of all hair points was not practicable; those that remain are probably still having some effect on r_A. Re-drawn from data of Proctor (1980a).

neighbouring plants. It is thus difficult to make relevant laboratory measurements and much field measurement is needed (cf. the field potometer of Blaikley, 1932), but this could be a fertile field of investigation in terms of ecological understanding.

The life-forms of bryophytes are strikingly correlated with habitat (e.g. Gimingham and Birse, 1957; Gimingham and Smith, 1971; Meusel, 1935; Mägdefrau, 1969); the physical basis for this correlation lies in their capillary properties on the one hand, and relations with the atmospheric environment on the other. Bryophyte mats will often hold 5–10 times their dry weight of water, sometimes more, notably in *Sphagnum* and *Leucobryum* (Mägdefrau and Wutz, 1951; Coufalová, 1957) Mägdefrau and Wutz found the thick moss cover on the floor of coniferous forests capable of absorbing up to about 9 mm of rainfall (with one figure of 14.7 mm from a *Sphagnum capillifolium*-dominated sample). Pócs (1980) found that the profuse epiphytic bryophytes of an elfin forest in the Uluguru Mountains, Tanzania, intercepted 50 000 l ha^{-1} (equivalent to 5 mm rainfall) during a single period of rain. Life-form largely determines the capacity for storing water, the nature of the evaporating surface, and the length and capacity of the conducting path between them. These relationships merit detailed quantitative study.

It was pointed out above that eddy diffusion in the turbulent surrounding air is very much more rapid than the molecular diffusion in the immediate proximity of a bryophyte. Nevertheless, the rate of eddy diffusion is finite. Gradients of humidity, temperature and windspeed do develop, and they are important in the microclimate of bryophyte habitats, but their scale is measured in metres or tens of metres rather than in the fractions of a millimetre or millimetres of laminar boundary layers. Many measurements have been published of microclimatic variables in notable bryophyte habitats (e.g. Billings and Anderson, 1966; Zehr, 1977), but detailed physical analysis on the lines of current work on agricultural, forest or urban micrometeorology is a field yet to be exploited (Lowry, 1969; Monteith, 1975, 1976; Oke, 1978).

10.4 ENERGY BALANCE AND TEMPERATURE

Evaporation and temperature are closely interrelated. The temperature of a bryophyte is determined by the energy balance at its surface (Gates, 1968, 1980; Gates and Papian, 1971; Monteith, 1973). By convention, fluxes towards a surface are regarded as negative, and away from it positive. Then:

$$R_a + M + \epsilon\sigma T^4 + C + \lambda E + J + G = 0 \qquad (10.7)$$

where R_a = absorbed radiation, M = energy produced or absorbed by metabolism (negligible for plants), ϵ = emissivity (c. 0.95 at the relevant wavelengths), σ = Stefan-Boltzmann constant (5.57×10^{-8} W m^{-2} K^{-4}), T = absolute temperature (K), C = flux of sensible heat by convection, λ = latent heat of evaporation, E = rate of evaporation (or condensation) of water, J = change of stored heat, G = conduction from or to the environment.

Generally speaking, M can be ignored, and J and G are also likely to be small, though the last may be important at times for bryophytes growing on rock surfaces. Under most conditions, absorbed radiation must be balanced by some combination of radiative heat loss ($\epsilon\sigma T^4$), convective heat transfer, and latent heat exchange by evaporation or condensation. The physics of convective heat loss and evaporation are closely related (Monteith, 1965, 1973, 1981; Gates, 1980). Qualitatively, some of the consequences of these relations are obvious. In full sun (where the absorbed radiation may approach 1000 W m^{-2}) a bryophyte will be hotter than the surrounding air – and much hotter if it is dry than if it is moist. A moist bryophyte in strong sun can readily reach temperatures between 35 and 40°C, but temperatures much higher than this are uncommon. They are most likely to arise when the ambient humidity is high, the evaporating surface large, and the windspeed low, giving a high boundary-layer resistance. These conditions may be realized in large sunflecks in woods, or in *Sphagnum* carpets. Rudolph has recorded temperatures up to about 44°C on the surface of *Sphagnum* in a raised bog in Schleswig (Ellenberg, 1978, p. 464). Dry bryophytes in the sun are often much hotter than this; temperatures in the range 55–65°C are common (Table 10.4).

If the net gain or loss of energy by radiation is small (as it may be in a shady wood) a moist bryophyte will generally be somewhat cooler than its surroundings, but the temperature difference will seldom be large because in such situations both saturation deficit and windspeed are likely to be low, so limiting evaporation. In open situations at night, net loss of radiation to the sky may be 100 W m^{-2} or more, and much of this energy may be supplied by condensation of dew: calculation shows that dewfall is seldom likely to exceed about 0.5 mm of water in the course of a night, and it will often be much less. Dew formation has been shown to be important for the

Table 10.4 Some temperatures measured in bryophytes in the field under high irradiance in South-West England. See also measurements quoted by Lange (1955) from Central Europe.

Species and locality	Irradiance (W m^{-2})	Air temp. (°C)	r.h. (%)	Windspeed (m s^{-1})	Bryophyte temp. (°C)	
					dry	wet
Tortula intermedia Limestone rocks, Chudleigh, Devon, 6.76.	1000	30	25	c. 1.5	65	39
Pleurochaete squarrosa Short limestone turf, Chudleigh, Devon, 6.76.	1000	30	25	c. 1.5	62	36
Tortula intermedia Ledge of Cenomanian limestone, Branscombe, Devon, 6.76.	875	24.5	48	c. 2	71	—
Anomodon viticulosus Open turf on steep chalk cliff slope, Branscombe, Devon, 6.76.	875	24.5	48	c. 2	48.5	—
Tortula ruralis Thin fixed-dune turf, Braunton Burrows, Devon, 8.75.	—	31	—	—	64.5	—
Polytrichum piliferum S.W.-facing hornfels outcrop, Dunsford, Devon, 7.75.	—	24.5	—	—	57.5	—
Mnium hornum Acid oakwood, Stoke Woods, Exeter, Devon, 5.76. Max. temp. in sunflecks.	—	20	55	0.35–1.0	—	39
Fontinalis squamosa Boulder in R. Dart, Holne, Devon, 6.76.	650	23.5	56	0.5–1.0	44	—
	725	23.5	56	0.5–1.0	—	28

growth of lichens in the Negev Desert of Israel, especially on northerly aspects (Lange *et al.*, 1970; Kappen *et al.*, 1980), and this may be generally so for saxicolous bryophytes in open situations as well.

Used quantitatively, the energy balance equation provides an alternative means of estimating water loss if the other terms of the energy budget can be evaluated. Hoffman and Gates (1970) applied this method to estimate water loss from *Reboulia hemispherica*. In most cases, it is probably easier to measure water loss directly, but the energy budget approach is potentially useful in some field situations.

Spot readings of irradiance and temperature can be made easily enough, but continuous recording is time-consuming and expensive. Much useful information on the general features of the radiation climates of bryophyte habitats can be obtained from hemispherical ('fish-eye') photographs (Proctor, 1980b), which show how far the sky, and the tracks of the sun across it, are obscured by hills, trees and other obstructions. The interpretation of hemispherical photographs is discussed by Anderson

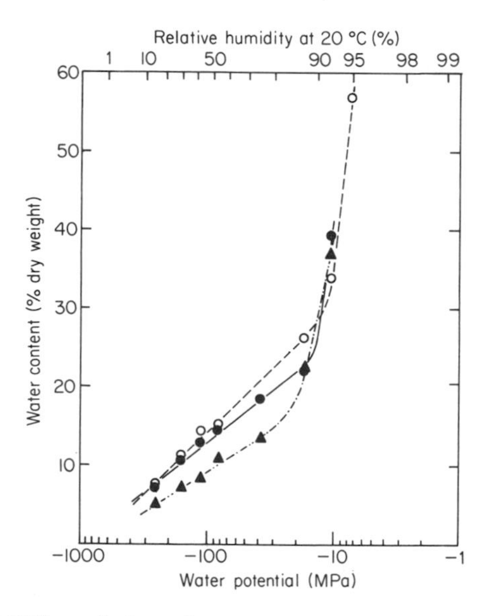

Fig. 10.7 The relation of water content to water potential for *Tortula ruralis* (——●——), *Racomitrium lanuginosum* (— — —O— — --), and *Homalothecium lutescens* (—·—·▲—·—·). The curve for *Rhytidiadelphus loreus* lies very close to that for *R. lanuginosum* within the range of this diagram, which corresponds largely with the region A to B of Fig. 10.8. Re-drawn from Dilks and Proctor (1979).

(1964, 1971). Data on irradiance in unobstructed sites useful in conjunction with hemispherical photographs are given by Cowley (1978), Day (1961) and Harding (1979); Page (1978) provides regression equations for estimating direct and diffuse radiation totals from sunshine figures (Meteorological Office, 1976; see also Gates, 1980).

10.5 WATER CONTENT AND WATER POTENTIAL OF BRYOPHYTES

The water content of bryophytes can vary within very wide limits. Some species under wet conditions reach water contents of 1000% or more of their dry weight. Dry bryophytes in equilibrium with the air in sunny weather may have water contents of 5% or less. The water associated with a bryophyte shoot can be divided into (a) *apoplast water*, held within the cell walls, (b) *symplast water*, held within the cells themselves, and (c) *external capillary water*. Curves relating water content to water potential or relative humidity have been published for a number of bryophytes (e.g. Plantefol, 1927; Willis, 1964; Dilks and Proctor, 1979; see also Fig. 10.7). In the past, most sets of data have been based on relative humidity (r.h.). This is related to the more fundamental quantity *water potential* by the equation:

$$\psi = \frac{RT \ln \frac{e}{e_s}}{\overline{V}_w}, \tag{10.8}$$

where ψ = water potential (Pa),
R = gas constant (8.31 J mol^{-1} K^{-1}),
T = absolute temperature (K),
$\overline{V}_w$ = partial molal volume of water (1.8×10^{-5} m^3),
e = partial vapour pressure of water in air,
e_s = saturated vapour pressure of water at temperature, T.

Essentially, the same principles apply to the water relations of bryophyte and flowering plant cells (Slatyer, 1967; Kramer, 1969; Nobel, 1974), but much of the interest in bryophyte water relations is at water potentials far below those most relevant in flowering plant physiology. Most of the change in the water content of the cells takes place between the osmotic potential of the turgid cell (discussed further below, p. 355) and about −20 MPa. At lower water potentials than this, much of the change is in the apoplast water associated with the cell wall materials. In this range, the relation of water content to water potential is very similar to that found in wood shavings or dry seeds. It provides a useful – indeed often the only – practical method of estimating the water potential of bryophytes in the field. At water potentials above −0.1 MPa most of the change is in the external capillary water, e.g. in spaces between papillae (c. −0.1 MPa), between sheathing leaf bases or in hyalocysts of Pottiales, Leucobryaceae

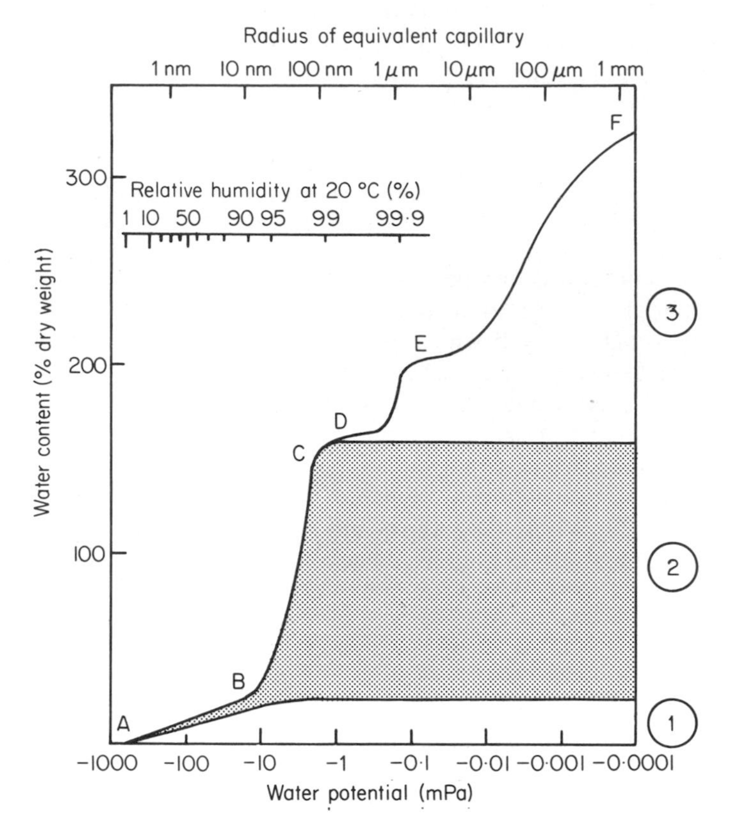

Fig. 10.8 Schematic relation of water content to water potential for a bryophyte. From A to B, most water is within the cell walls and is closely associated with surfaces and relatively immobile. At B, a free liquid phase becomes important, and from B to C increase of water content is mainly within the cells and is reflected by the changing osmotic potential of the cell solution. C is the point of incipient plasmolysis; from C to D the water content increases only slightly with the build-up of wall pressure and the first appearance of external capillary water associated with fine surface roughness and striation of the cell wall. From D to F the form of the curve will vary from species to species depending on the detailed architecture of the shoot. Within this region, water content can vary widely with only small variation in water potential, and virtually no change in cell water content. Region 1 in the diagram represents cell-wall (apoplast) water, region 2 symplast water (shaded), and region 3 external capillary water. The positions of the major points on the diagram are reasonably typical, from available measurements, but their exact location will vary from species to species. Vascular plant physiology is generally concerned with the region of the diagram to the right of point B (and largely to the right of C), and (because there is little or no external water) below a horizontal line through D. In poikilohydric bryophytes with external capillary water storage and conduction the remaining parts of the diagram are physiologically relevant and important. Re-drawn from Dilks and Proctor (1979).

or *Sphagnum* (c. −0.01 MPa) or in the concavities of leaves etc. (c. −0.1 to −1 kPa; Fig. 10.8).

In many bryophytes adapted to wet habitats most water is held within the vacuoles of the large cells (i.e. within region 2 of Fig. 10.8); loss of water quickly leads to loss of turgor. Apoplast water and external capillary water are relatively much more important in ectohydric species of dry habitats. In these, the symplast water within the cells is a relatively small fraction of the total water associated with the plant under most conditions. A large part of the water can thus be evaporated with only a small change in water potential, and with negligible physiological effect on the cells. However, once drying begins to affect the cells, metabolism quickly comes to a standstill.

This contrast, which represents the extremes of a continuous range of variation, is well illustrated by the water content–water potential curves for *Polytrichum commune* and *Racomitrium lanuginosum* (Fig. 10.9; Bayfield, 1973). It has important consequences for the response of photosynthesis and respiration to water content. Thus, in *Pellia epiphylla* (Fig. 10.10a), photosynthesis increases progressively up to a water content approaching 1000% of dry weight. Respiration, on the other hand, changes little over a wide range of water content. The highest rates occur at water contents around 200%, below which respiration falls steeply to low values. By contrast, in *Tortula intermedia* (Fig. 10.10b), photosynthesis rises to a well-marked maximum at a water content around 200% of dry weight, beyond which it declines again, probably owing to the restriction of gas exchange by the superincumbent film of water (p. 341). In *T. intermedia*, the

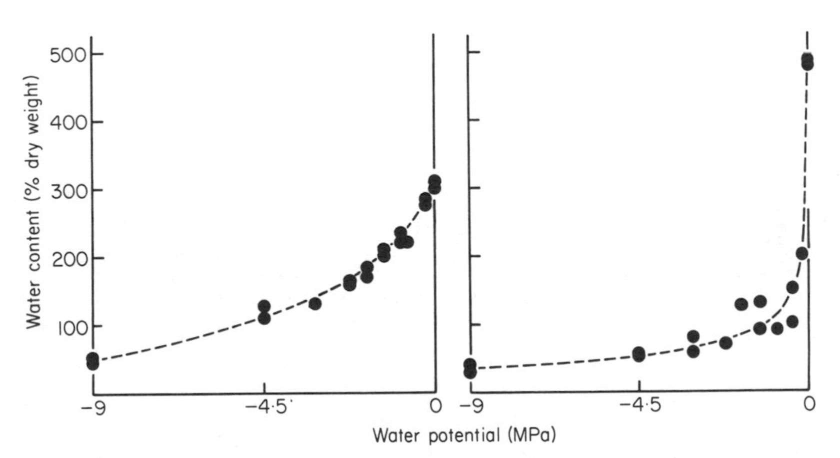

Fig. 10.9 The relation of water content to water potential for (a) *Polytrichum commune*, and (b) *Racomitrium lanuginosum*. These graphs correspond to the part of Fig. 10.8 to the right of B, but the right hand end of the linear water-content scale here is in that diagram greatly expanded. Re-drawn from Bayfield (1973).

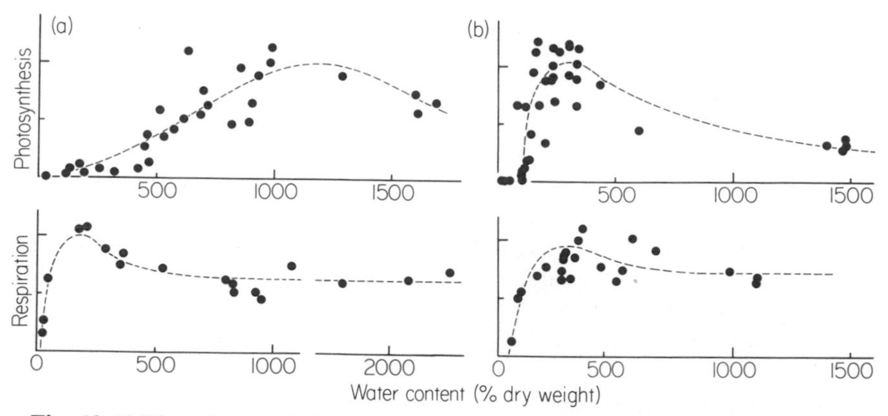

Fig. 10.10 The relation of photosynthesis and respiration to water content in (a) *Pellia epiphylla* and (b) *Tortula intermedia*. In (a) the maxima of respiration and photosynthesis are widely separated. In (b) the steep portions of the photosynthesis and respiration curves are nearly coincident; the peaks of both curves are probably broadened by variation of water content within the material. Re-drawn from Dilks and Proctor (1979).

steep regions of the photosynthesis and respiration curves at low water contents are not noticeably separated. Both patterns can be paralleled by other species. *Conocephalum conicum* (Slavik, 1965; Ensgraber, 1954), *Hookeria lucens* and *Plagiothecium undulatum* (Dilks and Proctor, 1979) behave essentially like *Pellia epiphylla,* while *Anomodon viticulosus, Homalothecium lutescens, Pleurochaete squarrosa* (Dilks and Proctor, 1979), *Homalothecium sericeum* (Romose, 1940) and *Racomitrium lanuginosum* (Tallis, 1959) behave like *Tortula intermedia.* Many species fall somewhere between the two extremes (Stålfelt, 1937; Seidel, 1976; Busby and Whitfield, 1978; Dilks and Proctor, 1979; Tobiessen *et al.*, 1979).

The responses of different species appear to be very much more alike in terms of water *potential* than of water *content.* Slavik (1965) found *Conocephalum conicum* extremely sensitive to drought stress, photosynthesis falling to zero at only −1.28 MPa, but this seems to be unrepresentative of bryophytes in general. Lange (1969a) found measurable positive net assimilation following absorption of water from air at 98% r.h. (−3.0 MPa) in 10 out of 14 species he investigated (the exceptions being *Rhizomnium punctatum* and three thalloid liverworts); a number of species still showed positive net assimilation at 94% r.h. (−8.5 MPa). In osmotic stress experiments, Dilks and Proctor (1979) found progressive decline of photosynthesis over a wide range of water content to negligible rates at around −5 to −10 MPa in *Hookeria lucens,* c. −10 MPa in *Anomodon viticulosus* and between −10 and −20 MPa in

Homalothecium lutescens. Respiration was usually somewhat stimulated by mild water stress (to c. −2 MPa) and then fell progressively but slowly down to water potentials around −20 MPa or lower. It is difficult to determine the water potential at which respiration ceases, because physical adsorption and release of gases by the bryophyte material interfere with measurements of metabolic gas exchange. In lichens, Cowan *et al.* (1979) found that incorporation of tritium into respiratory intermediates, amino acids and polyols from labelled water vapour was taking place rather freely down to −40 MPa, and detectable at −100 MPa; incorporation into photosynthetic intermediates ceased altogether below −10 MPa. Bryophytes are likely to behave in broadly the same way.

Osmotic potentials of turgid bryophyte cells vary widely (Table 10.5); values for many species are listed by Renner (1932), Patterson (1946), Will-Richter (1949), Ochi and Yonehara (1954) and Hosokawa and Kubota (1957). Thalloid liverworts of wet habitats tend to have rather low osmotic potentials, commonly in the range −0.5 to −1.5 MPa. Many bryophytes from mesic to dry habitats have osmotic potentials between −1.5 and −2.5 MPa, while some poikilohydric species of dry habitats such as *Frullania tamarisci* and *Radula complanata* may reach −3 to −6 MPa. Ochi and Yonehara and Hosokawa and Kubota showed seasonal changes of osmotic potential in various species, and such changes are probably general.

How do these values of osmotic potential and water potential relate to field conditions? A bryophyte with an osmotic potential of 6 MPa could maintain turgor in equilibrium with air to c. 96% r.h. at 20°C. However, humidities as high as this are unusual in field situations unless rain is actually falling. The saturation vapour pressure of water varies by about 7% per °C at normal temperatures, so small differences of temperature are often more likely to be important in determining the distribution of water than differences of a few MPa in osmotic potential. Similarly, the osmotic potential will only have a substantial effect on the water vapour deficit ($\chi_S - \chi_A$), and hence on evaporation, if the ambient humidity is very high. At an ambient relative humidity of 50% at 20°C, an osmotic potential of 6 MPa would reduce the rate of evaporation by only about 8%, though the rate of evaporation would drop progressively as the plant dried out and its water potential approached equilibrium with that of the air.

These considerations suggest that high osmotic potentials in bryophyte leaves are not primarily related either to retarding water loss or to taking up water from moist air. It may be that they reflect adaptation to maintain high water-potential gradients for water movement (osmotic potential often increases towards the apex of a shoot or leaf; Renner, 1932; Will-Richter, 1949) or that they are incidental to some other aspect of desiccation tolerance.

Table 10.5 Osmotic potentials of bryophytes. The selection of figures given here have been re-calculated from published data expressed in molar concentrations of various solutes. Unless otherwise indicated, values for mosses are from Patterson (1946) and values for liverworts from Will-Richter (1949). Sources: [1]Patterson (1946); [2]Will-Richter (1949), mean summer values; [3]Ochi and Yonehara (1954); [4]Hosokawa and Kubota (1957), autumn values. *Limits of seasonal variation.

Species	Osmotic potential (MPa)	Species	Osmotic potential (MPa)
Anthoceros punctatus	0.38	*Polytrichum formosum*	1.63–2.65[3]*
Conocephalum conicum	0.55, 0.89[1]	*Atrichum undulatum*	1.55
Pellia epiphylla	0.63, 0.67[1]	*Ceratodon purpureus*	1.77
Metzgeria furcata	3.53	*Dicranum scoparium*	1.98
Fossombronia wondraczekii	1.11[1]	*Dicranoloma fragiliforme*	3.60[4]
Trichocolea tomentella	1.45, 1.55[1]	*Boulaya mittenii*	3.78[4]
Bazzania trilobata	3.17, 2.64[1]	*Tortula papillosa*	2.21
Lophozia ventricosa	1.31	*Bryum argenteum*	1.77
Barbilophozia barbata	2.21	*Mnium marginatum*	1.98
Solenostoma cordifolium	1.59	*Rhizomnium punctatum*	1.55
Nardia scalaris	1.26	*Aulacomnium palustre*	2.42
Gymnomitrion concinnatum	2.38	*Philonotis fontana*	1.33
Mylia taylori	2.06	*Ulota crispula*	4.12[4]
Plagiochila asplenioides	2.38, 1.98[1]	*Hedwigia ciliata*	3.46–6.60[3]*
Lophocolea bidentata	2.21	*Neckera pennata*	2.42
Cephalozia bicuspidata	1.04, 1.55[1]	*Anomodon attenuatus*	2.42
Diplophyllum albicans	1.90	*Anomodon giraldii*	3.16[4]
Scapania undulata	1.84	*Thuidium cymbifolium*	2.75[4]
Radula complanata	5.57	*Thuidium delicatulum*	1.55
Porella platyphylla	2.68	*Amblystegium riparium*	2.42
Lejeunea cavifolia	2.21	*Rhynchostegium riparioides*	2.42
Frullania tamarisci	4.68	*Plagiothecium sylvaticum*	1.98

In field situations in dry weather equilibrium water potentials are often low, and sometimes very low. Field water-content measurements and calculation from air humidity and temperature measurements both suggest that values between −20 and −40 MPa are common in woods or shaded grassland turf in spells of dry weather (Klepper, 1963; Morton, 1977; Dilks and Proctor, 1979; C. Nichols, unpublished measurements from Maiden Castle, Dorset). In exposed places −100 MPa is commonplace, and ψ may fall to −200 MPa or less, especially if the ground (and bryophyte) are strongly heated by the sun.

10.6 RESPONSES TO DESICCATION

How low a water content (or water potential) can bryophytes tolerate, and for how long? Abel (1956) exposed detached leaves of some 60 species of mosses (collected around Vienna) to a wide range of relative humidities for 24–48 h, and tested their ability to plasmolyse after gradual remoistening; Clausen (1952) carried out similar experiments with 38 species of liverworts in Denmark. In these experiments, the plants fell into three fairly clear categories. Some species of moist habitats were always killed by even slight drying, e.g. *Schistostega pennata, Bryum duvalii, Philonotis calcarea, Fontinalis antipyretica, Scorpidium scorpioides, Cephalozia media, Cladopodiella fluitans, Solenostoma triste* and *Calypogeia trichomanis.* Many mosses from a variety of more or less mesic habitats survived immediate drying to equilibrium with 50–90% r.h., though killed by more intense desiccation, but became much more desiccation-tolerant if they were kept initially for 24 h at 96% r.h. Species in this category included various Mniaceae, *Climacium dendroides, Bryum capillare, Calliergon cuspidatum, Ceratodon purpureus, Plagiothecium* spp., *Hylocomium splendens, Atrichum undulatum, Pogonatum aloides* and *Rhytidiadelphus* spp. Clausen's data for liverworts are less detailed, but suggest similar behaviour in, e.g., *Lophocolea heterophylla* and *Lepidozia reptans.* A third group of species proved very desiccation-tolerant under all conditions. Many of these plants are corticolous or saxicolous but some grow on dry soil; they include, e.g. *Neckera crispa, Racomitrium canescens, Hedwigia ciliata, Bryum pendulum, Pterogonium gracile, Tortula ruralis, Frullania dilatata, Cephaloziella divaricata* and *Porella platyphylla.*

Many of the more tolerant bryophytes not only tolerate drying to very low water contents, but can survive dry for very long periods. Thus, *Racomitrium lanuginosum* and *Andreaea rothii* survived for more than a year at 32% r.h. at 20°C (Dilks and Proctor, 1974), and *Tortula ruralis* quickly recovered normal protein synthesis after 10 months air dry (Bewley, 1973a). Fourteen species from a Japanese beech forest survived for 102 days at various relative humidities down to 30% (Hosokawa and Kubota, 1957). Keever (1957) reported recovery of dried herbarium specimens of *Grimmia laevigata* after 7–10 years. A noteworthy point is that some highly tolerant species appear to survive better at low than at higher humidities, e.g. *Boulaya mittenii, Ulota crispula* (Hosokawa and Kubota, 1957), *Racomitrium lanuginosum, Tortula ruralis* (Dilks and Proctor, 1974). This is worth emphasizing, because it parallels the common condition in seeds (Roberts and Abdalla, 1968; Roberts, 1975), it may be a general feature of highly desiccation-tolerant bryophytes, and it is contrary to the implicit assumption on which many studies of bryophyte desiccation-tolerance have been based.

The typical responses of photosynthesis and respiration following remoistening after various periods of desiccation are illustrated in Figs. 10.11–10.13; the behaviour of bryophytes is closely paralleled by that of

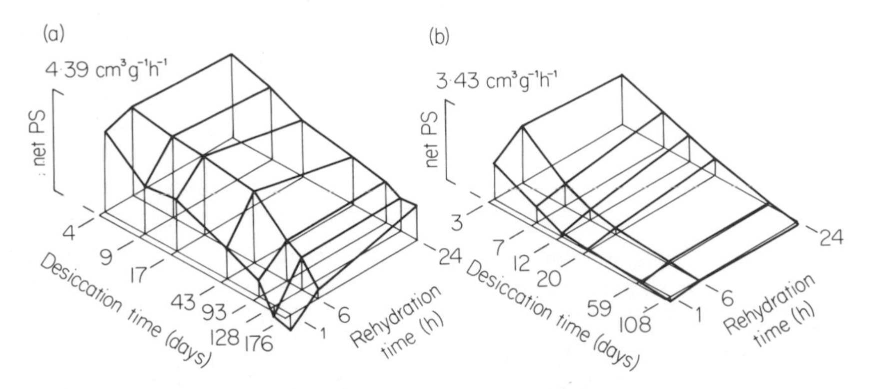

Fig. 10.11 Response surfaces for recovery on re-moistening after various periods of desiccation at 32% r.h. and 20°C for (a) *Ulota crispa* (Drewsteignton, Devon, March 1974) and (b) *Myurium hochstetteri* (Meall Sanna, Ardnamurchan, Argyll, Sept. 1973). From unpublished manometric measurements by T. J. K. Dilks.

lichens (Ried, 1960; Lange, 1969b; Smith and Molesworth, 1973; Farrar, 1976b; Farrar and Smith, 1976), intertidal seaweeds (Ried, 1969) and poikilohydric pteridophytes (e.g. *Selaginella lepidophylla,* Eickmeier, 1979). On remoistening, there is an initial physical displacement of adsorbed gas, and respiration rapidly builds up to a level substantially above the normal rate of dark respiration; the corresponding phenomena in lichens have been called the *wetting burst* and *resaturation respiration* (Smith and Molesworth, 1973; Farrar and Smith, 1976). Photosynthesis recovers more slowly than respiration, but the compensation point is reached within minutes or hours. Photosynthesis and respiration then gradually return to normal over a period which may typically be 12–24 hours. Following remoistening there is thus an initial stage (which may be prolonged in a sensitive species or after long drying; Fig. 10.12b) of net loss of assimilate which must be made good after compensation has been reached (Stålfelt, 1937; Hinshiri and Proctor, 1971; Dilks and Proctor, 1976b). This is obviously one factor which limits the frequency with which a bryophyte can tolerate drying and remoistening. Another is the leakiness of the cell membranes immediately after remoistening, which allows leaching of nutrients and soluble metabolites (Dhindsa and Bewley, 1977; Gupta, 1976, 1977a; Brown and Buck, 1979); this may be a major cause of harmful effects of rapid remoistening (e.g. Clausen, 1952). With increasing duration of desiccation, recovery is progressively slower and less complete. If desiccation is sufficiently prolonged, remoistening initiates a rapid build-up of respiration, followed within a day or two by death. It has been suggested that this rapid respiration is due to associated micro-organisms (Gupta, 1977b), but the regularity of the effect suggests that this cannot always be so, and the question remains to be resolved.

The response-surface of Fig. 10.11 can also be considered in the direction of the desiccation-time axis. Fig. 10.13 presents curves of this kind for a recovery time of 24 h. If time is plotted on a logarithmic scale, loss of viability is often well represented by a sigmoid cumulative Normal curve; this becomes a straight line if photosynthesis is expressed as a percentage of the initial (control) value and transformed to a probit scale (Bliss, 1967). Responses of this kind are commonplace in toxicology; loss of viability by stored seeds follows a similar course (Roberts and Abdalla, 1968; Roberts, 1975). The behaviour of a particular species can be specified by two parameters, one defining the time for 50% loss of

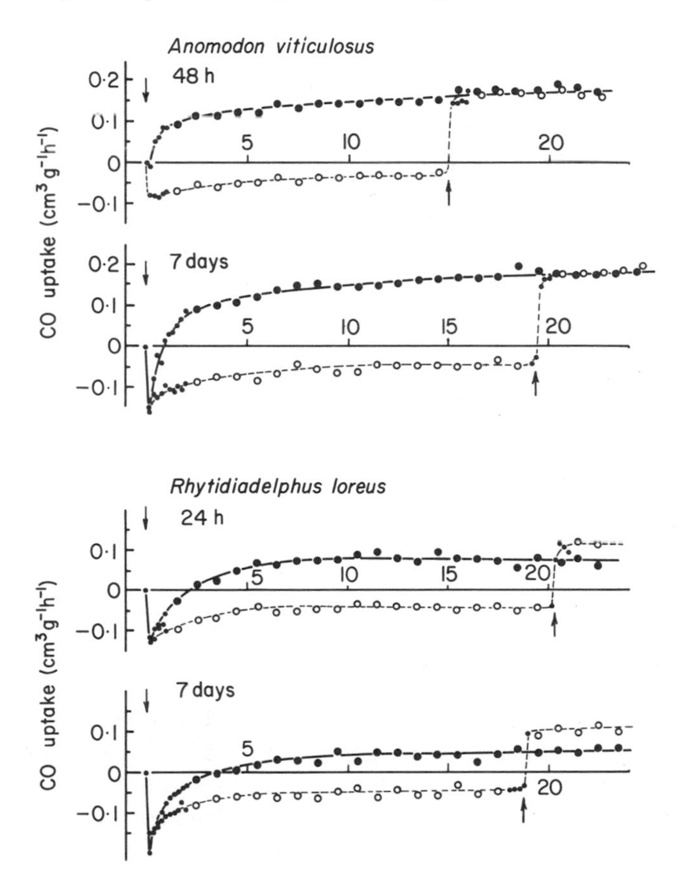

Fig. 10.12 Recovery of net photosynthesis and respiration after different periods of desiccation: (a) *Anomodon viticulosus*, (b) *Rhytidiadelphus loreus*. In *A. viticulosus* recovery takes place quickly, and equally in light or dark; recovery of *R. loreus* is slower, and less complete in the high irradiance conditions of the experiment (photon flux c. 250 μmol.m^{-2} s^{-1}). Infra-red gas analyser measurements, reproduced with permission from Dilks and Proctor (1976b). (Horizontal axis shows rehydration time (h)).

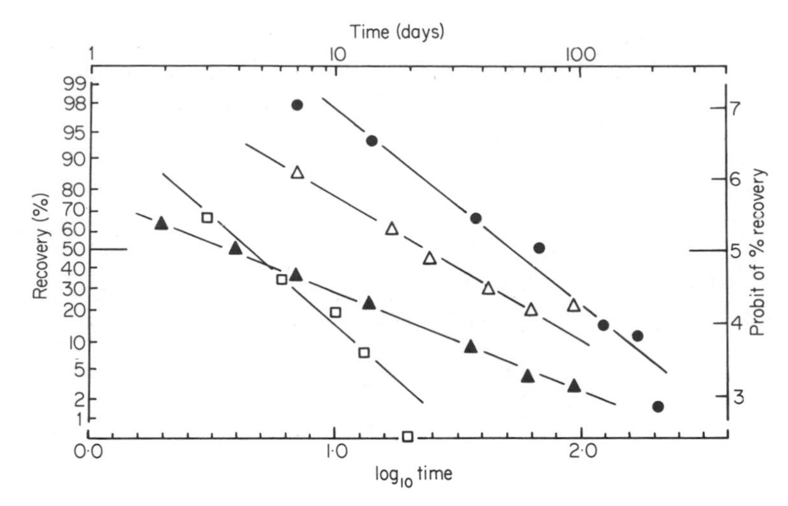

Fig. 10.13 The relation of photosynthesis after 24 h moist recovery to length of preceding desiccation period. Loss of viability follows a cumulative log normal curve, transformed to a straight line by plotting time on a logarithmic scale, and performance on a scale of probits (normal deviate+5). *Neckera crispa*, 32% r.h. (●); *Plagiothecium undulatum*, 76% r.h. (△); *Rhytidiadelphus loreus*, 32% r.h. (▲); *Plagiochila spinulosa*, 32% r.h. (□). Re-drawn from Dilks and Proctor (1974).

photosynthetic capacity, and another defining the standard deviation of the cumulative Normal curve – a measure of the relative spread of time over which the loss of viability takes place. There is very wide variation in the first parameter (Table 10.6), and evidence for a good deal of variation in the second (Dilks and Proctor, 1974, 1976a; Hearnshaw and Proctor, 1982). Obviously, photosynthesis after a 24 h moist period is only a limited measure of recovery; bryophyte shoots that have lost a substantial part of their photosynthetic capacity may have their functional integrity irreparably damaged, so that long-term survival may be from apical or basal innovations or secondary protonemata. Evidence of the relative tolerance of young and old shoots is somewhat conflicting (e.g. Abel, 1956), but my own experience is that the young growth seems in general to survive longest. Intraspecific differences in desiccation tolerance have been demonstrated in *Calliergon cuspidatum, Climacium dendroides,* and *Hypnum cupressiforme* (Lee and Stewart, 1971), and are probably common.

The desiccation tolerance of bryophytes typically varies from season to season, and with environmental conditions in the immediately preceding period. In many species, drought-hardening takes place under sub-critical water stress. This is clearly shown by the much enhanced tolerance of many liverworts and mosses to low water potentials after a period of relatively mild water stress (e.g. 24 h at 96% r.h.; Höfler, 1946, 1950; Abel, 1956; Gupta, 1978). Slow drying can have the same effect, and experiment shows better recovery after slow than rapid drying (*Tortula muralis,* Dhindsa and Bewley, 1977; *Cratoneuron filicinum,* Krochko *et al.*, 1978), and longer

Table 10.6 Half-viability times for bryophytes kept dry at 20°C. Most of the figures in this table are from manometric measurements of photosynthesis after 24 hours' moist recovery. They give an idea of the range likely to be encountered, but in general are liable to wide fluctuation with season and previous treatment. See also Table 10.9. Sources: (1) Hinshiri and Proctor, 1971; (2) Proctor, 1972; (3) Dilks and Proctor, 1974; (4) Dilks and Proctor, 1976a; (5) Dilks and Proctor, 1976b. *From logarithmic plots of original data.

Species	Month	Relative humidity (%)	Half-viability time (days)	Source
Andreaea rothii	throughout year	32	c.250–300	(4)
Racomitrium aquaticum	Oct. (min.); Mar.–July (max.)	32	c. 90– > 300	(4)
R. lanuginosum	Jan.	76	100	(3)
		54	150	(3)*
		32	> 334	(3)*
Tortula ruralis	Jan.	76	16	(3)*
		54	35	(3)*
		32	20	(3)*
	Oct. (min.); Mar.–May (max.)	32	> 40– > 300	(4)
Neckera crispa	April	32	55	(3)
Anomodon viticulosus	April	50	c. 30	(1)
	February	32	c. 20	(2)
	November	c.50	15	(Table 10.9)
Scorpiurum circinatum	June	32	4	(3)
	Jan. (min.); May–July (max.)	32	c. 8–50	(4)
Plagiothecium undulatum	Jan.	76	25	(3)*
		54	8	(3)*
		32	3	(3)*
Rhytidiadelphus loreus	Nov.	76	14	(3)*
		54	c. 14	(3)*
		32	4	(3)
Rhytidiadelphus loreus	July	32	15	(5)
Hylocomium splendens	April	32	10	(3)
	Oct. (min.); Mar.–July (max.)	32	c. 2–25	(4)
	July	32	c. 20	(5)
Plagiochila spinulosa	Jan.	32	5	(3)
	Oct. (min.); Mar.–May (max.)	32	< 1–c. 10	(4)
Saccogyna viticulosa	May	32	3	(3)
Porella platyphylla	April	50	c. 80	(1)

survival after repeated dry/wet cycles (*Rhytidiadelphus loreus,* Dilks and Proctor, 1976b). Desiccation tolerance declines again if a bryophyte is kept continuously moist; this 'de-hardening' may be the reason why material of *Tortula intermedia, Rhytidiadelphus loreus* and *Hylocomium splendens* kept wet and dry for alternate weeks performed relatively poorly in the intermittent desiccation experiments of Dilks and Proctor (1976b, see below).

Desiccation tolerance tends to be greatest at the times of year when water stress is greatest (either through drought or freezing) and growth least. In Britain, tolerance is generally least in autumn, increasing progressively through winter to a maximum in late spring and early summer; the details of the pattern vary from species to species (Dilks and Proctor, 1976a). The common pattern in Central Europe is for desiccation tolerance to be greatest during late spring and early summer, with a second peak during the coldest part of winter, and relatively low in early spring and especially during the main growing period in late summer and early autumn (Abel, 1956; Nörr, 1974a). In Japan, with a wet summer and dry winter, desiccation tolerance is apparently greatest during the latter season (Ochi, 1952; Hosokawa and Kubota, 1957). Drought hardening probably plays at least some part in these seasonal variations in tolerance, but other factors are probably at work too. In material of *Lunularia cruciata* from Israel, Schwabe and Nachmony-Bascomb (1963) found the change to the dormant and drought-tolerant summer phase determined primarily by photoperiod, but influenced also by temperature.

It is probably rare for bryophytes to have to withstand very long periods of unbroken desiccation in the field. Even prolonged drought is likely to be punctuated by dew on clear nights – and, paradoxically, it may be bryophytes of shaded habitats which suffer the longest drying, though not the most severe. A moist break of 24 hours allowed complete recovery of surviving cells from the effects of a preceding dry period in *Anomodon viticulosus, Tortula intermedia, Hylocomium splendens* and *Rhytidiadelphus loreus* (Proctor, 1972; Dilks and Proctor, 1976b). In the last two species, moist-break experiments showed that recovery was not complete in 6 hours, but the time-scale would no doubt vary from species to species. Evidently, provided the moist intervals are not below a critical length, the effects of successive periods of desiccation are not cumulative.

In general, bryophytes that are even moderately desiccation-tolerant are also tolerant of freezing. Bryophyte tissues are seldom more than a few cells thick so, at the slow rates of cooling experienced in the field, intracellular freezing will not occur (Mazur, 1969). Freezing brings about partial dehydration of the cells, which lose water until their water potential matches that of the ice outside (and the solutes in the cell depress its freezing point to the ambient temperature). Recovery curves of net photosynthesis after freezing and after drying are similar in form (Kallio and Kärenlampi, 1975), and the long-term effects of freezing and drying also appear to be similar (Dilks and Proctor, 1975). However, the parallel is not exact. Freezing in itself cannot produce very severe desiccation stress (Table 10.7), though obviously more intense stress can arise later by evaporation. Even desiccation-sensitive species may survive freezing well, provided they are cooled reasonably slowly. Pihakaski and Pihakaski (1979) found that material of *Pellia epiphylla,* collected frozen, showed good recovery after being maintained at −22°C.

Table 10.7 Saturation vapour density over water and ice (Slatyer, 1967), and water potential of ice.

Temperature °C	Saturated water vapour over water ($kg\ m^{-3} \times 10^{-3}$)	Saturated water vapour over ice ($kg\ m^{-3} \times 10^{-3}$)	Water potential of ice relative to water at same temperature (MPa)
0	4.85	4.85	0.0
−5	3.41	3.25	−5.96
−10	2.36	2.14	−11.85
−15	1.61	1.39	−17.55
−20	1.074	0.883	−22.89

Desiccation-tolerant bryophytes will survive very high temperatures in the dry state (Table 10.8). Survival time at a given temperature, and temperature tolerance for a given interval are two dimensions of the same response surface, and are hence closely related. Hearnshaw and Proctor (1981) found that the time for 50% loss of chlorophyll (measured after a period of remoistening as an index of viability) declined very steeply with temperature (Table 10.9). For most of the species investigated, the relation to temperature follows Arrhenius's equation (Fig. 10.14), and is similar to the relation found (over a more limited range of temperature) for loss of viability by dry seeds in storage (Roberts, 1975). Extrapolation suggests that these bryophytes should survive almost indefinitely at very low temperatures. Recovery of normal metabolism after short periods at 77 K (Bewley, 1973b) or 0.05 K (Becquerel, 1951) is thus unsurprising, but the form of the curve for *Racomitrium lanuginosum* should be a warning against taking extrapolation of the straight lines in Fig. 10.14 too seriously!

Obviously the responses of bryophytes to desiccation are complex, but they can usefully be thought of as divided into 'catastrophe' responses and 'housekeeping' responses. For species of normally moist habitats, with low desiccation tolerance, occasional periods of drought pose a catastrophic hazard, grossly interrupting normal metabolism and growth and liable to kill a greater or less part of the population before normally favourable

Table 10.8 Survival of dry bryophytes after treatment for 30 minutes at different temperatures (Lange, 1955). The bryophytes were scored after cultivation for 1 week on damp sand in petri dishes: *** no visible difference from control material; ** visible damage, but less than half of material dead; * severe damage, more than half material killed but living portions still visible; — all dead.

Species, date and locality	Temperature (°C)											
	65	70	75	80	85	90	95	100	105	110	115	120
Gymnomitrion obtusum, 4.53 Shaded block-scree, Meissner	***	**	*	*	—	—	—	—	—	—	—	—
Porella platyphylla, 5.53 L. Garda area, on *Populus*	***	***	***	*	—	—	—	—	—	—	—	—
Polytrichum formosum, 7.54 Shady oak-beech wood, Brittany	***	***	***	**	**	*	—	—	—	—	—	—
Mnium hornum, 4.53. Shady beech-wood, Meissner (650 m)	***	***	***	***	**	—	—	—	—	—	—	—
Brachythecium rutabulum, 6.54 Turf at side of ditch, Göttingen	***	***	***	***	**	*	—	—	—	—	—	—
Dicranum scoparium, 6–8.54 Woods: Göttingen, Hanover, Thuringer Wald	***	***	***	***	***	**	*	—	—	—	—	—
Polytrichum piliferum, 9.54 Sandy *Calluna* heath, Hanover	***	***	***	***	***	***	**	—	—	—	—	—
Leucodon sciuroides, 7.54 On oak, Brittany	***	***	***	***	***	***	***	*	—	—	—	—
Racomitrium lanuginosum, 4.53 Sunny block-scree, Meissner (700 m)	***	***	***	***	***	***	***	*	—	—	—	—

Table 10.8 Continued.

Species, date and locality	Temperature (°C)											
	65	70	75	80	85	90	95	100	105	110	115	120
Rhytidium rugosum, 9.54 Loiseleurietum, Hohe Tauern	***	***	***	***	***	***	***	*	—	—	—	—
Orthotrichum anomalum, 5.53 L. Garda area, sunny limestone	***	***	***	***	***	***	***	*	*	—	—	—
Schistidium maritimum, 7.54 Sunny coastal rocks, Brittany	***	***	***	***	***	***	***	*	*	—	—	—
Tortella tortuosa, 5.53 L. Garda area, sunny limestone	***	***	***	***	***	***	***	**	*	—	—	—
Tortula ruralis, 4–54 Garrigue, Montpellier	***	***	***	***	***	***	***	**	**	—	—	—
Campylopus flexuosus, 8.54 Shady siliceous rocks, Brittany	***	***	***	***	***	***	***	***	**	—	—	—
Ceratodon purpureus, 8.54 Berlin, sandy ground under *Pinus*	***	***	***	***	***	***	***	***	**	*	—	—
Pleurochaete squarrosa, 7.54 Dry calcareous grassland, Montpellier	***	***	***	***	***	***	***	***	***	**	—	—
Barbula gracilis, 7–54 Dry calcareous grassland, Montpellier	***	***	***	***	***	***	***	***	***	***	*	—

Table 10.9 Survival of dry bryophytes at different temperatures: time for chlorophyll content (measured after 3 days' remoistening) to fall to 50% of initial value; experiment with *Anomodon viticulosus* begun Nov. 1979, remaining mosses Jan. 1980, and the two liverworts Mar. 1980 (data from Hearnshaw and Proctor, 1982).

Species and locality	Temperature (°C)				
	20	37	60	80	100
Anomodon viticulosus, limestone rocks, Chudleigh, Devon	15 d	1.2 d	2.1 h	—	1.4 min.
Andreaea rothii, granite, Crockern Tor, Dartmoor, Devon	—	89 d	68 h	—	9.3 min.
Racomitrium lanuginosum, granite nr Wistmans Wood, Dartmoor	—	98 d	406 h	18 h	8.0 min.
R. aquaticum, Crockern Tor, Dartmoor, Devon	—	64 d	314 h	—	27 min.
Tortula intermedia, sunny limestone rocks, Chudleigh, Devon	—	68 d	39 h	—	14 min.
Frullania tamarisci, rocky turf, Hartland Point, Devon.	—	6.2 d	4.1 h	—	0.9 min.
Porella platyphylla, shady limestone, Chudleigh, Devon	—	43 d	20 h	—	7.1 min.

conditions return. For the most sensitive, the immediate question is 'how low a water potential can the plant survive?' Data of the kind presented by Clausen (1952), Abel (1956), Nörr (1974a), Johnson and Kokila (1970) and Brown and Buck (1979) are directly relevant to this kind of situation. For many rather tolerant species, which may withstand severe desiccation when drought-hardened, the duration of desiccation becomes important. The most obviously relevant laboratory experiments are those measuring survival time (e.g. Hosokawa and Kubota, 1957; Dilks and Proctor, 1974; Hearnshaw and Proctor, 1981), though short-term measurements such as those of Brown and Buck (1979) may still correlate usefully with survival time and habitat adaptation. Often, survival times measured in laboratory experiments are longer than seem likely ever to be required in the field. There are probably two main reasons for this. The first is that natural selection is primarily for 'housekeeping' survival rather than 'catastrophe' survival – for keeping the plant essentially undamaged after short periods of drying rather than for recovery from catastrophically long desiccation. The second is that bryophytes in exposed situations may reach temperatures of 60°C or more in hot sunshine, greatly accelerating desiccation damage. Nevertheless, for many bryophytes of dry habitats frequent severe desiccation is normal, and for these it is dynamic adaptation to maintain a positive carbon balance through repeated cycles of drying and

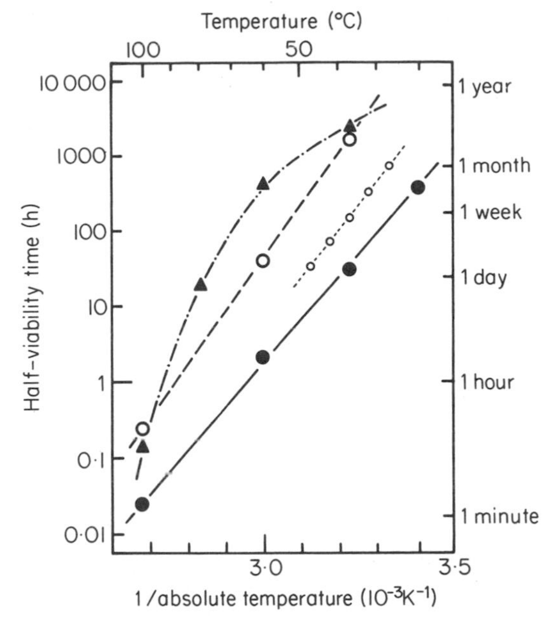

Fig. 10.14 Relation of half-viability time to temperature for *Anomodon viticulosus* (———●———), *Tortula intermedia* (– – – – O – – – –) and *Racomitrium lanuginosum* (—·—·▲—·—·). The small circles and dotted line are from data presented by Roberts (1975) for viability of rice in storage at a water content of 12% (wet weight). The data are presented as an 'Arrhenius plot'; points for a given species follow a straight line if the relation of loss of viability to temperature obeys Arrhenius's equation. Re-drawn from data of Hearnshaw and Proctor (1982).

wetting that is important; their ultimate desiccation tolerance is probably rarely, if ever, tested by the environment. For some at least, too-constant moisture is probably harmful, as it is for some lichens (Farrar, 1976a).

10.7 RESPONSES TO LIGHT, TEMPERATURE AND CARBON DIOXIDE: CARBON BALANCE

The physiological responses of photosynthesis in bryophytes to light, temperature and carbon dioxide concentration are essentially the same as in flowering plants. The effects of these three factors interact, so that a single response curve for one factor is in effect a two-dimensional slice of a four-dimensional response surface. Under field conditions, the carbon dioxide concentration in the air does not vary greatly and is usually around 320 parts per million by volume (v.p.m); we can then consider the effects of light and temperature only, as illustrated in Fig. 10.15.

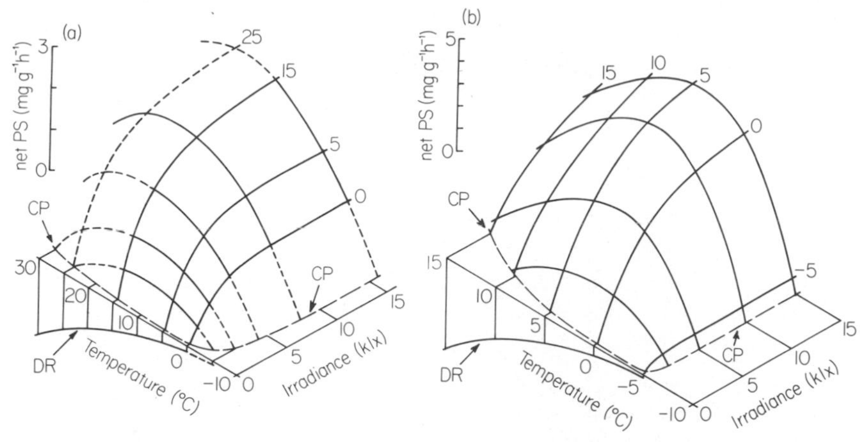

Fig. 10.15 Response surfaces of net photosynthesis to irradiance and temperature at normal atmospheric CO_2 content (c. 320 v.p.m.) for (a) *Hylocomium splendens*, based on the data of Stålfelt (1937), and (b) *Racomitrium lanuginosum* (S. Finland) based on the data of Kallio and Heinonen (1975). Smooth curves have been drawn through the experimental points or interpolated where appropriate; broken lines are extrapolated beyond the range of the data. The dark respiration curve (DR) is shown at zero irradiance. Irradiance curves intersect the line marked CP at the light compensation point for that temperature; temperature curves intersect the same line at upper and lower compensation points characteristic for a particular irradiance.

Photosynthesis consists of two sets of processes. The *light reactions* use the energy of absorbed light to generate the energy-rich compound adenosine triphosphate (ATP) and, by the splitting of water with the release of oxygen, reducing power as the reduced form of the coenzyme nicotinamide adenine dinucleotide phosphate ($NADPH_2$); the rate of the light reactions is limited by irradiance, and is little affected by temperature. The *dark reactions* utilize the ATP and $NADPH_2$ to fix carbon dioxide and convert the products to sugars and other metabolites; the dark reactions are temperature-dependent in the usual way of enzyme-controlled biochemical processes. At low temperatures, reaction-rate increases with temperature until an optimum is reached, beyond which the rate declines again. If we are considering net photosynthesis (and this is what is ecologically relevant) the effect of temperature on respiration must also be taken into account. Respiration reaches an optimum at a substantially higher temperature than photosynthesis, so that its effect is to depress the temperature optimum of net photosynthesis below that for gross photosynthesis. The greater the proportion that respiration is of gross photosynthesis, the greater the effect will be.

At low irradiance, the rate of photosynthesis is limited by the amount of light absorbed. At high irradiance, it is limited by the rate of the dark reactions; increasing irradiance then makes no difference to the rate of photosynthesis and the plant is said to be *light-saturated*. Evidently, light-saturation will in general be reached at a low irradiance if temperature is low and the dark reactions are proceeding slowly, but at progressively higher irradiances as temperature rises. The temperature optimum for net photosynthesis (Table 10.10) will be highest under saturating irradiance, falling to progressively lower temperatures as light becomes limiting.

Most bryophytes that have been investigated reach light-saturation at about 10–20 klx (or roughly 50–100 W m^{-2} PhAR or a photon flux between 400 and 700 nm of 200–400 $\mu mol\ m^{-2}\ s^{-1}$) at normal temperatures. This is much the same as for normal C_3 vascular plants. Higher saturation levels have been reported for a few species of exposed habitats (e.g. *Ulota crispa* (Miyata and Hosokawa, 1961), *Polytrichum juniperinum* (Bazzaz *et al.*, 1970), *Polytrichum alpinum* and *Calliergon sarmentosum* (Oechel and Collins, 1976), *Sphagnum capillifolium* (Grace and Marks, 1978) and S. *fuscum* (Silvola and Heikkinen, 1979). Mache and Loiseaux (1973) found that *Marchantia polymorpha* reached saturation at 2–3 klx, and Priddle (1980b) found that a species of *Drepanocladus* from an Antarctic lake reached saturation at about 3 W m^{-2} (total; about 300 lx) at 2°C. Similarly low levels would probably be found for other bryophytes growing in deep shade, particularly at low temperatures.

It can be seen from Fig. 10.15 that the light compensation point – the irradiance at which net photosynthesis is zero – depends on temperature. Thus, Seidel (1976) found compensation points for a range of common woodland species from 160–620 lx at 5°C, but 740–1930 lx at 25°C. Bryophytes of more open situations generally have higher compensation points; values of 1–2 klx have been found for *Homalothecium sericeum* (Romose, 1940), 2.7 klx and 3.8 klx respectively for *Polytrichum alpinum* and *Calliergon sarmentosum* (Oechel and Collins, 1976), and 3–6 klx at 15°C (but only 200–400 lx at −5°C) for *Dicranum elongatum* and *Racomitrium lanuginosum* (Kallio and Heinonen, 1975). Bryophytes growing in deep shade (e.g. in woods or in deep water) may have much lower compensation points. Priddle (1980b) found compensation points at 0.11 and 0.64 W m^{-2} respectively (total irradiance; c. 10 and 60 lx) for deep-water Antarctic material of an unidentified *Drepanocladus* species and *Calliergon sarmentosum* at 2°C; at 20°C the compensation point of the latter was c. 220 lx (2.2 W m^{-2}).

Generally, measurable net photosynthesis takes place at 0°C (in the temperate species that have been studied), and the lower temperature compensation point is reached between −5°C and −10°C. This is reasonably in line with what would be expected from the degree of water-stress produced by freezing. Upper temperature compensation points generally lie between 30°C and 40°C at saturating irradiance.

Table 10.10 Temperature optima for net photosynthesis of bryophytes. All measurements were made at normal atmospheric CO_2 concentration, and most at saturating irradiances (or nearly so); the optima quoted by Seidel (1976) are at an irradiance (5 klx) substantially below saturation but appropriate to conditions in shady woodland. *Limits of seasonal variation or acclimatization to preceding conditions.

Species and origin	Optimum temperature (°C)	Author
Sphagnum girgensohnii (Sweden)	18	Stålfelt, 1937
S. magellanicum (Germany)	18	Rudolph, 1968
S. fuscum (Finland)	15	Silvola and Heikkinen, 1979
S. capillifolium (N. England)	> 20	Grace and Marks, 1978
Polytrichum formosum (S. Germany)	10	Seidel, 1976
P. juniperinum (U.S.A.)	10	Bazzaz *et al.*, 1970
P. alpestre (Signy I., Antarctica)	5–15*	Collins, 1977
P. alpinum (Alaska)	10–12	Oechel and Collins, 1976
Atrichum undulatum (S. Germany)	15	Seidel, 1976
Dicranum elongatum (Finland)	5	Kallio and Heinonen, 1975
D. elongatum (Finland)	10	Kallio and Kärenlampi, 1975
D. elongatum (Alaska)	13–21*	Oechel, 1976
D. fuscescens (Alaska)	12–19*	Oechel, 1976
D. fuscescens (Quebec)	c. 5–15	Hicklenton and Oechel, 1976
D. scoparium (S. Germany)	12	Seidel, 1976
D. majus (S.W. England)	20	M.C.F. Proctor (unpublished)
Leucobryum glaucum (S. Germany)	12–15	Seidel, 1976
Racomitrium lanuginosum (see text)	c. 5	Kallio and Heinonen, 1973
Tetraphis pellucida (U.S.A.)	c. 18–21	Forman, 1964 (growth measurements)
Plagiomnium cuspidatum (S. Germany)	12	Seidel, 1976
Fontinalis antipyretica	8–20*	Harder, 1925
Anomodon viticulosus (S.W. England)	c. 18	M.C.F. Proctor (unpublished)
Drepanocladus uncinatus (Signy I., Antarctica)	15	Collins, 1977
Calliergon sarmentosum (Alaska)	11–19*	Oechel, 1976
Homalothecium sericeum (Denmark)	10	Romose, 1940
Eurhynchium striatum (S. Germany)	15	Seidel, 1976
Orthothecium rufescens (N. England)	c. 22	M.C.F. Proctor (unpublished
Pleurozium schreberi (Sweden)	18–19	Stålfelt, 1937
P. schreberi (S. Finland)	15	Kallio and Kärenlampi, 1975
P. schreberi (Kevo, N. Finland)	5	Kallio and Kärenlampi, 1975
Rhytidiadelphus squarrosus (Sweden)	18–19	Stålfelt, 1937
R. triquetrus (Sweden)	15–16	Stålfelt, 1937
R. loreus (S. Germany)	12–15	Seidel, 1976
Hylocomium splendens (Sweden)	18–19	Stålfelt, 1937
H. splendens (S. Germany)	12–15	Seidel, 1976
Herbertus aduncus (W. Ireland)	c. 20	M.C.F. Proctor (unpublished
Plagiochila asplenioides (S. Germany)	15	Seidel, 1976
P. spinulosa (S.W. England)	c. 18	M.C.F. Proctor (unpublished
Scapania gracilis (S.W. England)	c. 18	M.C.F. Proctor (unpublished
Porella platyphylla (S.W. England)	c. 19	M.C.F. Proctor (unpublished

Response curves for net photosynthesis at high carbon dioxide concentrations show higher optimum temperatures than the corresponding curves at normal atmospheric carbon dioxide concentrations. These curves, like those for uptake of $^{14}CO_2$ are probably of more physiological than ecological interest, though their optima and other features may show useful correlations with distribution and behaviour in the field. Carbon dioxide concentration interacts with temperature in its effect on net photosynthesis in a similar way to light (Fig. 10.16). At low concentrations, photosynthesis is steeply dependent on carbon dioxide concentration; the concentration at which photosynthesis balances respiration so that net photosynthesis is zero, is the *CO_2 compensation point* (Γ). This varies with temperature, but at normal temperatures is typically about 50–100 v.p.m. (Dilks, 1976), rising (obviously) to 320 v.p.m. at the upper temperature compensation point. In general values of Γ for terrestrial bryophytes are of little direct ecological interest, but they provide one of several pieces of evidence that bryophytes fix carbon dioxide by the C_3 pathway normal for most vascular plants (see also Rundel *et al.*, 1979). Diffusion resistances of terrestrial bryophytes are probably rarely, if ever high enough to influence photosynthesis (see Nobel, 1977), though a few mosses, notably Leucobryaceae and such sphagna as *S. magellanicum* and *S. compactum,* look peculiarly ill-adapted to uptake of carbon dioxide from the air; it seems possible that these draw up much of

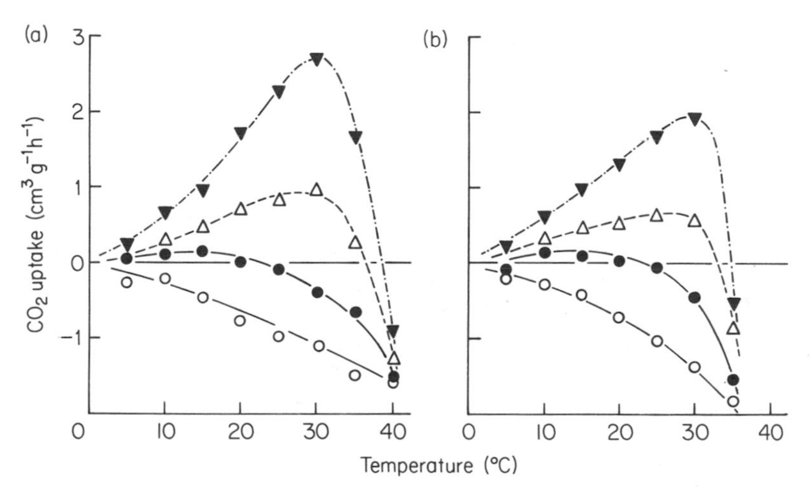

Fig. 10.16 Response curves of net photosynthesis to temperature at different carbon dioxide concentrations; irradiance (photon flux) c. 250 μmol.m^{-2} s^{-1}.
(a) *Hylocomium splendens,* (b) *Porella platyphylla.* CO_2 concentration c. 300 v.p.m., ——●——; 2000 v.p.m., ———△———; 10000 v.p.m., —·—▲—·—; dark respiration, ———O———. From unpublished manometric measurements by T. J. K. Dilks.

their carbon dioxide in solution like any other nutrient. The carbon dioxide relations of aquatic bryophytes are considered in the next section.

Lethal limits for moist bryophytes are generally between 40°C and 50°C. Ewart (1896) found limiting temperatures of 40–45°C for *Dicranum scoparium* and *Bryum caespiticium,* Dircksen (quoted in Nörr, 1974b) found 41–43°C for *Plagiochila asplenioides,* 41–48°C for *Preissia quadrata,* 43–45°C for *Mnium hornum* and 45–47°C for *Dicranum scoparium,* and Biebl (1967) found values between 42°C and 50°C for two liverworts and three mosses from tropical forest in Ceylon. Nörr (1974b) found the lethal limit generally between 42°C and 47°C, with seasonal changes of up to about 3°C; *Racomitrium lanuginosum* stood out from the other species with a limit rising to 51°C in summer. Antropova (1971) reported a reversible temperature-hardening effect of similar magnitude.

Generally, a single physiological measurement or response curve can only reflect in a very inadequate way the habitat adaptations of a bryophyte. Thus, light compensation points will probably correlate with habitat but are of little direct ecological significance in themselves; what matters is whether the plant can attain a positive carbon balance over 24 hours. Hosokawa and Odani (1957) found good correlation between habitat and daily compensation period – the time needed to recoup the carbon dioxide lost by respiration during the night – in a Japanese beech forest. Similarly, temperature responses need to be seen in the context of microclimatic variation from point to point (pp. 348–351) and the daily march of irradiance and temperature.

Ideally, for modelling the behaviour of a bryophyte under varying environmental conditions, or for comparing the responses of two species, a knowledge is needed of the whole response surface and of the nature and extent of seasonal variation and acclimitization to preceding environmental conditions (Hicklenton and Oechel, 1976; Oechel, 1976). Very few published sets of data even approach this ideal, and for many purposes much less will suffice. In comparing isolated response curves it is important to bear in mind the values of the other factor(s). Temperature-response curves obtained at atmospheric carbon dioxide concentration and saturating irradiance should be reasonably comparable; so should light-response curves at or near the optimum temperature.

Some promising results have been achieved in predicting net photosynthesis or growth under field conditions from measured environmental factors, using models based on response curves from laboratory measurements. Thus, *Polytrichum alpestre* showed reasonable agreement between carbon fixation computed from physiological measurements of light and temperature response and actual values of dry-matter production on Signy Island, Antarctica (Collins, 1977); water content was probably non-limiting throughout. In a Canadian population of *Dicranum fuscescens,*

Hicklenton and Oechel (1977) found that temperature accounted for most of the day-to-day variation in net photosynthesis, but variation from hour to hour was controlled by both temperature and light (see also Kellomäki *et al.*, 1978). Miller *et al.* (1978) demonstrated excellent prediction of the detailed course of net photosynthesis of *Calliergon sarmentosum, Polytrichum alpinum, Dicranum angustatum* and *D. elongatum* at Barrow, Alaska, from hourly measurements of irradiance, air temperature and water vapour density, windspeed, surface temperature and daily precipitation. *C. sarmentosum* showed the highest rate of photosynthesis at high soil water potential, but the lowest at low water potential; correspondingly this species and *Dicranum angustatum* photosynthesized most rapidly early in the season (late June), while *Polytrichum alpinum* reached a peak in mid-July and *Dicranum elongatum* in late July and early August. For bryophytes in many situations, availability of water is the most important single factor influencing carbon balance. Busby *et al.* (1978) found growth of *Homalothecium nitens* closely correlated with precipitation ($r = 0.92$), less so with mean daily irradiance ($r = 0.84$) and still less so with mean temperature ($r = 0.70$); *Hylocomium splendens* showed practically no correlation of growth with either mean temperature or mean amount of precipitation, but a very good correlation ($r = 0.93$) with the number of days the moss was wet.

On the whole, attempts to relate geographical distribution limits of bryophytes to physiological temperature responses have been rather disappointing. Thus, temperature response curves of *Racomitrium lanuginosum* from sites in Spitzbergen, Finland, Austria, Wales, Ireland and the sub-Antarctic island of South Georgia are remarkably similar (Kallio and Heinonen, 1973), as are the responses of growth to temperature of polar, temperate and tropical populations of *Bryum argenteum* (Longton, 1979). T.J.K. Dilks and M.C.F. Proctor investigated the temperature responses of a number of liverworts of the 'mixed hepatic mat' community of the western Scottish Highlands (Ratcliffe, 1968) and lowland hepatics growing in comparable situations, but found only a few differences that could even tentatively be related to altitudinal or geographical distribution limits. Probably, temperature often operates indirectly on bryophytes through its effects on evaporation and water balance.

10.8 AQUATIC BRYOPHYTES

Because of the high heat capacity and thermal conductivity of water, aquatic bryophytes live in a more constant and predictable temperature environment than terrestrial species. On the other hand, while the solubility of carbon dioxide in water is near unity, the rate of gaseous diffusion is slower in water than in air by a factor of about 10^4 (Table

Table 10.11 Some physical properties of air and water at various temperatures: kinematic viscosity (ν) and diffusion coefficient of carbon dioxide (D_C) for air and water, and solubility of carbon dioxide in water.

Temperature (°C)	Air		Water		Solubility of CO_2 in water (v/v)
	(ν/m^2 s$^{-1}\times 10^{-5}$)	(D_C/m^2 s$^{-1}\times 10^{-5}$)	(ν/m^2 s$^{-1}\times 10^{-6}$)	(D_C/m^2 s$^{-1}\times 10^{-9}$)	
0	1.33	1.29	1.79		1.713
5	1.37	1.33	1.52		
10	1.42	1.38	1.31	1.46	1.194
15	1.46	1.42	1.14		1.019
20	1.51	1.47	1.00	1.77	0.878
25	1.55	1.51	0.89	2.00	0.759
30	1.60	1.56	0.80		0.665

(Compiled from various sources.)

10.11). This means that the external boundary-layer resistance is likely to be a major component of the total resistance in the diffusion path of carbon dioxide to the photosynthesizing cells (cf. Nobel, 1977).

Reynolds number (p. 345) depends on kinematic viscosity (ν); the value for air is about 10–15 times that for water at normal temperatures (Table 10.11), so in terms of boundary-layer geometry a flow-rate of 10 cm s^{-1} in water is equivalent to 100–150 cm s^{-1} in air.

The velocity of a stream can be measured using a propeller-type current meter, or by timing a float between two marks; small pieces of dog biscuit make good floats for this purpose, and are biodegradable. The average velocity is given roughly by a current meter reading at 0.6 of the depth of the stream. Average stream velocity can be estimated approximately from the empirical relationship of Manning's formula (Simons, 1971).

$$V = (K/n)R^{2/3}S^{1/2}, \tag{10.9}$$

where V = average velocity of a stream (m s^{-1}),
K = a constant depending on units used, about 1.0 in SI units,
R = hydraulic radius (area of cross-section divided by wetted perimeter), approximately equal to depth in channels shallow in proportion to width,
S = slope of channel (m m^{-1}),
n = 'Manning's roughness coefficient', a parameter depending on the character of the channel, for many natural channels between about 0.03 (moderately smooth, clear channels) and 0.05 (channels with bouldery beds and/or substantial growth of vegetation).

This underlines that there is often little difference in mean velocity between stony mountain streams and lowland rivers (both commonly in the range from a few dm s^{-1} to rather over 1 m s^{-1}), but turbulence and the velocity gradient close to the stream bed are often much greater in mountain streams, giving the appearance of faster flow. Small streams in flat lowland country may flow at only a few cm s^{-1}, and effective flow rates close to surfaces that can support bryophytes in large lowland rivers may be equally low. A stream that normally flows at much over 1 m s^{-1} in summer is likely to be too fast for anything beyond sparse bryophyte growth.

All aquatic bryophytes that have been investigated require free carbon dioxide as an inorganic carbon source for photosynthesis, and have carbon dioxide compensation points between about 40 and 160 v.p.m. (1.5–6 μmol l^{-1}; Ruttner, 1947; Steeman-Nielsen, 1947; Bain and Proctor, 1980). This is in contrast to many aquatic vascular plants and algae of nutrient-rich waters, which are capable of using bicarbonate, and can deplete the carbon dioxide concentration to 1 v.p.m. or less raising the pH to between 10 and 11 in doing so. Thus, in general, bryophytes cannot compete with vascular plants or algae in highly productive lakes. Their métier is in situations where lack of other nutrients limits growth (probably the situation in some calcareous dune-slacks or the shallow calcareous lakes of the Burren, Co. Clare, Ireland, and in many acid oligotrophic pools), and in well-aerated fast-flowing streams. River and stream ecosystems typically depend largely on terrestrial detritus for their source of energy and carbon. The carbon dioxide concentration in river water is often above equilibrium with air, and carbon dioxide derived from breakdown of organic matter may often be important for aquatic bryophyte growth.

From the limited available evidence, it appears that at air-equilibrium carbon dioxide concentrations diffusion becomes limiting below about 1

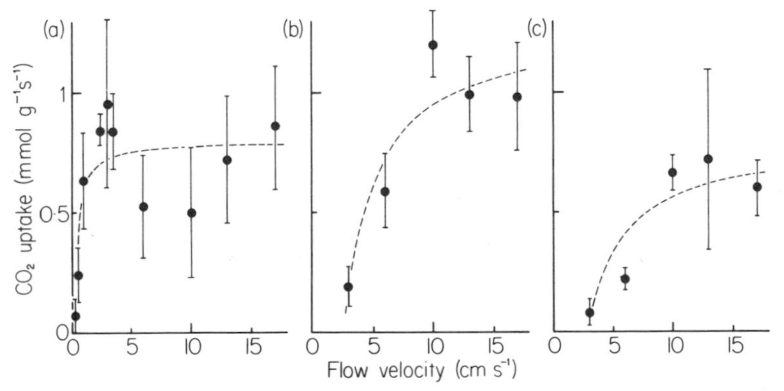

Fig. 10.17 Relation between rate of flow of water and photosynthetic CO_2 uptake in (a) *Fontinalis antipyretica,* (b) *Nardia compressa,* and (c) *Scapania undulata.* The first grows (and was exposed in the experiment) as long trailing stems, the last two as short turfs or mats covering boulders. From unpublished flow-tank measurements by Janet T. Bain.

cm s^{-1} for the trailing shoots of *Fontinalis antipyretica*, and below about 10 cm s^{-1} for the dense mats of *Scapania undulata* and *Nardia compressa* (Fig. 10.17). Calculation suggests that free convection becomes significant at flow rates below a few mm s^{-1} if irradiance is substantial, as in shallow still pools. In deep lakes, photosynthesis will generally be light-limited (e.g. Bodin and Nauwerck, 1968; Priddle, 1980a), and the carbon dioxide concentration will be enhanced by microbial breakdown of organic matter. Even in apparently still water there will generally be some movement due to turbulent mixing brought about by the wind.

10.9 HETEROTROPHIC NUTRITION

In most of this chapter it has been assumed implicitly that the carbon balance of a bryophyte depends on the relation of photosynthesis and respiration. However, *Cryptothallus mirabilis* is well known as a saprophytic liverwort, and *Aneura pinguis* and several of the smaller *Riccardia* species have similar orchid-type endotrophic mycorrhizas (Harley, 1969). These could be contributing to carbon nutrition, but this has not been experimentally tested. *Buxbaumia* species are also commonly regarded as saprophytic. An ability to utilize external organic carbon sources has been demonstrated for a number of bryophytes. Thus, *Funaria hygrometrica, Riccia fluitans* and various Splachnaceae will grow in the dark given glucose or suitable amino-acids as carbon sources, but *Pellia endiviifolia, Climacium dendroides* and *Calliergon stramineum* will only grow in the light (Keilová-Klecková, 1959). Simola (1969) found that *Sphagnum capillifolium* did not grow in culture without an organic carbon source, but that it would grow with or without light in the presence of sucrose, glucose or mannose. Organic nutrient requirements may be rather widespread among bryophytes, and could be ecologically important.

REFERENCES

Abel, W.O. (1956), *Sber. öst. Akad. Wiss. math.-naturw. Kl.* Abt. 1, **165,** 619–707.
Anderson, M.C. (1964), *J. Ecol.*, **52,** 27–41.
Anderson, M.C. (1971), Radiation and crop structure. In: *Plant Photosynthetic Production: Manual of Methods* (Šesták, Z., Čatský, J. and Jarvis, P.G., eds), pp. 412–66. Junk, The Hague.
Antropova, T.A. (1971), *Bot. Zhurn.*, **56,** 1681–6.
Bain, J.T. and Proctor, M.C.F. (1980), *New Phytol.*, **86,** 393–400.
Bayfield, N.G. (1973), *J. Bryol.*, **7,** 607–17.
Bazzaz, F.A., Paolillo, D.J. and Jaegels, R.H. (1970), *Bryologist,* **73,** 579–85.
Becquerel, P. (1951), *C. R. hebd. Séanc. Acad. Sci., Paris,* **232,** 22.
Bewley, J.D. (1973a), *Can. J. Bot.*, **51,** 203–6.
Bewley, J.D. (1973b), *Plant Sci. Letters,* **1,** 303–8.

Biebl, R. (1967), *Flora, Jena,* **157,** 25–30.
Billings, W.D. and Anderson, L.E. (1966), *Bryologist,* **69,** 76–95.
Blaikley, N. (1932), *Ann. Bot.,* **46,** 289–300.
Bliss, C.I. (1967), *Statistics in Biology,* Vol. 1. McGraw-Hill, New York.
Bodin, K. and Nauwerck, A. (1968), *Schweiz. Z. Hydrobiol.,* **30,** 318–52.
Bopp, M. and Stehle, E. (1957), *Z. Bot.,* **45,** 161–74.
Bopp, M. and Weniger, H.-P. (1971), *Z. Pflanzenphysiol.,* **64,** 190–8.
Bowen, E.J. (1931), *Ann. Bot.,* **45,** 175–200.
Bowen, E.J. (1933), *Ann. Bot.,* **47,** 401–22; 635–61; 889–912.
Brown, D.H. and Buck, G.W. (1979), *New Phytol.,* **82,** 115–25.
Buch, H. (1945, 1947), *Commentat. biol.,* **9(16),** 1–44; **9(20),** 1–61.
Busby, J.R., Bliss, L.C. and Hamilton, C.D. (1978), *Ecol. Monogr.,* **48,** 95–110.
Busby, J.R. and Whitfield, D.W.A. (1978), *Can. J. Bot.,* **56,** 1551–8.
Castaldo, R., Ligrone, R. and Gambardella, R. (1979), *Rev. bryol. lichénol.,* **45,** 345–60.
Clausen, E. (1952), *Dansk. bot. Ark.,* **15,** 1–80.
Clee, D.A. (1937), *Ann. Bot.,* **1,** 325–8.
Clee, D.A. (1939), *Ann. Bot.,* **3,** 105–11.
Clee, D.A. (1943), *Ann. Bot.,* **7,** 185–93.
Collins, N.J. (1977), The growth of mosses in two contrasting communities in the maritime Antarctic: measurement and prediction of net annual production. In: *Adaptations within Antarctic Ecosystems* (Llano, G.A., ed.), pp. 921–33. Smithsonian Institution, Washington.
Coufalová, E. (1957), *Biológia,* **12,** 255–65.
Cowan, D.A., Green, T.G.A. and Wilson, A.T. (1979), *New Phytol.,* **82,** 489–503.
Cowley, J.P. (1978), *Met. Mag.,* **107,** 357–73.
Day, G.J. (1961), *Met. Mag.,* **90,** 269–83.
Dhindsa, R.S. and Bewley, J.D. (1977), *Plant Physiol.,* **59,** 295–300.
Dilks, T.J.K. (1976), *J. exp. Bot.,* **27,** 98–104.
Dilks, T.J.K. and Proctor, M.C.F. (1974), *J. Bryol.,* **8,** 97–115.
Dilks, T.J.K. and Proctor, M.C.F. (1975), *J. Bryol.,* **8,** 317–36.
Dilks, T.J.K. and Proctor, M.C.F. (1976a), *J. Bryol.,* **9,** 239–47.
Dilks, T.J.K. and Proctor, M.C.F. (1976b), *J. Bryol.,* **9,** 249–64.
Dilks, T.J.K. and Proctor, M.C.F. (1979), *New Phytol.,* **82,** 97–114.
Eickmeier, W.G. (1979), *Oecologia,* **39,** 93–106.
Ellenberg, H. (1978), *Vegetation Mitteleuropas mit den Alpen in ökologischer Sicht,* 2nd edn. Ulmer, Stuttgart.
Ensgraber, A. (1954), *Flora, Jena,* **141,** 432–75.
Ewart, A.J. (1896), *J. Linn. Soc. Bot.,* **31,** 344–461.
Farrar, J.F. (1976a), *New Phytol.,* **77,** 93–103.
Farrar, J.F. (1976b), *New Phytol.,* **77,** 105–13.
Farrar, J.F. and Smith, D.C. (1976), *New Phytol,* **77,** 115–25.
Forman, R.T.T. (1964), *Ecol. Monogr.,* **34,** 1–25.
Gates, D.M. (1968), *A. Rev. Pl. Physiol.,* **19,** 211–38.
Gates, D.M. (1980), *Biophysical Ecology,* Springer, Berlin.
Gates, D.M. and Papian, L.E. (1971), *Atlas of Energy Budgets of Plant Leaves,* Academic Press, London and New York.

Gimingham, C.H. and Birse, E.M. (1957), *J. Ecol.*, **45**, 533–45.
Gimingham, C.H. and Smith, R.I.L. (1971), *Br. Ant. Surv. Bull.*, **25**, 1–21.
Grace, J. and Marks, T.C. (1978), Physiological aspects of bog production at Moor House. In: *Production Ecology of British Moors and Mountain Grasslands* (Heal, O.W. and Perkins, D.F., eds), pp. 38–51. Springer, Berlin.
Green, T.G.A. and Clayton-Greene, K. (1977), *Bull, Br. Bryol. Soc.*, **29**, 5–6.
Grubb, P.J. (1970), *New Phytol.*, **69**, 303–26.
Gupta, R.K. (1976), *Biochem. Physiol. Pfl.*, **170**, 389–95.
Gupta, R.K. (1977a), *Can. J. Bot.*, **55**, 1196–00.
Gupta, R.K. (1977b), *Can. J. Bot.*, **55**, 1195–200.
Gupta, R.K. (1978), *Ind. J. exp. Biol.*, **16**, 350–3.
Haberlandt, G. (1886), *Jb. wiss. Bot.*, **17**, 359–498.
Harder, R. (1925), *Jb. wiss. Bot.*, **64**, 169–200.
Harding, R.J. (1979), *J. appl. Ecol.*, **16**, 161–70.
Harley, J.L. (1969), *The Biology of Mycorrhiza*, 2nd edn., Leonard Hill, London.
Hearnshaw, G.F. and Proctor, M.C.F. (1982), *New Phytol.*, **90**, 221–8.
Hébant, C. (1977), *The Conducting Tissues of Bryophytes*, Cramer, Vaduz.
Hicklenton, P.R. and Oechel, W.C. (1976), *Can. J. Bot.*, **54**, 1104–19.
Hicklenton, P.R. and Oechel, W.C. (1977), *Arct. Alp. Res.*, **9**, 407–19.
Hinshiri, H.M. and Proctor, M.C.F. (1971), *New Phytol.*, **70**, 527–38.
Hoffman, G.R. and Gates, D.M. (1970), *Bull. Torrey bot. Cl.*, **97**, 361–6.
Höfler, K. (1946), *Anzeiger der Akad. Wiss. Wien math.-naturw. Kl.*, **1945(3)**, 5–9.
Höfler, K. (1950), *Ber. dt. bot. Ges.*, **68**, 3–10.
Hosokawa, T. and Kubota, H. (1957), *J. Ecol.*, **45**, 579–91.
Hosokawa, T. and Odani, N. (1957), *J. Ecol.*, **45**, 901–15.
Isaac, I. (1941), *Ann. Bot.*, **5**, 339–51.
Johnson, A. and Kokila, P. (1970), *Bryologist*, **73**, 682–6.
Kallio, P. and Heinonen, S. (1973), *Reports from the Kevo Subarctic Research Station*, **10**, 43–54.
Kallio, P. and Heinonen, S. (1975), CO_2 exchange and growth of *Rhacomitrium lanuginosum* and *Dicranum elongatum*. In: *Fennoscandian Tundra Ecosystems* (Wiegolaski, F.E., ed.), pp. 138–48. Springer, Berlin.
Kallio, P. and Kärenlampi, L. (1975), Photosynthetic activity of multicellular lower plants (mosses and lichens). In: *Photosynthesis and productivity in different environments* (Cooper, J.P., ed.), pp. 393–423. Cambridge University Press.
Kappen, L., Lange, O.L., Schulze, E.-D., Buschbom, U. and Evenari, M. (1980), *Flora, Jena*, **169**, 216–29.
Keever, C. (1957), *Ecology*, **38**, 422–9.
Keilová-Klečková, V. (1959), *Preslia*, **31**, 166–78.
Kellomäki, S., Hari, P. and Koponen, T. (1978), *Bryophytorum Bibliotheca*, **13**, 485–507.
Klepper, B. (1963), *Bryologist*, **66**, 41–54.
Kramer, P.J. (1969), *Plant and Soil Water Relationships: a Modern Synthesis*, McGraw-Hill, New York.
Krochko, J.E., Bewley, J.D. and Pacey, J. (1978), *J. exp. Bot.*, **29**, 905–17.
Lange, O.L. (1955), *Flora, Jena*, **142**, 381–99.
Lange, O.L. (1969a), *Planta, Berl.*, **89**, 90–4.

Lange, O.L. (1969b), *Flora, Jena,* **B158,** 324–59.
Lange, O.L., Schulze, E.-D. and Koch, W. (1970), *Flora, Jena,* **159,** 38–62.
Laüchli, A. (1976), Apoplasmic transport in tissues. In: *Encyclopaedia of Plant Physiology. N.S.2. Transport in Plants. IIB.* (Lüttge, U. and Pitman, M.G., eds), pp. 3–34. Springer, Berlin.
Lee, J.A. and Stewart, G.R. (1971), *New Phytol.,* **70,** 1061–8.
Leyton, L. (1975), *Fluid Behaviour in Biological Systems,* Clarendon Press, Oxford.
Longton, R.E. (1979), Climatic adaptation of bryophytes in relation to systematics. In: *Bryophyte Systematics* (Clarke, G.C.S. and Duckett, J.G., eds), pp. 511–31. Academic Press, London.
Longton, R.E. (1981), Physiological ecology of mosses. In: *Mosses of North America* (Taylor, R.J. and Leviton, A.E., eds), pp. 77–113. Pacific Division, American Academy for the Advancement of Science, Washington.
Lowry, W.P. (1969), *Weather and Life: an introduction to bio-meteorology,* Academic Press, New York.
Mache, R. and Loiseaux, S. (1973), *J. Cell. Sci.,* **12,** 391–401.
Mägdefrau, K. (1935), *Z. Bot.,* **29,** 337–75.
Mägdefrau, K. (1969), *Vegetatio,* **16,** 285–97.
Mägdefrau, K. and Wutz, A. (1951), *Forstwiss. Zbl.,* **70,** 103–17.
Mazur, P. (1969), *A. Rev. Pl. Physiol.,* **20,** 419–48.
McConaha, M. (1939), *Am. J. Bot.,* **26,** 353–5.
McConaha, M. (1941), *Am. J. Bot.,* **28,** 301–6.
Meteorological Office (1964), *Hygrometric Tables. III 2nd edn. Aspirated psychrometric readings, degrees Celsius.* Met.0.265c. H.M.S.O., London.
Meteorological Office (1976), *Averages of bright sunshine for the United Kingdom,* 1941–1970. Met.0.844. H.M.S.O., London.
Meusel, H. (1935), *Nova Acta Leopoldina,* N.F. **3,** 123–277.
Milburn, J.A. (1979), *Water Flow in Plants,* Longman, London.
Miller, P.C., Oechel, W.C., Stoner, W.A. and Sveinbjörnsen, B. (1978), *Photosynthetica,* **12,** 7–20.
Miyata, I. and Hosokawa, T. (1961), *Ecology,* **47,** 766–75.
Monteith, J.L. (1965), Evaporation and environment. In: *The State and Movement of Water in Living Organisms* (Fogg, G.E., ed.), pp. 205–34, Symposia of the Society for Experimental Biology, No. 19. Cambridge.
Monteith, J.L. (1973), *Principles of Environmental Physics,* Arnold, London.
Monteith, J.L. (1975, 1976), *Vegetation and the Atmosphere,* Academic Press, London.
Monteith, J.L. (1981), The coupling of plants to the atmosphere. In: *Plants and their Atmospheric Environment* (Grace, J., Ford, E.D. and Jarvis, P.G., eds), pp. 1–29, 21st Symposium of the British Ecological Society. Blackwell, Oxford.
Morton, M.R. (1977), *Ecological Studies of Grassland Bryophytes,* Ph.D. thesis, University of London.
Nobel, P.S. (1974), *Introduction to Biophysical Plant Physiology,* Freeman, San Francisco.
Nobel, P.S. (1977), *Physiologia Pl.,* **40,** 137–44.

Nörr, M. (1974a), *Flora, Jena,* **163,** 371–87.
Nörr, M. (1974b), *Flora, Jena,* **163,** 388–97.
Ochi, H. (1952), *Bot. Mag., Tokyo,* **65,** 10–12.
Ochi, H. and Yonehara, S. (1954), *Bot. Mag., Tokyo,* **67,** 265–70.
Odu, E.A. (1978), *J. Bryol.,* **10,** 163–81.
Oechel, W.G. (1976), *Photosynthetica,* **10,** 447–56.
Oechel, W.C. and Collins, N.J. (1976), *Can. J. Bot.,* **54,** 1355–69.
Oke, T.R. (1978), *Boundary Layer Climates,* Methuen, London.
Page, J.K. (1978), *Methods for the Estimation of Solar Energy on Vertical and Inclined Surfaces.* Publ. BS 76, Department of Building Science, University of Sheffield.
Patterson, P.M. (1946), *Am. J. Bot.,* **33,** 604–11.
Pihakaski, K. and Pihakaski, S. (1979), *Ann. Bot.,* **43,** 773–81.
Plantefol, L. (1927), *Ann. Sci. Nat. (Bot.), Sér. 10,* **9,** 1–263.
Pócs, T. (1980), *Acta Bot. Acad. Hung.,* **26,** 143–67.
Priddle, J. (1980a), *J. Ecol.,* **68,** 141–53.
Priddle, J. (1980b), *J. Ecol.,* **68,** 155–66.
Proctor, M.C.F. (1972), *J. Bryol.,* **7,** 181–6.
Proctor, M.C.F. (1979a), Structure and eco-physiological adaptation in bryophytes. In: *Bryophyte Systematics* (Clarke, G.C.S. and Duckett, J.G., eds, pp. 479–509. Academic Press, London.
Proctor, M.C.F (1979b), *J. Bryol.,* **10,** 531–8.
Proctor, M.C.F. (1980a), Diffusion resistances in bryophytes. In: *Plants and their Atmospheric Environment,* (Grace, J., Ford, E.D. and Jarvis, P.G., eds), pp. 219–29, 21st Symposium of the British Ecological Society. Blackwell, Oxford.
Proctor, M.C.F. (1980b), *J. Bryol.,* **11,** 351–66.
Proctor, M.C.F. (1981), Physiological ecology of bryophytes. In: *Advances in Bryology,* (Schultze-Motel, W., ed.). Cramer.
Ratcliffe, D.A. (1968), *New Phytol.,* **67,** 365–439.
Raven, J.A. (1977), *Adv. bot. Res.,* **5,** 153–219.
Renner, O. (1932), *Planta, Berl.,* **18,** 215.
Ried, A. (1960), *Biol. Zbl.,* **79,** 657–78.
Ried, A. (1969), *Ber. dt. bot. Ges.,* **82,** 127–41.
Rifot, M. and Barrière, G. (1974), *Rev. bryol. lichénol.,* **40,** 45–52.
Roberts, E.H. (1975), Problems of long-term storage of seed and pollen for genetic resources conservation. In: *Crop Genetic Resources for Today and Tomorrow* (Frankel, O.H. and Hawkes, J.G., eds), pp. 269–96. Cambridge.
Roberts, E.H. and Abdalla, F.H. (1968), *Ann. Bot.,* **32,** 97–117.
Romose, V. (1940), *Dansk. bot. Ark.,* **10,** 1–134.
Rudolph, H. (1968), *Planta, Berl.,* **79,** 35–43.
Rundel, P.W., Stichler, W., Zander, R.H. and Ziegler, H. (1979), *Oecologia,* **44,** 91–94.
Ruttner, F. (1947), *Öst. bot. Z.,* **94,** 265–94.
Schönherr, J. and Ziegler, H. (1976), *Planta, Berl.,* **124,** 51–60.
Schwabe, W. and Nachmony-Bascomb, S. (1963), *J. exp. Bot.,* **14,** 353–78.
Seidel, D. (1976), *Flora, Jena,* **165,** 163–96.

Silvola, J. and Heikkinen, S. (1979), *Oecologia,* **37,** 273–83.
Simola, L.K. (1969), *Physiologia Pl.,* **33,** 1079–84.
Simons, D.B. (1971), Open channel flow. In: *Introduction to physical hydrology,* (Chorley, R.J., ed.), pp. 131–52. Methuen, London.
Slatyer, R.O. (1967), *Plant-Water Relationships,* Academic Press, London.
Slavik, B. (1965), The influence of decreasing hydration level on photosynthetic rate in the thalli of the hepatic *Conocephalum conicum.* In: *Water Stress in Plants* (Slavik, B., ed.), pp. 195–201. Junk, The Hague.
Smith, D.C. and Molesworth, S. (1973), *New Phytol.,* **72,** 525–33.
Smithsonian Institution (1951), *Smithsonian Meteorological Tables,* 6th edn. Washington.
Stålfelt, M.G. (1937), *Planta, Berl.,* **27,** 30–60.
Steeman-Nielsen, E. (1947), *Dansk bot. Ark.,* **12(8),** 5–71.
Tallis, J. (1959), *J. Ecol.,* **47,** 325–50.
Tobiessen, P.L., Slack, N.G. and Mott, K.A. (1979), *Can. J. Bot.,* **57,** 1994–8.
Trachtenberg, S. and Zamski, E. (1979), *New Phytol.,* **83,** 49–52.
Troughton, J.H. and Sampson, F.B. (1973), *Plants: a scanning electron microscope study.* Wiley, Sydney.
Will-Richter, G. (1949), *Sber. öst. Akad. Wiss. math.-naturw. Kl.* Abt. I **158,** 431–542.
Willis, A.J. (1964), *Trans. Br. Bryol. Soc.,* **4,** 668–83.
Zacherl, H. (1956), *Z. Bot.,* **44,** 409–36.
Zehr, D.R. (1977), *Bryologist,* **80,** 571–83.

Chapter 11

Mineral Nutrition

D.H. BROWN

11.1 INTRODUCTION

Research into the effects of physical factors such as temperature, light intensity and relative humidity on bryophyte growth has achieved much success in explaining the distribution and habitat preferences of certain species (see Chapter 10). The dominance shown by particular physical factors, especially those impinging on the plant's water economy, has tended to relegate work on chemical factors to a subordinate role. In the past, many workers have either assumed that the nutrient supply has been adequate and therefore of minor importance or have entirely failed to consider the possible relevance of the plant's mineral nutritional requirements. Such attitudes may partly account for Richards' (1959) comment that 'The mineral economy of bryophytes is a subject on which so little is known that a connected discussion is hardly possible'. Subsequent research has somewhat rectified this situation and improved analytical techniques have undoubtedly contributed to this.

In recent years there has been increasing interest in the use of bryophytes, and also lichens, in monitoring the aerial distribution of heavy metals, gaseous pollutants and the radio-isotopic products of nuclear explosions as well as in biogeochemical prospecting (see, for example, Brooks, 1971; LeBlanc and Rao, 1974; Rao *et al.*, 1977; Shacklette, 1965a and Chapter 12). While much of this work has dealt with elements of no known nutritional value, it has greatly increased our knowledge of the responses of bryophytes to mineral elements, their sources of supply and the factors influencing availability and uptake. Naturally, caution has to be shown in extrapolating from 'heavy metals' (see Nieboer and Richardson, 1980 for a discussion of the use of this term) to more nutritionally useful elements and dangers exist in generalizing from observations on a limited range of species and situations.

Three approaches to the study of bryophyte mineral nutrition will be considered here: (a) plant chemical analyses as a measure of requirement and availability, (b) responses to controlled nutrient supply in the laboratory and the field and (c) observations on certain, particularly mineral-rich, habitats. The coverage of the literature is intended to be illustrative and critical rather than wholly comprehensive. An extended treatment is provided because bryophyte mineral nutrition has previously only been reviewed as part of more wide-ranging considerations of physiology or ecology (see for example Barkman, 1958; Kenworthy, 1978; Longton, 1980; Puri, 1973; Richardson, 1981; Streeter, 1970; Watson, 1971).

11.2 CHEMICAL ANALYSIS OF ELEMENTS IN BRYOPHYTES

As a measure of element availability to bryophytes, analysis of the plant itself would appear to be highly instructive. A number of surveys have been made of a relatively limited number of bryophyte species with at least 68 elements tested for, although not all have been above the detection limits of the techniques used (e.g. Erämetsä and Yliruokanen, 1971a, b; Hancock and Brassard, 1974; Lounemaa, 1956; Rastorfer, 1972, 1974; Sainsbury *et al.*, 1966; Shacklette, 1965a, 1967). Such total analyses of element contents of bryophytes have also been implied to provide data on their nutrient requirements. How far these assumptions are correct requires consideration of the source of the chemical and its final location within the plant and its cells.

It is frequently suggested that the atmosphere, rather than the substratum, is the major, if not the sole, source of all mineral elements for bryophytes. This is particularly true of work associated with pleurocarpous species, work involved in monitoring aerial pollution patterns or work concerned with plants growing on or under trees. These situations will be considered in more detail later. Absorption of minerals over the whole surface of the gametophyte is aided by (a), the large surface area to volume ratio and (b), low surface resistance to ion uptake from solution, due to limited cutin development. Because of the large number of exposed cation exchange sites in bryophytes, compared to higher plants, (Clymo, 1963; Knight *et al.*, 1961) many workers have attributed uptake and retention of cations, particularly heavy metals, exclusively to such physical binding processes. Whether these generalizations are equally valid for all situations is the subject of this section.

11.2.1 Element location studies

The introduction of techniques such as atomic absorption

spectrophotometry and X-ray fluorescence has meant that total analyses of bryophytes may be readily performed. For nutritional studies, however, it is important to know what proportion and amount of a particular element in a plant is present within the cytoplasm of the cell and therefore directly capable of influencing metabolism.

Elements associated with bryophytes may be located in five situations as follows: (1) dissolved in the solution bathing the exterior of the plant and within the matrix of the cell wall, (2) as particles of relatively refractory material trapped by the plant leaves, branches, rhizoids etc., (3) ions bound to exchange or chelating sites located external to and including the cell's plasma-membrane, (4) soluble within the limits of the cell's plasma-membrane in the cytoplasm and vacuoles and (5) insoluble material within the same location as (4). Uptake to the first three sites is the result of passive physico-chemical processes while the latter two reflect incorporation into the cell through a selectively permeable membrane by biological process. Incorporation of material into these fractions will be considered in roughly this order.

(a) Passive uptake of particles and solutions without binding

Much use has been made recently of mosses in the monitoring of element dispersal patterns around emission sources by analysis of material either sampled directly from the field (e.g. Andersen *et al.*, 1978; Barclay-Estrup and Rinne, 1978; Burkitt *et al.*, 1972; Ellison *et al.*, 1976; Huckabee, 1973; Lötschert *et al.*, 1975; Wallin, 1976), following an exposure period after transplanting from similar, uncontaminated habitats (e.g. Goodman and Roberts, 1971; Pilegaard, 1979), or the use of 'moss bags' (e.g. Cameron and Nickless, 1977; Goodman and Roberts, 1971; Little and Martin, 1974; Mäkinen, 1977; Ratcliffe, 1975). Many regional surveys have attributed enrichment by heavy metals to generalized industrial and urban pollution (e.g. Glooschenko and Capobianco, 1978; Gorham and Tilton, 1978; Groet, 1976; Grodzińska, 1978; Pakarinen and Tolonen, 1976; Rinne and Barclay-Estrup, 1980; Rühling and Tyler, 1971; Solberg and Selmer-Olsen, 1978).

Survey work of the above kind originally emphasized the high cation exchange capacities of bryophytes as an explanation for the high trapping efficiencies observed. Later work showed that the majority of the emissions studied were primarily particulate in nature and mosses were therefore morphologically, rather than chemically, suitable for this kind of investigation. The high water-holding capacity of suspended moss bags reduces the chance of trapped particles being washed off except after periods of very heavy rainfall.

The technique of particle trapping by bryophytes can only be considered to be semi-quantitative as trapping depends on the contribution made, in dry deposition, by gravitational sedimentation, inertial impaction or eddy diffusion deposition, whose efficiency is determined by the particle size and wind velocity as well as the morphology, surface properties and hydration of the accepting surface (see Little, 1979; Roberts, 1975). Clough (1975) found a decreasing trapping efficiency from suspended moss bags to moss carpets to grass sward in wind tunnel experiments. Micro-environmental differences accounted for much of the difference in trapping efficiency between grass and moss bags. It is, therefore, unfortunate that comparisons of trapping efficiencies between bryophytes and lichens have sometimes compared material from different habitats or different parts of the same host (e.g. Folkeson, 1979; Groet, 1976; Steinnes, 1977 and see p. 396). Other studies have used material collected from very similar situations (Grodzińska, 1978; Pilegaard *et al.*, 1979). With artificial rain containing radio-isotope enriched particles Taylor and Witherspoon (1972) showed that bryophytes with closed sward characteristics *(Dicranum)* (see also Svensson and Lidén, 1965) retained more particles than did more open lichen swards *(Cladonia)* in which particle penetration occurred to a greater depth.

Where maximal values for deposition are required, washing before analysis is inappropriate. Many particles will be removed by washing but the proportion has rarely been quantified. Woollon (1975) attributed the 50% loss of Ca from *Fissidens cristatus* by water washing to removal of soil particles from the boat-shaped clasping leaves. Washing may also partially solubilize particulate matter and in the field or laboratory when the solution is acid (i.e. due to acid rain contaminated with sulphur or nitrogen oxide emissions) solubilization may be enhanced. Whether solubilization will result in extra uptake of ions or their release will be determined by a number of processes (see p. 388).

Analyses intended to demonstrate binding and incorporation of elements must exclude particulate matter and material in solution on the exterior of the plant (situations 1 & 2). Failure to wash maritime bryophytes will, for example, severely over estimate their Na and Mg contents (Bates, 1975a). A number of workers have emphasized that washing may cause loss of soluble ions from within the cells, especially if the material is senescent or has been damaged by environmental stress (e.g. Brown and Buck, 1979; Shacklette, 1965a; Streeter, 1965; Tamm, 1953). Brown and Buck (1979) studied the effect of desiccation stress and concluded that species from arid habitats lose very little intracellular material after desiccation but that species from wet habitats may become highly leaky after a similar stress. Gupta (1976, 1977) reported similar losses of organic molecules and showed release to be a non-selective process. Storage at high relative humidities after desiccation stress decreased or eliminated ion leakage from many, but not all, desiccation-sensitive bryophytes (Brown and Buck, 1979) and lichens (Buck and Brown, 1979) but may induce some leakage from desiccation-resistant species.

Shacklette (1965a) suggested that dry-ashed bryophyte samples which had ash weights greater than 10% of the original dry weight should be rejected as liable to be contaminated by soil particles. Such judgements cannot be used for example where species actively accumulate $CaCO_3$ in the form of tufa (Glenn, 1931) or in hepatics which have high, and biologically significant, levels of silica associated with leaf surface sculpturing (Duckett and Soni, 1972). Heavy metal-rich particles may only slightly alter the ash weight but contribute appreciably to the content of these elements. In addition to particles derived from industrial or automobile emissions, soil particles, especially at times when crop fields are being prepared, may contaminate inadequately washed samples (Gorham and Tilton, 1978). In one of the few reports in which microscopical observations were used to check for contamination, Willis (1964) found sand grains in preparations of extensively mechanically cleansed *Tortula ruralis* ssp. *ruraliformis* from sand dunes.

Brown and Buck (1978a) used various extractants on both soil and *Brachythecium rutabulum* and measurements of the Ca : Sr ratio to support the suggestion that Sr in this species was not due to soil contamination in an Sr-rich region. Lee *et al.* (1977) recalculated chromium values in *Aerobryopsis longissima* growing on *Homalium guillainii* when ash weights exceeded 10% on the assumption that this represented wind-blown soil contamination. The validity of such an approach appears highly suspect. Element enrichment in plant ash compared to soil concentrations may indicate some biological enrichment process but selective leaching of elements from soil particles during pedogenesis may create false enrichments.

Mitchell (1960) suggested that soil contamination may be assessed by the titanium content of the sample because most soils, except limestone, are rich in this element, it is barely soluble in the soil and plants do not appear to actively accumulate it. The high Ti concentrations in Shacklette's (1965a) data (average Ti = 297 p.p.m. in bryophytes, 15 p.p.m. in vascular plants) indicated that the bryophytes were significantly contaminated with soil particles and this is supported by high values for other barely soluble elements found in soils (e.g. Al, Be, Fe, Si and Zr). Nieboer *et al.* (1978) showed that the Fe : Ti ratio in lichens, over a wide range of concentrations was relatively constant (7.0 ± 0.2), closely resembled the proportion in rock (weighted average for the Canadian Shield = 7.2) and the Ti, being only extracted from the plant by drastic treatments, was probably in the form of oxides. A similar relationship has been reported for *Dicranum* (Richardson, 1981) but with a different Fe : Ti ratio (9.7), perhaps due to a combination of sampling near to mining and milling activities and greater uptake of soluble Fe (in proportion to the amount supplied in the particles) than with the lichens. Higher total Fe and Ti values were reported for *Dicranum,* again reflecting the greater retention of particles by a closed moss sward compared to the more open lichen carpet. Reasonable agreement was shown between the observed Cr, V and Ni levels in the lichen and those predicted on the basis of their ratios to Ti in the parent rock, while biologically active elements (e.g. K, S and Zn) did not. While this approach is a valuable method of indicating contamination by soil-derived elements in unwashed samples it should not be taken to mean, without other evidence, that all such elements are still in the original particles.

.Scanning electron microscopy and electron microprobe studies (Garty *et al.*, 1979; Lawrey, 1977) have shown heavy metal-rich particles between the cells within the thalli of lichens. In *Rhytidiadelphus squarrosus* exposed to lead-rich automobile exhaust, electron-dense, lead-containing regions have been demonstrated, within the plasma-membrane, in vesicles, vacuoles, chloroplasts and nuclei as well as in the cell wall (Gullvåg *et al.*, 1974; Ophus and Gullvåg, 1974; Skaar *et al.*, 1973). No intracellular lead was found after exposure to lead salts in the laboratory (Gullvåg *et al.*, 1974). The authors concluded that pinocytotic incorporation of particles occurred but it is also possible that secondary deposition of lead on polyphosphate bodies, known to be present in bryophyte organelles (Keck and Stich, 1957; Tanaka and Sato, 1977), may have occurred as Ophus and Gullvåg (1974) reported both elements to be the major constituents of the particles they observed. Only the cell wall was found to contain electron-dense regions when field-polluted material of *Hylocomium splendens* was studied (Gullvåg, *et al.*, 1974). The authors concluded that the thicker cell wall in this species acted as a barrier to the entry of lead, but whether this was in the form of soluble or particulate lead was not considered.

(b) Binding of elements by exchange and chelation

Information on the high, but not unique, cation exchange capacities (CEC) of bryophytes is of extreme importance in understanding their mineral nutrition and the value and relevance of chemical analyses. It is not appropriate here to give detailed discussion of the physical chemistry of ion-binding to such sites but rather to highlight those features of major relevance to ion uptake by bryophytes. Useful discussions will be found in Clymo (1963), Nieboer and Richardson (1980), Nieboer *et al.* (1978) and Rühling and Tyler (1970).

The cation exchange capacity of living bryophyte cells has been shown to be highly correlated, at least in *Sphagnum* (Clymo, 1963; Spearing, 1972), with the presence of polygalacturonic acid, or related molecules (Craigie and Maass, 1966; Painter and Sørensen, 1978; Schwarzmaier and Brehm, 1975) present within the cell wall. As lichens are known to have similar, but lower CEC's (Clymo, 1963; Knight *et al.*, 1961), but have not been conclusively shown to contain polyuronic acid molecules (Brown, 1976; Tuominen and Jaakkola, 1974), it is likely that other negatively-charged molecules may contribute to the total CEC of bryophytes. Burton and Peterson (1979a) showed 33% of cell wall-bound Zn could be released by incubation with pronase and suggested protein binding. Bryophytes also show anion exchange properties, but the capacity may be a hundred-fold less than the CEC (Clymo, 1963). Proteins are potentially suitable molecules for both cation and anion binding and these are present within the cell wall and plasma-membrane; the exterior of the latter may act in physically binding ions irrespective of its ability to selectively transport ions and molecules into the cell interior. Rupture of the plasma-membrane permits ions access to previously unexposed, mainly proteinaceous, binding sites which may account for the enhanced cation retention observed by some workers with dead compared to live plants (Buck and Brown, 1978; Pickering and Puia, 1969). With *Sphagnum* the major part of the dry weight of living plants is represented by dead hyaline cells which, coupled with their high water- and particle-holding capacity, accounts for their suitability and frequent use in pollution monitoring. It must be remembered that all soluble, passively retained ions are not held exclusively by ion exchange but that complex and chelate formation may also be involved (Nieboer *et al.*, 1978).

Whether a particular ion becomes bound to an exchange site depends on a number of factors related to its relative affinity for the acceptor group compared with other competing ions also present in the solution and to the nature of the ions already occupying the acceptor site. Thus carboxyl groups of uronic acid molecules will be fully protonated and un-ionized at low pH values. Acidifying a bryophyte will release ions due to suppression of ionization and to their replacement by protons on the carboxyl groups. Conversely, supplying mineral cations to a partially or fully protonated plant will result in their uptake and a decrease in the pH of the bathing solution due to the exchange of cations for protons. Mineral cations with high affinity for the exchange sites will displace other mineral cations depending on their relative affinity for the site and their relative concentrations. As protons compete for acceptor sites with mineral cations the pH of the bathing solution may critically effect the uptake of the latter. However, at neutral or alkaline pHs, as frequently experienced in nature, proton competition may be slight.

It has been found experimentally that divalent cations have higher affinities for exchange sites compared to monovalent ions and that, generally, ions with higher atomic weight displace lighter elements. Thus, Rühling and Tyler (1970) established, experimentally, the following sorption and retention series: Cu, Pb $>$ Ni $>$ Co $>$ Zn, Mn for *Hylocomium splendens* and this has been repeatedly observed by other workers. Such an affinity sequence is obtained, irrespective of the anion used, with solutions containing single elements or mixtures, but in the latter case the quantities bound will vary depending on the ratio of the cations and their affinities for the acceptor sites. Although cation binding in bryophytes is dominated by uronic acid acceptors the nature of other acceptor molecules may also be important because different ligands have different affinities for metal atoms. Thus, class A metals (e.g. Na, K, Ca, Mg) have preference for metal binding donor atoms in ligands in the order $O > N > S$ while class B metals (e.g. Ag, Au, Hg) have the reverse order. Divalent borderline metal ions have increasingly class B preferences in the order $Mn < Zn < Ni < Fe \simeq Co < Cd < Cu < Pb$ (Nieboer and Richardson, 1980).

It is extremely important to realize that, excluding particulate systems, any soluble ion presented to a bryophyte cell, whether in the laboratory or in the field as rain or soil solution, will firstly establish an equilibrium with the binding sites on the cell exterior. When an equilibrium has been established, and this is a very rapid physical process, the remaining atoms are available for uptake into the cell (Pickering and Puia, 1969). Equally, it must be remembered that any alteration in the concentration or composition of the bathing solution must establish a new equilibrium position with material on the cell-wall exchange sites. Hence, any cation bound to the cell wall is liable to be displaced and the quantity present in this extracellular location is a reflection of the recent past history of the plant and not, as has been sometimes suggested, a representation of the ions acquired by the plant during its whole existence. A continual build-up of elements is only likely to occur with the trapping of particles or the incorporation of material into the interior of the cell. Extracellular binding of ions is therefore a purely physico-chemical process with any selection of ions reflecting passive exchange and not the result of the immediate metabolic activity of the cell, although past activity, in the sense of metabolically controlled synthesis of cell wall constituents, is involved.

Brehm (1968) showed that the apparent CEC increased with a decrease in the weight of *Sphagnum* in a constant volume of salt solution or when the volume of the solution decreased and the weight of *Sphagnum* remained constant. Brown and House (1978), working with small and variable weights of *Solenostoma crenulatum,* used the expression 'apparent copper concentration' to represent the effective concentration of copper in solution, this being the actual copper concentration divided by the dry weight of plant material used.

(c) Analysis of the cellular location of cations

Brehm and Brown and co-workers have attempted to establish the proportions of different cations which are bound to extracellular cation exchange sites and are also present within the plasma membrane of the cell. Brehm (1968, 1970) used dilute (N/100) mineral aids to displace cations from exchange sites and then added formaldehyde to rupture cells and release soluble ions, finally displacing the remaining cations with normal acid. Brown and co-workers used a technique of sequentially eluting bryophytes as follows: discharge of most cations from extracellular exchange sites being achieved by treatment with 1000 p.p.m. Sr (Bates and Brown, 1974) or Ni (Brown and Buck, 1978a, b; 1979), release of soluble ions by boiling and recovery of residual material by total digestion in concentrated HNO_3. This latter, total digest, fraction has been shown to contain material bound within the cell, but may also include ions which were formerly soluble within the cell but which, due to their high affinity for newly exposed sites, become bound during the process of destroying the cell membranes by boiling (Brown and Buck, 1979).

Using these techniques, both groups generally found Na and K in solution within the cell, while Ca was mainly in an extracellular exchangeable form. Elements such as Mg and Zn show intermediate patterns, the proportion present in each fraction being variable and dependent on the presumed cellular requirements for the element and the environmental supply. Alcoholic extraction and differential centrifugation of aqueous extracts of a range of aquatic bryophytes were used by Burton and Peterson (1979a) to show that no more than 20% of th Zn was soluble. Similar conclusions were reached by Pickering and Puia (1969) who studied uptake of radioactive Zn by *Fontinalis antipyretica*. Although McLean and Jones (1975) considered Zn uptake in *Scapania undulata* and *Fontinalis squamosa* was active, saturating at 10 p.p.m. Zn, it is more probable that this was due to saturation of extracellular binding sites and that additional uptake above 25–30 p.p.m. Zn was due to rupture of cell membranes exposing intracellular exchange sites. Brown and Bates (1972) considered that Pb was exclusively bound to the cell walls of *Grimmia donniana* growing on Pb slag. Buck and Brown (1978) pointed out that the technique of acid displacement of Pb used by Brown and Bates (1972), ruptured cell membranes so that intracellular Pb could have been displaced. Their own data, using Ni to displace Pb from exchange sites, failed to fully release all endogenous or supplied Pb. Subsequently, it was appreciated that a concentration greater than the 1000 p.p.m. Ni used was needed to remove an element such as Pb which had a very high affinity for the exchange sites (Brown, 1980; Buck, 1980). It is probable that the reported failure of Rühling and Tyler (1970) to displace one third of the Cu supplied to *Hylocomium splendens* by adding $MgCl_2$ was due to the greater affinity for the exchange site shown by the Cu relative to the Mg at the concentration used. Brown and House (1978) found that 1000 p.p.m. Ni only removed 67% of native Cu from *Solenostoma crenulatum* grown on Cu mine debris, whereas a similar concentration of Pb removed 93% of the Cu.

(d) The significance of extracellular cation binding

It is possible that the cell wall exchange sites may act as a buffer against the presentation of excessive concentrations of toxic elements to the plasma membrane surface by binding incoming ions. Thus, the toxicity of certain media to *Physcomitrella patens* protoplasts (Stumm *et al.*, 1975) may be due to the absence of a cell wall to reduce the concentration of certain elements to tolerable levels. Species from Ca-rich habitats, due to the abundance of Ca in the environment, have the bulk of this element bound in an exchangeable form. As Bates (1982) has shown a greater Ca-binding capacity in such species compared to those from Ca poor habitats, it is possible that this is a method of decreasing the concentrations of this element to an acceptable level at the plasma membrane. Brown and Bates (1972) suggested that the similar Pb-binding efficiency of *Grimmia donniana* obtained from habitats with high or low Pb levels indicated a general ability to withstand this element. Briggs (1972), however, showed that growth of *Marchantia polymorpha* gemmae obtained from habitats contaminated with Pb from automobile exhaust were more tolerant to Pb in culture than was material obtained from unpolluted sites. Similarly, Brown and House (1978) found that, while samples of *Solenostoma crenulatum* taken from a copper waste tip and a lead mining region were indistinguishable in their exchangeable Cu-binding capacity over a wide range of values, membrane integrity and photosynthetic reactions of the former material were substantially less damaged by high applied Cu concentrations. This suggested that Cu tolerance may only have evolved in the Cu mine population and that this did not involve a change in the Cu-binding capacity of the plant as has been reported for certain Cu-tolerant grasses (see Antonovics *et al.*, 1971).

Brehm (1971) reported that living *Sphagnum* cells appeared to be able to maintain their intracellular cation concentrations within narrow limits while major changes in the ion concentration of adhering water and the exchange sites was occurring. Fluctuations in exchangeable concentrations were greatest in Na and K and least with Ca in the field. Brehm (1970) and Skene (1915) have shown that *Sphagnum* treated with a nutrient solution and then transferred to distilled water survived and grew for a longer time than material pretreated with dilute acid to displace cations. Skene concluded from this that cations moved from the exchange sites into the new living cells. However, it is also possible that the filled exchange sites retained their cations and that ions lost from dying cells were released into the adhering transpiration stream and drawn towards the growing apex by evaporation (Brehm, 1971) avoiding binding to the exchange sites because of the ions already present. Exchange of individual atoms would occur but total release of cations is improbable unless adequate production of soluble anions occurred. Although Ramaut (1955a, b) reported production of soluble organic acids by *Sphagnum*, the quantities were insufficient to account for the low pH maintained around *Sphagnum* plants (Clymo, 1966). This low pH appears to be maintained by the continual production of polyuronic acid molecules and the release of protons by cation exchange. The latter process means that unlimited cycling of cations between living cells is unlikely, due to the need to divert sufficient cations to new exchange sites in order to maintain a suitable pH for growth.

There is no direct evidence with any bryophyte species to show that the exchange sites can act as cation reservoirs for the intracellular uptake of ions. Presentation of high divalent cation concentrations may displace monovalent ions and temporarily enhance their concentration in the intercellular solution from which intracellular ion uptake may occur. Brown and Buck (1979) have shown that rewetting after desiccation causes loss of soluble ions from desiccation-sensitive species. High concentrations of K may be released directly into the medium but soluble divalent cations, such as Mg, are frequently bound to the exchange sites and may therefore displace other ions already present (Brown and Buck, 1979). It should be noted, however, that there is no evidence that the latter can then be incorporated into the cell.

With natural rainfall, the solution arriving contains ions (see p. 410) and, as the volume presented would not cause so great a dilution of ions as occurs in laboratory experiments, a temporary increase in intercellular ion concentration can occur. The first rainfall, tending to wash ions from the atmosphere, would be richest in ions, as would the intercellular solution during the last stages of drying; both situations are potentially capable of causing ion release from exchange sites. Acidic rainwater, due to solution of oxides of sulphur and nitrogen, will cause cations to be displaced from exchange sites (Fairfax and Lepp, 1976; Lepp and Fairfax, 1976). Prolonged rain, with wash-off of solution, would deplete the plant's total cation content and would cause an altered mixture of ions to be presented to the cell in the intercellular solution. Both situations are potentially harmful to the cell and may prove to be toxic although changes in the environmental pH or uptake of SO_2 or NO_x may more directly alter metabolism and thereby induce cell death.

(e) Distribution of elements in different parts of bryophytes

Some attention has been paid to the distribution of elements along the length of gametophytes and therefore related to the different ages of the plant segments. It has generally been found that monovalent cations are concentrated towards the growing apex while the concentration of divalent elements increases, with segment age, towards the base (e.g. Bates, 1979; Brown and Buck, 1978a; Rühling and Tyler, 1970; Tamm, 1953). This pattern generally reflects the presence of monovalent elements in solution in the living cells and the continued passive binding of divalent cations to exchange sites where senescent or dead cells may have more of these exposed to cations in the bathing solution. The youngest segments frequently also contain most N and P (Dézsi and Simon, 1976; Pakarinen, 1978a, b; Tamm, 1953) as they are usually acquired as anions and are present intracellularly in forms associated with cell metabolism. Exceptions to this pattern occur; for example N may increase with age, possibly due to microbial growth (Allen *et al.*, 1967; Grubb *et al.*, 1969), and brown segments usually have high ash contents due to Fe and Al from soil particles (Gorham and Tilton, 1978). Higher heavy metal concentrations occur in stems than in leaves (Lötschert *et al.*, 1975). Washing causes greater loss of ions from older parts of the gametophyte (Smith, 1978; Woollon, 1975), but it is not clear whether this is the result of release of ions from cells or removal of contaminating particles.

Brown and Buck (1978a) showed that Sr : Ca and Sr : K ratios of different segments of *Brachythecium rutabulum,* growing in a Sr-rich area, decreased from the base to the growing tip. They suggested that this reflected the poorer upward movement of soil-derived Sr, due to its greater affinity for the available exchange sites, compared to the other elements analysed. The observation that analyses of growing tips of acrocarpous species showed higher Sr concentrations compared to pleurocarpous species from the same habitats, was explained (Brown and Buck, 1978a) by the shorter distance for Sr to move from the soil to the apex of acrocarpous species. Although Ca concentrations ($\mu g\ g^{-1}$ air dried moss) increased with increasing segment age in *Pleurozium schreberi,* Bates (1979) showed that the content per segment (μg per segment) initially increased and then subsequently decreased (see also Tamm (1953) for *Hylocomium splendens,* p. 406). The change in concentration may only reflect a change in the ratio of weight of cell wall material: weight of protoplast. The initial increase in Ca content and decrease in concentration is an indication that the developing apex had not attained its full ion content at the time of analysis.

In a study of cation changes during the development of sporophytes and gametophytes of *Funaria hygrometrica,* Brown and Buck (1978b) showed that the K content of the sporophyte increased rapidly during development, while the content of the less-mobile Ca rose more slowly. During gametophyte senescence, loss of K occurred but Ca and Mg contents increased, attributable to a transfer of K from the senescing gametophyte to the developing sporophyte and uptake of Ca and Mg to newly exposed exchange sites, together with enhanced water flow through the gametophyte to the expanding sporophyte. Such interpretations are applicable to many other situations where changes in cell physiology and movement of ion-containing solutions may occur. Czarnowska and Rejment-Grochowska (1974), using material from the field, reported *Atrichum undulatum* sporophytes contained a tenth of the Fe concentration of gametophytes while Mn, Zn and Cu were the same or slightly more concentrated in the gametophyte.

Although element concentrations of *Sphagnum* vary in relation to bog-water chemistry (Pakarinen and Tolonen, 1977a), Pakarinen has pointed out that differences in element concentrations in gametophytes, where aerial inputs are presumed to be comparable, can be affected by the volumetric density and rate of plant growth. Thus, comparison of the green parts of hummock- and hollow-inhabiting *Sphagnum* species showed significantly lower concentrations of extracellular Ca, Pb, Fe and Mn and higher concentrations of intracellular K, Mg and especially N and P in the faster growing hollow species (Pakarinen, 1978a) and in profiles of ombrotrophic *Sphagnum* peat (Pakarinen and Tolonen, 1977b). Similarly, the more productive *Sphagnum girgensohnii* and *Polytrichum commune* from wet microsites in *Picea abies* forests had lower concentrations of Pb, Fe and Mn than did *Dicranum majus* or the 'feather mosses' *Hylocomium splendens* and *Pleurozium schreberi,* although the accumulation values (amount retained in new phytomass per unit area of ground) were the same in *S. girgensohnii* and *Pleurozium* (Pakarinen and Rinne, 1979).

While growth rates strongly influence element accumulation, it must also be noted that CEC's are greater per unit weight in species of *Sphagnum* sect. *Acutifolia* inhabiting drier habitats than in species of the sect. *Cuspidata,* which inhabit relatively wet sites (Puustjärvi, 1959). Clymo (1963) and Spearing (1972) showed a correlation between CEC of *Sphagnum* species and the height of the optimum habitat above the water table, both increasing above the water table. Other data (Damman, 1978) show increases in Fe, Al, Pb and Zn in both hummock and hollow species in the zone of water level fluctuations. Some of this build-up may result from alternation of peat aeration from anaerobic and reducing when water-logged, with the formation of soluble sulphides of Fe and Al and insoluble sulphides of Pb and Zn, to aerobic, oxidizing conditions on drying, causing precipitation of the Fe and Al and the partial solubilizations of Pb and Zn in the form of sulphates. Both processes will result in the accumulation of these elements at the water table but it should be realized that it has yet to be shown that such elements are in an exchangeable form. Damman (1978) also showed that the amount of N, Na, K and probably P contained in annual growth increment exceeded the input due to precipitation, while only a quarter of the Ca and half of the Mg in precipitation was retained. This again implies relocation of N, Na, K and P from decaying zones to the apex. In addition, a greater volume of solution will flow over hummock species due to transpiration, tending to draw ions released from dead parts of the plant toward the apex (Brehm, 1971) and this is reflected in the steeper gradient in Pb, Zn, Fe, Cu and Mn from green to brown tissues in hummock compared to hollow species (Pakarinen, 1978c). All of these factors combine to suggest that comparisons between species based on dry weight alone are of extremely limited value.

Studies of *Pleurozium schreberi* contaminated in the field by particles containing ^{137}Cs, derived from fall-out from nuclear weapon testing, showed that the mean residence time was c. 4 years compared to c. 10–20 years in the lichen *Cladonia stellaris* (Mattsson and Lidén, 1975), the difference being partly related to differences in growth rates. It was also observed that Cs was mainly present in the green part of the moss, even in periods of reduced Cs deposition, suggesting some Cs relocation from older parts of the plant, even though the Cs:K ratio increased with segment age. In *C. stellaris*, ^{137}Cs increased at the apex in dry conditions and declined during periods of rainfall (Mattsson, 1975a). With artificial ^{137}Cs additions to live and dead *Dicranum scoparium*, Witkamp and Frank (1967) were able to show that only 4–8% of the total uptake was by biological processes but that, probably due to loss of leaves etc., loss was greater from dead than live plants. Other, less readily soluble, radionucleides (e.g. ^{155}Eu and ^{125}Sb) are less mobile and in lichens with apical growth are retained in progressively lower segments corresponding to the annual growth increment of the year of initial deposition (Mattsson, 1975b).

Many of the above movements of ions probably occur in the surface film of water with absorption and deposition on exchange sites. Shimwell and Laurie (1972) suggested that the Pb-rich crystalline deposit observed on *Dicranella varia* carpets, growing on heavy metal rich mine waste, after dry weather was the result of an excretion process, implying internal conduction and release. It is more probable that evaporation of the Pb-rich surface water film could have caused such an accumulation. They also reported that the myxohydric *Philonotis fontana* have lower Zn and Pb concentrations, especially at the apex, attributable to uptake only being from the soil and not over the entire surface as in *Dicranella varia.* Salt crusts occurred on *Bryum antarcticum* from dry habitats but not on *Bryum argenteum* from wetter habitats (Schofield and Ahmadjian, 1972).

Grubb (1961, 1965) initially failed to show any selectivity in the passive transfer of nutrients through the gametophyte and to the sporophyte of *Polytrichum formosum,* suggesting passive movement in the transpiration stream passing through the hydrome. He later concluded that this was possibly due to damage to the rhizome during washing and/or the use of inappropriate salt mixtures (Grubb, 1968). He observed that rates of K, N and P redistribution were considered too fast for diffusion and might involve the symplastic pathway through the lumen of cells of the leptome. In *Polytrichum juniperinum,* Pb originally supplied as the EDTA complex but finally visualized by electron microscopy as a lead sulphide precipitate, can penetrate the lumen of the cytoplasm-free hydroids of the hydrome as well as being recovered in their cell walls and those of the leptome (Trachtenberg and Zamski, 1979; Zamski and Trachtenberg, 1976). When Pb was supplied as the nitrate, i.e. in an ionic not chelated form (Trachtenberg and Zamski, 1978), it was only recovered in the cytoplasm of the leptoids and in the cell walls of the leptoids and cortical tissue. Recovery of Pb in the hydroids was presumed to show movement of the intact Pb–EDTA complex and that in this form it could not enter living cells. The latter observation is in accord with the evidence (Coombes and Lepp, 1974) that a 500-fold greater concentration of the EDTA complexes of Cu and Pb were required to produce the same reduction in growth of *Marchantia* gemmae and *Funaria* spores as the free ions. Trachtenberg and Zamski (1978) reported little entry of lead nitrate into the rhizome, however the solution was presented, and showed with Pb–EDTA (Trachtenberg and Zamski, 1979) that absorption of water was more efficient through the aerial surfaces than through the rhizome.

Chevallier *et al.* (1977) found the gametophyte of *Funaria hygrometrica* acted as a partial barrier to the movement of added radioactive phosphate to the sporophyte. They observed a decline in the rate of P translocation with maturation of the sporophyte, co-incident with the stage of development at which leptoids degenerate (Schulz and Weincke, 1976). This suggests that initially some active transport in the symplastic, leptome, pathway occurs but that later stages permit only apoplastic movement in the hydroids. Although inorganic phosphate was supplied to the gametophyte, Brown and Buck (1978b) suggested that early transport in the leptome to the developing sporophyte may be partly in the form of organic phosphates because Chevallier *et al.* (1977) observed that when phosphate was supplied directly to the seta (without the possibility of conversion to organic compounds by the gametophyte) the amount transported to the capsule was greater and high rates of transport were sustained to later developmental stages, consistent with movement of inorganic phosphate in the hydroids.

(f) Uptake and location of heavy metals following aerial pollution

The form in which heavy metals are supplied to bryophytes remains unclear but emissions from industrial, urban and vehicular sources are probably particulate. Apart from direct trapping without further modification, dissolution can also occur from which uptake to exchange sites or into the cell may occur. As studies have generally involved healthy material, entry of heavy metals into the cell to toxic levels has been avoided.

Barclay-Estrup and Rinne (1978) considered that analyses of washed

Pleurozium schreberi and *Hylocomium splendens* showed contents of sorbed Pb and Zn. They showed higher levels of both elements in urban areas, with Pb > Zn in such regions and Zn > Pb in rural areas. Rinne and Barclay-Estrup (1980) found higher concentrations of heavy metals in organic soils under *Pleurozium* but attributed this to concentration of heavy metals present in decaying moss rather than as an indication of the route of heavy metal entry into the plant. The concentration of all metals, except Mn, increased with increasing tree cover as might be predicted if the source of heavy metals was the atmosphere (see p. 411). The apparent under-representation of Mn may reflect competition for exchange sites between soluble heavy metal cations, amongst which Mn has the lowest affinity for the available sites. The implication is that both Mn and some of the other heavy metals have interacted with the receiving moss plant in a soluble form, even though the initial emission was probably in a particulate form. Grodzińska (1978) reported a similar apparent inverse relationship between Mn and heavy metal contents in *Pleurozium* and *Hylocomium* following a large-scale regional study of Polish national parks.

In a regional study of heavy metal recovery in *Leucobryum* and other mosses, Groet (1976) found distribution patterns of (a), Cd and Zn and (b), Pb, Cu, Ni and Cr probably reflecting different emission sources. The pattern of Cd recovery was unlike the other elements and, as with Mn above, Groet (1976) suggested that this was due to Cd being a poor competitor for exchange sites, exacerbated by the low concentrations of Cd available for binding. A positive correlation was found between altitude and Cd and Zn concentration, suggesting that precipitation may be important both for removal of heavy metal particles from the atmosphere and also their solubilization. The negative relationship between Ni and altitude, and therefore precipitation (Groet, 1976) may indicate the involvement of dry deposition but may again reflect competition between soluble heavy metals for binding sites. When a range of bryophytes sampled in 1951 were compared with more recently collected samples Rasmussen (1977) observed a decrease in Ca, Mg, Cd and Mn and an increase in Cr, Cu, Fe, Ni, Pb, V and Zn for which replacement of the former group by the latter on available exchange sites appears a reasonable explanation.

Not all reports from heavy metal surveys show competition between certain elements. Ellison *et al.* (1976) showed a significant positive correlation between Mg and Mn in *Hypnum cupressiforme* near an iron works, but there was no negative correlation between either metal and any of the heavy metals studied. Johnsen and Rasmussen (1977) found *Pterogonium gracile* to be enriched fifty-fold with Mn, relative to the bulk precipitation, compared to a factor of about ten for other elements, and concluded that this was due to deposition of Mn-enriched soil derived dust.

They also observed a decrease in Mn and Fe values during the 30-year period studied and attributed this to the growth of *Acer* adjacent to the single beech tree studied, thereby shielding it from soil-derived dust. As proportions of heavy metals were similar both in the moss and in precipitation in the relatively unpolluted area studied, Johnsen and Rasmussen (1977) concluded that no competition for exchange sites had occurred, although it should be noted that absence of release from particulate deposits would achieve the same result. In a highly polluted site, Pilegaard (1979) showed heavy metal uptake was proportional to time of exposure and for individual elements there was a significant linear relationship between concentration in *Dicranoweisia cirrata* (and the lichen *Hypogymnia physodes*) and in the bulk precipitation, suggesting particle trapping only. Partial exchange uptake may have occurred because statistical analyses showed that the two species did not behave identically.

11.3 CULTURE EXPERIMENTS

Many valuable insights into the mineral nutrition of bryophytes have been obtained from controlled culture experiments in the laboratory under rigorously controlled conditions of light, temperature and, more rarely, water availability. Useful information is also obtained with experiments in which areas of natural vegetation are treated with more or less defined fertilizer mixtures. However, in all of these studies care must be taken to avoid unjustifiable extrapolations and to be aware of the limitations of the methods employed.

One problem with culture experiments is to decide what form of propagule may be involved in the natural situation (Black, 1974; Seltzer and Wistendahl, 1971; Studlar, 1980; Tallis, 1959). Few studies have compared the response of both spores and gametophytic tissue although fragmentation of gametophytes may be an important means of propagation (Miller and Ambrose, 1976; Odu, 1979). Egunyomi (1978) found germination and gametophyte production was different if cultures were initiated from spores or gemmae of *Octoblepharum albidum*. Although spores and gemmae represent convenient and apparently reproducible inocula, failure to control their stage of maturity and period of storage will influence their germination frequency and response to culture conditions (Berrie, 1975; Hoffman, 1970; Paolillo and Kass, 1973; Tenge, 1959). With the use of gametophyte fragments, hormonal control of regeneration means that the size of the fragment and its location on the original plant may apparently modify its ability to respond to nutritional conditions (e.g. see Wilmot-Dear, 1980). In all cases, a source of nutrients will be introduced with the propagule and may even be sufficient to avoid or delay the onset of deficiency symptoms (Chevallier, 1973, 1975a; Dietert, 1979).

Chalaud (1937) considered that spores of *Scapania subalpina* germinated in leaf folds of adult plants rather than directly on soil. The suggestion that many pleurocarpous species are normally unconnected to their substratum (e.g. Tamm, 1953; Watson, 1960) need not mean that they are independent of their substratum chemistry throughout their life and experiments must be made on the appropriate propagule to determine the influence of nutrients during establishment. The observation of changes in pH from the surface of a moss hummock to its substratum (e.g. Briggs, 1965; Clymo, 1963) means that, although once established the plant is able to modify its environment, initially it was subject to the soil chemical environment. Geldreich (1948b) reported that *Leucolejeunea* maintained the pH of agar media whereas the pH of uninoculated agar fell, due to water loss.

11.3.1 Laboratory experiments

Recipes for many of the culture media used in bryophyte studies have been compiled by Basile (1975) and Geldreich (1949). It is unfortunate that much of the early, detailed, work on such species as *Marchantia polymorpha* (Voth, 1943; Voth and Hamner, 1940), *Funaria hygrometrica* (Hoffman, 1966a, b) and *Leucolejeunea clypeata* (Fulford *et al.*, 1947) used the too complex technique of cation and anion 'triangles' in which the concentrations of more than one cation or anion was varied at any one time. The latter authors commented 'it has not been possible to ascribe specific reactions to the effects of specific combinations of the nutrients, for the variables are too numerous and the gaps in the proportions of the nutrients too great'. Nevertheless, they were able to find many changes in amount and form of growth and, with adequate culture periods, showed deficiencies of macro- and some micro-nutrients known to be essential for other plant groups. As might be expected, because *Funaria* and *Marchantia* grow in a wide range of habitats, adequate growth occurred in culture in a wide range of nutrient concentrations and compositions. A more specific example of deficiency is given by *Funaria* spores which contain low concentrations of Mn (Chevallier, *et al.,* 1969), which is depleted during germination (Chevallier, 1973). Although subsequent growth and the development of functional chloroplasts requires additional Mn (Chevallier, 1973, 1974, 1975a, b, 1976), no evidence that it is essential, in the sense that the *Funaria* life cycle cannot be completed in its absence, has yet been provided.

Improved growth of bryophytes at reduced nutrient concentrations (Buckholder, 1959; Dietert, 1979; Griggs and Ready, 1934; Southorn, 1977; Voth, 1943) has often been used to support the suggestion that they have low nutrient requirements and can therefore grow in habitats nutritionally too poor for more demanding plants (see p. 404). However, strong nutrient solutions may have inappropriate osmotic potentials. Cells of *Marchantia* were smaller in more concentrated media (Voth, 1943) and more reproductive structures were formed. *Funaria* protonemata were found to be more tolerant than gametophytes of high osmotic potentials (Hoffman, 1966b). Dilute solutions often stimulate the production of rhizoids and rhizoidal filaments (see Allsopp and Mitra, 1958; Kofler, 1959).

Water availability in culture experiments may be important, but growth in liquid culture has not been extensively compared with growth on more solid substrata. Agar is frequently added to solidify nutrient solutions but in deficiency experiments it must be realized that agar can contribute a range of nutrients, some in substantial concentration (Araki, 1953; Bridson and Brecker, 1970) and will also chelate or otherwise reduce the availability of certain heavy metals (Ramanovsky and Kushner, 1975). Coombes and Lepp (1974) showed chelating agents substantially reduced the toxicity of Cu to *Marchantia* and *Funaria*. Schneider *et al.* (1967) attributed differences in growth on glass cloth, vermiculite and perlite to differences in their ion exchange properties. Saxena and Rashid (1980) found that charcoal inhibited the transition from chloronema to caulonema and bud formation while Schofield and Ahmadjian (1972) found it suppressed protonemata but not shoot growth. Both groups of workers attributed the changes observed to removal of growth substances. Conversely, growth of aquatic species in wholly inorganic media may not adequately represent the field situation where organic matter may combine with many inorganic elements, especially heavy metals (Strumm and Morgan, 1970). Boatman and Lark (1971) germinated *Sphagnum* spp. on cellophane discs, but only observed growth when nutrient solutions were supplied, none occurring with natural oligotrophic waters. Boatman's (1978) demonstration that the cellophane discs used adsorbed ions suggests that natural water chemistry was modified to an unacceptable composition by the disc rather than demonstrating the possible failure of spores to germinate in nature as suggested earlier.

(a) Calcium

Calcium is an essential element with, amongst other functions, a role in maintaining membrane integrity (see p. 414) and in cellular adhesion. Fulford *et al.* (1947) reported that cell separation occurred in new growth of *Leucolejeunea* when maintained on a Ca-free medium. Geldreich (1948a) only observed rounding up and bulging of cells which he attributed to replacement of calcium pectate by the magnesium derivative, creating a less rigid binding between cells. Fulford *et al.* probably observed an extreme case of this phenomenon.

Although rhizoid development may be associated more with attachment (e.g. Odu, 1978) than with nutrient absorption, Nehira (1973) showed, but only for *Marchantia,* that Ca was required for their differentiation and accumulated at the rhizoid base. Chen and Jaffee (1979) suggested rhizoid formation from *Funaria* spores occurred at the point of maximum Ca entry. Iwasa (1965) observed growth of *Funaria* protonema in the apparent absence of Ca and found ammonium oxalate to inhibit bud formation while calcium oxalate promoted false bud formation. Use of the effective Ca-chelating agent EDTA, even in the presence of Ca, decreased both growth and bud formation indicating that Ca was required for both processes.

(b) Nitrogen

Nitrate is generally assumed to be the most readily available nitrogen source in neutral or alkaline soils and ammonium in acidic environments. Growth of cultured bryophytes is frequently similar on both nitrate and ammonium salts (e.g. Buckholder, 1959). Schuler *et al.* (1955) used ^{15}N-labelled ammonium ions, in the form of ammonium nitrate, to show equal uptake of both inorganic nitrogen compounds by *Sphaerocarpos texanus.* Ammonium ions were considered (Sironval, 1947) to cause degeneration of *Funaria* caulonema and morphological abnormalities were reported for *Funaria* (Southorn, 1977) and *Scapania* (Killian, 1923).

Failure to show satisfactory growth on ammonium salts can often be ascribed to the fall in pH which occurs during ammonium assimilation in unbuffered media. Machlis (1962) suggested that the small size of male plants of *Sphaerocarpos donnellii* found in the field may be related to the demonstration, in culture, of their readier adsorption of ammonium ions compared to female plants, resulting in a pH drop which may suppress growth. Growth on potassium showed no difference between the two sexes and no pH change. Whether the sex difference was due to differences in the utilization of ammonium ions was not demonstrated by Machlis.

Schwoerbel and Tillmanns (1974) showed ammonium ions were preferentially assimilated by *Fontinalis antipyretica* due to the repression of nitrate reductase by the presence of ammonium ions. Nitrate reductase was only formed in the light in *Fontinalis* (Schwoerbel and Tillmanns, 1974), which may account for the failure of Fries (1945) in obtaining only slight growth of *Leptobryum pyriforme* and none of *Funaria hygrometrica* on nitrate in the dark. However, Dietert (1979) showed nitrate to be the best nitrogen source for *Funaria* and *Weissia controversa* protonemal growth in the light. She also showed nitrate could support good growth of both species but ammonium ions were poor even in buffered media. Growth on nitrate requires its conversion to ammonium ions before assimilation. Light may be required to supply the necessary reductant because preliminary observations (Brown, unpublished) have shown that, although many bryophytes can reduce nitrate to nitrite in the dark, the conversion of nitrite to ammonium is either an obligate light reaction or is substantially stimulated by light. Lee and Stewart (1978) reported adequate activities of the enzymes, glutamine synthetase, ferredoxin-linked glutamate synthase and glutamate dehydrogenase, required for the further assimilation of ammonium ions into amino acids and Montague *et al.* (1969) described transaminase activity in *Sphaerocarpos texanus,* implying that the usual nitrogen assimulation pathways occur in bryophytes.

It is possible that growing apices of species with poor or non-existent access to inorganic nitrogen sources in the substratum, may acquire nitrogen by transfer of organic molecules from senescent or dead cells in other parts of the plants. The ability of amino acids, dipeptides and other organic nitrogenous compounds to support growth in a range of bryophytes is therefore an important contribution to our understanding of their nitrogen nutrition (Buckholder, 1959; Simola, 1975, 1979). A number of amino acids, particularly arginine and alanine, have been shown (Simola, 1975) to support growth of *Sphagnum nemoreum* almost as well as ammonium salts. Nitrate was a poor N source and some amino acids were not utilized at all, e.g. leucine, lysine and methionine. Probably because of poor translocation and hydrolysis of dipeptides, only glycylaspartate and glycylglutamate supported reasonable growth of *Sphagnum fimbriatum* and even these were not as good as ammonium nitrate (Simola, 1979). Tryptone (a mixture of amino acids) spared the utilization of ammonium ions in *Sphaerocarpos texanus* (Schuler, *et al.*, 1955) and produced more typical plants than those grown exclusively on ammonium nitrate.

While Simola (1975) reported no morphological abnormalities when amino acids were utilized by *Sphagnum,* Dunham and Bryan (1968) found that while amino acids were non-toxic in the presence of inorganic nitrogen sources, alone they caused a number of highly specific morphological changes to developing *Marchantia polymorpha* gemmalings. In a series of papers (see Basile and Basile, 1980 for references) hydroxyproline has been shown to have dramatic morphogenetic effects on a number of leafy liverworts, in particular the induction of amphigastria (underleaves). These changes have been attributed to an increase in hydroxyproline in cell wall-bound proteins and a decrease in the total amount of such protein. Although Basile and Basile (1980) found that nitrate had no such morphogenetic effect, ammonium ions, even in the presence of nitrate, could cause alterations to the structure of *Gymnocolea inflata,* even though changes in composition and distribution of hydroxyproline-containing proteins were not as previously reported for other species directly subjected to hydroxyproline. Hydroxyproline had no effect on *Sphagnum nemoreum* studied by Simola (1975). However, it, and certain other amino acids, were inhibitory to leafy shoot development, though less so to protonemata of *Atrichum,* and induced atypical development in *Sphagnum squarrosum* (Buckholder, 1959). Buckholder (1959) also found that uracil, in the presence of ammonium nitrate, induced a somewhat thalloid form of *S. squarrosum,* which also occurred in media with aspartic acid. It is not clear whether this thalloid form was the same as that demonstrated by Bold (1948) and found by Boatman and Lark (1971) to decline in importance with decreasing solute concentration.

11.3.2 Field experiments

Field fertilizer experiments have shown that, in many situations, bryophyte growth may be suppressed by the more vigorous growth of higher plants (Jeffrey and Pigott, 1973; Willis, 1963). Mickiewicz (1976) found *Brachythecium albicans* and *Eurhynchium swartzii* to be least affected by NPK fertilization, but total bryophyte biomass was inversely related to the amount of fertilizer supplied. Alteration in the nutrient balance of lawns to favour grasses has been used to eradicate bryophytes but drainage and addition of semi-persistent toxic chemicals, especially mercurous chloride (calomel), have also been employed (Blandy, 1954; Booer, 1951; Dawson, 1969; Jackson, 1961; Woolhouse, 1972). On other occasions, bryophyte cover or vigour has been enhanced by fertilization, but this may, in part, be achieved by the complete (O'Toole and Synnott, 1971; Southorn, 1976, 1977) or partial (Miles, 1973) removal of competing phanerogams during the experiment.

Miles (1973) stripped the surface layers of Callunetum from different soils and compared recolonization with or without the addition of fertilizer containing N, P, K, Ca and Mg. Although phanerogams re-established the Callunetum rapidly, the frequency of recolonizing cryptogams was unlike that of the surrounding vegetation and fertilized plots contained some species absent from the surrounding vegetation, *Bryum argenteum, Ceratodon purpureus* and *Pohlia nutans* were particularly successful in covering unoccupied ground. Fertilization of stripped tundra soils with compound (NPK) fertilizers enriched with extra N or P enhanced growth of mosses (primarily *Polytrichum* spp.), had little effect on the cover of native grasses and sedges, but failed to support growth of introduced grasses (Chapin and Chapin, 1980). O'Toole and Synnott (1971) found *Marchantia polymorpha* and *Funaria hygrometrica* rapidly colonized turned peat plots treated with lime ($CaCO_3$) and phosphate, becoming replaced later by *Bryum inclinatum* and *Ceratodon purpureus.* Nitrogen and potassium had no effect on this pattern and could have been supplied from the peat in adequate amounts. Substantial growth of *Polytrichum longisetum* occurred when N, P and K were added to the absence of lime (i.e. without an increase in soil pH) and *Campylopus pyriformis* grew well without lime, or initially in the presence of lime but the absence of K. No bryophyte growth occurred without the addition of fertilizer.

While the addition of lime to a soil may correct Ca deficiency, it may also cause changes in the vegetation due to an increase in the soil pH. Invasion of *Bryum bornholmense* into heathland plots treated with calcium carbonate was noted by Miles (1968) and was considered to be due to a pH change (from 3.7 to 7.8) as no invasion occurred when calcium sulphate or calcium dihydrogen phosphate (which did not change the soil pH) were added. Adjustment of pH has frequently been used to eliminate acid-loving bryophytes from lawns (Dawson, 1968).

Application of NPK fertilizer to a herb-rich *Calluna* heath over ultrabasic rock (Ferreira and Wormell, 1971) caused the loss of *Racomitrium lanuginosum* and acid-soil phanerogams, but resulted in an increase in grass cover and bareground bryophyte species (e.g. *Pogonatum urnigerum* and *Barbula rigidula*), although the latter may have been of temporary occurrence.

The permanence of any of these changes needs to be considered, especially as Al-Mufti *et al.* (1977) commented on the 'marked amplitude of seasonal variation in the abundance of bryophytes, expansion of which coincided with the moist cool conditions of spring and autumn' in a range of herbaceous vegetation. Hence, the time of year at which comparisons are made between treated and untreated plots may influence the relative amounts of bryophytes and phanerogams.

11.4 STUDIES OF SPECIFIC NUTRIENT REGIMES

In order to understand bryophyte mineral nutrition in a particular habitat it is important to know the sources of mineral elements and their potential and actual availability. A number of investigators have considered some of these aspects but a fully comprehensive picture of any habitat it still unavailable.

11.4.1 Comparison of aerial and substratum inputs

(a) Direct aerial inputs

Studies with suspended moss bags clearly show patterns of heavy metal dispersal from urban and industrial sources (see p. 385 and Chapter 12). It is still not clear how much of the metal burden acquired by moss bags arrives as dry deposition and how much dissolved in rainwater. Many analyses of natural vegetation for heavy metals have assumed a similar source of supply and the many valuable studies of distribution patterns apparently vindicate this assumption. Because heavy metals are readily immobilized at the soil surface the chances of secondary uptake from this source are reduced. Huckabee and Janzen (1975) showed that, although *Dicranum scoparium* took up ^{203}Hg from solution, it was unable to remove Hg when added to the soil, even though Hg-leaching occurred.

Studies of particulate radio-isotope fallout have shown that, following the cessation of atmospheric testing of nuclear weapons, such isotopes were found in progressively lower layers of the moss or lichen carpet due in part to apical growth in the absence of additional inputs (see p. 396). Such work emphasizes the atmosphere rather than the soil as the source of nutrients for many species, especially those lacking any apparent internal conducting tissue and not obviously attached to the substratum. Support for this view comes from studies made on the growth and mineral contents of material from unpolluted sites, especially those associated with leaching of nutrients from other plants. In this respect, the work of Tamm (1953) on *Hylocomium splendens* has been particularly influential.

(b) The importance of input from leachates

Tamm (1953) studied *Hylocomium splendens* from a number of coniferous woodland sites ranging from humid western Norway to drier eastern Sweden. Analyses were made of changes of weight and chemical composition of the readily-dated shoot segments with time of year and position relative to tree canopies.

Analyses of segments away from the growing apex showed a decline in K and P concentrations. For segments up to about three years old this decline was due to increasing dry weight of the segment as the amount per segment increased. The decline in both concentration and content of these soluble intracellular elements from segments older than three years was related to the decline in segment growth and photosynthetic capacity (Callaghan, 1978; Kallio and Kärenlampi, 1975) and the expected degeneration of cells at about this age. Nitrogen, although an intracellular element, did not show such a marked decline in concentration, probably reflecting its combination in organic molecules such as proteins. As much of the Ca may be assumed to be located in an extracellular exchangeable location (see p. 414), its increasing content with segment age, even after the fourth year when segment dry weight began to decline, reflects physico-chemical, rather than biologically controlled, uptake to increasing numbers of exposed exchange sites. The concentration of Ca, Mn, Fe and Al increased with segment age, although the first two were often slightly elevated in the bud region.

It is important to appreciate the differences in location and the nature of the processes controlling the uptake and incorporation of these elements (see p. 392) when considering the validity of Tamm's conclusions. Thus, although he was able to show that Ca could be removed from solutions at concentrations comparable to those in rainwater, presumably by equilibration with available exchange sites, this does not provide evidence that the biologically incorporated elements N, P and K can be acquired from the same solutions. Tamm also showed release of K and Ca from *Hylocomium* by shaking in distilled water. Much of this loss certainly represented intercellular rather than intracellular material, but the latter could also be involved if membranes had been stressed either by desiccation (Brown and Buck, 1979) or the use of ion-deficient solutions over a long period of time. Rieley *et al.* (1979) leached *Rhytidiadelphus loreus* and *Polytrichum formosum* in the field with the actual throughfall from sessile oaks (thereby avoiding the latter criticism) and showed uptake of nitrate and Ca by the moss and loss to the leachate of Mg and Na. Changes in K and ammonium ions were variable and P showed slight release. Substantial increases in length and branching occurred during these experiments but it is not possible to relate these increases to the annual dry weight increases quoted by Rieley *et al.* (1979). Lang *et al.* (1976) reported some uptake of N and loss of Ca, Mg and, initially, K after leaching lichens with artificial rainwater. The value of such experiments is that they show that changes in the composition of the available nutrient solution can cause loss as well as gain of elements from plants and that no element is completely removed from such solutions.

Statements to the effect that mosses can take up ions from dilute solutions and that the amounts of elements in rainfall and throughfall are of the same order of magnitude or equal to that taken up by mosses are misleading over-simplifications. Many factors, such as solution concentration, frequency of wetting and drying and physiological activity of the moss, will determine the proportion of the available nutrient solution actually incorporated into the plant. Madgwick and Ovington (1959) demonstrated that the concentration of K, Ca and probably Na declined with increasing duration of rain. Higher frequencies of rain reduce the concentration of heavy metals in subsequent rainfall (Lazrus *et al.*, 1970).

(c) Growth of species in relation to tree canopies

Abolin' (1974) showed there was a highly significant correlation between total moss cover and the percentage of rainfall penetrating tree canopies and reported maximum coverage of *Pleurozium* in the 'windows' between tree stands and *Sphagnum* in the zone of run-off from the surface of tree crowns. He failed to find any relationship between height increases of individual mosses and rainfall which he attributed to limitation by other factors such as light intensity.

Tamm (1953) showed that productivity of *Hylocomium* individuals was related to their location relative to tree canopies. Annual dry weight increments increased with distance from the tree trunk to the canopy margin, which Tamm related to the increasing light intensity up to a maximum of

50% of that experienced in forest clearings. Beyond the edge of the canopy the light intensity increased but, after a canopy margin transition zone, the plant dry weight did not.

Tamm considered that the failure of *Hylocomium* to respond to higher illumination by increased dry weight increments was due to nutrient limitations, because the percentage N, P, K and Ca concentration in the moss were low. This was compared to the high element concentrations, especially N, observed in samples from under the tree canopy where growth was limited by light intensity. In addition, Tamm provided a very few analyses of rainfall from under the tree canopy and in the open which showed enrichment of all the elements studied under the canopy. The quantities present in canopy leachates were calculated to be approximately sufficient for the annual input of K, P and Ca (Tamm, 1953, 1964).

In spite of the substantially lower nutrient content of the rainfall in the open, *Hylocomium* plants in the open did not show proportionally lower nutrient contents when compared to samples from under the canopy where nutrients were considered adequate. This suggests that, in the open, other sources of nutrient supply, besides precipitation, must exist to satisfy the *Hylocomium* plants nutrient demand and there is no reason to think that they would not also exist for plants under the tree canopy. Hence, although numerically there may be sufficient potential nutrient input from canopy leachate, it does not follow that it is actually and exclusively used.

It is to be regretted that no nutrient addition experiments were performed by Tamm in order to directly demonstrate whether nutrient limitation occurred. Oechel (unpublished) showed fertilization experiments failed to increase the productivity of *Hylocomium* and *Pleurozium* but showed an increase in tissue nitrogen content in the latter, possibly representing 'luxury' or uncontrolled consumption. *Sphagnum nemoreum* showed a similar increase in N concentraton but also increased productivity which demonstrated that *Sphagnum,* but not *Hylocomium* or *Pleurozium* (possibly due to light limitation of growth), was nutrient-limited.

The limitation of *Hylocomium* growth away from the tree canopy observed by Tamm (1953) need not be caused by nutrient limitation as dry weight increments in plants at the edge of the canopy, with enhanced nutrient input, were less, not greater, than those just outside the canopy. Growth in the open may be limited by, for example, the greater evaporation possible in such situations compared to under the tree canopy, reducing the time during which the plant is sufficiently hydrated to support photosynthesis (Busby and Whitfield, 1978). Conversely, under high nutrient and low light intensity conditions near the tree trunk, the higher nutrient concentrations in the moss with reduced weight increments may be due to etiolation-like processes increasing the cytoplasm volume to cell wall weight ratio and favouring the incorporation of elements usually found soluble within the cell, e.g. K and N.

Weetman and Timmer (1967), working in similar habitats to those of Tamm, have similarly shown that, with decreasing light intensities, the dry weight of green shoots of *Pleurozium schreberi* decreased, the concentration of N, P, K and, to a limited extent, Ca and Mg increased, while the amount of these elements (except Ca which decreased) showed no trend related to light intensity or precipitation volume. Both Tamm and Weetman and Timmer emphasized that the species they studied lacked internal structural modifications suitable for nutrient translocation, had very limited capacity to compensate for moisture loss by capillary rise of water from below and they considered little redistribution of nutrients occurred within the plant. Callaghan *et al.* (1978) failed to show transfer of radioactive photosynthetic products from the apical region to older segments of *Hylocomium splendens* but as (a) older adjacent segments were shown to be fully photosynthetically competent and (b) transfer of mineral elements in the reverse direction need not be a rapid process, such data may not truly reflect the plants capacity to redistribute such material.

Mnium hornum has been extensively studied from a number of woodland habitats (Thomas, 1970). This species is in intimate contact with the substratum and displays both external and internal water conduction, the latter becoming proportionally more important with increasing desiccation stress (cf. Bayfield, 1973). Field and laboratory culture experiments showed that increasing light intensities increased the sward density and shoot weight while lower relative humidities decreased the sward height, and that growth was probably light-limited under natural conditions. As Thomas (1970) pointed out, concentrations of N and P in the plant were negatively correlated with light intensity in laboratory grown plants. Culture data showed this to be the result of differences in light intensity and not nutrient supply implying, both here and with other species, uptake in excess of requirements ('luxury consumption'). Cultured *Mnium hornum* showed greater Ca, K and Mg concentrations at higher light intensities, which may reflect (a) increased transpiration at high light intensities due to increased temperature or (b) a shorter path length to the apex of shorter plants (cf. Brown and Buck, 1978a).

Culture of *Mnium hornum* was possible when water and nutrients were only supplied from below but while Ca and Mg concentrations in the plant were equal to those in field grown material, K and P were lower, perhaps indicating that another source was also required for these elements. Longton and Greene (1979) reported that cultured *Pleurozium schreberi* showed nutrient deficiency symptoms unless extra nutrients were supplied to the leaves; watering the substratum was ineffective. However, Oechel (unpublished) concluded that precipitation and litterfall were insufficient to supply Ca, Mg and K requirements of *Pleurozium* and this species, therefore, probably drew on forest floor reserves of these elements. On the other hand, Oechel estimated that the *Polytrichum, Hylocomium* and *Sphagnum* he studied required, respectively, only 4%, 26% and 27% of the nutrients provided in precipitation and litterfall to satisfy their requirements.

(d) Modification of solution chemistry by trees

Because bryophytes appear capable of controlling the entry of N, P and K into their cells, it is impossible to discriminate between different sources of these elements simply on the basis of their relative abundance. Binding of cations to exchangeable sites means that these will be derived, in proportion, from the available sources. Thus, Thomas (1970) showed Mn enrichment in *Mnium hornum* growing under larch and spruce where the litter and possibly leachates, but not the soil, were rich in Mn. Rieley *et al.* (1979) indicated a strong positive correlation between Na and Mg and the volume of throughfall, consistent with rainfall being the main source of these elements. As the site studied was relatively coastal, both elements probably derived originally from seawater and this has frequently been suggested as the cause of elevated Na (Rasmussen, 1977) and Mg (Ellison *et al.*, 1976; Rühling and Tyler, 1973) levels in mosses. Ca was less significantly correlated with throughfall volume and Rieley *et al.* (1979) suggested that decomposing litter might be an important source of this element. Tamm (1953) emphasized dust, mainly derived from soil, as another source for Ca and also Fe, neither of which is readily leached from leaves. K and P are readily leached from leaves and the latter, due to its poor mobility in soil, is liable to be only slightly available from the soil. N in rainfall is frequently depleted by passage through the canopy (Carlisle *et al.*, 1967; Szabó and Csortos, 1975). Release of N and P has been associated with caterpillar infestation and throughfall enriched with these elements in organic form in their excreta (Szabó and Csortos, 1975).

A number of uncertainties exist in analyses of rainfall. Adsorption of ions can occur in collecting vessels where the build-up of microbial populations in the solution will change the distribution of ions between soluble and particulate. Few authors dealing with bryophytes have compared amounts in solution with that removed by filtration. Weetman and Timmer (1976) demonstrated that about 60% of the organic, 30% of the nitrate and 80–90% of the ammonium nitrogen was removed from throughfall collections by filtration through filter paper which would not exclude bacteria. Whether the particulate material was insect debris and excreta, dust particles or plant parts was not indicated, not was its solubility. Calculations of the amount of nitrogen in throughfall based on the soluble material showed that this was insufficient to account for the annual input of N to *Pleurozium schreberi* (Weetman and Timmer, 1976). Tamm (1953), who did not analyse for the more abundant nitrate nitrogen, failed to find sufficient ammonium nitrogen in throughfall and postulated direct ammonium absorption from the air.

Increased concentrations of elements in throughfall compared to rainfall can be achieved by processes other than leaching. Particles trapped by foliage can be washed down by rainfall. The volume of throughfall can be increased by interception of aerosols and cloud or fog droplets. Schlesinger and Reiners (1974) reported a 4.5-fold volume increase when buckets with artificial, plastic, conifer-like foliage were suspended over rain gauges. They observed, inspite of much contamination, a significant increase in the concentration of Ca and Mg at a very exposed site. Tamm (1953) and Abolin' (1974) reported increased volumes of water recovered at the canopy margin due to run-off and possibly surface interception. The morphology of the canopy will be important in determining whether precipitation is enhanced at the canopy margin, e.g. centrifugal types such as *Betula, Picea* and *Tilia* due to run-off at the branch tips, or increased stem-flow, e.g. centripetal types such as *Acer, Fagus* and *Fraxinus* or intermediate such as *Alnus* and *Quercus*. Barkman (1958) quoted percentage rainfall reaching the tree bole, e.g. *Fagus* 8–22%, *Acer* 6%, *Quercus* 6%, *Picea* 1%. Rasmussen (1978) pointed out that initially oaks are somewhat centripetal and age to a more centrifugal type. Nihlgård (1970) showed that 19% of the incident rainfall was intercepted by beech canopies, 70% recovered as throughfall and 11% as stem-flow with the comparable figures for spruce being 39%, 58% and 3%. The concentration of ions in stem-flow is frequently higher than in throughfall (Carlisle *et al.*, 1967; Eaton *et al.*, 1973; Nihlgård, 1970) but the precise amounts depend on leaching of material from leaves and bark and uptake and loss by any epiphytes present.

Decreased pH of rainfall increases cation leakage from foliage, which may be an exchange reaction because the hydrogen ion content of the throughfall may be reduced tenfold below that of the rain (Eaton *et al.*, 1973; Lepp and Fairfax, 1976). However, an acidic solution may also reduce the cation content of the receiver plants (Fairfax and Lepp, 1976; Lepp and Fairfax, 1976). These authors concluded that the changes they observed experimentally with *Dicranum scoparium* and *Hypnum cupressiforme* were caused both by competition for exchange sites between hydrogen ions and other cations and also by possible membrane damage. Unfortunately, the complex and confusing data they obtained may have been erratically modified by unequal desiccation stress experienced by the mosses.

(e) Epiphytes

Barkman's (1958) major work on cryptogamic epiphytes provided a detailed introduction to possible sources of nutrients to epiphytes based, unfortunately, on very little data. Consideration was given to the nature and quantity of nutrient elements in bark and how their solubility might be increased by acidic rain. No detailed correlations were presented between chemical factors and epiphytes growing on different tree species but comment was provided on the relationship between the nutrient richness of bark, often as total electrolyte content, and bryophyte communities.

The trapping of dust particles was related by Barkman (1958) to bark texture with consequent changes in bark pH and nutrient supply. For example, Barkman suggested that the growth of the terrestrial calciphile *Barbula recurvirostra* on the otherwise acid bark of oak or birch was due to trapping of calcareous dust. Differences between the epiphytic flora of trees with otherwise similar bark chemistries were attributable to differences in bark rugosity and dust trapping. The failure of authors to estimate the degree of dust contamination of bark may possibly account for the variability found between reported bark pH values as a result of calcareous dust raising the pH measured. Bark pH tends to decrease with (a) tree age, often due to the build-up of acidic tannins, (b) acidity of the rainfall and (c) aspect (e.g. Young (1938) reported a lower pH on the south side of trees and fewer mosses) and to increase with (a) deposition of dust and sea spray and (b) location within woodlands, due to foliage interactions with hydrogen ions (see p. 411). Buffer capacity of bark was related to the concentration of Ca and Mg in the bark and Barkman suggested it was an underrated feature in comparing tree species. Coker (1967) reported a decrease in ion exchange and buffer capacity with increasing gaseous SO_2 pollution. As this pollutant is acidic Lötschert and Köhm (1978), for example, used the decrease in bark pH and increase in bark S content as a measure of pollution while using Ca content as a measure of dust contamination.

In a study of cations in bark, moss and bulk precipitation, Rasmussen and Johnsen (1976) suggested that the differences in the epiphytic flora of *Fraxinus* and *Fagus* were probably caused by differences in light intensity, humidity and bark surface characteristics rather than pH, conductivity, buffer capacity and mineral composition. They used solutions in contact with the surface of bark discs, to represent readily available elements present in stem-flow, rather than the use of suspensions of bark powder, which more closely represents the maximum potential bark element content. They considered that, because the cation content of epiphytic *Hypnum cupressiforme* growing on the two trees was less variable than the bark composition, epiphytic mosses probably acquired their cations from the atmosphere and not from the bark, except possibly for K. Rasmussen (1978) drew similar conclusions from a study of the same species growing on a range of trees but also pointed out that certain relationships did exist. This was in part due to Rasmussen (1978) using a more drastic extraction procedure which increased the apparent bark element content compared to values reported by Rasmussen and Johnsen (1976). Extreme values in the bark were frequently reflected in the moss but it must be realized that as no estimates were made of moss productivity, it is not possible to determine whether such enrichment or deficiency has any nutritional relevance.

Pitkin (1975) related changes in growth rate of corticolous epiphytes to periods of wetness (especially rainfall minus potential evapotranspiration) but did not investigate changes in nutrient content of the available water or the bryophytes measured. From their detailed analyses of N, P, Ca, K and Na in species from outer branches *(Ulota)*, large branches and upper parts of the trunk *(Hypnum, Dicranum)* and lower trunk *(Isothecium)*, Grubb *et al.* (1969) concluded that the lower concentrations of N, K and Na in *Ulota* indicated that such species can live in regions of the tree with more restricted nutrient supplies. However the supply of nutrients was never directly measured and luxury uptake of nutrients may have occurred in regions of the tree where growth may be restricted by light intensity. Grubb *et al.* (1969) observed that levels of N and P were higher in *Ulota* from parts of the country where pollution was presumed to cause restricted growth.

In an experiment in which ^{137}Cs was introduced into stems of *Liriodendron* (Hoffman, 1972), it was shown that leaching of the radio-isotope occurred, being greatest in the throughfall on an area basis but being more concentrated in the stem-flow. On the weight of Cs recovered in lichens and bryophytes, 60% was in main-stem material, 4% in the canopy, 9% in the tree-base bryophytes and 27% in the terricolous bryophytes. These values were related to the biomass of cryptogams. On a concentration basis, tree-base bryophytes had the highest values, lichens showed decreasing concentrations with increasing height and a similar pattern was shown by the *Liriodendron* bark. As fructicose lichens on the tree trunk have less Cs than foliose species it is probable that stem-flow in equilibrium with the bark was the source of Cs rather than a higher incidence of throughfall near the tree trunk as implied by Tamm (1953). This work unambiguously demonstrates uptake of leachate from throughfall by terricolous species and stem-flow by epiphytes but does not exclude uptake of other elements from different sources.

(f) Epiphylls

Some information is available on the mineral nutrition of epiphyllous liverworts. Thus, Berrie and Eze (1975) demonstrated some penetration of leaf cells of the host by rhizoids of *Radula flaccida* and reported movement of water and radioactive phosphate from the host leaf to the epiphyll. The lower osmotic potential of the epiphyll cells relative to those of the host leaf is suitable for such water movement but rhizoid penetration did not support the interchange of photosynthetic products between the two plants (Eze and Berrie, 1977). Olarinmoye (1974) related growth of a number of epiphylls to frequency rather than total amount of rainfall. He found growth did not stop, but was severely reduced, during dry periods. After a dry period, subsequent growth was high, even with relatively little rainfall, which Olarinmoye suggested was due to enhanced supply of nutrients from dust washed down in the first rainfall. However, in culture, *Radula flaccida* grew best on low nutrient concentrations (Olarinmoye, 1975), possibly reflecting growth in deep shade with minimal dust deposition. As Olarinmoye (1975) reported artificial and natural establishment of *Radula flaccida* on bark, as well as leaves, substrate specificity may be slight, indicating relative nutritional independence from the substratum.

11.4.2 The importance of calcium

The use of chemical analyses of bryophytes in biogeochemical prospecting lead Shacklette (1965a) to comment that 'The content of an element in a bryophyte generally is related to the amount contained in the supporting substratum only if the substrate has a greater than average amount of the element'. It is, therefore, not surprising that Nagano (1972), Bates (1978) and others reported elevated Ca concentrations in bryophytes from calcareous substrata, especially as passive extracellular binding by exchange is probably responsible for most Ca binding (Brown and Buck, 1979).

Bates (1982) found that the 16–17-fold greater Ca concentration found in species from calcareous habitats (calcicoles) compared to those from comparable Ca-poor habitats (calcifuges) (Bates, 1978) is partly reflected in a 3–4-fold greater Ca-exchange capacity in the calcicoles. Experiments suggested that the calcicole mosses, rather than using the cell wall to reduce the Ca concentration at the cell membrane, may require Ca enrichment around their cells in order to maintain membrane integrity (Bates, 1982). It may be that calcicoles have inherently leakier membranes than calcifuges when maintained at the same Ca concentration. Thus, active influx of K from dilute solutions has been shown (Jefferies, 1969; Jefferies *et al.*, 1969) to be maximal for *Cephalozia connivens* (a calcifuge) at 0.1 mM Ca and pH 4 and for *Leiocolea turbinata* (a calcicole) at 3.0 mM Ca and between pH 4 and 8, but K efflux was unaffected by Ca concentration. The calcifuge species had a higher affinity for K than the calcicole which is, like the Ca response, related to the amount present in their respective environments. Ca is always required for membrane integrity and Patterson (1946) noted that a K:Ca ratio of 49:1 was required when KCl was used to test osmotic potentials of bryophytes. Too often, studies on ionic requirements of bryophytes neglect the Ca requirement.

A number of workers have established that, with calcicole and calcifuge species, there is a relationship between the amounts of Ca in their normal environment and the amount of Ca required for growth in culture experiments. Vaarama and Tarén (1963) showed that protonemal growth of certain calcifuge species was depressed by high Ca concentrations but occurred in the apparent absence of Ca. In a comparison of *Scapania aspera* (calcicole) with *S. gracilis* (calcifuge), Harrington (1966a, b) considered growth of the former to be limited by low Ca and, while germination of the latter was possible in the absence of added Ca, differentiation of the calcifuge was inhibited at Ca concentrations which supported germination. *S. gracilis* was more tolerant of Mg than Ca in both the field and the laboratory. Vaarama and Tarén (1963) suggested that sufficient Ca for growth of calcifuges might be derived from the glassware used but it is also possible that Ca in the initial inoculum (Borysławski, 1978) or present in other reagents would be sufficient for the modest requirements of calcifuges in short-term growth experiments.

It is probable that other elements may partially spare Ca by binding to sites otherwise uselessly occupied by Ca or even by substituting for Ca in some essential process (Bell and Lodge, 1963; Borysławski, 1978). Bell and Lodge (1963), basing their comments on analyses of water from habitats in which *Cratoneuron commutatum* grew, suggested that low Mg was not inhibitory and that, particularly with the var. *falcatum,* low Ca was tolerated if Mg was present or, if Mg was also low, by K and Na in place of Mg. The var. *falcatum* appeared to show most calcifuge features in that it grew with high Fe, low Ca and low pH while the var. *commutatum* was more calcicole by growing in water rich in Ca, low in Fe and with a high pH. Although *Scorpidium scorpioides* did not sustain growth when cultured in the absence of Ca, and is generally found in areas rich in Ca, Borysławski (1978) pointed out that such areas are rich in other elements also and this species can tolerate low Ca regions in nature probably by the substitution of other bases for the Ca requirement, i.e. this species was basicolous rather than calcicolous. Although *Scorpidium turgescens* is frequently found in base-rich habitats, analyses by Birks and Dransfield (1970) from one habitat showed low cation concentrations and the authors suggested that the rate of nutrient supply in running water might compensate for the apparent nutrient-deficient conditions as demonstrated for *Sphagnum* (Clymo, 1973).

Inability to tolerate high concentrations of available Fe, Mn and Al at low pH has been suggested as a factor restricting calcicole species to high pH environments. Mayer and Gorham (1951) commented that mosses contained higher Fe and Mn than cultivated plants but Bates (1978), although finding low values of Fe and Al in saxicolous calcicoles, did not find consistently high concentrations in calcifuges. Woollon (1975), using unchelated solutions, showed that above 0.1 mg l^{-1} Fe growth of *Fissidens cristatus* was increasingly inhibited. She also reported optimal growth at pH 8 and 10 mg l^{-1} Ca, but adequate growth above these values, and a broad optimum for Mg up to 75 mg l^{-1}. Tallis (1959) considered that basal segments of *Racomitrium lanuginosum* were able to solubilize Fe under waterlogged (anaerobic) conditions, accounting for its high Fe content in acidic habitats and its frequent occurrence in Fe-rich areas. However, a direct Fe requirement has not been established for this species.

A number of studies have been made of changes in cation concentration occurring during the year for a range of bryophytes from calcareous habitats (Beaumont, 1968; Streeter, 1965; Thomas, 1970; Woollon, 1975). In grassland habitats, bryophyte cation concentrations, especially K, were higher under shrubs and bushes and as the greatest differences occurred with the highest rainfall (Streeter, 1965) and precipitation was shown to be enriched with appropriate cations (Beaumont, 1968), leaf leachates were considered as the source of these extra cations. Woollon (1975) suggested that, because *Fissidens cristatus* grows in intimate contact with the substratum in relatively open communities, high rainfall may supply cations by down-wash ions derived from the soil rather than vegetation. Beaumont (1968) noted that *Neckera crispa* inhabited similar open habitats and contained low K and high Ca concentrations. Its absence from more closed communities may be due to poor competitive ability rather than an inappropriate nutrient supply. On the other hand, *Pseudoscleropodium purum,* although a characteristic species of calcareous turf (Watson, 1960), was shown (Beaumont, 1968) to grow optimally in culture at low Ca (1 p.p.m.), had a pH optimum of 8 and grew better above pH 5 with high K supplies. Growth of this species between grasses, having minimal contact with the calcareous soil, (Watson, 1960), would result in a greater K supply by leaching and lower Ca availability.

Comments to the effect that calcareous areas may be inimical to certain species because high Ca may antagonize K uptake have been apparently supported by plant analyses (Beaumont, 1968; Woollon, 1975). However, earlier discussion (see p. 391) has shown that these two elements are located in different parts of the cell. Thus, high Ca will reduce uptake of K to exchange sites (Woollon, 1975) but what significance this has to the cells metabolism is not clear from simple chemical analyses of total element concentrations. Such an interaction is supported by the observation (Hoffman, 1966b) that *Funaria* grown with low Ca and K showed symptoms of Ca deficiency not noticed in the absence of K. Hoffman

(1966a), however, showed that growth of *Funaria* protonemata at low concentrations of K was best with high Ca and at high K with lower Ca, which does not support the suggestions of Ca antagonizing K. Studies of the presumed calcifuge *Mnium hornum* (Thomas, 1970) showed that, on calcareous substrata, growth was positively correlated with soil depth and negatively correlated with soil Ca and pH. Soil depth is probably related to leaching of Ca and it is noticeable that growth can occur directly on siliceous rocks but only occurs on calcareous rock when at least 5 mm of soil is present as a Ca buffer. In culture experiments, Ca depressed shoot growth at lower concentrations than those depressing protonemal growth. High Ca produced short stocky plants and high P produced slender plants. It was suggested that P substantially decreased Ca uptake but not vice versa. However, this may be an analytical artifact as $CaCO_3$ or limestone added to soil increased the soil-extractable P, but also increased the plant Ca content. As $CaCO_3$ reduced growth but $CaSO_4$ did not, it is probable that the elevated pH of the former rather than the Ca concentration was responsible for reduced growth. Statistical analysis (Thomas, 1970) failed to show a correlation between pH and productivity and, although Mn contents decreased with increasing pH, this was related more to the nature of the tree leachates than availability from the soil.

Experiments like those of Thomas (1970) and Miles (1968) (see p. 404) show that an alkaline pH rather than elevated Ca concentrations may be critical for terricolous species. The high acidity of strip mining areas was considered by Lawrey (1978) to be the cause of low nutrient uptake by bryophytes, probably as a result of protons competing for exchange sites. Difference in cation contents may be explained by such a mechanism but it does not imply any effect on growth. *Sphagnum* species maintain environmental pH values at those favourable for growth by the production of unesterified polyuronic acid in their cell walls (see p. 393). The unequal sensitivity of *Sphagnum* species to elevated Ca concentrations in the environment (Skene, 1915) is only demonstrated after the exchange sites have been saturated with Ca. The damage caused by the remaining alkaline solution depends on how far the pH has been displaced from the preferred value. Growth was shown (Skene, 1915) to be sensitive to direct addition of dilute alkali while dilute acid sometimes caused an increase in growth. Clymo (1973) showed most *Sphagnum* species grew well at low pH and low Ca, nearly as well at high pH or high Ca but the combination of high pH and high Ca was lethal to most species.

11.4.3 Aquatic habitats

In aquatic situations, conditions which favour other macrophytes may eliminate bryophytes by competition. Bryophytes must therefore tolerate more extreme conditions to survive. Unlike competing macrophytes, aquatic bryophytes utilize free carbon dioxide and not bicarbonate ions (Bain and Proctor, 1980; Raven, 1970). Entz (1961) related decreasing bryophyte importance downstream in a Karst-spring fed stream to decreasing free carbon dioxide. In lower regions, bryophytes occurred in shaded areas (decreased photosynthesis of other plants) or in waterfalls and rapids (increased aeration) due to increased free CO_2 concentrations in such areas. Where acid waters from an abandoned lignite mine entered an otherwise neutral or alkaline water system, macrophyte diversity and abundance decreased, being replaced by bryophyte communities because in acid conditions free CO_2 is the dominant form of available carbon, thereby favouring bryophytes (Sand-Jensen and Rasmussen, 1978). In calcareous (alkaline) conditions Bain and Proctor (1980) suggested that bryophytes only survive in competition with other plants where high aeration provides adequate free CO_2 or where other nutrients are insufficient to support growth of other plants. Tufa formation occurs in calcareous areas (Glenn, 1931; Parihar and Pant, 1975), probably because bryophytes act as a surface from which Ca-laden water can evaporate leaving a deposit of $CaCO_3$, plant growth being sufficient to avoid complete coverage. Equivalent iron-rich ochre deposits occur on bryophytes in acidic regions (Huckabee *et al.*, 1975; Sand-Jensen and Rasmussen, 1978).

Pollution of aquatic habitats causes many problems for bryophyte growth (see for example Empain, 1976, 1977, 1978; Lewis, 1974; Wattez, 1976; Whitton, 1972; Whitton and Say, 1975). Hargreaves *et al.* (1975) observed that only protonemata of two mosses survived in extremely acid (below pH 3) streams, but it was not established whether this was due to a direct effect of low pH or the increased solubility and availability of potentially toxic heavy metals (Whitton, 1972). Use of aquatic bryophytes as monitors of heavy metal pollution (subject to all the problems of availability, chemical nature and cellular location discussed for aerial deposition on p. 397) generally assumes tolerance of elevated levels of such elements (Burton and Peterson, 1979b; Dietz, 1973; Empain, 1976, 1977, 1978; Whitehead and Brooks, 1969). Bryophyte distribution patterns associated with sources of complex pollution mixtures implies differential responses to heavy metal stress (Benson-Evans and Williams, 1976; Empain, 1977; McLean and Jones, 1975; Wattez, 1976; Whitton and Say, 1975). Distribution patterns are also affected by physical factors such as amount of silt for the rooting of higher plants, decreased light intensities due to suspended matter, abrasive action of mineral particles as well as damage due to high water velocity (e.g. Holmes and Whitton, 1977; Lewis, 1973). The presence of organic matter can substantially modify the availability of both nutrient and toxic elements (Prosi, 1979; Skogerboe *et al.*, 1979; Strumm and Morgan, 1970; Whitton and Say, 1975). The relative binding capacities of plants and associated silt need to be considered when using bryophytes to monitor heavy metal burdens of rivers (Foulquier and Hébrard, 1976). The complexity of the aquatic environment has meant that, so far, no specific bryophyte distribution patterns have been precisely related to aquatic chemistry.

11.4.4 Saxicolous habitats

Very little has been written on the mineral nutrition of bryophytes in saxicolous habitats. Studies have either considered species composition in relation to physical factors on a single rock type (Bates, 1975b; Foote, 1966; Yarranton, 1967, 1970; Yarranton and Beasleigh, 1968, 1969) or compared species compositon on different rock types without chemical analysis of the substratum (Nagano, 1969; Shacklette, 1965b). Many authors have either quoted total analyses of the underlying rock (Lounemaa, 1956; Shacklette, 1965a) or used values obtained from other sources (Pentecost, 1980), neither approach providing any insight into the availability of nutrients. Bates (1978) analysed water or solutions of the chelating agent EDTA which had been bubbled over freshly exposed rock surfaces and found similarities between the cation concentration in the latter extract and in plants growing on a range of rock types. Direct transfer of ^{90}Sr from artificially enriched rocks to planted *Grimmia orbicularis* clumps was shown by Hébrard *et al.* (1974). These authors found maximum radioactivity in the plant correlated with periods of maximal rainfall, although ^{90}Sr declined in the plant with prolonged rainfall. Mosses and lichens can actively alter the mineral form and hence the chemistry of the underlying rock (e.g. Dormaar, 1968).

11.4.5 Salinity

In coastal regions it is apparent from changes in the composition of higher plant and bryophyte communities (Bates, 1975a, b) that sea water probably adversely affects many species. Shacklette (1961) and Adam (1976) have reported a range of moss species found regularly inundated by seawater, the latter author relating the occurrence of species in different zones of salt marshes to their frequency of tidal inundation. Shacklette (1961) suggested that *Scapania undulata* was unique among leafy liverworts in having been found submerged in seawater, but Adam (1976) reported 10 species from salt marshes and references to others in maritime habitats.

The highly characteristic maritime species *Schistidium maritimum*, is normally confined to siliceous rocks just above the high tide mark, although it can tolerate immersion (Bates, 1975a; Dalby, 1966; Shacklette, 1961). Anchorage problems may determine its lower limit on siliceous rocks (Dalby, 1966), its absence from calcareous rocks and its occurrence in sheltered Scottish salt marshes (Adam, 1976). Other species tolerant of maritime conditions may also be found inland on a range of substrata. Those which are found on both acidic and basic maritime rocks, but only on calcareous rocks inland, may be considered as basiphiles where the divalent cation requirements may be satisfied by either Ca from the substratum or Mg from seawater (high levels of Na appearing to be non-toxic). Although Barkman (1958) suggested that the occurrence of *Ulota phyllantha* on maritime rocks and inland *Ulmus* was due to the latter's bark K/Na ratio being low, and therefore resembling that in seawater, this species also grows on *Alnus* (Shacklette, 1961) and *Acer* (Bates, 1975a) which have higher K/Na ratios. Bates and Brown (1975) found differences between the sensitivity to seawater of epiphytic and saxicolous populations of *U. phyllantha* suggesting there may be two distinct nutritional ecotypes.

On maritime Antarctic islands cryptogams derive nutrients from three sources: rock breakdown (K and Ca), sea bird and mammal excreta (N and P) and sea spray (Na and Mg) (Allen *et al.*, 1967; Smith, 1978). Smith (1978) showed that, unrelated to the distribution of individual species, the sodium concentration in bryophytes (a) decreased with horizontal and vertical distance from the coast, (b) increased, up to four-fold, in the summer (as a result of the sea being frozen in winter and the snow therefore being Na-poor) and (c) increased with wetness of the habitat and increased plant water content. It is likely that these changes in sodium concentration reflect differences in the extracellular exchangeable fraction. Potassium values were also related to the hydration of the habitat but, in this case, it is more likely that the increases occurred in intracellular soluble K as a result of the greater cytoplasmic volume to cell wall weight ratio expected for plants from wetter habitats. Smith considered that Na and Ca concentrations in mosses (the latter derived from underlying substratum) were a response to availability of these ions in the immediate environment. Substantial increases in exchangeable Na and Mg and decreases in Ca occur when maritime or inland mosses are washed with artificial sea water (Bates and Brown, 1974). It is interesting to note that these authors found such changes occurred with *Schistidium maritimum,* indicating that it was not normally in equilibrium with seawater, but that the changes were substantially less than those observed with inland species.

It is not known whether the fluctuations in moss Na concentration noted by Smith (1978) were sufficient to influence metabolism but Bates and Brown (1975) showed that photosynthesis, protein synthesis and chlorophyll concentrations were severely decreased in inland, compared to maritime, mosses when treated with artificial seawater. The toxicity of seawater to inland species is probably caused by the observed increase in intracellular Na uptake and loss of K affecting the tertiary structure, and therefore function, of proteins (Evans and Sorger, 1966). Maritime species failed to show similar Na and K changes even when treated with double strength artificial seawater. However, when NaCl alone was supplied to maritime species, Na was taken up and K lost and this was partially offset, in proportion to concentration, by the addition of Ca. Seawater contains more Mg (56 mM) than Ca (10 mM) but was not as effective in preventing NaCl-induced changes. The calcifuge tendency of *S. maritimum* may account for it being responsive to low levels of Ca. The inland calcicole *Grimmia pulvinata* responded to added Ca by reducing K loss but stimulating Na uptake from NaCl, while Mg reduced Na uptake but slightly stimulated K loss (Bates, 1976). Spore germination and protonemal growth experiments with similar manipulations of ionic balance showed that growth was influenced by these deviations from the normal intracellular cation composition (Bates, 1975).

From studies with metabolic inhibitors Bates (1976) concluded that *S. maritimum* maintained low intracellular Na levels by a low cation permeability, permitting Na entry by passive diffusion, and removal by an active Na-efflux pump. *G. pulvinata* appeared to take up Na actively, but lacked an active Na-efflux mechanism. How far this is generally true for all inland species is not clear, as Sinclair (1967) reported active Na exclusion from the non-maritime *Hookeria lucens,* although the material was collected from an Na-rich site. Bates (1976) considered K retention by both maritime and inland species to be under metabolic control.

Bates and Brown (1975) showed that maritime species were not permanently damaged by immersion in seawater, but inland species were, and suggested that inhibition during saline treatment might be due to osmotic stress or competition between photosynthesis and active Na-efflux pumps for available energy. As Brown (unpublished) observed loss of K when *Fontinalis antipyretica* was treated with mannitol solutions of equal osmotic potential to that of seawater, it remains to be seen whether maritime species should be considered as halotolerant or osmotolerant.

The close proximity to the sea of coastal sand dunes might suggest that their characteristic flora was either halophilic or at least halotolerant. Bates (1975) reported that sand cultures of saxicolous maritime species survived seawater treatment but inland species did not. From a comparison of the effect of seawater misting of inland and dune populations of *Polytrichum commune, Aulacomnium palustre* and *Ceratodon purpureus,* Boerner and Forman (1975) concluded that only the latter showed any tolerance to salt spray and then only when periodically watered with non-saline solutions. Moore and Scott (1979) sprayed a range of sand dune species with up to 30% NaCl, although as shown above the use of NaCl alone is not an adequate substitue for seawater, and found the relative sensitivity of species to roughly reflect their distribution in the sand dune sequence relative to the presumed salinity gradient. Strong salinity gradients exist across sand dunes (Sluet van Oldruitenborgh and Heeres, 1969) but, because strong onshore winds are often associated with high rainfall and sandy soils are readily leached, build up of high salt concentrations in the substratum may not occur. Oliver (1971) studied rhizoid production in cultures treated with soil extracts from different dune zones and concluded that, because rhizoids are produced in distilled water, differences observed were due to the production of inhibitory substances in the older dunes. As undiluted nutrient solutions were frequently too concentrated to support rhizoid production it is likely that high salt concentrations do not build up on the fore-dunes closest to the sea, where rhizoid formation was observed, again emphasizing the steepness of the salinity gradient.

Vegetation changes occurring during land reclamation from the sea was studied by Joenje and During (1977) who observed initial colonization by 'weedy' bryophyte species such as *Funaria hygrometrica, Marchantia polymorpha* and *Physcomitrium pyriforme* in areas where salt concentrations were rapidly reduced. In coarse sandy areas, *Funaria* grew in the wet winter months but was killed during the summer when desiccation caused increased salt contents at the soil surface. In silty sand, salinity declined slowly and the first colonist was *Pottia heimii,* which is frequently found in salt marshes (Adam, 1976). It was considered that the arrival of particular species was related more to their availability in the surrounding regions than to any particular tolerance of salinity. Flowers (1933) reported *Funaria hygrometrica* from salt lakes growing on up to the equivalent of 0.5% NaCl, while general values from Joenje and During (1977) implied concentrations below 0.8% NaCl.

Salinization can occur by other chemicals than NaCl and Mizra and Shimwell (1977) reported on the bryophyte flora of alkaline industrial waste where Mg rather than Na was the dominant cation. A *Barbula–Funaria* community grew on soils with a Mg/K ratio of 52/1 while a more flushed community rich in *Puccinellia* and *Pohlia proligera* had a ratio of 11/1. In the former community, Mg concentrations exceeded the high Ca concentrations but, as the plants always contained more Ca, there was probably sufficient to counteract the potential toxicity of high levels of Mg, as was shown by Bates and Brown (1974) for maritime species. *Barbula tophacea* and *Funaria hygrometrica* appeared to be enriched four-fold with Ca and Mg compared to *Pohlia proligera,* although the soil values were not so extreme, suggesting a sequestering process, probably to extracellular exchange sites, permitting their growth in otherwise unfavourable cation-rich environments.

11.4.6 'Copper mosses'

The occurrence of so-called 'copper mosses' represents a situation where substratum chemistry has been considered of paramount importance in determining the distribution of certain species. The scarcity and limited distribution of such plants has resulted in frequent reports of their occurrence and regular reviews of the possible conditions responsible for their being confined to particular habitats (see for example: Brooks, 1971; Coker, 1971; Persson, 1956; Schatz, 1955; Shacklette, 1967; Wilkins, 1977). The coincidence of such species as *Mielichhoferia elongata, M. mielichhoferi, Merceya (Scopelophila) ligulata* and *Gymnocolea acutiloba* with rocks rich in Cu (often pyrites) has been used to suggest Cu as an element controlling their distribution. As pointed out by Schatz (1955), their growth being limited to such habitats implies not only tolerance to Cu, but a specific requirement for this element.

Only a few chemical analyses have been made of rock, soil or detritus beneath 'copper mosses'. These generally involved estimates of total element content and, although confirming that high Cu concentrations can occur, have shown 'copper mosses' growing at Cu concentrations known to be lower than those tolerated by more widespread species (Persson, 1956). Marked discontinuities in the Cu content of such soils have been reported (Mårtensson, 1956; Persson, 1956; Shacklette, 1967). As total analyses overestimate the amount available to the moss, Wilkins (1977) used ammonium acetate (pH 7) to estimate exchangeable Cu, but failed to detect any at 'copper moss' sites. It is likely that available Cu is in a chelate form and EDTA or other complexing agents would be better extractants.

Schatz (1955) speculated that the term 'sulphur mosses' might be a more appropriate term as substrates were generally rich in S and certain specimens of *Merceya* have been found around S-rich fumaroles and hot springs from which Cu was presumed absent (Noguchi, 1956; Persson, 1948). It is regrettable that Schatz's suggestion that reduced S compounds might be utilized as reductant donors for photosynthesis by these species has never, apparently, been tested experimentally. However, as some *Merceya* specimens have been reported growing in the run-off from Cu-covered roofs, where S enrichment was improbable (Noguchi, 1956) and Persson (1956) reported a number of sites without elevated sulphur values, the term 'sulphur mosses' is equally inappropriate and unsubstantiated.

Only Coker (1971), Shacklette (1967) and Wilkins (1977) have provided direct analyses of 'copper mosses' and none of these reported substantial uptake of copper. Many other workers (Boyle, 1977; Brooks *et al.*, 1973; Brown and House, 1978; Dykeman and DeSousa, 1966; Kendrick, 1962; Wilkins, 1977) have reported substantially higher copper concentrations in other bryophyte species, many from copper-rich swamps or areas where the substratum, but not the atmosphere, was enriched with copper. In *Solenostoma crenulatum,* more than 90% of the copper was bound to extracellular exchange sites (Brown and House, 1978), in agreement with the known high affinity of the element for such sites (Rühling and Tyler, 1970). It is therefore likely that similar distribution will be found with 'copper mosses' and hence very little copper may enter the cell. In addition, Url (1956) demonstrated, on the basis of plasmolysis with copper sulphate solutions, greater cytoplasmic resistance of 'copper mosses' to this element. The copper-tolerant ecotypes of *Solenostoma crenulatum* detected by Brown and House (1978) did show enhanced photosynthesis in the presence of Cu, suggesting some degree of Cu requirement.

Low pH is another feature common to soils bearing 'copper mosses' but many other acidic sites lack such plants, indicating that pH alone is not the controlling factor. *Merceya latifolia* is exceptional in being found in Cu-rich habitats at pH 7.63 (Persson, 1956). Although germination and growth of *Merceya ligulata* and *Mielichhoferia mielichhoferi* spores occurred at pH 2.02 and 2.62, they also grew at pH 8.20 and 7.20 and therefore showed no obligate requirement for acidic conditions (Nagano and Shimizu, 1972). Studies of spore germination are of limited value as sporophytes rarely occur in nature and spread is by vegetative methods. Studies of *M. ligulata* and *M. gedeana* showed greater regeneration from leaves and greater spore germination at low pH values, but also showed that copper sulphate solutions from 0.1% to 0.8% totally inhibited the former species but not the latter (Noguchi and Furuta, 1956). The copper tolerance of *M. gedeana* is the more surprising because the Cu concentrations used were greatly in excess of the ionic Cu concentrations likely to occur in 'copper moss' habitats. It must be remembered that chelation reduces Cu toxicity to *Marchantia* and *Funaria* (Coombes and Lepp, 1974). Copper chelation may account for the growth of angiosperms in Cu-rich swamps (Boyle, 1977; Dykeman and DeSousa, 1966; Fraser, 1961; Kendrik, 1962) and Usui *et al.* (1975) found *Athyrium yokoscense* growing on a copper mine only on the remains of *Nardia sieboldii*. *Pohlia nutans* grows and acquires a very high Cu content in the more mineralized parts of such Cu swamps. *Pohlia nutans* has not been associated with typical 'copper moss' habitats and grows in the absence of Cu enrichment (Persson, 1956). Boyle (1977) suggested that slightly chlorotic *P. nutans* might indicate the presence of higher than usual Cu concentrations. Pigott (1958) and Wilkins (1977) stressed that dark green plants of *M. elongata* grew in moist, shaded habitats and became chlorotic in more exposed conditions. Such habitats may reduce excessive evaporation-driven uptake of Cu by capillarity and part of the success of *M. elongata* may be its ability to tolerate conditions supporting low photosynthetic activity.

Much remains to be done in the study of 'copper mosses' but future work should more carefully consider the availability of Cu in such habitats as well as analysis of a range of other elements and determine the cellular location of the limited amount of copper taken up by these plants. Whether

Cu will be found to be associated with specific exchange sites (and we do not yet know the cation exchange capacity of these plants) or even with inorganic sulphides needs to be considered. It is very likely that there is no single explanation for the occurrence of all species of 'copper moss'.

11.4.7 Association of bryophytes with other organisms

(a) Phosphate metabolism

Higher plants in phosphate deficient-habitats are frequently infected with mycorrhizal fungi in a mutualistic association in which a beneficial transference of phosphate occurs from the soil, via fungal hyphae, to the host plant. Sporadic reports exist of such fungi in bryophytes. Stahl (1949) observed structures within thalloid liverwort cells similar to those of functional vesicular-arbuscular (VA) endomycorrhizas and fungi, derived from a range of liverworts, induced typical VA infections of the higher plant genus *Coprosma* (Johnson, 1977). Johnson also obtained infection from mats of *Lembophyllum,* without observing mycorrhizas in them and Rabatin (1980) achieved infection of *Poa* using *Pogonatum* homogenates. Sporocarps of *Endogone pisiformis* are frequently associated with *Sphagnum* (Seymour, 1929) and *Glomus epigaeus* with *Funaria hygrometrica* (Daniels and Trappe, 1979). Daniels and Trappe (1979) considered the high humidity in the moss carpet promoted sporocarp formation without a mycorrhizal association being developed with the moss. Neither they nor Rabatin (1980) observed any fungal penetration of moss cells, whereas Butler (1939) found hyphae and vesicles in decomposing *Sphagnum.* Parke and Linderman (1980) found similar infections in *Funaria hygrometrica,* but only when pot cultures also contained infected 'companion' plants of asparagus. There is, therefore, no evidence to suggest that mycorrhizal infection may occur in bryophytes and improve their phosphate status although Parke and Linderman (1980) quoted an unpublished report of Pokorny and Hendrix that *Philonotis glaucescens* was more prolific when a mixture of VA mycorrhizal fungi was added to their planting mixture. Whether this system has anything to do with the phosphate metabolism of the moss remains to be seen, as Hildebrand *et al.* (1978) suggested that the growth of liverworts under acidic conditions may be promoted by fungal exudates.

In nature, it is likely that phosphate is a major, if not the major, element limiting the growth of many bryophytes. However, there have been very few direct studies of this problem which have considered bryophytes. A positive correlation has been shown between biomass of a *Hydrogonium* community and the available phosphate content of the soil (Datta Munshi, 1977). Fertilizer experiments (see p. 404) have shown increased productivity after phosphate applications. An increase in the P concentration of

Pleurozium schreberi and *Sphagnum nemoreum* after field fertilization with Hoagland's solution (Oechel, unpublished) apparently indicated a phosphate deficiency in nature. With *Sphagnum,* increased fertilizer concentration increased shoot productivity and photosynthesis, but in *Pleurozium* and *Hylocomium splendens* productivity decreased and photosynthesis was unchanged. In *Pleurozium,* the increased tissue P concentration may therefore only reflect unaltered phosphate uptake to a lower weight of plant material. It remains to be seen how far these responses represent differences in absolute nutrient requirements between species or differences in retention capacity of hummock or feather-moss morphologies or limitations imposed by other environmental factors such as light intensity. In both *Pleurozium* and *Sphagnum,* Oechel apparently observed luxury uptake of nitrogen.

Microscopical observations of cultured protonemata have demonstrated polyphosphate bodies in chloroplasts which suggests that under these conditions there was sufficient phosphate to permit its storage (Keck and Stich, 1957; Tanaka and Sato, 1977). Tanaka and Sato showed loss of polyphosphate bodies in response to transfer of protonemata to phosphate-deficient media.

Another type of fungal infection, unrelated to phosphate metabolism, has been observed in the non-photosynthetic thalli of *Cryptothallus mirabilis.* These mycorrhizal fungi are of the orchid type in which typical intracellular coiled fungal hyphae (pelotons) are digested by the liverwort as a source of organic matter (Williams, 1950). Benson-Evans (1960) suggested that failure of cultured *Cryptothallus mirabilis* to grow beyond the 20–30 celled stage was due to an unspecified requirement for fungal infection products because carbohydrates were supplied in the medium.

(b) Nitrogen metabolism

Nitrogen fixation has been regularly demonstrated in a range of *Sphagnum* species from a variety of habitats. In coniferous forests low heterotrophic, light-independent, activity has been ascribed to *Azotobacter*-like bacteria (Basilier, 1979). Substantially higher, light-dependent, nitrogen-fixing activity has been found with *Sphagnum* from nutrient-rich minerotrophic mires. In these habitats epiphytic cyanobacteria (blue-green algae) are present, particularly on the surface of submerged plants (Basilier, 1979; Blasco and Jordan, 1976; Granhall and Selander, 1973), where, in spite of similar heterocyst frequencies to free-living cyanobacteria, they showed enhanced nitrogen fixation (Basilier, 1980). Plants emergent above the water table have cyanobacteria within the cavity of hyaline cells (endophytic) (Basilier, 1979).

The electron microscopical studies of Granhall and Hofsten (1976) did not show any substantial change in endophytic cyanophyte morphology.

They also reported the presence of green algae and bacteria (including *Methanosarcina*-like bacteria) within the hyaline cells. There is no evidence to suggest any obligate relationship between endophytic cyanobacteria and *Sphagnum* nor the hypothesis of an interrelationship between them and bacteria, including methane bacteria, suggested by Granhall and Hofsten (1976). The existence of highly active nitrogen-fixing cyanobacteria, epiphytic on *Drepanocladus* from similar habitats, and the reports of high nitrogen fixation by epiphytically infected *Sphagnum* plants (Basilier *et al.*, 1978; Granhall and Selander, 1973) imply that endophytic associations are not significantly more effective. Lower nitrogen fixation and cyanobacterial colonization is associated with the growing apex and non-green parts of the host plant (Basilier *et al.*, 1978). Lack of cyanobacteria at the apex may be caused by delayed colonization of the more rapidly growing moss tissue, production of antimicrobial substances or inappropriate physical and chemical conditions (Basilier *et al.*, 1978; Broady, 1977, 1979).

Nitrogen fixation by endophytic cyanobacteria is strongly light-dependent and shows maximum fixation at noon and in the middle of the *Sphagnum* growth season (Basilier *et al.*, 1978). In the laboratory, nitrogen fixation by the symbiotic system is barely affected by pH values from 4.3 to 6.8 although free-living and epiphytic cyanobacteria prefer values above 5 (Basilier *et al.*, 1978; Lambert and Reiner, 1979). Blasco and Jordan (1976) considered that leaf exudates might make the hyaline cell pH more alkaline and Granhall and Hofsten (1976) suggested that acidification of the environment by *Sphagnum* could cause an alkaline reaction within the hyaline cells thereby buffering the cyanobacterial environment against extreme pH conditions. They suggested that lack of cyanobacterial colonization observed in *Sphagnum* from acidic communities (pH 3.8 or below) was due to insufficient pH-buffering capacity with the hyaline cells of such plants. Heterotrophic bacterial nitrogen-fixing symbioses appear to tolerate pH's below 4 (Lambert and Reiners, 1979). Nitrogen fixation by cyanobacteria requires much phosphate and comparisons between communities showed that low phosphate concentrations were associated with low nitrogen fixation by *Sphagnum* symbioses (Basilier *et al.*, 1978).

Within communities there is a general correlation between *Sphagnum* growth and nitrogen-fixing capacity which Basilier *et al.* (1978) considered to be a causal relationship and, because no such relationship was established with individual *Sphagnum* plants, suggested that nitrogen was released into the medium from where it could be acquired by other *Sphagnum* plants or angiosperms (see also p. 428). Basilier (1980), using ^{15}N as a tracer, showed the rapid transfer of nitrogen from the cyanobacteria to the apex of the *Sphagnum* plant and presumed it occurred in the water moving towards the exposed apex. The occurrence of nitrogen-fixing organisms along the length of a *Sphagnum* plant and the demonstrated movement of nitrogen may account for the similar nitrogen concentration along the length of *Sphagnum* plants (see p. 395).

Cyanobacteria did not appear to be early colonizers of volcanic lava on Surtsey Island, Iceland (Brock, 1973) in spite of the low soil nitrogen and high phosphate content (Henriksson and Henriksson, 1974), possibly due to the high mobility (Brock, 1973), low water-holding capacity and high salt content of ash (Henriksson *et al.*, 1972). Mosses were amongst the earliest colonists but were generally closely associated with free-living cyanobacteria (Brock, 1973; Schwabe, 1974) which were capable of nitrogen fixation (Henriksson and Henriksson, 1974). Using mixtures of cultured cyanobacteria and *Funaria,* Rodgers and Henriksson (1976) showed no reduction in nitrogen fixation rate by the presence of the moss, an increase in cyanobacterial and moss growth and an increase in the moss gametophyte nitrogen content when cyanobacteria were present. Because gametophyte numbers were stimulated in the presence of cyanobacteria, even when nitrate was added, Rodgers and Henriksson (1976) suggested that cyanobacteria might produce moss-stimulating growth substances. Cyanobacteria benefited by the improved water regime at the soil surface but beneficial moss exudates may also exist. In the Antarctic, however, Horne (1972) found better development of the cyanobacterium *Nostoc* away from moss banks, possibly due to the higher nitrate content of water running through moss banks.

Cyanobacteria have been reported as epiphytes with other mosses besides *Sphagnum* and *Funaria.* Alexander and Schell (1973) reported nitrogen fixation in *Polytrichum commune* associated with *Nostoc* colonization in the arctic tundra and Lambert and Reiners (1979) showed significant nitrogen-fixation with *Plagiomnium cuspidatum,* in addition to *Sphagnum,* in subalpine zones of New Hampshire. Vlassak *et al.* (1973) demonstrated nitrogen fixation in *Ceratodon purpureus* in grassland but, although cyanobacteria were present in its phyllosphere, they could not eliminate a contribution from heterotrophic bacterial nitrogen fixation. On granite outcrops, Snyder and Wullstein (1973) reported high nitrogen fixation rates in the soil below *Grimmia leavigata* which they attributed to *Azotobacter.* More distant associations between nitrogen-fixing cyanobacteria and mosses have been shown, with the use of ^{15}N tracer experiments by Horne (1972) for Antarctic moss banks, Jones and Wilson (1978) with *Gymnostomum recurvirostrum* on temperate sugar limestone and Stewart (1967) with *Bryum algovicum* var. *rutheanum* on temperate sand dunes. In none of these studies was information provided on the nature of the nitrogenous compound(s) moving from the cyanobacterium to the moss.

Under normal conditions nitrogen-fixing cyanobacterial colonies are present in mucilage-containing cavities in gametophytes of *Anthoceros* and *Blasia* and may also occur in a range of thalloid liverworts (Duckett *et al.*, 1977; Rodgers and Stewart, 1977). Preformed cavities become infected with cyanobacteria, enlarge and multicellular filamentous cells proliferate from the cavity margin throughout the cyanobacterial mass. These proliferations are assumed to assist in the interchange of metabolites between cyanobacterial and host cells. Older reports of cyanobacteria consuming host-derived mucilage within the cavity can be discounted because of the absence of any structures normally associated with mucilage production in the host cells (Duckett *et al.*, 1977).

Cyanobacteria symbiotic with *Anthoceros* and *Blasia* have higher heterocyst frequencies, compared to free-living material, higher rates of nitrogen fixation in the light, particularly under microaerophilic conditions, are depleted of nitrogen-storing phycobilin pigments and have a minimal capacity to evolve oxygen or fix carbon dioxide photosynthetically (Rodgers and Stewart, 1977). When cultured away from *Blasia* the cyanobacteria can grow heterotrophically on a range of carbohydrates and, when in symbiosis, are supplied with carbon by transfer of photosynthate from the host. In *Anthoceros,* illuminated, photosynthetic, sporophytes can sustain nitrogen fixation in darkened gametophytes (Stewart and Rodgers, 1977). Symbiotic systems have higher nitrogen contents than cyanobacterial-free plants, which can grow in the presence of combined nitrogen (Rodgers and Stewart, 1977). Because freshly isolated cyanobacteria primarily release nitrogen in the form of ammonium ions, it is presumed that this is the form in which nitrogen is transferred from the cyanobacteria to the host cells (Stewart and Rodgers, 1977, 1978).

As mentioned above, volcanic lava and ash is nitrogen deficient. Substantial colonies of the liverworts *Lophozia bicrenata* and *Cephaloziella divaricata* were observed on the primary ash from the 1912 Katmai eruption (Griggs, 1933). *Cephaloziella* appeared to be an efficient nitrogen scavenger as it grew in nitrogen-depleted cultures, although deficiency symptoms did develop in 9 months (Griggs and Ready, 1934). This ability is related to the observed low levels of nitrogen in the ash colonized by liverworts (Griggs, 1933). Where ash was colonized by other plants it contained higher levels of nitrogen and impure cultures of *Cephaloziella* readily became overgrown when nitrate was supplied, indicating a poor competitive ability. Unlike observations on Surtsey (see p. 428), no cyanobacteria were associated with the liverworts in the field or the laboratory. Some growth was probably at the expense of nitrogen in the original inoculum as is seen with other species such as *Funaria* (Dietert, 1979).

At the opposite extreme, some bryophytes are capable of growing in areas enriched with nutrients from excreta. Certain genera, e.g.

Splachnum, are mainly confined to fresh dung, Whitmore (1965) reporting them to be overgrown by *Sphagnum* and *Polytrichum* as the pH declined. Lists of species have been published from sewage treatment plants (Cooke, 1953), bird-cliffs (e.g. Eurola and Hakala, 1977; Grønlie, 1948) and penguin rookeries and seal wallowing grounds in the maritime Antarctic (Gimingham and Smith, 1970; Smith and Gimingham, 1976). The high nutrient concentrations in areas contaminated with penguin or seal excreta, characteristically richer in P and N than other areas (Allen *et al.,* 1967), often represses sexual development (Grønlie, 1948), alters morphology, viz *Bryum siplei* appearing like *Bryum argenteum* near bird colonies (Schofield and Ahmadjian, 1972) or is directly lethal. Nitrogen is mainly in organic form as uric acid, with birds, and urea, with seals (Schofield and Ahmadjian, 1972). Growth of *Bryum antarcticum* has been shown with, in decreasing order of effectiveness, ammonium, nitrate, urea, uric acid, xanthine and allantoin. As this species appears to posses an extracellular urease, growth may still be responding directly to ammonium ions (Schofield and Ahmadjian, 1972).

11.4.8 Post-fire communities

Initial plant colonization following vegetation burning frequently consists of a highly characteristic assemblage of bryophyte species. Because burning increases the nutrient content of the soil surface, attempts have been made to correlate the entry of particular species with their nutrient requirements. *Funaria hygrometrica* is regularly, but not invariably, reported from burnt sites. Southorn (1976) included *Ceratodon purpureus, Bryum argenteum* and tuberous species of *Bryum* as 'bonfire species' but emphasized the rapidly achieved dominance of *Funaria.* Although often associated with *Funaria, Marchantia polymorpha* is a conspicuous initial colonizer of the wetter habitats (Cremer and Mount, 1965; Graff, 1936; Skutch, 1929) while *Polytrichum* species have been reported from the driest sites (Maikawa and Kershaw, 1976; Skutch, 1929). *Polytrichum* species are also conspicuous in lightly burnt sites where they were initially present in the vegetation, probably due to regeneration from rhizomes surviving the low temperature burn. Heathland and moorland fires, where slow re-colonization by angiosperms occurs, are dominated initially by acrocarpous species from the surrounding vegetation, pleurocarpous species only entering with the angiosperm cover (Gimingham, 1978; Rawes and Hobbs, 1979; Southorn, 1976).

Colonization by spores was suggested by Cremer and Mount (1965) and Southorn (1976), the latter measuring the increase in protonema frequency in soil samples. Gemmae (Cremer and Mount, 1965) and detached fragments (Benson and Blackwell (1926) for *Campylopus flexuosus*) have been considered for later colonists. The initial plant cover reduces temperature variation at the soil surface (Richardson, 1958) and improves moisture retention (Raison, 1979) favouring the entry of more competitive species. Skutch (1929) regularly found *Marchantia* remains beneath colonizing *Polytrichum* sp. and Richardson (1958) found that *Holcus lanatus* colonized through carpets of *Pohlia nutans* on cooling pit heaps. Benson and Blackwell (1926) observed that carpets of *Campylopus flexuosus* retarded the spread of later fires, but were destroyed by drought.

Burning causes changes in the soil chemistry due to direct heating effects as well as due to the addition of soluble salts from the breakdown of surface vegetation (see Raison (1979) for references and a detailed discussion of burning-induced changes). Grassland fires (Daubenmire, 1968) do not stimulate bryophyte colonization but it is not clear whether this is due to the lower and shorter elevation of surface temperatures, release of only partially destroyed organic matter which may be toxic, lesser addition of soluble salts or smaller increase in soil pH due to the release of fewer basic salts (Raison, 1979). Forest fires, with a greater bulk of inflammable material, and bonfires produce hotter burning conditions and cause more extreme changes (Raison, 1979; Southorn, 1976). Increased initial alkalinity, decreasing when other species colonize, has been stressed by certain authors (Cremer and Mount, 1965; Marshall and Averill, 1928; Skutch, 1929) partly because other factors were not measured in early work. Entry of *Polytrichum* later in the succession may require a drop in pH (Cremer and Mount, 1965), as may *Ceratodon purpureus*; Summerhayes and Williams (1926) related this to salt loss by leaching. High initial salt concentrations may inhibit angiosperm colonization (Graff, 1936; Southorn, 1976). Bryophytes perhaps avoid this, possibly osmotic, problem by growing in the most leached surface layers, i.e. *Funaria* rhizoids in the top 5 mm, *Marchantia* down to 5 cm and *Polytrichum* to 7.5 cm (Cremer and Mount, 1956; Hoffman, 1966b), and colonizing after leaching the initial excessive salt concentrations (Hoffman, 1966b; Southorn, 1976). Hoffman (1966b), using humus-poor C horizon subsoil, found growth of *Funaria* was depressed when cultured on soil heated to up to 150°C, heating to between 200 and 300°C stimulated growth, but above this temperature growth was non-existent. Southorn (1977), however, observed high germination frequencies up to 700°C except between 150 to 250°C when no growth was observed, probably due to release of excessive quantities of soluble material from the greater amount of organic matter initially present. While Southorn used the more appropriate surface soil, Hoffman's use of growth rather than germination is a better assay for comparison with field observations. Future work on factors controlling bryophyte growth after fire must consider the role of organic matter as potential toxins and chelating compounds.

It is far from clear how important the addition of salts from plant ash is to the colonization of burnt areas. Southorn (1976) removed ash from a bonfire site and found recolonization by 'bonfire species' but in different proportions (*Funaria* never dominating), while ash added to unburnt soil had the same species but in lower than usual abundance. Hoffman (1966b) added wood ash to heated subsoil and found only slight stimulation of *Funaria* growth. While Cremer and Mount (1956) found *Funaria* growth was stimulated by the addition of blood or bone fertilizer, Southorn (1976) failed to find this species after similar fertilizer treatment, although some other mosses did grow. O'Toole and Synnott (1971) fertilized freshly exposed peat and reported that *Funaria* and *Marchantia* grew when calcium carbonate and phosphorus levels were elevated, being replaced by *Bryum inclinatum* and *Ceratodon purpureus* after the first year. *Polytrichum* and *Campylopus* developed in the absence of lime when phosphorus was applied. Responses to addition of nitrogen and potassium were poor and no mosses grew on unfertilized peat.

Southorn's (1976) study of bryophyte colonization provides a description of the chemical changes occurring following experimental burning, compared to a weeded area, and confirmed that initial increases occurred in pH, exchangeable K, Mg and Ca, extractable P and soluble organic matter, all of which subsequently decline (see also Raison (1979) for further details). In an 80-week period the organic matter, K and Mg values approached those of the control site, while the others remained relatively high. Unfortunately, only limited analyses of nitrate and ammonium ions were provided from another burnt site, but did show an initial increase. Rapid-fire sites, uncolonized by bryophytes, showed no increase in extractable P, Ca or N. Changes in soil chemistry need to be considered immediately after burning to understand colonization. Similarly, because Southorn (1976) observed that the bryophyte community was maintained after the second growing season only if the experimental area was weeded free of angiosperms, alterations in soil chemistry which are maintained for some time are important in understanding continued growth of 'bonfire species'. Soil analyses indicate that elevated levels of Ca, P and N are maintained for some time. Hoffman (1966b) concluded from a combination of the changes observed in the chemistry of heated soil and culture experiments on *Funaria* that elevation of N and P were the most important features controlling growth of this species on burnt sites.

It is perhaps unfortunate that *Funaria* (Dietert, 1979; Hoffman, 1966a, b) and *Marchantia* (Voth and Hamner, 1940) are capable of growing in a wide range of nutrient mixtures such that culture experiments fail to provide any insight into specific nutritional requirements satisfied by burnt areas. Voth and Hamner (1940) did note that *Marchantia* grew well on low K but not on high P and this may explain its growth on wetter burnt areas where leaching may have occurred. The suggestion (Crum, 1972) that weedy species, including *Funaria,* require relatively high levels of K and N is not supported by culture experiments which only emphasize the latter element.

During colonization, the nature of the nitrogen source may be important as Dietert (1979), Kofler (1959) and Southorn (1977) found ammonium ions inhibited *Funaria* growth and ammonium ions are usually present in high concentrations immediately after burning when bryophyte growth is minimal. Leaching and the restoration of an appropriate nitrifying microflora are probably essential stages in developing a suitable supply of nitrate (Raison, 1979; Southorn, 1976, 1977). However, Dietert (1979) considered that, because Southorn (1976) showed soil N and P declined to near control levels by the end of the observation period, these elements were of limited significance in stimulating bryophyte growth. Instead Dietert emphasized Ca and pH, on the basis of O'Toole and Synnott's (1971) fertilizer experiments. Ca was dismissed as probably unavailable to *Funaria* (without supporting evidence) and therefore she singled out high pH as the most important factor in the continued survival of *Funaria* in burnt areas. In support of this hypothesis is the observation that, although reported from habitats with widely varying pH's (e.g. Apinis and Lacis, 1936; Hoffman, 1966b; Montgomery, 1931), optimal growth of *Funaria* in nature is apparently between pH 7 and 8 (Ikenberry, 1936; Hoffman, 1966b). Further support is given by culture experiments which show that while protonemal growth of *Funaria* is possible over a wide range of pH values (e.g. Apinis, 1939; Ikenberry, 1936) it increases with increasing pH between 3.5 and 7.8 (Armentano and Caponetti, 1972).

No unambiguous evidence has yet been presented to show that, after burning and an initial leaching period, colonization of burnt sites by bryophytes is due to the production of a specifically favourable nutrient environment. The initial chemical regime may be more inhibitory to higher plant colonization than bryophytes. Amongst the chemicals which may be of importance, and which should be considered in more detail in future studies, are organic molecules which may have inhibitory or stimulatory effects on growth of all plants liable to colonize such sites.

11.5 CONCLUDING REMARKS

Our understanding of bryophyte mineral nutrition has been much improved by recent research but, as can be seen from the data quoted here, there are still many areas of uncertainty. Some of these are due to insufficient information on the sources from which bryophytes acquire their nutrients. Too frequently it has been assumed that the atmosphere is the sole source of supply. Future work must clarify how far this is correct for a range of habitats and morphologies. Where the substratum provides the potential for a major contribution, i.e. in heavy metal- or calcium-rich areas, more information is required on the availability of elements to bryophytes; it is an oversimplification to consider only the inorganic components when investigating mineral nutrition. Total analyses of inputs, whether to terricolous, epiphytic or aquatic habitats, generally contain unknown proportions of unavailable material. Analyses of organic matter may be of importance because of (a), ion binding by exchange or chelation and, particularly during establishment, (b), utilization either as metabolites or as positively or negatively acting growth substances.

As an aid to understanding mineral nutritional requirements, direct analysis of plant material has apparently much to contribute. However, future research will require more detailed studies of the influence physical factors have on the uptake of ions both in the laboratory and when investigating natural populations. Greater emphasis should be placed on growth studies and the effects elements have on different propagules. These will assist our comprehension of the influence of minerals on plant productivity as will short-term studies of their effects on specific physiological processes. Interpretation of plant element concentration data are only fully explicable if growth rates and location relative to the metabolically active interior of the cell are known. How far the cell wall cation exchange sites act as a buffer against intracellular uptake of potentially toxic elements or as a reservoir of essential elements needs clarification.

REFERENCES

Abolin', A.A. (1974), *Ekologiya,* **5,** 51–6.

Adam, P. (1976), *J. Bryol.,* **9,** 265–74.

Alexander, V. and Schell, D.M. (1973), *Arctic Alpine Res.,* **5,** 77–88.

Allen, S.E., Grimshaw, H.M. and Holdgate, M.W. (1967), *J. Ecol.,* **55,** 381–96.

Allsopp, A. and Mitra, G.C. (1958), *Ann. Bot.,* **22,** 95–115.

Al-Mufti, M.M., Sydes, C.L., Furness, S.B., Grime, J.P. and Band, S.R. (1977), *J. Ecol.,* **65,** 759–92.

Andersen, A., Hormand, M.F. and Johnsen, I. (1978), *Env. Poll.,* **17,** 133–51.

Antonovics, J., Bradshaw, A.D. and Turner, R.G. (1971), *Adv. Ecol. Res.,* **7,** 1–85.

Apinis, A. (1939), *Acta Horti. Bot. Univ. Latv.,* **11–12,** 1–14.

Apinis, A. and Lacis, L. (1936), *Acta Horti. Bot. Univ. Latv.,* **9–10,** 1–100.

Araki, C. (1953), *Mem. Fac. Ind. Arts Kyoto Tech. Univ. Japan,* **2,** 17.

Armentano, T.V. and Caponetti, J.D. (1972), *Bryologist,* **75,** 147–53.

Bain, J.T. and Proctor, M.C.F. (1980), *New Phytol.,* **86,** 393–400.

Barclay-Estrup, P. and Rinne, R.J.K. (1978), *Oikos,* **30,** 106–8.

Barkman, J.J. (1958), *Phytosociology and Ecology of Cryptogamic Epiphytes,* Van Gorcum & Comp. N.V., Assen.

Basile, D.V. (1975), *Bryologist,* **78,** 403–13.
Basile, D.V. and Basile, M.R. (1980), *Am. J. Bot.,* **67,** 500–7.
Basilier, K. (1979), *Lindbergia,* **5,** 84–8.
Basilier, K. (1980), *Oikos,* **34,** 239–42.
Basilier, K., Granhall, U. and Stenström, T.-A. (1978), *Oikos,* **31,** 236–46.
Bates, J.W. (1975a), Ph.D. Thesis, University of Bristol.
Bates, J.W. (1975b), *J. Ecol.,* **63,** 143–62.
Bates, J.W. (1976), *New Phytol.,* **77,** 15–23.
Bates, J.W. (1978), *J. Ecol.,* **66,** 457–82.
Bates, J.W. (1979), *J. Bryol.,* **10,** 339–51.
Bates, J.W. (1982), *New Phytol.,* **90,** 239–52.
Bates, J.W and Brown, D.H. (1974), *New Phytol.,* **73,** 483–95.
Bates, J.W. and Brown, D.H. (1975), *Oecologia,* **21,** 335–44.
Bayfield, N.G. (1973), *J. Bryol.,* **7,** 607–17.
Beaumont, E.A. (1968), Ph.D. Thesis, University of Sussex.
Bell, P.R. and Lodge, E. (1963), *J. Ecol.,* **51,** 113–22.
Benson, M. and Blackwell, E. (1926), *J. Ecol.,* **14,** 120–37.
Benson-Evans, K. (1960), *Trans Br. Bryol. Soc.,* **3,** 729–35.
Benson-Evans, K. and Williams, P.F. (1976), *J. Bryol.,* **9,** 81–91.
Berrie, G.K. (1975), *J. Bryol.,* **8,** 443–54.
Berrie, G.K. and Eze, J.M.O. (1975), *Ann. Bot.,* **39,** 955–63.
Birks, H.J.B. and Dransfield, J. (1970), *Trans. Br. Bryol. Soc.,* **6,** 129–32.
Black, V.J. (1974), Ph.D. Thesis, University of Aberdeen.
Blandy, R.V. (1954), *J. Sci. Food Agric.,* **5,** 397–400.
Blasco, J.A. and Jordan, D.C. (1976), *Can. J. Microbiol.,* **22,** 897–907.
Boatman, D.J. (1978), *Bull. Br. Bryol. Soc.,* **31,** 10.
Boatman, D.J. and Lark, P.M. (1971), *New Phytol.,* **70,** 1053–9.
Boerner, R.E. and Forman, R.T.T. (1975), *Bryologist,* **78,** 57–73.
Bold, H.C. (1948), *Bryologist,* **51,** 55–63.
Booer, J.R. (1951), *Ann. App. Biol.,* **38,** 334–47.
Borystwski, Z.R. (1978), *Acta Soc. Bot. Pol.,* **47,** 15–23.
Boyle, R.W. (1977), *J. Geochem. Exploration,* **8,** 495–527.
Brehm, K. (1968), *Planta,* **79,** 324–45.
Brehm, K. (1970), *Beitr. Biol. Pflanzen,* **47,** 91–116.
Brehm, K. (1971), *Beitr. Biol. Pflanzen,* **47,** 287–312.
Bridson, E.Y. and Brecker, A. (1970), Design and formulation of microbial culture media. In: *Methods in Microbiology* (Norris, J.R. and Ribbons, D.W., eds), Vol. 3A, pp. 229–95, Academic Press, London.
Briggs, D. (1965), *J. Ecol.,* **53,** 69–96.
Briggs, D. (1972) *Nature,* **238,** 166–7.
Broady, P.A. (1977), *Br. Antarct. Surv. Bull.,* **45,** 47–62.
Broady, P.A. (1979), *Br. Antarct. Surv. Bull.,* **47,** 13–29.
Brock, T.D. (1973), *Oikos,* **24,** 239–43.
Brooks, R.R. (1971), *N.Z.J. Bot.,* **9,** 674–7.
Brooks, R.R., Yates, T.E. and Ogden, J. (1973), *N.Z.J. Bot.,* **11,** 443–8.
Brown, D.H. (1976), Mineral uptake by lichens. In: *Lichenology: Progress and Problems* (Brown, D.H., Hawksworth, D.L. and Bailey, R.H., eds), pp. 419–39, Academic Press, London.

Brown, D.H. (1980), *Abstract FESPP II Congress, Spain,* **1980,** 248–9.
Brown, D.H. and Bates, J.W. (1972), *J. Bryol.,* **7,** 187–93.
Brown, D.H. and Buck, G.W. (1978a), *J. Bryol.,* **10,** 199–209.
Brown, D.H. and Buck, G.W. (1978b), *Ann. Bot.,* **42,** 923–9.
Brown, D.H. and Buck, G.W. (1979), *New Phytol.,* **82,** 115–25.
Brown, D.H. and House, K.L. (1978), *Ann. Bot.,* **42,** 1383–92.
Buck, G.W. (1980), Ph.D. Thesis, University of Bristol.
Buck, G,W, and Brown, D.H. (1978), *Bryophytorum Bibliotheca,* **13,** 735–50.
Buck, G.W. and Brown, D.H. (1979), *Ann. Bot.,* **44,** 265–77.
Buckholder, P.R. (1959), *Bryologist,* **62,** 6–15.
Burkitt, A., Lester, P. and Nickless, G. (1972), *Nature,* **238,** 327–8.
Burton, M.A.S. and Peterson, P.J. (1979a), *Bryologist,* **82,** 594–8.
Burton, M.A.S. and Peterson, P.J. (1979b), *Env. Poll.,* **19,** 39–46.
Busby, J.R. and Whitfield, D.W.A. (1978), *Can. J. Bot.,* **56,** 1551–8.
Butler, E.J. (1939), *Trans. Br. Mycol. Soc.,* **22,** 274–301.
Callaghan, T.V., Collins, N.J. and Callaghan, C.H. (1978), *Oikos,* **31,** 73–88.
Cameron, A.J. and Nickless, G. (1977), *Water, Air & Soil Poll.,* **7,** 117–25.
Carlisle, A., Brown, A.H.F. and White, E.J. (1967), *J. Ecol.,* **55,** 615–27.
Chalaud, G. (1937), *Rev. Gen. Bot.,* **49,** 111–21.
Chapin, F.S. and Chapin, M.C. (1980), *J. App. Ecol.,* **17,** 449–56.
Chen, T.H. and Jaffe, L.F. (1979), *Planta,* **144,** 401–6.
Chevallier, D. (1973), *Physiol. Vegetal.,* **11,** 461–73.
Chevallier, D. (1974), *Soc. Bot. Fr. Coll. Bryologie,* **121,** 179–89.
Chevallier, D. (1975a), *Bryologist,* **78,** 194–9.
Chevallier, D. (1975b), *Physiol. Plant.,* **34,** 216–20.
Chevallier, D. (1976), *Biologia Plant.,* **18,** 132–9.
Chevallier, D., Fourcy, A. and Kofler, L. (1969), *C.R. Acad. Sci. Paris,* **D268,** 1879–92.
Chevallier, D., Nurit, F. and Pesey, H. (1977), *Ann. Bot.,* **41,** 527–31.
Clough, W.S. (1975), *Atmos. Env.,* **9,** 1113–9.
Clymo, R.S. (1963), *Ann. Bot.,* **27,** 309–24.
Clymo, R.S. (1966), Control of cation concentrations, and in particular of pH, in *Sphagnum*-dominated communities. In: *Chemical Environment in the Aquatic Habitat* (Golterman, H.L. and Clymo, R.S., eds), pp. 273–84, North Holland, Amsterdam.
Clymo, R.S. (1973), *J. Ecol.,* **61,** 849–69.
Coker, P.D. (1967), *Trans. Br. Bryol. Soc.,* **5,** 341–7.
Coker, P.D. (1971), *Trans. Br. Bryol. Soc.,* **6,** 317–22.
Cooke, W.B. (1953), *Bryologist,* **56,** 143–5.
Coombes, A.J. and Lepp, N.W. (1974), *Bryologist,* **77,** 447–52.
Craigie, J.S. and Maass, W.S.C. (1966), *Ann. Bot.,* **30,** 153–4
Cremer, K.W. and Mount, A.B. (1965), *Aust. J. Bot.,* **13,** 303–22.
Crum, H. (1972), *J. Hattori bot. Lab.,* **35,** 269–98.
Czarnowska, K. and Rejment-Grochowska, I. (1974), *Acta Soc. Bot. Pol.,* **43,** 39–44.
Dalby, D.H. (1966), *Fld. Studies,* **2,** 283–301.
Damman, A.W.H. (1978), *Oikos,* **30,** 380–95.

Daniels, B.A. and Trappe, J.M. (1979), *Can. J. Bot.*, **57**, 539–42.
Datta Munshi, J. (1977), *Geobios. Jodhpur*, **4**, 54–6.
Daubenmire, R. (1968), *Adv. Ecol. Res.*, **5**, 209–66.
Dawson, R.B. (1968), *Practical Lawn Craft and Management of Sports Turf*, Crosby Lockwood, London.
Dézsi, L. and Simon, T. (1976), *Acta Bot. Acad. Sci. Hung.*, **22**, 17–28.
Dietert, M.F. (1979), *Bryologist*, **82**, 417–31.
Dietz, F. (1973), The enrichment of heavy metals in submerged plants. In: *Advances in Water Pollution Research* (Jenkins, S.H., ed.), 6th Congress, pp. 53–62, Pergamon Press, Oxford.
Dormaar, J.F. (1968), *Can. J. Earth Sci.*, **5**, 223–30.
Duckett, J.G., Prasad, A.K.S.K., Davis, D.A. and Walker, S. (1977), *New Phytol.*, **79**, 349–62.
Duckett, J.G. and Soni, S.L. (1972), *Bryologist*, **75**, 583–6.
Dunham, V.L. and Bryan, J.K. (1968), *Am. J. Bot.*, **55**, 745–52.
Dykeman, W.R. and DeSousa, A.S. (1966), *Can. J. Bot.*, **44**, 871–8.
Eaton, J.S., Likens, G.E. and Bormann, F.H. (1973), *J. Ecol.*, **61**, 495–508.
Egunyomi, A. (1978), *J. Hattori bot. Lab.*, **44**, 25–30.
Ellison, G., Newham, J., Pinchin, M.J. and Thompson, I. (1976), *Env. Poll.*, **11**, 167–74.
Empain, A. (1976), *Mem. Soc. R. Bot. Belg.*, **7**, 141–56.
Empain, A. (1977), Thèse doct. sc. Bot., Université de Liège.
Empain, A. (1978), *Hydrobiologia*, **60**, 49–74.
Entz, B. (1961), *Ver. Int. Verein. theor. angew. Limnol.*, **14**, 495–8.
Erämetsä, O. and Yliruokanen, I. (1971a), *Suomen Kemi.*, **B44**, 121–8.
Erämetsä, O. and Yliruokanen, I. (1971b), *Suomen Kemi.*, **B44**, 372–4.
Eurola, S. and Hakala, A.V.K. (1977), *Aquilo ser bot.*, **15**, 1–18.
Evans, H.J. and Sorger, G.J. (1966), *Ann. Rev. Pl. Physiol.*, **17**, 47–76.
Eze, J.M.O. and Berrie, G.K. (1977), *Ann. Bot.*, **41**, 351–8.
Fairfax, J.A.W. and Lepp, N.W. (1976), The effects of some atmospheric pollutants on the cation status of two woodland mosses. In: *Proceedings of the Kuopio Meeting on Plant Damages Caused by Air Pollution* (Kärenlampi, L., ed.), pp. 26–36, Kuopion korkeakoulu & Kuopion Luonnon Ystäväin Yhdistys, Kuopio.
Ferreira, R.E.C. and Wormell, P. (1971), *Trans. Bot. Soc. Edinb.*, **41**, 149–54.
Flowers, S. (1933), *Bryologist*, **36**, 34–43.
Folkeson, L. (1979), *Water, Air and Soil Poll.*, **11**, 253–60.
Foote, K.G. (1966), *Bryologist*, **69**, 265–96.
Foulquier, L. and Hébrard, J.P. (1976), *Oecol. Plant.*, **11**, 267–76.
Fraser, D.C. (1961), *Econ. Geol.*, **56**, 951–62.
Fries, N. (1945), *Bot. Not.*, **1945**, 417–24.
Fulford, M., Carroll, G. and Cobbe, T. (1947), *Bryologist*, **50**, 131–46.
Garty, J., Galun, M. and Kessel, M. (1979), *New Phytol.*, **82**, 159–68.
Geldreich, E.E. (1948a), *Bryologist*, **51**, 218–29.
Geldreich, E.E. (1948b), *Bryologist*, **51**, 229–35.
Geldreich, E.E. (1949), *Ohio J. Sci.*, **69**, 191–4.
Gimingham, C.H. (1978), *Vegetatio*, **36**, 179–86.

Gimingham, C.H. and Smith, R.I.L. (1970), Bryophyte and lichen communities in the maritime Antarctic. In: *Antarctic Ecology* (Holdgate, M.W., ed.), Vol. 2, pp. 752–85, Academic Press, London.
Glenn, G.G. (1931), *Proc. Indiana Acad. Sci.*, **40,** 87–101.
Glooschenko, W.A. and Capobianco, J.A. (1978), *Water, Air and Soil Poll.*, **10,** 215–20.
Goodman, G.T. and Roberts, T.M. (1971), *Nature,* **231,** 287–92.
Gorham, E. and Tilton, D.L. (1978), *Can. J. Bot.*, **56,** 2755–9.
Graff, P.W. (1936), *Bull. Torrey bot. Club,* **63,** 67–74.
Granhall, U. and Hofsten, A.V. (1976), *Physiol. Plant.*, **36,** 88–94.
Granhall, U. and Selander, H. (1973), *Oikos,* **24,** 8–15.
Griggs, R.F. (1933), *Am. J. Bot.*, **20,** 92–113.
Griggs, R.F. and Ready, D. (1934), *Am. J. Bot.*, **21,** 265–77.
Grodzińska, K. (1978), *Water, Air and Soil Poll.*, **9,** 83–97.
Groet, S.S. (1976), *Oikos,* **27,** 445–56.
Grønlie, A.M. (1948), *Nytt. Mag. Naturvidensk,* **86,** 117–243.
Grubb, P.J. (1961), *Trans. Br. Bryol. Soc.*, **4,** 184.
Grubb, P.J. (1965), *Trans. Br. Bryol. Soc.*, **4,** 900.
Grubb, P.J. (1968), *Trans. Br. Bryol. Soc.*, **5,** 654.
Grubb, P.J., Flint, O.P. and Gregory, S.C. (1969), *Trans. Br. Bryol. Soc.*, **5,** 802–17.
Gullvåg, B.M., Skaar, H. and Ophus, E.M. (1974), *J. Bryol.*, **8,** 117–22.
Gupta, R.K. (1976), *Biochem. Physiol. Pflanzen,* **170S,** 389–95.
Gupta, R.K. (1977), *Can. J. Bot.*, **55,** 1186–94.
Hancock, J.A. and Brassard, G.R. (1974), *Can. J. Bot.*, **52,** 1861–5.
Hargreaves, J.W., Lloyd, E.J.H. and Whitton, B.A. (1975), *Freshwater Biol.*, **5,** 563–76.
Harrington, A.J. (1966a), *Trans. Br. Bryol. Soc.*, **5,** 212.
Harrington, A.J. (1966b), PhD. Thesis, University of London.
Hébrard, J.-P., Foulquier, L. and Grauby, A. (1974), *Bull. Soc. Bot. France,* **121,** 235–50.
Henriksson, E., Henriksson, L.E. and Pejler, B. (1972), *Surtsey Res. Prog. Reps,* **6,** 66–8.
Henriksson, L.E. and Henriksson, E. (1974), *Surtsey Res. Prog. Reps,* **7,** 30–44.
Hildebrand, R., Kottke, I. and Winkler, S. (1978), *Beitr. Biol. Pflanzen,* **54,** 1–12.
Hoffman, G.R. (1966a), *Bryologist,* **69,** 182–92.
Hoffman, G.R. (1966b), *Ecol. Monogr.*, **36,** 157–80.
Hoffman, G.R. (1970), *Bryologist,* **73,** 634–5.
Hoffman, G.R. (1972), *Bot. Gaz.*, **133,** 107–119.
Holmes, N.T.H. and Whitton, B.A. (1977), *Freshwater Biol.*, **7,** 43–63.
Horne, A.L. (1972), *Br. Antarct. Surv. Bull.*, **27,** 1–18.
Huckabee, J.W. (1973), *Atmos. Env.*, **7,** 749–54.
Huckabee, J.W., Goodyear, C.P. and Jones, R.D. (1975), *Trans. Am. Fisheries Soc.*, **104,** 677–84.
Huckabee, J.W. and Janzen, S.A. (1975), *Chemosphere,* **4,** 55–60.
Ikenberry, G. (1936), *Am. J. Bot.*, **23,** 271–8.
Iwasa, K. (1965), *Pl. Cell. Physiol.*, **6,** 421–9.
Jackson, N. (1961), *J. Sports Turf. Res. Inst.*, **37,** 264–75.

Jefferies, R.L. (1969), *Trans. Br. Bryol. Soc.*, **5**, 901.
Jefferies, R.L., Laycock, D., Stewart, G.R. and Sims, A.P. (1969), The properties of mechanisms involved in the uptake and utilisation of calcium and potassium by plants in relation to an understanding of plant distribution. In: *Ecological Aspects of the Mineral Nutrition of Plants* (Rorison, I.H., ed.), pp. 281–307, Blackwell Scientific Publications, Oxford.
Jeffrey, D.W. and Pigott, C.D. (1973), *J. Ecol.*, **61**, 85–92.
Joenje, W. and During, H.J. (1977), *Vegetatio*, **35**, 177–85.
Johnsen, I. and Rasmussen, L. (1977), *Bryologist*, **80**, 625–9.
Johnson, P.N. (1977), *New Phytol.*, **78**, 161–70.
Jones, K. and Wilson, R.E. (1978), *Ecol. Bull.*, **26**, 158–63.
Kallio, P. and Kärenlampi, L. (1975), Photosynthesis in mosses and lichens. In: *Photosynthesis and Productivity in Different Environments* (Cooper, J.P., ed.), pp. 393–423, Cambridge University Press, Cambridge.
Keck, K. and Stich, H. (1957), *Ann. Bot.*, **21**, 611–9.
Kendrick, W.B. (1962), *Can. J. Microbiol.*, **8**, 639–47.
Kenworthy, J.B. (1978), *Trans. Bot. Soc. Edinb.*, Suppl. **2**, 43–50.
Killian, C. (1923), *Soc. Biol. Paris C.R.*, **88**, 746–8.
Knight, A.H., Crooke, W.M. and Inkson, R.M.E. (1961), *Nature*, **192**, 142–3.
Kofler, L. (1959), *Rev. bryol. lichenol.*, **28**, 1–202.
Lambert, R.L. and Reiners, W.A. (1979), *Arctic Alpine Res.*, **11**, 325–33.
Lang, G.E., Reiners, W.A. and Heier, R.K. (1976), *Oecologia*, **25**, 229–41.
Lawrey, J.D. (1977), *Mycologia*, **69**, 855–60.
Lawrey, J.D. (1978), *Bull. Torrey bot. Club*, **105**, 201–4.
Lazrus, A.L., Lorange, E. and Lodge, J.P. (1970), *Env. Sci. Technol.*, **4**, 55–8.
LeBlanc, F. and Rao, D.N. (1974), *Soc. bot. Fr. Coll. Bryologie*, **121**, 237–55.
Lee, J., Brooks, R.R., Reeves, R.D. and Jaffre, T. (1977), *Bryologist*, **80**, 203–5.
Lee, J.A. and Stewart, G.R. (1978), *Adv. Bot. Res.*, **6**, 2–43.
Lepp, N.W. and Fairfax, J.A.W. (1976), The role of acid rain as a regulator of foliar nutrient uptake and loss. In: *Microbiology of Aerial Plant Surfaces* (Dickinson, C.H. and Preece, T.F., eds), pp. 107–18, Academic Press, London.
Lewis, K. (1973), *Freshwater Biol.*, **3**, 251–7.
Lewis, K. (1974), Ph.D. Thesis, University of Wales.
Little, P. (1979), *Agric. Avi.*, **20**, 129–44.
Little, P. and Martin, M.H. (1974), *Env. Poll.*, **6**, 1–19.
Longton, R.E. (1980), Physiological ecology of mosses. In: *Mosses of North America* (Taylor, R.J. and Leviton, A.E., eds), pp. 77–113, Pacific Division, American Academy Advancement of Science, San Francisco.
Longton, R.E. and Greene, S.W. (1979), *J. Bryol.*, **10**, 321–38.
Lötschert, W. and Köhm, H.-J. (1978), *Oecologia*, **37**, 121–32.
Lötschert, W., Wandtner, R. and Hiller, H. (1975), *Ber. Deutsch. Bot. Ges.*, **88**, 419–31.
Lounemaa, K.J. (1956), *Annal. Bot. Soc. Zoo. Bot. Fenn. Vanamo*, **29**, 1–196.
Machlis, L. (1962), *Physiol. Plant.*, **15**, 354–62.
McLean, R.O. and Jones, A.K. (1975), *Freshwater Biol.*, **5**, 431–44.
Madgwick, H.A.I. and Ovington, J.D. (1959), *Forestry*, **32**, 14–27.

Maikawa, E. and Kershaw, K.A. (1976), *Can. J. Bot.*, **54**, 2679–87.
Mäkinen, A. (1977), *Suo*, **28**, 79–88.
Marshall, R. and Averill, C. (1928), *Ecology*, **9**, 533.
Mårtensson, O. (1956), *K.V.A. avh. i. Naturskydds.*, **14**, 1–321.
Mattsson, L.J.S. (1975a), *Health Phys.*, **28**, 233–48.
Mattsson, L.J.S. (1975b), *Health Phys.*, **29**, 27–41.
Mattsson, L.J.S. and Lidén, K. (1975), *Oikos*, **26**, 323–7.
Mayer, A.M. and Gorham, E. (1951), *Ann. Bot.*, **15**, 247–63.
Mickiewicz, J. (1976), *Polish Ecol. Stud.*, **2**, 57–62.
Miles, J. (1968), *Trans. Br. Bryol. Soc.*, **5**, 587.
Miles, J. (1973), *J. Ecol.*, **61**, 399–412.
Miller, N.G. and Ambrose, L.J.H. (1976), *Bryologist*, **79**, 55–63.
Mitchell, R.L. (1960), *J. Sci. Food Agric.*, **10**, 553–60.
Mizra, R.A. and Shimwell, D.W. (1977), *J. Bryol.*, **9**, 565–72.
Montague, M.J., Otero, J.G., Evans, L.W., Irland, J.M. and Swartz, I.J. (1969), *Bryologist*, **72**, 53–5.
Montgomery, C.E. (1931), *Bot. Gaz.*, **91**, 225–51.
Moore, C.J. and Scott, G.A.M. (1979), *J. Bryol.*, **10**, 291–311.
Nagano, I. (1969), *J. Hattori bot. Lab.*, **32**, 155–203.
Nagano, I. (1972), *J. Hattori bot. Lab.*, **35**, 391–8.
Nagano, I. and Shimizu, Y. (1972), *J. Saitama Univ.*, **8**, 29–44.
Nehira, K. (1973), *Mem. Fac. Gen. Ed. Hiroshima Univ. III*, **7**, 1–5.
Nieboer, E. and Richardson, D.H.S. (1980), *Env. Poll. B*, **1**, 3–26.
Nieboer, E., Richardson, D.H.S. and Tomassini, F.D. (1978), *Bryologist*, **81**, 226–46.
Nihlgård, B. (1970), *Oikos*, **21**, 208–17.
Noguchi, A. (1956), *Kumamoto J. Sci. Ser. B, Sect. 2*, **2**, 239–57.
Noguchi, A. and Furuta, H. (1956), *J. Hattori bot. Lab.*, **17**, 32–44.
Odu, E.A. (1978), *J. Bryol.*, **10**, 163–81.
Odu, E.A. (1979), *Ann. Bot.*, **44**, 147–52.
Olarinmoye, S.O. (1974), *J. Bryol.*, **8**, 275–89.
Olarinmoye, S.O. (1975), *J. Bryol.*, **8**, 357–63.
Oliver, R.L.A. (1971), *Trans, Br. Bryol. Soc.*, **6**, 296–305.
Ophus, E.M. and Gullvåg, B.M. (1974), *Cytobios*, **10**, 45–58.
O'Toole, M.A. and Synnott, D.M. (1971), *J. Ecol.*, **59**, 121–5.
Painter, T.J. and Sørensen, N.A. (1978), *Carbohydrate Res.*, **66**, C1–C3.
Pakarinen, P. (1978a), *Ann. Bot. Fenn.*, **15**, 15–26.
Pakarinen, P. (1978b), *Bryophytorum Bibliotheca*, **13**, 751–62.
Pakarinen, P. (1978c), *Ann. Bot. Fenn.*, **15**, 287–92.
Pakarinen, P. and Rinne, R.J.K. (1979), *Lindbergia*, **5**, 73–83.
Pakarinen, P. and Tolonen, K. (1976), *Ambio*, **5**, 38–40.
Pakarinen, P. and Tolonen, K. (1977a), *Lindbergia*, **4**, 27–33.
Pakarinen, P. and Tolonen, K. (1977b), *Suo*, **28**, 95–102.
Paolillo, D.J. and Kass, L.B. (1973), *Bryologist*, **76**, 163–8.
Parihar, N.S. and Pant, C.B. (1975), *Current Sci.*, **44**, 61–2.
Parke, J.L. and Linderman, R.G. (1980), *Can. J. Bot.*, **58**, 1898–904.
Patterson, P.M. (1946), *Am. J. Bot.*, **33**, 604–11.

Pentecost, A. (1980), *J. Ecol.*, **68**, 251–67.
Persson, H. (1948), *Revue bryol. lichénol.*, **17**, 75–8.
Persson, H. (1956), *J. Hattori bot. Lab.*, **17**, 1–18.
Pickering, D.C. and Puia, I.L. (1969), *Physiol. Plant.*, **22**, 653–61.
Pigott, C.D. (1958), *Trans. Br. Bryol. Soc.*, **3**, 382.
Pilegaard, K. (1979), *Water, Air Soil Poll.*, **11**, 77–91.
Pilegaard, K., Rasmussen, L. and Gydesen, H. (1979), *J. Appl. Ecol.*, **16**, 843–53.
Pitkin, P.H. (1975), *J. Bryol.*, **8**, 337–56.
Prosi, F. (1979), Heavy metals in aquatic organisms. In: *Metal Pollution in the Aquatic Environment* (Förstner, U. and Wittmann, G.T.W., eds), pp. 271–323, Springer–Verlag, Berlin.
Puri, P. (1973), *Bryophytes: A Broad Perspective*, Atma Ram & Sons, Delhi.
Puustjärvi, V. (1959), *J. Sci. Agr. Soc. Finland*, **31**, 103–19.
Rabatin, S.C. (1980), *Mycologia*, **72**, 191–5.
Raison, R.J. (1979), *Plant Soil*, **51**, 73–108.
Ramanovsky, S. and Kushner, D.J. (1975), *Microbiol Ecl.*, **2**, 162–76.
Ramaut, J. (1955a), *Bull. Acad. Roy. Belgique 5th ser.*, **41**, 1137–55.
Ramaut, J. (1955b), *Bull. Acad. Roy. Belgique 5th ser.*, **41**, 1168–99.
Rao, D.N., Robitaille, G. and LeBlanc, F. (1977), *J. Hattori bot. Lab.*, **42**, 213–39.
Rasmussen, L. (1977), *Env. Poll.*, **14**, 37–45.
Rasmussen, L. (1978), *Lindbergia*, **4**, 909–18.
Rasmussen, L. and Johnsen, I. (1976), *Oikos*, **27**, 483–7.
Rastorfer, J.R. (1972), Comparative physiology of four west antarctic mosses. In: *Antarctic Terrestrial Biology* (Llano, G.A., ed.), pp. 143–61, American Geophysical Union, Washington.
Rastorfer, J.R. (1974), *Ohio J. Sci.*, **74**, 55–9.
Ratcliffe, J.M. (1975), *Atmos. Env.*, **9**, 623–9.
Raven, J.A. (1970), *Biol. Rev.*, **45**, 167–221.
Rawes, M. and Hobbs, R. (1979), *J. Ecol.*, **67**, 789–807.
Richards, P.W. (1959), Bryophyta. In: *Vistas in Botany* (Turrill, W.B., ed.), Vol. I, pp. 387–420, Pergamon Press, London.
Richardson, D.H.S. (1981), *The Biology of Mosses*, Blackwell Scientific Publications, Oxford.
Richardson, J.A. (1958), *J. Ecol.*, **46**, 537–46.
Rieley, J.O., Richards, P.W. and Bebington, A.D.L. (1979), *J. Ecol.*, **67**, 497–527.
Rinne, R.J.K. and Barclay-Estrup, P. (1980), *Oikos*, **34**, 59–67.
Roberts, T.M. (1975), *Symp. Proc. Int. Conf. Heavy Metals in the Environment, Toronto*, **2**, 503–32.
Rodgers, G.A. and Henriksson, E. (1976), *Acta Bot. Isl.*, **4**, 10–15.
Rodgers, G.A. and Stewart, W.D.P., (1977), *New Phytol.*, **78**, 441–58.
Rühling, Å. and Tyler, G. (1970), *Oikos*, **21**, 92–7.
Rühling, Å. and Tyler, G. (1971), *J. appl. Ecol.*, **8**, 497–507.
Sainsbury, C.L., Hamilton, J.C. and Huffman, C. (1966), *U.S. Geol. Survey Bull.*, **1242F**, 1–42.
Sand-Jensen, K. and Rasmussen, L. (1978), *Bot. Tidsskr.*, **72**, 105–12.
Saxena, P.K. and Rashid, A. (1980), *Z. Pflanzenphysiol.*, **99**, 373–7.
Schatz, A. (1955), *Bryologist*, **58**, 113–20.

Schlesinger, W.H. and Reiners, W.A. (1974), *Ecology,* **55,** 378–86.
Schneider, M.J., Voth, P.D. and Troxler, R.F. (1967), *Bot. Gaz.,* **128,** 169–74.
Schofield, E. and Ahmadjian, V. (1972), Field observations and laboratory studies of some Antarctic cold desert cryptogams. In: *Antarctic Terrestrial Biology* (Llano, G.A., ed.), pp. 97–141, American Geophysical Union, Washington.
Schuler, J.F., Diller, V.M., Fulford, M. and Kerstein, H.J. (1955), *Pl. Physiol.,* **30,** 478–82.
Schulz, D. and Weincke, C. (1976), *Flora,* **165,** 47–60.
Schwabe, G.H. (1974), *Surtsey Res. Prog. Reps.,* **7,** 22–5.
Schwarzmaier, U. and Brehm, K. (1975), *Z. Pflanzenphysiol.,* **75,** 250–5.
Schwoerbel, J. and Tillmanns, G.C. (1974), *Arch. Hydrobiol.,* **47,** 282–94.
Seltzer, R.C. and Wistendahl, W.A. (1971), *Bryologist,* **74,** 28–32.
Seymour, A.B. (1929), *Host index of the fungi of North America,* Harvard University Press, Cambridge.
Shacklette, H.T. (1961), *Bryologist,* **64,** 1–16.
Shacklette, H.T. (1965a), *U.S. Geol. Survey Bull.,* **1198D,** 1–21.
Shacklette, H.T. (1965b), *U.S. Geol. Survey Bull.,* **1198C,** 1–17.
Shacklette, H.T. (1967), *U.S. Geol. Survey Bull.,* **1198G,** 1–18.
Shimwell, D.W. and Laurie, A.E. (1972), *Env. Poll.,* **3,** 291–301.
Simola, L.K. (1975), *Physiol. Plant.,* **35,** 194–9.
Simola, L.K. (1979), *J. Hattori bot. Lab.,* **46,** 49–54.
Sinclair, J. (1967), *J. exp. Bot.,* **18,** 594–9.
Sironval, C. (1947), *Bull. Soc. r. Bot. Belgique,* **79,** 48–78.
Skaar, H., Ophus, E. and Gullvåg, B.M. (1973), *Nature,* **241,** 215–6.
Skene, M. (1915), *Ann. Bot.,* **29,** 65–87.
Skogerboe, R.K., Hartley, A.M., Vogel, R.S. and Keirkyohann, S.R. (1979), Monitoring lead in the environment. In: *Lead in the Environment* (Boggess, W.R. and Wixson, B.G., eds), pp. 33–70, Castle House Publications Ltd.
Skutch, A.F. (1929), *Ecology,* **10,** 177–89.
Sluet van Oldruitenborgh, C.J.M. and Heeres, E. (1969), *Acta bot. neerl.,* **18,** 315–24.
Smith, R.I.L. (1978), *J. Ecol.,* **66,** 891–909.
Smith, R.I.L. and Gimingham, C.H. (1976), *Br. Antarct. Surv. Bull.,* **43,** 25–47.
Snyder, J.M. and Wullstein, L.H. (1973), *Bryologist,* **76,** 196–9.
Solberg, Y. and Selmer-Olsen, A.R. (1978), *Bryologist,* **81,** 144–9.
Southorn, A.D.L. (1976), *J. Bryol.,* **9,** 63–80.
Southorn, A.D.L. (1977), *J. Bryol.,* **9,** 361–73.
Spearing, A.M. (1972), *Bryologist,* **75,** 154–8.
Stahl, M. (1949), *Planta,* **37,** 103–48.
Steinnes, E. (1977), *Kjeller Report,* **KR-154,** 1–29.
Stewart, W.D.P. (1967), *Nature,* **214,** 603–4.
Stewart, W.D.P. and Rodgers, G.A. (1977), *New Phytol.,* **78,** 459–71.
Stewart, W.D.P. and Rodgers, G.A. (1978), *Ecol. Bull.,* **26,** 247–59.
Streeter, D.T. (1965), *Trans. Br. Bryol. Soc.,* **4,** 818–27.
Streeter, D.T. (1970), *Science Progrees,* **58,** 419–34.
Strumm, W. and Morgan, J.J. (1970), *Aquatic Chemistry,* Wiley, New York.

Studlar, S.M. (1980), *Bryologist,* **83,** 301–13.
Stumm, I., Meyer, Y. and Abel, W.O. (1975), *Plant Sci. Letters,* **5,** 113–8.
Summerhayes, V.S. and Williams, P.H. (1926), *J. Ecol.,* **14,** 203–43.
Svensson, G.K. and Lidén, K. (1965), *Health Phys.,* **11,** 1033–42.
Szabó, M. and Csortos, Cs. (1975), *Acta Bot. Acad. Sci. Hung.,* **21,** 419–32.
Tallis, J.H. (1959), *J. Ecol.,* **47,** 325–50.
Tamm, C.O. (1953), *Meddn. Stat. Skogsforskinst.,* **43,** 1–140.
Tamm, C.O. (1964), *Bryologist,* **67,** 423–6.
Tanaka, R. and Sato, S. (1977), *Cytologia,* **42,** 383–7.
Taylor, F.G. and Witherspoon, J.P. (1972), *Health Phys.,* **23,** 867–9.
Tenge, F.-K. (1959), *Z. Bot.,* **47,** 287–305.
Thomas, H. (1970), Ph.D. Thesis, University of Bristol.
Trachtenberg, S. and Zamski, E. (1978), *J. exp. Bot.,* **29,** 719–27.
Trachtenberg, S. and Zamski, E. (1979), *New Phytol.,* **83,** 49–52.
Tuominen, Y. and Jaakkola, T. (1974), Absorption and accumulation of mineral elements and radioactive nuclides. In: *The Lichens* (Ahmadjian, V. and Hale, M.E., eds), pp. 185–223, Academic Press, New York.
Url. W. (1956), *Protoplasma,* **46,** 768–95.
Usui, H., Arihara, T. and Shimada, R. (1975), *Bull. Coll. Agric. Utsunomiya Univ.,* **9,** 25–36.
Vaarama, A. and Tarén, N. (1963), *J. Linn. Soc. (Bot.),* **58,** 297–304.
Vlassak, K., Paul, F.A. and Harris, G.C. (1973), *Plant Soil,* **38,** 637–49.
Voth, P.D. (1943), *Bot. Gaz.,* **104,** 591–601.
Voth, P.D. and Hamner, K.C. (1940), *Bot. Gaz.,* **102,** 169–205.
Wallin, T. (1976), *Env. Poll.,* **10,** 101–14.
Watson, E.V. (1960), *J. Ecol.,* **48,** 397–414.
Watson, E.V. (1971), *The Structure and Life of Bryophytes,* 3rd edn. Hutchinson, London.
Wattez, J.R. (1976), Les bryophytes aquatiques et sub-aquatiques, bioindicateurs de la pollution des eaux douces. In: *La pollution des eaux continentales* (Pesson, P., ed.), pp. 173–82, Gauthier-Villars, Paris.
Weetman, G.F. and Timmer, V. (1967), *Pulp & Paper Res. Inst. Canada Tech. Rep.,* **503,** 1–38.
Whitehead, N.E. and Brooks, R.R. (1969), *Bryologist,* **72,** 501–7.
Whitmore, R. (1965), *Bryologist,* **68,** 342–3.
Whitton, B.A. (1972), *Symp. Zool. Soc. Lond.,* **29,** 3–19.
Whitton, B.A. and Say, P.J. (1975), Heavy metals. In: *River Ecology* (Whitton, B.A., ed.), pp. 286–311, Blackwell Scientific Publications, Oxford.
Wilkins, P. (1977), *Bryologist,* **80,** 175–81.
Williams, S. (1950), *Trans. Br. Bryol. Soc.,* **1,** 357–66.
Willis, A.J. (1963), *J. Ecol.,* **51,** 353–74.
Willis, A.J. (1964), *Trans. Br. Bryol. Soc.,* **4,** 668–83.
Wilmot-Dear, C.M. (1980), *J. Bryol.,* **11,** 145–60.
Witkamp. M. and Frank, M.L. (1967), *Health Phys.,* **13,** 985–90.
Woolhouse, A.R. (1972), *J. Sports Turf Res. Inst.,* **48,** 102–7.
Woollon, F.B.M. (1975), *J. Bryol.,* **8,** 455–64.
Yarranton, G.A. (1967), *Can. J. Bot.,* **45,** 249–58.

Yarranton, G.A. (1970), *Can. J. Bot.*, **48**, 1387–404.
Yarranton, G.A. and Beasleigh, W.J. (1968), *Can. J. Bot.*, **46**, 1591–9.
Yarranton, G.A. and Beasleigh, W.J. (1969), *Can. J. Bot.*, **47**, 959–74.
Young, C. (1938), *Proc. Indiana Acad. Sci.*, **47**, 106–15.
Zamski, E. and Trachtenberg, S. (1976), *Isr. J. Bot.*, **25**, 168–73.

Chapter 12

Responses of Bryophytes to Air Pollution

DHRUVA N. RAO

12.1 INTRODUCTION

The study of the response of plants as biological systems to air pollution is a convenient method for recognizing the presence and identity of airborne contaminants. With increasing concern about air pollution as a factor in ecosystem disturbance, there is a need to assess the effects and relative sensitivity of the various groups of plants to different air pollutants.

Bryophytes, which are the simplest of green land plants, lack a vascular system and are simple both morphologically and anatomically. The growth potential in bryophytes is not as highly polarized as in vascular plants. Often, in mosses dormancy or death of the growing point may stimulate the development of dormant buds lower down the stem. The plants can reproduce vegetatively with great ease; any living cell is regeneratively totipotent and can produce a new plant (Steere, 1970).

Bryophytes grow in a variety of habitats but especially in moist places on soil, rocks, trunks and branches of trees and fallen logs. They obtain many of their nutrients from substances dissolved in the ambient moisture. Some substances are probably absorbed directly from the substrate by diffusion through the cells of the gametophyte. Like all land plants, bryophyte diversity is a function of the climate and the substrate where they grow. There are species with habitats so circumscribed that their presence reveals precise ecological conditions: *Leskea polycarpa* (in Europe) and *Leskea nervosa* (in America) on tree bark subjected to flood water; *Funaria hygrometrica* on burnt soil where the substrate has a high pH and a high nutrient content, especially potash; cuprophile communities or 'copper mosses', chiefly *Mielichhoferia, Dryptodon* and *Merceya* species, on substrates rich in copper (Persson, 1956; Shacklette, 1961, 1966). It has been suggested that some bryophytes could serve as a useful tool for foresters or prospectors in determining the condition and character of the land (Watson, 1947). There is ample evidence that some bryophytes can be used as reliable indicators of air pollution (LeBlanc and Rao, 1974, 1975). Taoda (1973a) suggested the use of bryophytes as bryo-meters, instruments for measuring phytotoxic air pollution.

12.2 BRYOPHYTES VERSUS AIR POLLUTION

Ecologists in several countries have noted the impoverishment of bryophytic communities in and around cities and industrial areas (Barkman, 1961; LeBlanc, 1961; Rao and LeBlanc, 1967; Gilbert, 1968; Daly, 1970). Many bryophytes have become extinct in urban industrial environments because of their sensitivity to polluted air (LeBlanc and Rao, 1973a). According to Barkman (1969), within the last century the Dutch flora has lost 15% of its terrestrial bryophytes and 13% of its epiphytic bryophytes, and in Amsterdam alone 23 species of bryophytes which occurred in the year 1900 are now extinct. Delvosalle *et al.* (1969) observed that, among some 600 species of bryophytes which were indigenous to Belgium in 1850, nearly 114 have disappeared and 34 are now very rare. They attribute this change to direct and indirect effects of human activities, especially to air pollution. LeBlanc and De Sloover (1970) reported that some bryophyte species which were common on Mount Royal in the city of Montreal about 50 years ago are now extremely rare and some species have disappeared altogether. Similarly, many of the most common bryophytes in Northumberland, England are totally absent from a large part of the lower Tyne Valley where scattered colliery towns, burning pit heaps, and the huge conurbation of Newcastle-upon-Tyne are sources of air pollution (Gilbert, 1968).

Transplantation experiments have shown that bryophytes die within a short period of time, depending on the level of pollution, when transferred along with their substrates from unpolluted to polluted areas in a city or around a factory (LeBlanc and Rao, 1966; Daly, 1970; LeBlanc *et al.*, 1971, 1976).

The major air pollutants in metropolitan areas are usually smoke, containing sulphur dioxide, soot and fly ash, and automobile exhaust fumes containing carbon monoxide, hydrocarbons, nitrogen oxides, sulphur dioxide, aldehydes, lead and other compounds. These pollutants can affect organisms either individually or in combination of two or more. When in combination, their effect can be antagonistic, synergistic or neutral. The nature of emissions from an industrial plant depend upon the fuel burned and the product manufactured; for example, sulphur dioxide is emitted from thermal power plants or iron sintering plants, and hydrogen fluoride from phosphate fertilizer plants or aluminium factories.

Gaseous pollutants so far investigated with respect to bryophyte sensitivity are sulphur dioxide (SO_2), hydrogen fluoride (HF) and ozone (O_3). The presence of particulate pollutants, such as air-borne minerals including heavy metals and radioactive materials, have been detected using bryophytes.

Since urban and industrial environments comprise a series of habitats with a variety of substrates and moisture regimes, and are subject to varying levels of pollution, it will be useful to consider the substrate, life-form and reproductive potential of species while assessing their responses to air pollution.

12.2.1 Substrate and life-form affecting species response

Bryophytes occupying certain substrates appear to be more sensitive to air pollution than others. Species diversity in a polluted area varies not only with the distance from the source of pollution but also with the type of substrate (Table 12.1). Generally, the order of increasing sensitivity is from terricolous to saxicolous and corticolous species. Species growing on trees are generally far more sensitive than those growing on other substrata.

While considering pollution toxicity to bryophytes, attention should be given also to the effects induced in them via the polluted substratum as, for example, the bark in the case of epiphytes. Earlier this was ignored due to a preconception that epiphytes depend primarily on the air environment for their metabolic needs of gases, moisture and minerals, and that they do not derive any nutritional benefit from the underlying bark. In this connection, it should be remembered that pollutants are regularly adsorbed on the bark surface, that epiphytes are occasionally flooded with rainwater charged with pollutants washed out from the air, from the tree canopy and trunk, and that vegetation propagules, such as gemmae, protonemata, etc., after lodging in bark crevices, are greatly influenced by the physico-chemical properties of the bark during ecasis. Therefore, air pollutants, either in a gaseous state mixed with air or in a liquid state affected by dew, rain or snow, will be noxious to bryophytes attached to the bark.

Robitaille *et al.* (1977) while studying the factors contributing to the paucity of epiphytic cryptogams in the vicinity of a copper smelter in Canada, noted that the factors determining the sensitivity of an epiphytic community to SO_2 pollution, in order of their importance are: ambient SO_2 concentration, stem-flow pH, bark pH, epiphyte pH, buffer capacity of bark and relative percentages of SO_2 derivatives (bisulphite ions and sulphurous acid) in and around the bryophyte plant. According to them, all these factors linked together would provide a sound basis for ascertaining the overall effects of SO_2 pollution on epiphytes.

From growth and distribution study of bryophytes in Christchurch, New

Table 12.1 Minimum distances (km) from the centre of Newcastle-upon-Tyne at which bryophytes were recorded on different substrates (modified after Gilbert, 1970a).

	Asbestos roofs	Sandstone (wall tops)	Ash trees (base)	Short grassland
Ceratodon purpureus	1.6	1.6		1.6
Bryum argenteum	1.6	1.6		
B. capillare	3.2	1.6		
Tortula muralis	3.2	1.6		
Grimmia pulvinata	6.4	11.2		
Orthotrichum diaphanum	8.0		17.6	
Tortula ruralis	11.2			
Orthotrichum anomalum	12.8			
Hypnum cupressiforme	14.4	8.0	16.0	4.8
Schistidium apocarpum	14.4	16.0		
Homalothecium sericeum	14.4	9.6		
Brachythecium rutabulum	16.0	8.0		1.6
Funaria hygrometrica		1.6		
Leptobryum pyriforme		1.6		
Barbula convoluta		1.6		
Dicranoweisia cirrata		11.2	17.6	
Barbula cylindrica		11.2	17.0	
Rhynchostegium murale		11.2		
Eurhynchium praelongum		12.8		1.6
Mnium hornum		12.8		11.2
Lophocolea bidentata		12.8		8.0
Pohlia nutans		12.8		4.8
Encalypta streptocarpa		14.4		
Orthotrichum affine			17.6	
Rhytidiadelphus squarrosus				8.0
Atrichum undulatum				12.8
Calliergon cuspidatum				12.8
Fissidens taxifolius				4.8
Pseudoscleropodium purum				14.4
Plagiomnium undulatum				14.4
Thuidium tamariscinum				12.8

Zealand, Daly (1970) observed that bryophyte species on stone walls can tolerate higher levels of pollution than those on tree trunks (Table 12.2). In the case of epiphytes, species growing on tree bases can cope with the pollution conditions better than those on tree trunks (Table 12.3). It has been suggested that shelter, by reducing the level of pollution, and a high pH of the substrate, by increasing the degree of ionization and the rate of oxidation of acidic pollutants, provide favourable conditions for survival of species (Gilbert, 1968). Perhaps this explains why, in a city ecosystem, the shady and alkaline niches are more preferred by bryophytes.

Table 12.2 Maximum concentrations of sulphur dioxide ($\mu g\ SO_2\ m^{-3}$ of air) tolerated by bryophytes on tree trunks and stone walls in Christchurch, N.Z. (modified after Daly, 1970).

	Tree trunks	Stone walls
Eucalypta vulgaris	5	
Triguetella papillata		40
Bryum blandum		40
Brachythecium rutabulum	10	50
Tortula princeps	10	100
Bryum rubrum	50	75
Campylopus sp.	50	50
Lunularia sp.		100
Bryum laevigatum		100
Rhynchostegiella muricata	75	100
Lophocolea sp.	75	100
Hypnum cupressiforme	100	75
Grimmia pulvinata		110
Tortula muralis	100	125
Ceratodon purpureus	125	125
Pohlia cruda	125	125
Bryum argenteum	125	125
Pottia macrophylla		125
Funaria hygrometrica		125

It is reported that under a regime of nutrient flushing of the habitat, certain species, such as *Brachythecium rutabulum, Grimmia pulvinata, Orthotrichum diaphanum, Bryum capillare, B. argenteum, Tortula muralis, Rhynchostegium confertum* and *Hypnum cupressiforme* show enhanced survival and luxuriant growth in polluted environments (Gilbert, 1970a).

Besides habitat, the life-forms of bryophytes can to a certain extent modify the effects of pollution. According to Gilbert (1970a), the general trend of increasing resistance to pollution among life-forms would be in the order of (i) rough mats, tall turfs, wefts, large cushions and leafy liverworts, (ii) smooth mats and small cushions, and (iii) short turfs and thalloid forms. This sequence, however, only suggests a general tendency. The moss protonemata, precursors of the leafy gametophores, appear to be more sensitive than the mature gametophyte (Gilbert, 1969).

12.2.2 Reproductive potential and plant response

It is possible that the reproductive potential of a species determines its degree of success in a polluted environment. Gilbert (1971b) comments that in polluted urban areas, the secret of survival of some bryophytes,

Table 12.3 The percentage occurrence (*) of epiphytic bryophytes on base (B) and trunk (T) of trees in five pollution zones in Wawa, Ontario (modified after Rao and LeBlanc, 1967).

Soil Sulphate (meq 100 g^{-1})	>1.4		0.9–1.4		0.7–0.9		0.4–0.7		<0.4	
Pollution zone	I		II		III		IV		V	
	B	T	B	T	B	T	B	T	B	T
Dicranum flagellare			1		1		2	1	1	1
Drepanocladus uncinatus			1		1	1	1	1	1	1
Ptilidium pulcherrimum			1		2	1	2	1	2	2
Tetraphis pellucida			1		1		1		1	
Blepharostoma trichophyllum					1		1	1		
Brachythecium oxycladon					1					
B. reflexum					1				1	
B. salebrosum					1		1		1	1
Dicranum fulvum					1					
D. montanum					1		1		1	
Fissidens cristatus					1					
Frullania eboracensis						1		1		1
Hypnum pallescens					1		1	1	2	1
Jamesoniella autumnalis					1		1			
Lophocolea heterophylla					1		1		1	
Plagiomnium cuspidatum					1		1	1	3	1
Plagiochila asplenioides					1				1	
Plagiothecium denticulatum					1		1		1	1
P. sylvaticum					1				1	
Platygyrium repens						1	1	1		
Pohlia nutans					1		1		1	
Pylaisia polyantha					1	1	1	1		1
Amblystegium serpens							1			
Campylium chryrophyllum							1		1	
Barbilophozia barbata							1			
Dicranum polysetum							1			
D. scoparium							1			
Heterophyllium haldanianum							1			1
Plagiomnium drummondii							1			
Orthotrichum obtusifolium								1		1
Pleurozium schreberi							1		1	
Platydictya subtile										1
Eurhynchium pulcherrimum									1	
Lepidozia reptans									1	
Lophocolea minor									1	
Oncophorus wahlembergii									1	
Orthotrichum sp.									1	1
Paraleucobryum longifolium										1
Ulota crispa										1
Total number of species (39)	0	0	4	0	20	5	23	10	20	15

*The percentage occurrence of 1–20%, 21–40%, 41–60%, 61–80% and 81–100% are expressed by numbers 1, 2, 3, 4 and 5, respectively.

such as *Bryum argenteum, Ceratodon purpureus, Dicranella heteromalla, Funaria hygrometrica, Leptobryum pyriforme, Lunularia cruciata, Marchantia polymorpha* and *Pohlia proligera* lies in their high reproductive capacity and fast subsequent growth. According to him, these species produce spores and/or gemmae on a very large scale. These propagules, when sown on tap-water agar (pH 5), germinate in less than a day, producing fast growing colonies of protonemata which soon give rise to protonemal buds. The vegetative propagation of these species from stem and leaf fragments would be even faster.

It is now generally accepted that air pollution inhibits sexual reproduction in bryophytes. De Sloover and LeBlanc (1970) stated that fertility is related to vitality; as the pollution level goes down, the percentage frequency of species goes up, subsequently increasing the fertility percentage. However, this situation may vary from species to species depending on the prevailing climatic conditions in the area. Though a decrease in fertility under pollution stress seems to be common to all bryophytes, the rate of decrease may vary in different species (Table 12.4). In *Leskea polycarpa*, there is an increase in percentage presence and of fertility from the most polluted to the least polluted area in Montreal but in *Platygyrium repens,* though there is an increase in percentage presence with decreasing levels of pollution, this species never reproduced sexually even in the least polluted areas. Perhaps fertilization occurs more readily in the monoecious

Table 12.4 Relative frequency of gametophyte (presence) and sporophyte (fertility) of some mosses in different IAP zones in Montreal (modified after LeBlanc and De Sloover, 1970).

IAP zone		I	II	III	IV	V
Number of stations studied		63	48	69	108	61
Brachythecium salebrosum	Presence (%)		4	10	42	32
	Fertility (%)		0	0	2	9
Leskea polycarpa	Presence (%)	3	19	57	89	95
	Fertility (%)	0	0	21	60	76
Orthotrichum obtusifolium	Presence (%)	2	2	1	19	43
	Fertility (%)	0	0	0	15	27
Orthotrichum ohioense	Presence (%)					2
	Fertility (%)					100
Orthotrichum pumilum	Presence (%)				3	7
	Fertility (%)				33	75
Orthotrichum pusillum	Presence (%)					2
	Fertility (%)					100
Platygyrium repens	Presence (%)	2	2	10	12	12
	Fertility (%)	0	0	0	0	0
Pylaisia polyantha	Presence (%)			1	2	3
	Fertility (%)			0	0	50

Leskea than in the dioecious *Platygyrium*. In the latter this may be due partly to the relatively longer exposure to air pollutants of the spermatozoids, leading to failure of fertilization.

12.3 SULPHUR DIOXIDE POLLUTION STUDIES

During the last 15 years serious efforts have been made to study the effects of SO_2 pollution on bryophytes. In these studies the methods that have been employed are mainly ecophysiological and phytosociological. The former method includes fumigation and transplantation and the latter field survey and pollution mapping techniques. By these methods it is possible to relate the external and internal injury symptoms and ecological parameters of species to the level of pollution.

12.3.1 Ecophysiological and bioassay methods

(a) Fumigation techniques

Following the techniques of Rao and LeBlanc (1966), Coker (1967) fumigated the epiphytic bryophytes *Orthotrichum lyellii, O. diaphanum* and *Radula complanata* with 5, 10, 30, 60 and 120 p.p.m. of SO_2. Microscopic examination of leaves of these SO_2-exposed plants showed brownish spots on the chloroplasts at all SO_2-levels, and plasmolysis in cells of leaves exposed to concentrations above 5 p.p.m. Most likely, the necrosis of chloroplasts, which eventually stops further assimilation even when the source of pollution is removed altogether, and the irreversible plasmolysis in cells contributed to the ultimate death of the epiphytes. Studies show that SO_2 absorbed in the plant tissue causes degradation of chlorophyll *a* by increasing the concentration of free H^+ ions, which subsequently displace the Mg^{2+} ions from the chlorophyll molecule, converting it into phaeophytin *a*. The chemical reactions involved in this process are suggested to be as follows (Rao and LeBlanc, 1966; Coker, 1967):

$$SO_2 + H_2O \rightleftharpoons H_2SO_3^{2-}$$
$$H_2SO_3^{2-} \rightleftharpoons HSO_3^- + H^+$$
$$2HSO_3^- + O_2 \rightleftharpoons 2\ SO_4^{2-} + 2H^+$$
$$\text{Chlorophyll } a + 2H^+ \rightarrow \text{Phaeophytin } a + Mg^{2+}$$

Since SO_2 within the plant body acts as sulphurous acid, the level of moisture in the tissue will determine the rate of chlorophyll degradation. Syratt and Wanstall (1969) studied the effects of 5 p.p.m. SO_2 at 0, 20, 40, 80 and 100% relative humidities on chlorophyll breakdown of *Dicranoweisia cirrata, Metzgeria furcata, Hypnum cupressiforme* var. *filiforme, Frullania dilatata, Neckera pumila* and *Ulota crispa*. Damage to chlorophyll showed dependence on the humidity at which the plant was exposed to SO_2. However, under similar conditions, *Dicranoweisia cirrata* was able to withstand higher levels of SO_2 than the other species. Perhaps, its high concentration of chlorophyll (2.897 O.D. units 100 mg^{-1} dry wt.) and its high efficiency in converting SO_3^{2-} into SO_4^{2-} (521.3 μg SO_4^{2-} 100 mg^{-1} dry wt.) enable this moss to cope with the SO_2 pollution better than others.

Respiration rates of some of the above-mentioned species were determined after exposing them for various lengths of time to 5 p.p.m. SO_2 at 100% relative humidity. In *Dicranoweisia* and *Metzgeria,* respiration continued to rise, at least during the experiment, whereas in *Hypnum* it rose to a peak in about 24 hours and then dropped (Syratt and Wanstall, 1969). It is likely that the energy produced by the enhanced rate of respiration is used by the plants in the conversion of SO_3^{2-} into SO_4^{2-}.

According to Taoda (1973b) the toxicity of H_2SO_3 remains stronger than H_2SO_4 at the same pH. Bryophytes treated in neutralized H_2SO_4 remain uninjured while those in neutralized H_2SO_3 are severely injured. High pH did not decrease the toxic effect of H_2SO_3 and he also observed that bryophytes in a dry condition were much more tolerant to SO_2 than in a wet condition.

Comeau and LeBlanc (1971) observed that the regenerative power of *Funaria hygrometrica* leaves exposed to 0.5, 1, 5 and 10 p.p.m. of SO_2 for 4, 6 and 8 h duration, was inversely proportional to pollution dose, that is the concentration of $SO_2 \times$ duration of exposure.

(b) Transplantation technique

Bryophyte species are transplanted *in situ,* that is along with their substrates, from unpolluted to ecologically more or less similar, but SO_2-polluted, sites to obtain evidence of injury caused by pollution. Following the technique of Brodo (1961), LeBlanc and Rao (1966, 1973b) transplanted bark discs bearing mosses, such as *Orthotrichum obtusifolium, Platygyrium repens, Pylaisia selwynii, P. polyantha* and *Ulota crispa,* in the region of Sudbury, Ontario, where the environment is polluted with SO_2. They observed that after one year of exposure the epiphytes were either seriously injured or dead, depending on the levels of pollution to which they were exposed. The common symptoms of injury were plasmolysis and chlorophyll degradation in the leaf cells.

Gilbert (1968) transplanted four species of mosses, *Homalothecium sericeum, Grimmia pulvinata, Hypnum cupressiforme* and *Tortula muralis,* in Newcastle-upon-Tyne and observed that all transplants at sites with high levels of SO_2, showed a rapid breakdown of chlorophyll, with less than 10% of it remaining after 10 weeks; their respiration rates increased slightly in the first few weeks and then fell rapidly. Daly (1970) made similar observations on *Brachythecium* sp. and *Hypnum* sp. transplanted in the city of Christchurch, New Zealand.

The typical response of mosses to SO_2 pollution starts as a loss of colour at the tips of the more exposed leaves and this gradually extends down the leaves and down the shoots until they have lost all their chlorophyll. The last parts to become bleached are those deep down in the mat or cushion. Usually, completely bleached specimens fail to recover when replaced in their original habitat, but some acrocarps or primarily erect mosses, are capable of producing basal regenerative shoots and thus completely renewing the moss cushion (Gilbert, 1968).

It has been suggested that SO_2 is toxic to bryophytes more than any other pollutant (Taoda, 1972) and that most bryophytes cannot exist when the average winter concentration of SO_2 exceeds 50 $\mu g\ m^{-3}$ or 0.017 p.p.m. (Gilbert, 1968; Daly, 1970). On the basis of their transplantation studies in the Sudbury area, Ontario, LeBlanc and Rao (1973b) concluded that in relatively humid areas long exposure periods of one year or more to SO_2 above 0.03 p.p.m. can cause acute injury, between 0.006 and 0.03 p.p.m. can cause chronic injury, and below 0.002 p.p.m. may cause no obvious injury to epiphytic mosses.

12.3.2. Field survey and phytosociological methods

On approaching a large city or an industrial complex, one finds that there is gradual decrease in number of species, reduction in coverage, loss of vitality and inhibition of sexual reproduction of bryophytes before most of them are finally eliminated from the city centre or from the areas of high levels of pollution. Since the above phytosociological changes are directly related to the changes in the levels of pollution, bryophytes have been successfully used, either singly or collectively, for delimiting pollution zones (Barkman, 1958, 1963; Gilbert, 1968; Rao and LeBlanc, 1967; LeBlanc and De Sloover, 1970; Daly, 1970). Taoda (1972) divided the metropolitan area of Tokyo into five zones on the basis of the decline, both in luxuriance and in the number of species of epiphytic bryophytes caused by phytotoxic air pollutants.

To analyse the pollution-induced phytosociological changes, bryophytes growing on a particular substrate, say tree bark, are examined along a line or belt transect having an air pollution gradient. Studies from a number of transects radiating in all directions from the source of pollution are needed for preparing a pollution map of an area. It makes relatively little difference whether one uses a single but well-selected species or a number of species for this purpose, because in either case the results obtained are more or less similar (Gilbert, 1970a; LeBlanc and De Sloover, 1970; Granger, 1972).

Phytosociological data have also been used in the quantification and delineation of areas affected by pollution. De Sloover and LeBlanc (1970) have shown that an index of atmospheric purity (IAP) determined on the basis of number, frequency-coverage and resistance-factor of species, can provide a fair picture of the long-range effect of pollution of a site. Consequently, on the basis of the IAP values of a large number of sites studied on transects radiating in all directions from a given pollution source, the area around it can be divided and mapped into different pollution zones. It has been shown that maps prepared on the basis of IAP values and on actual SO_2-levels are more or less similar in their outline (LeBlanc *et al.*, 1972a).

From his observation on the distribution of bryophytes in Newcastle-upon-Tyne, England, where the SO_2 pollution levels are known, Gilbert (1970b) listed all the species from the very sensitive to the most resistent ones with respect to their substrates (sandstone, limestone, asbestos and tree bark) and related them to the ambient SO_2 levels which they were able to tolerate. He could thus prepare a biological scale which may be used for estimating the average concentration of SO_2 in ecologically similar areas by studying the bryophytes occurring in these areas.

12.3.3 Species susceptibility

In air pollution studies, it is primarily the sensitive species which are useful as bio-indicators of pollution. However, it is not always easy to ascertain, especially in the field, the degree of sensitivity and resistance of species. Rao and LeBlanc (1967) assessed the extent of SO_2 pollution in Wawa region, Ontario on the basis of SO_4^{2-} concentration in soil. They observed that the epiphytes, both with respect to their number and percentage occurrence, were responsive to the level of pollution prevailing in the area. From the data presented in Table 12.3, it is evident that out, of the 39 bryophyte species recorded in the investigated area, none were present in zone I, four in zone II, 22 in zone III, 25 in zone IV and 28 in zone V; also in any particular zone, the number of species growing on tree bases (a protected niche) was always more than that growing on tree trunks (an exposed niche). In Wawa, the epiphytes *Platydictya subtile, Orthotrichum* spp., *Paraleucobryum longifolium* and *Ulota crispa* were found to be highly sensitive to SO_2 pollution as inferred from their presence on tree trunks in the relatively unpolluted air of zone V only.

Similarly, in Newcastle-upon-Tyne, Gilbert (1970b) observed that *Homalothecium sericeum, Frullania dilatata, Metzgeria furcata, Orthotrichum affine, Tortula laevipila* and *Zygodon viridissimus* could grow on tree trunks only in relatively pure air. Daly (1970) noted a striking parallel on the reduction of air pollution and the reappearance of sensitive bryophytes such as *Hypnum* and *Lophocolea* in Christchurch, New Zealand. According to Taoda (1972) the epiphytic bryophytes *Frullania muscicola, Haplohymenium sieboldii, Herpetineuron toccoae, Trocholejeunea sandvicensis, Macromitrium japonicum, Eurohypnum lepto-*

thallum, Schwetschkea matsumurae, Lophocolea minor and *Cheilolejeunea* spp. occur in the relatively unpolluted air of rural areas around Tokyo, where the SO_2 concentration remains below 0.01 p.p.m.

The presence or absence of certain species under uniform habitat conditions in nature can provide a relative picture of their sensitivity or resistance to pollution. For example, the epiphytic bryophytes in Wawa (Table 12.3) can be arranged into four groups in the order of their increasing sensitivity, represented by species able to survive in zones II, III, IV and V, respectively. The species such as *Dicranum flagellare, Drepanocladus uncinatus, Ptilidium pulcherrimum* and *Tetraphis pellucida,* present in the second zone are apparently the most resistant ones, while those appearing gradually at decreasing pollution levels in zones III, IV and V represent species which are more and more sensitive to SO_2 pollution.

Taoda (1973b), while studying the SO_2 tolerance of epiphytic bryophytes in Japan, observed that most bryophytes were injured by 0.8 p.p.m. SO_2 in 10–40 h or by 0.4 p.p.m. in 20–80 h of total exposure; at 0.2 p.p.m., acute injury, such as discolouration of shoots, did not occur even after 100 h exposure but chronic injury, such as growth retardation, was noticed.

Certain species of bryophytes are known to be toxiphilous because they can thrive even under severe conditions of SO_2 pollution. According to Daly (1970), *Bryum argenteum, Ceratodon pupureus* and *Pohlia cruda* remain unaffected by the city conditions of Christchurch, New Zealand, and their coverage even seems to increase with pollution. Similarly, Gilbert (1970b) noted that *Bryum argenteum* and *Ceratodon pupureus* on asbestos roofs and *Funaria hygrometrica* and *Tortula muralis* on sandstone, never reached their extinction point even in the central part of Newcastle-upon-Tyne, where the average annual concentraton of SO_2 was more than 170 $\mu g\ m^{-3}$ or 0.06 p.p.m. Rao and LeBlanc (1967) noted that three terricolous species *Dicranella heteromalla, Pohlia nutans* and *Ceratodon purpureus* showed good growth on soils with a pH of 3.4 and a SO_4^{2-} concentration of 4.3 meq 100 g^{-1}, in the SO_2-polluted areas of Wawa, Ontario, where the other species could not survive. Taoda (1972) observed that *Hypnum yokohamae* var. *kusatsuensis* and *Glyphomitrium humillimum* could grow on trees in wooded parks in the centre of Tokyo and were able to tolerate SO_2 concentrations of 0.04 to 0.05 p.p.m.

All SO_2-resistant bryophytes show a common feature of fast growth rate. In the case of mosses, the sensitive protonema is usually short-lived and soon initiates protonemal buds which develop rapidly into more resistant gametophytes (Gilbert, 1970a). Perhaps the active metabolism and high reproductive capacity of certain bryophytes enable them to colonize urban areas successfully (Gilbert, 1971b). The apparent ecological success of terricolous and saxicolous bryophytes in polluted areas may be attributed to the buffering action of their substrates (Barkman, 1958), to the physical protection provided by snow cover during winter when the level of pollution is maximum (LeBlanc, 1961), and to the minimum contact with pollution-laden winds by virtue of the location of species on or near the surface of the ground (Geiger, 1965).

12.4 FLUORIDE POLLUTION STUDIES

The responses of bryophytes to air-borne fluorides have been studied mainly by ecophysiological methods, including fumigation and transplantation techniques, and phytosociological methods, including field survey and pollution-mapping techniques.

12.4.1 Ecophysiological and bioassay methods

(a) Fumigation technique

During fumigation, plants are exposed to fluorides under controlled conditions in which only the HF concentration and exposure duration are variable while other conditions remain constant.

Comeau and LeBlanc (1972) exposed gametophytes of *Funaria hygrometrica* to gaseous HF concentrations of 13 (1.07 $\mu g\, m^{-3}$), 65 (5.33 $\mu g\, m^{-3}$) and 130 (10.66 $\mu g\, m^{-3}$) p.p.b. for 4, 8 and 12-h durations and to 13 p.p.b. for 36, 72 and 108-h durations. When these HF-exposed plants were examined with respect to certain morphological and chemical changes, it was found that when the dose or exposure factor (concentration$\times$time) was low, the overall injury to plants was minimal and the rates of fluoride-accumulation was low. The recuperation rate or the loss of accumulated fluoride from the plant body in the post-fumigation period was high. Conversely, when the exposure factor was high, the foliar injury was severe, the F concentration was high and the recuperation rate was low. They observed that at an exposure factor of 780 (65 p.p.b. $\times$ 12 h), *Funaria* leaves showed apical necrosis, disintegration of chloroplasts and plasmolysis in their cells and, after a recovery period of 3 weeks, the leaf F concentration was reduced by 26–36% of that accumulated during fumigation.

The above results indicate that the dose–response relationships of bryophytes, especially mosses, are similar to those of higher plants. Once the HF is absorbed on the plant surface, it moves towards the tips of leaves, causing a distinct pattern of injury which remains proportional to the exposure factor. It is suggested that during the post-fumigation period there is detoxification or loss of the absorbed fluoride through weathering or leaching, and reduction in concentration through subsequent plant growth (Treshow, 1971).

(b) Transplantation technique

This technique involves *in situ* transplantation of species from an unpolluted to a polluted habitat. LeBlanc *et al.* (1971) transplanted bark discs on which *Orthotrichum obtusifolium* and *Pylaisia polyantha* were growing onto trees at various distances from an aluminium factory in Quebec, Canada. After 4 and 12 months of exposure the transplants from the control and polluted sites were compared with respect to colour, external morphology, degree of cell plasmolysis, loss of chlorophyll, reaction to Molisch phase test, absorption spectra of chlorophyll and fluoride concentration. Results indicated that, depending on the duration of exposure and distance and direction from the factory the moss colour changed from golden green to brown or dark brown and the apical and marginal portions of leaves became more plasmolysed, necrotic and fragile than the basal portion. Chlorophyll extracts of pollution-affected moss transplants with respect to their control were only weakly positive to the Molisch phase test. It was observed that, after 12 months of exposure to fluoride contaminated air, the transplants near the factory showed complete destruction of chlorophyll, plasmolysis and other cellular abnormalities and high levels of fluoride accumulation. The F concentration in *Orthotrichum obtusifolium,* transplanted at 1.0 km north of the factory was 600 p.p.m., while in the control at 40 km distance it was 20 p.p.m. only. These injury symptoms clearly indicate the sensitivity of bryophytes to fluoride pollution.

12.4.2 Field survey and phytosociological methods

In these methods the bryophyte community of an area having a point source of pollution is analysed with respect to different ecological parameters. Gilbert (1971a) observed the absence of bryophytes from several habitats around an aluminium plant in Scotland. He found that samples of *Racomitrium lanuginosum* collected from exposed sandstone at 1.6, 2.0, 3.0, 7.0 and 10.5 km downwind from the factory, had 146, 101, 48, 38 and 25 p.p.m. soluble fluoride, respectively. At a site 1.2 km upwind the F concentration in the moss was only 30 p.p.m.

According to Gilbert (1971a), among the fluoride-sensitive epiphytic bryophytes, *Orthotrichum lyellii* and *O. affine* are more sensitive than *Ulota crispa* and *Frullania dilatata* while *Pohlia nutans* and *Aulacomnium androgynum* are rather resistant species; among the terricolous bryophytes, *Sphagnum* spp., *Polytrichum commune, Leucobryum glaucum, Rhytidiadelphus squarrosus* and numerous small hepatics and acrocarpous mosses can resist fluoride pollution.

LeBlanc *et al.* (1972b) established a relationship between the epiphytic vegetation and the level of fluoride pollution around an aluminium factory in Quebec. They determined indices of atmospheric purity (IAP) on the basis of phytosociological parameters, such as number, frequency-coverage and ecological index of species. They delineated the polluted area into six IAP zones lying one outside the other at increasing distances from the factory. The increase in frequency-coverage of epiphytes from zone I to VI, suggests dependence of this parameter on the atmospheric purity *vis-a-vis* the fluoride level in the area.

12.5 OZONE POLLUTION EFFECTS ON BRYOPHYTE REGENERATION

Ozone (O_3) is a secondary pollutant formed by the action of sunlight on nitrogen dioxide and on certain hydrocarbons. It is considered to be more phytotoxic than the primary pollutants involved in its formation. Depending on the plant species, gas concentration and microclimatic factors, O_3 uptake by leaves of plants often results in acute injury, premature ageing and senescence (Heggestad, 1968). However, little is known regarding the phytotoxicity of this photochemical pollutant on bryophytes.

Comeau and LeBlanc (1971) studied the regenerative ability of O_3-exposed *Funaria hygrometrica* leaves. Leaves from the apex (upper third) and from the base (lower third) of *Funaria* plants exposed to 0.25, 0.5, 1 and 2 p.p.m. O_3 concentrations for 4, 6 and 8-h periods as well as those of the control, were grown on solid agar with a mineral base in petri dishes kept in a phytotron. In each case the percentage of leaves developing protonemal structures on their margins were determined. The results indicated that O_3 could stimulate growth when administered at low concentrations for short periods; that the apical leaves of the gametophyte possessed a higher regeneration potential than the basal ones. Also the percentage of regeneration was inversely proportional to the exposure factor, that is concentration$\times$time. However, further studies are needed to confirm this report and to understand the mechanism of growth stimulation. Bryophytes, being easily cultivable, should prove excellent materials for such experiments.

12.6 HEAVY METAL AND RADIOACTIVE POLLUTION STUDIES

The problem of environmental pollution by heavy metals results mostly from industrial operations involved in the production of metallic products. In addition, widespread lead pollution takes place from automobile exhaust. Metals are non-degradable and once released into the environment they become an integral part of the habitat. Air-borne metal particles are removed from the atmosphere by rainwater, gravitational settling and impaction on various surfaces, especially vegetation, and may also be actively or passively accumulated by plants. The concentration of airborne elements in plants is determined by their concentration in the air, the inherent ability of plants to absorb these elements, the ratio of plant surface to total plant mass, the ion exchange capacity of plants and the length of time of exposure. Studies by Rühling and Tyler (1968), Hutchinson and Whitby (1974) and others have shown that plants can be used as indicators of aerial fallout of heavy metals. The growing literature on heavy metals, recently reviewed by Rao *et al.* (1977) shows beyond doubt that bryophytes are efficient absorbers of these elements.

Bryophytes are able to concentrate heavy metals in large amounts, usually greatly surpassing the absorbing capacity of vascular plants (Table 12.5). Chemical analysis of carpet-forming bryophytes has proved to be a rapid and inexpensive method for surveying heavy metal deposition in the

Table 12.5 Metal concentration in vascular plants *Picea mariana* (a) and *Clintonia borealis* (b) and in bryophytes *Hylocomium splendens* (c) and *Pleurozium schreberi* (d) collected in IAP zones I to V with decreasing levels of air pollution in the Murdochville copper mine area in Canada (modified after LeBlanc *et al.*, 1974).

IAP zone	Plant species	Metal concentration (p.p.m. dry wt.)				
		Pb	Cd	As	Zn	Cu
I	a	349.5	1.6	15.8	84.9	23.1
	b	458.5	6.2	13.0	157.0	93.7
	c	17320.0	8.4	4.0	304.0	724.0
	d	695.0	3.0	1.5	229.55	752.5
II	a	39.5	0.4	1.0	98.5	3.0
	b	66.0	3.1	2.0	37.0	11.0
	c	1283.0	5.1	4.0	274.0	341.0
	d	820.0	2.6	4.5	198.0	324.0
III	a	14.5	0.4	2.2	60.7	4.0
	b	58.0	1.1	2.0	35.0	18.3
	c	651.0	4.0	5.7	187.0	253.0
	d	403.3	1.6	3.5	166.0	246.0
IV	a	20.2	0.4	2.0	66.0	4.0
	b	40.0	0.98	2.0	25.4	13.6
	c	459.0	4.0	5.2	136.8	228.6
	d	201.0	1.2	3.0	103.2	137.4
V	a	11.0	0.4	5.0	44.5	6.0
	b	42.0	0.9	3.0	40.0	116.0
	c	196.5	1.5	4.0	130.0	346.0
	d	62.5	0.5	2.5	89.0	236.5
Control site	a	5.0	0.7	1.0	43.0	10.0
	b	26.0	0.4	4.0	24.0	54.0
	c	141.0	1.1	4.0	86.0	81.0
	d	15.0	0.5	3.0	46.0	45.0

terrestrial ecosystem (Rühling and Tyler, 1968, 1970, 1971; Tyler, 1971; Goodman and Roberts, 1971; LeBlanc *et al.*, 1974).

Moss gametophytes are known to accumulate iron 5–10 times more readily than do vascular plants (Czarnowska and Rejement-Grochowska, 1974). Mayer and Gorham (1951) analysed the content of 19 moss species from the lake district of England and they found that these plants accumulated between 0.005 and 0.4% manganese dry weight. Shacklette (1965b) determined the element contents of 29 species of bryophytes and found that the concentrations of Al, Ba, Cr, Cu, Fe, Ga, Ni, Pb, Ag, Ti, V, Zn and Zr were higher in bryophytes than those in angiosperms. Also, the bryophyte species were able to concentrate rare earth elements even when these were below the detection level in the substrate. Certain elements, very rarely found in plants, were also detected; Bi in the thallose liverworts *Concephalum conicum* and *Marchantia polymorpha,* Sn in the saxicolous mosses *Grimmia laevigata* and *Hedwigia ciliata,* and Ag in *Atrichum angustatum* and *Polytrichum commune. Thelia asprella,* growing on *Juniperus virginiana,* contained eight elements that were not found in the bark of the tree on which it was growing. It is likely that these elements were obtained either from windblown materials lodged in the moss mats or from the solutes in the stemflow rainwater. Also, Brassard (1969) showed that Cu, Pb and B in the substrate of *Mielichhoferia* collected in the Canadian High Arctic were 70, 50 and 770 p.p.m., respectively, while the average amounts of the respective elements in soils were 20, 10 and 10 p.p.m. only.

Though the above observations prove that mosses are efficient absorbers of elements from their substrates, it may not always be so. Nagano (1972) studied the relationship between the quantity of five macro-elements (N, Mg, Ca, K and P) present in 28 species of mosses and the quantity present in their substrate rocks in the Chichibu mountain area north west of Tokyo City. He did not find definite relationships between the amounts of these elements in the mosses with those in their substrates, except in the case of calcium. A very close correlation existed between the Ca content of calciphilous mosses and their substrates. Generally, mosses of the family Neckeraceae contained a high percentage of Ca and those of the families Grimmiaceae, Polytrichaceae, and Dicranaceae had a low percentage.

12.6.1 Factors affecting absorption and accumulation of metals

(a) Meteorology, topography and pollution-source proximity

The effects of meteorological factors and topography on the spatial distribution of airborne heavy metals are more or less similar to those of the fallout of gaseous pollutants. Lee (1972) studied Pb pollution from a factory manufacturing anti-knock compounds consisting of tetra-ethyl and tetra-methyl lead. He found that the Pb content in the moss species *Ceratodon purpureus, Eurhynchium praelongum, Plagiomnium cuspidatum* and *Calliergon cuspidatum* growing at various distances from the factory declined rapidly in the first 400 m and then slowly in the next 400 m distances; at 800 m, however, the amount of Pb in the mosses was still two to three times greater than that in the species collected from rural areas remote from the factory. Lee also observed that the Pb content in *Ceratodon purpureus* was about 320 p.p.m. at a distance of 2 m and 185 p.p.m. at a distance of 20 m away from a road in central Manchester. In addition, Briggs (1972) found high lead values in *Marchantia polymorpha* even at great distances from roads, with the Pb contents corresponding to the respective traffic intensities and without significant correlation to the lead content of the soil.

Rühling and Tyler (1969, 1970) and Tyler (1971) found an increase in the heavy metal content (Pb, Cd, Cu, Cr, Ni, Zn, Co and Mn) in the mosses *Hypnum cupressiforme* and *Hylocomium splendens* with increasing precipitation and decreasing distance from industrialized areas. Yeaple (1972) has shown that terrestrial mosses accumulate Hg in inverse proportion to their distances from pollution centres. Huckabee (1973) found that the Hg content in mosses was a function of the distance from smoke stacks, mean values ranging from 1.3 p.p.m. in polluted areas to 0.066 p.p.m. in remote areas.

Burkitt *et al.* (1972) and LeBlanc *et al.* (1974) found high concentrations of Zn, Cd and Pb around metal smelters within the zone of prevailing winds, although there was a rapid fall in metal concentrations as the distance from the smelters increased. As a general rule, the concentrations of airborne materials decreases with distance from the source of pollution and there may be a wide variation in metal accumulation from species to species and from habitat to habitat under different microclimatic conditions (Nieboer *et al.*, 1972).

(b) Acid rain

Acid rain, that is concurrent SO_2 emission with rain, enhances the solubility of metal salts and increases their uptake by bryophytes. Robitaille *et al.* (1977) have shown that acid rain produces drastic changes in the chemical properties of both epiphytes and their bark substrates by reducing the pH of stemflow, increasing the proportion of toxic bisulphite (HSO_3^{2-}) ions in stemflow, decreasing the buffer capacity of the bark, decreasing the internal pH of epiphytes, and increasing the metal concentration and chlorophyll loss in epiphytes.

In the strip mine habitat under conditions of low pH, heavy metals are easily solubilized and thus made available for uptake by terrestrial species of bryophytes, especially from runoff water. It is suspected that small amounts of heavy metals accumulated by bryophytes are also removed by low pH rain water (Lawrey and Rudolph, 1975).

Therefore, accumulation of heavy metals could occur concomitantly with their mobilization and release during periods of acid rain. In conclusion, one may say that under conditions of acid precipitation and reduced buffer capacity, the heavy metal absorption is invariably increased, in other words a weakly acid medium favours accumulation of heavy metals by bryophytes.

(c) Substrate specificity, life-form and lifespan of species

Certain bryophytes thrive on substrates rich in a certain metals. For instance, some species of mosses are known as indicators of sites of copper enrichment in the substrate. These plants have commonly been designated 'copper mosses' in the literature and the topic is dealt with in Chapter 11, pp. 422–5. Certain species of bryophytes can successfully grow on soil derived from serpentine rock which weathers to form a soil containing as much as 4000 p.p.m. Ni and 900 p.p.m. Cr. Shacklette (1965a) reports that shaded serpentine rock may be colonized with small patches of *Schistidium strictum, S. tenerum, Hedwigia ciliata, Thuidium abietinum, Rhytidium rugosum* and *Weissia controversa,* but fully exposed rocks may be colonized by dense mats of *Racomitrium lanuginosum.*

In view of their substrate specificity, it is possible to use certain bryophytes for biogeochemical prospecting. However, their use must be based on the total status of metal composition of the species and enough care should be taken to avoid contamination of samples by the substratum.

Among mosses the profusely branched and ramifying pleurocarps and the densely packed acrocarps are generally more efficient entrappers and absorbers of metal particles than the unbranched and erect acrocarps. Huckabee (1973) found a greater accumulation of mercury in *Dicranum scoparium* than in *Polytrichum commune* and he attributed this difference to their life-forms. Surface cuticle can also affect the absorption and accumulation rates of dissolved materials. Shimwell and Laurie (1972) noted that the 'ectohydric' moss *Dicranella varia,* which lacks cuticle and absorbs water over the whole of its surface, accumulates higher contents of heavy metals than does the 'myxohydric' moss *Philonotis fontana* which has a more or less continuous cuticle and absorption taking place mostly towards the base of the gametophyte. Tamm (1953) analysed the moss *Hylocomium splendens* for changes in the P, K, Na and Ca contents associated with the increasing life-span of that plant. He observed that the monovalent elements Na and K accumulated in the young moss and then their concentration decreased, Na more quickly and K at a slower rate. The bivalent and trivalent metals showed a successive accumulation from year to year in most segments of the moss plant, with or without previous accumulation in the bud. Tamm claimed that the annual growths of this moss are nutritionally isolated, both from each other and from the soil.

Further, according to Czarnowska and Rejement-Grochowska (1974) the mineral composition of mosses differ in the haploid and diploid phases. For example, in the moss *Atrichum undulatum,* they observed that the respective concentrations of Fe, Mn, Zn and Cu were 2000, 97, 45 and 4.5 p.p.m. in the gametophyte or haploid phase and 215, 93, 76 and 7.4 p.p.m. in the sporophyte or diploid phase.

(d) Cation exchange and chelation

Metal ions are retained by organic matter, either by means of simple ion exchange or by complexing or chelation with specific organic groups. The moss gametophytes readily entangle and retain particulate contaminants amongst the leaves and they possess remarkable ion-exchange properties. Certain mosses, especially *Sphagnum,* when dry absorb and retain rain water to a considerable extent, which facilitates ionic exchange between the soluble metal and moss cell walls (Clymo, 1963).

Results of cation exchange experiments with the moss *Hylocomium splendens* by Rühling and Tyler (1970) indicate that Ni accumulation is by passive ion exchange, at least when large amounts of Ni are supplied; for Cu, however, the process is not exclusively ion exchange because only two-thirds of the copper content is exchanged by repeated leaching with $MgCl_2$. The sorption and retention of the heavy metals studied by Rühling and Tyler were in the order: Cu, Pb > Ni > Co > Zn, Mn. The order was shown to be independent of whether or not the ions were supplied in pure or in mixed solution. According to them the natural carpet of *H. splendens* exhibits a continuous uptake of Mn, Fe and Cu from young to old tissues; the mor-layer beneath the moss carpet does not enrich the heavy metals above the concentrations measured in the old tissue.

12.6.2 Metal tolerance and detoxification mechanisms of species

The ability of plants to colonize heavy metal contaminated sites may suggest the evolution of tolerant genotypes or the development of tolerant metal ecotypes as a result of reduced competition there (Antonovics *et al.*, 1971). Coombes and Lepp (1974) observed morphological changes in Cu- and Zn-stressed *Funaria hygrometrica* and *Marchantia polymorpha* and they suggested that these changes could be possible metal tolerance mechanisms. *Funaria* develops 'capsule cells' and 'brood cells' when subjected to high doses of Cu and Zn respectively, thereby reducing the surface area of the protonema in contact with Cu and Zn. In Cu-stressed *Marchantia,* a buffering zone of outer dead cells may immobilize Cu within their walls, thus preventing toxic concentrations from reaching the inner cells.

Detoxification through excretion of heavy metals external to the cell wall has been seen in some bryophytes. Shimwell and Laurie (1972) observed that in *Dicranella varia* (an ectohydric moss devoid of cuticle) most of the Pb and Zn are excreted from the gametophyte and in periods of summer drought forms a powdery crust on the moss carpet. An analysis of the precipitate from *Dicranella* carpets on two shale soil heaps produced the abnormally high values for Pb of 42 000 and 60 450 p.p.m. respectively. The method of excretion by *Dicranella* appears to begin with an exudation at the tips of the leaves where crystals of complex heavy metal sulphates form and gradually spread out over the surface of the leaves to form a white crust. The crust disappears rapidly into solution in the periods of normal precipitation and the moss carpet remains healthy. If the period of drought and desiccation exceeds three or four weeks some areas of the carpet die and rapidly disintegrate to form raw humus, returning the Pb and Zn to the top layers of the soil. Theoretically, the absorption and precipitation of heavy metals by *Dicranella* provide a possible natural method of detoxification of spoil heap soils.

Skaar *et al.* (1973), in their electron microscopic studies of the leaves of the moss *Rhytidiadelphus squarrosus* cultured in the laboratory by watering once a day for 3 weeks with lead acetate solution and of samples collected in a Pb-polluted area, observed lead accumulation within the nuclei of moss leaf cells. The binding of Pb within the nuclear membrane, as a non-diffusible complex, has been suggested as the mechanism whereby the cytoplasmic concentration of diffusible Pb substances within the cell can be kept below a level that would otherwise be toxic to the mitochondrial or other lead-sensitive functions of the cytoplasm. Lead in slag specimens of *Grimmia donniana* was shown to be ionically bound to an extracellular site, the cell wall, thus preventing toxic amounts of lead from reaching or penetrating the cytoplasm (Brown and Bates, 1972).

Little is known of the maximum tolerance levels of bryophytes for individual metals. It appears that the majority of mosses have an innate tendency for the accumulation of at least some metals, such as Fe, Pb, Zn and Ni, from their substrate and the atmosphere and that they resemble higher plants in their behaviour towards certain metals (Lounamaa, 1956).

Besides being a significant sink of heavy metals bryophytes are efficient absorbers of radioactive materials. Gorham (1959), in a comparative study of lower and higher plants as accumulators of radioactive fallout, found that mean counts per minute in mosses and angiosperms were 152 and 63 respectively. The capacity of moss carpets to sorb radioactive fallout from nuclear bomb tests in the atmosphere has been demonstrated by Svensson and Leden (1965). They found an almost quantitative retention of the fission products $^{95}Zr+^{95}Nb$ and $^{140}Ba+^{140}La$ in carpets of *Pleurozium schreberi* from southern Sweden. Hoffman (1972) suggests that though bryophytes are only minor components of forest ecosystems, they represent an important sink in the cycling and accumulation of ^{137}Cs and the epiphytic, tree-base and terrestrial bryophytes constitute major pathways of ^{137}Cs from foliage of tagged trees.

Steere (1970) is of the opinion that perhaps because of their small nuclei, which range in size from 4 to 150 μm^3, the bryophytes are more resistant to ionizing radiations than higher plants. According to him, leafy hepatics are more sensitive than mosses and *Fissidens garberi* is perhaps the most resistant moss because it can survive within 10 m of a caesium source and even within 3 m in somewhat shielded habitats.

12.6.3 Bryophytes as temporal and spatial indicators of metal pollution

Bryophytes have been used as a bio-assay technique for detecting and measuring metal pollution over long periods of time. The first step in evaluating the potential significance of bryophytes to metal pollution is to establish the burden and location of various metals on or in these plants. By analysing the moss specimens *Hylocomium splendens, Pleurozium schreberi* and *Hypnum cupressiforme* that were preserved in the Botanical Museum in Lund, Sweden and that had been collected at intervals from 1860–1968 from the same location, Rühling and Tyler (1968) were able to record the effect of human activity historically on the concentration of atmospheric Pb. They found that from values of 20 p.p.m. (dry wt.) in the years 1860–1875 the Pb concentration was more than doubled between 1875–1900. During the first half of the twentieth century no measurable changes were observed but after 1950 there was a new increase to a present average of 80–90 p.p.m.

Rao *et al.* (1977) made a historical study of accumulation of Zn, Pb, Cu, Ni, Cd and Cr in herbarium specimens of certain mosses collected from Mount Royal in Montreal, Canada, between the years 1905 and 1971 (Table 12.6). During this 66-year period, they found an increase of Zn concentration from 88 to 440 p.p.m. in *Plagiomnium cuspidatum,* from 87 to 330 p.p.m. in *Brachythecium salebrosum,* from 100 to 120 p.p.m. in *Heterophyllium haldanianum,* from 90 to 160 p.p.m. in *Anomodon rostratus* and from 53 to 145 p.p.m. in *Atrichum angustatum.* When the average concentration of a metal in the mosses was plotted against the year of sampling it became evident that the concentration of metals, with the exception of copper, is slowly but inexorably increasing in the Montreal area. This trend is perhaps an indicator of the steadily increasing heavy metal pollution in the area through urbanization and industrialization.

Rühling and Tyler (1968, 1969, 1971, 1973) and Tyler (1971) conducted studies of the regional distribution of other airborne heavy metals by analysing samples of mosses and they found this method to be effective in evaluating aerial metal pollution. From the increasing concentrations of

Table 12.6 Metal accumulation in some species of mosses collected from Mount Royal, Montreal, P.Q., Canada, between the years 1905–1971 (after Rao *et al.*, 1977).

Species	Metal	Concentration (p.p.m. dry wt.)					
		1905/06	1910/11	1916	1925/26/28	1937	1971
Plagiomnium cuspidatum (wet soil)	Zn	88			120	205	440
	Pb	75			56	105	85
	Cu	37			69	25	45
	Ni	35			28	13	21
	Cd	1			6	1	5
	Cr	23			36	21	27
Brachythecium salebrosum (tree base)	Zn	87	105		155		330
	Pb	54	94		55		70
	Cu	25	37		33		27
	Ni	6	29		14		105
	Cd	1	1		7		6
	Cr	17	55		20		43
Heterophyllium haldanianum (rotten logs)	Zn	100	170		105		180
	Pb	20	50		85		110
	Cu	21	45		75		33
	Ni	14	31		27		14
	Cd	1	5		2		1
	Cr	21	37		36		36
Anomodon rostratus (rocks)	Zn	90		32	205		160
	Pb	13		4	55		34
	Cu	35		87	56		23
	Ni	21		16	30		3
	Cd	2		1	4		6
	Cr	63		7	61		19
Atrichum angustatum (dry sandy soil)	Zn	53	75		73		145
	Pb	9	18		24		37
	Cu	25	20		17		22
	Ni	22	54		25		37
	Cd	2	2		2		4
	Cr	24	24		26		27

metals in *Hylocomium splendens* from northern to southern parts of Scandinavia, Rühling and Tyler (1971) concluded that perhaps the much greater human activity over southern than northern Scandinavia, was responsible for the measured gradients in the deposition of heavy metals (Table 12.7).

LeBlanc *et al.* (1974) studied the spatial distribution of As, Cd, Cu, Pb and Zn emitted from a copper smelter in Canada by analysing their contents in two terrestrial mosses *Hylocomium splendens* and *Pleurozium schreberi* collected at various distances from the smelter. They observed

Table 12.7 Gradient concentration (p.p.m. dry wt.) of certain heavy metals in *Hylocomium splendens* from northern to southern parts of Scandinavia (after Rühling and Tyler, 1971).

Metal	Regional concentration gradient from N to S							
	North →							South
Fe	250	500	1000	1500				
Zn	20	40	60	80	100	120		
Pb	10	20	40	60	80	100	120	140
Cu	4	6	8	10	12	15	20	
Ni	1	2	4	6	8	10	12	
Cr	1	2	4	6	8	10		
Cd	0.063	0.125	0.25	0.5	0.75	1.0	1.5	
Hg	0.05	0.1	0.2	0.4	0.6			

that the average concentration of metals in the two mosses were negatively correlated with the index of atmospheric purity (IAP) values determined on the basis of ecological and phytosociological characters of epiphytic bryophytes growing on *Abies balsamea.* There existed a distinct relationship between the plant metal contents and the reciprocal of the distance from the smelter. This study suggests that the IAP method can be used for mapping the spatial distribution and for the diagnosis of heavy metal pollution. It is feasible to use bryophytes, especially the epiphytic species, to substantiate the presence of heavy metals in the environment. These organisms would serve as general long-term indicators of metal pollution.

12.7 CONCLUSIONS

The investigations with bryophytes in relation to different air pollutants prove their potential as bio-indicators of air pollution. Due to their habitat diversity, structural simplicity, totipotency and rapid rate of multiplication, bryophytes appear to be ideal organisms for pollution studies both under field and laboratory conditions. Phytosociological and ecophysiological studies indicate that the decline and absence of bryophyte populations, especially epiphytic ones, in urban-industrial areas, is a phenomenon primarily induced by air pollution caused by different gaseous and particulate pollutants. These plants can be reliable indicators and monitors of air pollution provided the metabolic responses of the sensitive species of bryophytes to various levels and kinds of pollutants under different ecological conditions are clearly deciphered and properly understood through controlled experiments. It may be reiterated that the First European Congress on the Influence of Air Pollution on Plants and Animals held at Wageningen in 1968 rightly resolved that cryptogamic epiphytes should be strongly recommended for general use as biological pollution indicators, because they are so easy to handle and they show a vast range of specific sensitivity to air pollutants greatly exceeding that of most higher plants.

REFERENCES

Antonovics, J., Bradshow, A.D. and Turner, R.G. (1971), Heavy metal tolerance in plants. In: *Advances in Ecological Research* (Cragg, J.B., ed.), pp. 1–80. Academic Press, New York.

Barkman, J.J. (1958), *Phytosociology and Ecology of Cryptogamic Epiphytes,* Assen, Netherlands.

Barkman, J.J. (1961), *Nature,* **58,** 141–51.

Barkman, J.J. (1963), *Verh. Kon. Ned. Akad. Weteusch.,* **54,** 1–46.

Barkman, J.J. (1969), The influence of air pollution on bryophytes and lichens. In: *Air Pollution: Proceedings of the First European Congress on the Influence of Air Pollution on Plants and Animals,* pp. 197–209. Wageningen.

Brassard, G.R. (1969), *Nature,* **222,** 584–5.

Briggs, D. (1972), *Nature,* **238,** 166–7.

Brodo, I.M. (1961), *Ecology,* **42,** 838–41.

Brown, D.H. and Bates, J.W. (1972), *J. Bryol.,* **7,** 187–93.

Burkitt, A., Lester, P. and Nickless, G. (1972), *Nature,* **238,** 327–8.

Clymo, R.S. (1963), *Ann. Bot.,* **27,** 309–24.

Coker, P.D. (1967), *Trans. Br. Bryol. Soc.,* **5,** 341–7.

Comeau, G. and LeBlanc, F. (1971), *Natur. Can.,* **98,** 347–58.

Comeau, G. and LeBlanc, F. (1972), *Can. J. Bot.,* **50,** 847–56.

Coombes, A.J. and Lepp, N.W. (1974), *Bryologist,* **77,** 447–52.

Czarnowska, K. and Rejement-Grochowska, I. (1974), *Acta Soc. Bot. Pol.,* **43,** 39–44.

Daly, G.T. (1970), *Proc. N.Z. Ecol. Soc.,* **17,** 70–9.

Delvosalle, L., Demaret, F., Lambinon, J. and Lawalree, A. (1969), *Serv. Rés. Nat. dom. Cons. Nat.,* **4,** 1–128.

De Sloover, J. and LeBlanc, F. (1970), *Bull. Acad. Soc. Lorraines Sci.,* **9,** 82–90.

Geiger, R. (1965), *The Climate Near the Ground,* Harvard University Press, Cambridge, Mass.

Gilbert, O.L. (1968), *New Phytol.,* **67,** 15–30.

Gilbert, O.L. (1969), The effects of SO_2 on lichens and bryophytes around Newcastle-upon-Tyne. In: *Air Pollution: Proc. 1st Eur. Cong. Influence Air Pollution Plants and Animals,* pp. 223–35. Wageningen.

Gilbert, O.L. (1970a), *New Phytol.,* **69,** 605–27.

Gilbert, O.L. (1970b), *New Phytol.,* **69,** 629–34.

Gilbert, O.L. (1971a), *Trans. Br. Bryol. Soc.,* **6,** 306–16.

Gilbert, O.L. (1971b), *Lichenologist,* **5,** 26–32.

Goodman, G.T. and Roberts, T.M. (1971), *Nature,* **231,** 287–92.

Gorham, E. (1959), *Can. J. Bot.,* **37,** 327–9.

Granger, J.M. (1972), *Sarracenia,* **5,** 43–84.

Heggestad, H.E. (1968), *Phytopathology,* **58,** 1089–97.

Hoffman, G.R. (1972), *Bot. Gaz.*, **133,** 107–18.
Huckabee, J.W. (1973), *Atmos. Env.*, **7,** 749–54.
Hutchinson, T.C. and Whitby, L.M. (1974), *Env. Conserv.*, **1,** 123–32.
Lawrey, J.D. and Rudolph, E.D. (1975), *Ohio. J. Sci.*, **75,** 113–17.
LeBlanc, F. (1961), *Rev. Can. Biol.*, **20,** 823–7.
LeBlanc, F., Comeau, G. and Rao, D.N. (1971), *Can. J. Bot.*, **49,** 1691–8.
LeBlanc, F. and De Sloover, J. (1970), *Can. J. Bot.*, **48,** 1485–96.
LeBlanc, F. and Rao, D.N. (1966), *Bryologist,* **69,** 338–46.
LeBlanc, F. and Rao, D.N. (1973a), *Bryologist,* **76,** 1–19.
LeBlanc, F. and Rao, D.N. (1973b), *Ecology,* **54,** 612–17.
LeBlanc, F. and Rao, D.N. (1974), *Soc. Bot. Fr. Coll. Bryologie,* **121,** 237–55.
LeBlanc, F. and Rao, D.N. (1975), Effects of air pollution on lichens and Bryophytes. In: *Responses of Plants to Air Pollution* (Mudd, J.B. and Kozlowski, T.T., eds). Academic Press, New York.
LeBlanc, F., Rao, D.N. and Comeau, G. (1972a), *Can. J. Bot.*, **50,** 519–28.
LeBlanc, F., Rao, D.N. and Comeau, G. (1972b), *Can. J. Bot.*, **50,** 991–8.
LeBlanc, F., Robitaille, G. and Rao, D.N. (1974), *J. Hattori bot. Lab.*, **38,** 405–33.
LeBlanc, F., Robitaille, G. and Rao, D.N. (1976), *J. Hattori bot. Lab.*, **40,** 27–40.
Lee, J.A. (1972), *Nature,* **238,** 165–6.
Lounamaa, K.J. (1956), *Ann. Bot. Soc. Zool. Fenn.*, **29,** 1–196.
Mayer, A.M. and Gorham, E. (1951), *Ann. Bot.*, **15,** 247–63.
Nagano, I. (1972), *J. Hattori bot. Lab.*, **35,** 391–8.
Nieboer, E., Ahmed, H.M., Puckett, K.J. and Richardson, D.H.S. (1972), *Lichenologist,* **5,** 292–304.
Persson, H. (1956), *J. Hattori bot. Lab.*, **17,** 1–18.
Rao, D.N. and LeBlanc, F. (1966), *Bryologist,* **69,** 69–75.
Rao, D.N. and LeBlanc, F. (1967), *Bryologist,* **70,** 141–57.
Roa, D.N., Robitaille, G. and LeBlanc, F. (1977), *J. Hattori bot. Lab.*, **42,** 213–39.
Robitaille, G., LeBlanc, F. and Rao, D.N. (1977), *Rev. bryol. Lichénol.*, **43,** 53–66.
Rühling, A. and Tyler, G. (1968), *Bot. Notiser,* **121,** 321–42.
Rühling, A. and Tyler, G. (1969), *Bot. Notiser,* **122,** 248–59.
Rühling, A. and Tyler, G. (1970), *Oikos,* **21,** 92–7.
Rühling, A. and Tyler, G. (1971), *J. appl. Ecol.*, **8,** 497–507.
Rühling, A. and Tyler, G. (1973), *Water, Air and Soil Pollution,* **2,** 445–55.
Shacklette, H.T. (1961), *Bryologist,* **64,** 1–16.
Shacklette, H.T. (1965a), *U.S. Geol. Surv. Bull.*, **1198-C,** 1–18.
Shacklette, H.T. (1965b), *U.S. Geol. Surv. Bull.*, **1198-D,** 1–21.
Shacklette, H.T. (1966), *U.S. Geol. Surv. Bull.*, **1198-G,** 1–18.
Shimwell, D.W. and Laurie, A.E. (1972), *Env. Poll.*, **3,** 291–301.
Skaar, H., Ophus, E.M. and Gullvåg, B.M. (1973), *Nature,* **241,** 215–16.
Steere, W.C. (1970), Bryophyte studies on the irradiated and control sites in the rainforest at El Verde. In: *Tropical Rain Forest: A Study of the Irradiation and Ecology at El Verde, Puerto Rico* (Odum, H.T., ed.), pp. D 213–D 225. U.S. Atomic Energy Commission, Washington, D.C.
Svensson, G.K. and Leden, K. (1965), *Health Physics,* **11,** 1033.

Syratt, W.J. and Wanstall, P.J. (1969), The effect of sulfur dioxide on epiphytic bryophytes. In: *Air Pollution: Proc. 1st Eur. Cong. Influence Air Pollution Plants and Animals,* pp. 79–85. Wageningen.
Tamm, C.O. (1953), *Medd. fran Stateus Skogforskning Institut,* **43,** 1–140.
Taoda, H. (1972), *Jap. J. Ecol.,* **22,** 125–33.
Taoda, H. (1973a), *Hikobia,* **6,** 224–8.
Taoda, H. (1973b), *Hikobia,* **6,** 238–50.
Treshow, M. (1971), *Ann. Rev. Phytopath.,* **9,** 21–44.
Tyler, G. (1971), Moss analysis – a method of surveying heavy metal deposition. In: *Proceedings of the Second International Clean Air Congress* (Englund, M. and Beery, W., eds), pp. 129–32. Academic Press, New York.
Watson, H. (1947), *Forestry Commission Booklet No. 1,* London.
Yeaple, D.S. (1972), *Nature,* **235,** 229–30.

Chapter 13

Quaternary Bryophyte Palaeo-ecology

H.J.B. BIRKS

13.1 INTRODUCTION

The Quaternary is the interval of geological time which covers the last 1–2 million years. It is unique in earth's history for its oscillating climates, alternating at the latitudes of Europe and North America between short (c. 10–20 × 10^3 years) temperate stages and long (c. 50–100 × 10^3years) cold stages within which glaciation commonly occurred. In addition, the Quaternary witnessed the evolution of man. The number of temperate stages is unknown but at least 20 have been identified in deep-sea cores covering the last 2 million years (Hays *et al.*, 1969; Shackleton and Opdyke, 1973, 1976). Nowhere on the continents have all these stages been identified. Indeed, the correlation between continental and marine sequences is a matter of considerable debate and uncertainty at present (see Kukla, 1977). In view of the great duration of the cold stages, the Quaternary should be viewed as a predominantly cold period interrupted periodically by warm phases – the brief interglacials with climates similar to those of today.

Palaeo-ecology is the ecology of the past. Although it can be studied in any period of earth's history in which there was life, it has the greatest relevance and strongest links with modern ecology in the Quaternary. This is because, in contrast to pre-Quaternary palaeo-ecology which deals predominantly with extinct taxa, the vast majority of Quaternary fossils can be identified with confidence and matched with modern genera and species whose ecology is often known. Quaternary palaeo-ecology is primarily concerned with the reconstruction of past populations, communities, and ecosystems for particular points in time and space (see Birks and Birks, 1980). For such reconstructions, all available evidence, both biological and geological is used to reconstruct past biotas and environments. Reconstruction of Quaternary terrestrial ecosystems is based largely on fossil remains of plants and animals. Of the various types of fossils preserved in Quaternary terrestrial sediments, those of plants are most abundant, both microscopic (pollen, spores, diatoms etc.) and macroscopic (seeds, fruits, leaves, wood, moss remains etc.), with the result that Quaternary palaeo-ecology is dominated by the study of plant remains.

Environmental reconstructions can use a few, selected species, the so-called indicator species or floristic approach, or the fossil assemblage as a whole, the so-called assemblage or vegetational approach (see Birks, 1973; Janssen, 1970). In both approaches, information is required about the present-day ecological and sociological tolerances and requirements of the taxa found as fossils. All these approaches assume methodological uniformitarianism (see Gould, 1965; Rymer, 1978), namely that modern-day observations and ecological tolerances can be used as a model for past conditions and, more specifically, that the relationships between plants and their environment have not changed with time, at least for the time period of the late Quaternary.

The pioneer studies of fossil plants in Quaternary sediments were based on plant macrofossils, including bryophyte remains (e.g. Mahony, 1869; Robertson, 1881; Reid, 1899). Such studies concentrated on obtaining species lists from different localities and demonstrated major changes in plant distribution in the past (e.g. Dixon, 1927; Gams, 1932; Tralau, 1963; Dickson, 1973). Although the standards of identification were high, the investigations were non-quantitative, the fossils were not placed in an exact stratigraphic context, and few palaeo-ecological insights were gained. When quantitative pollen analysis was developed by von Post (1916) and refined by Iversen (1954), macrofossils tended to be ignored in favour of this new, quantitative and stratigraphically valuable technique. However, the value of combining pollen and macrofossils was convincingly demonstrated by Jessen and Milthers (1928) for interglacial deposits in Denmark and by Jessen (1949) for late-glacial (c. 10–13 000 years ago) and post-glacial (0–10 000 years ago) deposits in Ireland. Macrofossils were not at that time sampled with the same stratigraphic precision as pollen. The first detailed stratigraphic macrofossil diagram was made by West (1957) from a British Ipswichian (last) interglacial deposit, and this approach has been extensively developed by W.A. Watts (e.g. Watts and Winter, 1966; Watts and Bright, 1968; Wright and Watts, 1969; Watts, 1970, 1978, 1979). The development of a quantitative, stratigraphic format has shown the value and importance of macrofossil studies in reconstructing past environments, particularly when studied in conjunction with pollen analysis. Plant macrofossils, including bryophyte remains, contribute primarily to the reconstruction of local limnic and wetland environments, whereas pollen and spores primarily reflect the regional, upland flora and vegetation (see H.H. Birks, 1980).

Bryophyte remains have only recently been studied in any detailed stratigraphic way (e.g. Walker and Walker, 1961; Tallis, 1964; de Vries and Bird, 1965; Rybniček and Rybničková, 1968, 1971; H.J.B. Birks, 1970, 1976; Birks and Mathewes, 1978; Aaby and Jacobsen, 1979). The aims of this chapter are to review recent Quaternary palaeo-ecological studies in which bryophyte remains have been studied stratigraphically, to show the value of bryophytes in palaeo-environmental reconstructions, and to discuss the contribution that bryophyte fossils have made to bryophyte ecology.

Plant nomenclature follows Clapham, Tutin and Warburg (1962) for vascular plants, Smith (1978) for mosses, and Paton (1965) for hepatics. Authorities are given for taxa not considered by these authors.

13.2 METHODS OF INVESTIGATING QUATERNARY BRYOPHYTES

Bryophyte remains are preserved in a variety of Quaternary sediments, including peats, lake muds, silts and clays, estuarine sediments and archaeological deposits. A thorough review of methods for extracting and studying plant macrofossils, including bryophytes, is given by Wasylikowa (1979).

The size of sample to be examined will vary with the nature of the investigation and the type of deposit. Samples of 50–100 cm^3 are preferable for stratigraphic studies. Smaller samples may be suitable for counts of *Sphagnum* leaves in peats (e.g. Green and Pearson, 1977). The sediment type and its volume should always be recorded. In the laboratory the sediment is dispersed as gently as possible, either by soaking in water or in either dilute alkali (NaOH, KOH, Na_2CO_3) or, if the sediment is calcareous, in dilute acid (HCl). The sediment is then washed with a gentle stream of water through sieves with mesh sizes of 0.5 and 0.2 mm diameter. The residue is transferred to jars for storage and a mixture of glycerine, alcohol, and formalin added to discourage fungal and bacterial growth. The residue is examined systematically in small quantities in water under a low-power binocular dissecting microscope. Remains of interest are picked out, identified, and counted.

Bryophyte remains can be identified in much the same way as modern bryophytes. Careful comparison with accurately determined modern reference material, vouched for by herbarium specimens, is essential. Ideally, a complete reference collection should be available, comprising all taxa which could be expected to have occurred within the study area. The collection should also encompass the full range of phenotypic variation exhibited by the species concerned. A bryophyte reference collection can be made by mounting shoots and detached leaves in gum chloral on microscope slides for ease of comparison with fossil remains. The vast majority of fossil bryophyte remains are stems, shoots, and leaves of gametophytes; very occasionally sporophytes (capsules, setae, opercula) are found (e.g. *Splachnum sphaericum* capsules; Conolly and Dickson, 1969). Archegonia and antheridia are also sometimes present. Unfortunately, records of sporophytes, archegonia and antheridia are so few that no information, as yet, is available on the former frequency of sporophyte production.

The number of shoots and stem fragments should be counted, and the number of detached leaves of any taxa not represented as shoot fragments should, if possible, be recorded. In many instances it is difficult to count all the bryophyte material and in such cases a scale of estimated abundance (e.g. rare, occasional, frequent, abundant, very abundant) can be used. The data can be presented in a similar way to a pollen diagram and they can be plotted stratigraphically as bryophyte concentrations (number of fragments in a known volume of sediment; e.g. H.J.B. Birks, 1970; Burrows, 1974), as estimated abundance (e.g. Birks and Mathewes, 1978; Hall, 1980; Aaby and Jacobsen, 1979), as presence and absence (e.g. de Vries and Bird, 1965; Rybniček and Rybničková, 1968; Birks, 1976), or as percentages of total bryophyte remains (e.g. Green and Pearson, 1977).

Several bryophytes produce characteristic and preservable spores (see Dickson, 1973; Boros and Jarai-Komlodi, 1975) but they are often produced in such small quantity and so close to the ground that they are rarely found and recognized by pollen analysts. The major exception is *Sphagnum,* which produces abundant spores in particular conditions. *Sphagnum* spores are often found in large numbers on slides prepared for pollen analysis. Fossil spores of *Anthoceros laevis, A. punctatus, Encalypta, Meesia triquetra,* and *Riccia* have also been identified from Quaternary sediments (Dickson, 1973).

13.3 ORIGIN OF BRYOPHYTES IN QUATERNARY SEDIMENTS

Bryophyte remains may be preserved in the place where the bryophytes lived and subsequently died. In this case they are termed autochthonous. Examples include *Sphagnum* spp. growing on bogs, *Calliergon* spp., *Drepanocladus* spp. and *Scorpidium scorpioides* growing in fens, and aquatic mosses such as *Drepanocladus exannulatus* growing submerged in clear-water lakes in arctic, antarctic, or alpine areas (see Light, 1975; Priddle, 1979). One of the most unusual types of autochthonous bryophyte assemblages known consists of pure *Aplodon wormskjoldii* remains, recorded by Brassard and Blake (1978) from the Carey Islands, northwest Greenland. Such autochthonous assemblages tend to be relatively poor in species. In contrast, bryophyte remains may be transported from where they lived to another locality by agencies such as water, wind and animals (including man), and fossilized in this new locality. Such fossils are termed allochthonous and the assemblages are frequently species-rich.

The proportion of autochthonous and allochthonous remains in a fossil bryophyte assemblage will clearly vary with the environment of deposition and will influence and limit the interpretation of the fossil assemblages in terms of past plant communities. For example, in a raised bog or fen the material forming the peat is closely related in time and space to the plant community or life assemblage on the mire surface. There is virtually no allochthonous input. All the bryophytes found as fossils can thus be assumed to have grown together *in situ* and past plant communities can be reconstructed with confidence (e.g. Rybniček and Rybničková, 1968, 1974; Jankovska, 1971; Birks and Birks, 1980), using the assemblage-approach for reconstruction. The fossil assemblages frequently have exact counterparts among contemporary bryophyte communities, and the history of particular mire communities can thus be directly reconstructed from the fossil record (Rybniček, 1973).

In a lake or riverine deposit, the remains preserved at a particular point may be mostly allochthonous, and consist of material originating in one part of the lake or river and transported by currents, of terrestrial material washed in by streams from the surrounding catchment, and of material blown onto the water surface by wind, particularly in storms. Interpretation of such fossil assemblages in terms of past plant communities is thus more difficult than that for autochthonous assemblages. The palaeo-ecologist invariably has to adopt the indicator-species approach to reconstruct the range of communities that were present in and around the site of deposition. In reconstructing any past community or environment from a fossil bryophyte assemblage, it is therefore essential to consider the mode of formation of the fossil assemblage or its taphonomy (see West, 1973; Lawrence, 1968; Birks and Birks, 1980).

The taphonomy of autochthonous bryophyte fossils is relatively straightforward. After death, decay, and fossilization, the life assemblage becomes the fossil assemblage which palaeo-ecologists subsequently study. Little is known, however, about decomposition processes and about differential preservation of bryophyte remains in sediments. Decomposition, even in bogs is high and may result in the loss of over 90% of the total annual productivity of the bog (Reader and Stewart, 1972). Preservation appears to be selective. Hepatics are poorly preserved, judging by their great rarity in fossil assemblages. Bog hepatics such as *Calypogeia fissa, Gymnocolea inflata, Lepidozia setacea,* and *Pleurozia purpurea* are unknown as fossils in British peat deposits. Similarly, several common mosses, such as *Dicranella palustris, Amblystegium riparium,* and *Leucobryum glaucum* are rarely, if ever, found as fossils, possibly also because of poor preservation. Clymo (1965) has shown that, within the genus *Sphagnum,* some species preserve more readily than others. This differential preservation may clearly result in a biassed fossil assemblage. There is a fascinating but as yet unexplored field of research on the decomposition of bryophyte leaves in relation to cell-wall chemistry and structure and to sedimentary environments and microbial activity (cf. Dickinson and Maggs, 1974).

Bryophyte fossils preserved in lake sediments may, in some instances, be derived from submerged bryophytes growing *in situ.* Light (1975) and Light and Lewis Smith (1976) have described assemblages of submerged bryophytes including *Drepanocladus exannulatus, Oligotrichum hercynicum, Polytrichum commune, Sphagnum auriculatum, Thuidium* cf. *T. tamariscinum, Cephalozia bicuspidata, Hygrobiella laxifolia, Nardia compressa, Scapania uliginosa* and *S. undulata* growing at depths of up to 20 m in clear, cold, oligotrophic high-altitude lakes in the Cairngorms. Bryophytes also appear to be the dominant submerged vegetation in some Antarctic (Light and Heywood, 1973, 1975; Priddle and Dartnall, 1978), and Arctic lakes (Persson, 1942, 1948; Rigler, 1972). Priddle (1979) has documented in detail the morphological variation of submerged *Calliergon sarmentosum* and *Drepanocladus* cf. *D. aduncus* from lakes on Signy Island, and has shown that the submerged forms have characteristic growth with large leaves and elongated internodes. H.J.B. Birks (1981) has recently interpreted a fossil assemblage from late-glacial lake sediments in Minnesota dominated by *Drepanocladus exannulatus* and *Calliergon sarmentosum* with identical growth forms to those described by Priddle (1979) as reflecting an *in situ* submerged bryophyte assemblage. Mosses probably formed the dominant aquatic vegetation from about 16000 to 13500 years ago in this lake.

Bryophyte remains in lake sediments may also be blown onto the lake surface in storms or they may be transported considerable distances by streams and rivers, particularly in floods (see Övstedal and Aarseth, 1975). They may also be washed in following soil erosion associated with, for example, forest clearance, agriculture, solifluction, and seasonal melting of snow. Remains carried by streams and rivers can also be deposited in calm backwaters of rivers, oxbow lakes, terrace sediments, and channel fills of streams and be fossilized in alluvial deposits. Many so-called 'full glacial' deposits in Britain are alluvial and are rich in macrofossils of both vascular plants (Bell, 1970; West, 1977) and bryophytes (Dickson, 1967, 1973). Deposits such as these are, by definition, laid down irregularly and are constantly subject to erosion. They are therefore not generally suitable for detailed stratigraphic studies, although they frequently provide valuable floristic information.

In arctic and alpine areas, dispersal of bryophyte fragments by wind, particularly in winter, appears to be an important mechanism of selective dispersal (Miller and Ambrose, 1976). Late snowbeds frequently accumulate plant debris blown onto the snow in winter from snow-free areas, such as wind-exposed ridges (see Warren Wilson, 1958; Bonde, 1969; Glaser, 1979). Snowbeds adjoining ponds or lakes may act as traps for wind-blown debris. When the snowbeds melt, the debris is deposited into the lake. Some plant remains sink, whereas others float and are carried ashore by winds. Glaser (1979) has analysed plant debris from snowbeds in Alaska and has shown that the diversity in the snowbed assemblage is much greater than in the adjoining pond sediments. This is because some plant remains are sufficiently buoyant to be swept to the opposite shore of the lake where they are deposited as flotsam. Thus, they never reach the sediments and they are not fossilized.

There is unfortunately no quantitative information on the mechanisms of dispersal and deposition of bryophyte remains in lake sediments and on how well modern assemblages of bryophytes reflect the vegetation around the lake, comparable to studies on modern pollen transport by Bonny (1976, 1978), Tauber (1967, 1977), and others. It would be of considerable interest to study modern deposition of bryophyte remains in lakes, both by trapping remains at various points on their way to the bottom sediments (e.g. in streams) and by examining remains in surficial sediments. With this knowledge, coupled with information about the present-day bryophytes of the catchment, new insights could be gained about the origin and interpretation of allochthonous bryophyte assemblages.

Gams (1932) and Dickson (1973) discuss the influence that habitat has on the probability of particular bryophytes being found as fossils. Many rupestral and epiphytic genera are rarely found as fossils, e.g. *Andreaea, Hedwigia, Seligeria, Cynodontium, Bartramia, Cryphaea, Orthotrichum,* and *Ulota.* Ephemeral genera such as *Ephemerum, Pottia, Phascum,* and *Weissia* are also rare or absent as fossils. Many common and ecologically ubiquitous species in Britain such as *Barbula convoluta, B. revoluta, B. cylindrica, B. unguiculata, Bryum argenteum, Fissidens taxifolius, Orthotrichum anomalum, Tortula muralis,* and *Ulota crispa* are not known as fossils in British Quaternary deposits (Dickson, 1973).

13.4 PALAEO-ECOLOGICAL STUDIES BASED ON BRYOPHYTE REMAINS

Any palaeo-ecological study requires the fossils to be identified to the lowest possible taxonomic level, so that the maximum amout of ecological information can be gleaned from the fossil record. Bryophyte remains can frequently be identified to species level. Although there are some difficult groups (e.g. *Bryum, Amblystegium, Drepanocladus, Racomitrium,*

Pohlia), nearly all bryophyte remains can be identified to a level below that of family, provided preservation is not bad. Bryophyte remains are invariably present in much lower concentrations in sediments than are pollen or spores. Larger volumes of sediment are therefore required to obtain adequate samples than for pollen analysis. However, because bryophytes can be identified in more detail than pollen, considerable information can be obtained about past environments. Bryophyte remains should ideally always be studied in conjunction with other plant macrofossils to provide a detailed picture of the local flora and vegetation, and with pollen analysis to provide a picture of regional vegetation. An independent chronology based on radiocarbon dating or some other radiometric technique is essential if sequences are to be compared and correlated. For palaeo-ecological interpretation, the modern ecological tolerances of the species found as fossils must be known so that ecological extrapolations can be made into the past. Although a large amount is known in a general, descriptive way about bryophyte communities and their ecology, less is known about the edaphic and climatic tolerances of bryophytes, about bryophyte population dynamics, and about community stability and integration.

Bryophytes are valuable palaeo-ecological indicators of past soil conditions, especially nutrients (particularly Ca), base-status, texture and moisture, of past water and peat chemistry and water depth, of snow-lie conditions and exposure, of local specialized habitats such as springs and gravel-flushes, and of local vegetation and vegetational change. Miller (1973, 1976a, 1980) has shown how bryophyte remains in late-glacial deposits in eastern North America can be used to reconstruct local habitats and their soil nutrient-status and moisture in considerably greater detail than is possible from either pollen or vascular plant macrofossils (cf. Birks, 1976; Watts and Winter, 1966).

13.4.1 Autochthonous assemblages

In autochthonous deposits, bryophyte remains can yield valuable insights into local vegetation, local changes in species abundance, and directions and rates of vegetational change, particularly as a result of hydroseral succession. Birks and Mathewes (1978) reconstructed in detail the hydroseral development of a channel bog in Abernethy Forest in eastern Scotland (see H.H. Birks, 1970) using bryophytes and other macrofossils. They showed the development, in early post-glacial times, from open-water submerged aquatic vegetation, through floating-leaved aquatics to reed- and sedge-swamps; the colonization at about 8500 BP* of the sedge-swamp by rich-fen with *Calliergon trifarium, Bryum pseudotriquetrum, Drepanocladus revolvens,* and *Campylium stellatum;* and the subsequent trans-

*Years before present.

formation to poor-fen with *Juncus effusus, Aulacomnium palustre,* and *Sphagnum*. This was, in turn, replaced by ombrotrophic bog at about 6000 BP with *A. palustre, Pohlia nutans, Polytrichum alpestre, Sphagnum* spp., *Vaccinium oxycoccos, Calluna vulgaris, Erica tetralix, Drosera rotundifolia,* and *Rhynchospora alba*.

Grosse-Brauckmann (1974, 1976) reconstructed the spatial and temporal patterns of hydroseral successions at Steinhuder Meer, Germany by means of pollen, macrofossils, and sediment stratigraphy from several sites within the basin. He showed by means of bryophyte remains that particular communities could be recognized in the fossil record. He demonstrated that the present vegetational zonation does not reflect in space the former hydroseral succession in time and that in different parts of the basin, different hydroseral successions had occurred. He was able to estimate the rate of sediment accumulation and the outward spread of vegetation, the duration of phases of stable vegetation at one place, the rate of change from one community to another, and changes in trophic status. He concluded that the hydrosere is not an autonomous, self-regulating system, but that it is affected by external changes in climate and hydrology and by human influence (cf. Walker, 1970).

Rybniček and Rybničková (1968, 1977) and Janssen *et al.* (1975) have reconstructed the direction of hydroseral change in mires in Czechoslovakia, Austria, and the Vosges mountains of France, respectively. Although not as detailed as Grosse-Brauckmann's study, the same conclusions about the non-autonomous nature of the hydrosere emerge from these studies. They all show how the palaeo-ecological record is important in testing ecological theories such as the hydrosere concept.

Another important ecological concept, namely the stability of communities in space and time, has been tested by Rybnicek (1973) in relation to central European mire communities. He could recognize 20 mire communities on the basis of macrofossil remains (including bryophytes), many of which could be related to modern communities at the association level. He showed the overall trend of occurrence of different communities during the post-glacial, from abundant eutrophic rich-fen communities in early post-glacial times to oligotrophic bog communities in late post-glacial times. Bog communities appear to have developed relatively recently, some, it appears, as a result of human disturbance. That such distinctive mire communities can evolve in response to recent environmental changes provides further support for the idea proposed by West (1964) that 'our present plant communities have no long history in the Quaternary, but are merely temporary aggregations under given conditions of climate, other environmental factors, and historical factors'. All the indications from the fossil record are that plant communities including those dominated by bryophytes, are not stable, permanent associations, but that they are formed by groups of plants best able to grow under particular environmental conditions, each species behaving individually to maintain itself (cf. Gleason, 1939).

Examination of bryophyte remains, particularly *Sphagnum* leaves, preserved in ombrotrophic peat has resulted in bryophyte palaeo-ecology making a third important contribution to ecology by providing a direct test of Osvald's (1923) 'regeneration-complex' theory of bog growth. According to this theory bog pools and hummocks alternate in a cyclical way with pools becoming hummocks and vice versa. If this theory is correct examination of *Sphagnum* remains in peat sections should reveal regular alternations between *S. papillosum* and *S. imbricatum* (hummock species) and *S. cuspidatum, S. pulchrum,* and *S. auriculatum* (pool species). Walker and Walker (1961) examined several peat faces 1–2 m deep in Irish raised bogs and showed from peat stratigraphy and detailed examination of *Sphagnum* remains that there was little evidence for any regeneration complex. Instead the evidence suggested that hummocks and pools were relatively permanent features, and that persistent large pools were associated with general rejuvenation of the bog surface resulting from short-term climatic changes. Bog growth appears, like the hydrosere, not to be an autonomous, self-regulating, and self-perpetuating system but is affected by external changes in climate and hydrology, particularly changes in precipitation (see Aaby and Tauber, 1975; Casparie, 1969).

Sphagnum-leaf analysis, with identification to species or section, can provide valuable information on recent changes in the composition of *Sphagnum* communities. For example, Green and Pearson (1977) showed that *S. recurvum* achieved its present dominance at Wybunbury and Chartley Mosses comparatively recently, probably as a result of drainage and nutrient (K, N, and P) enrichment of these basin 'schwingmoors' from the surrounding agricultural areas. Tallis (1973) has demonstrated that *S. recurvum*-dominated communities are characterized by high K, N, and P and he suggests that it is *S. recurvum*'s ability to tolerate and utilize such high nutrient levels that has enabled it to compete so successfully in recent years on lowland mires exposed to pollution and nutrient enrichment from agricultural land.

Tallis (1964) has analysed *Sphagnum* leaves in southern Pennine blanket-peats and has shown major changes in the *Sphagnum* flora of these bogs, particularly following human disturbance in the 15th century that resulted in the dominance of *Eriophorum vaginatum* and widespread peat erosion. The almost total absence of *Sphagnum* today on these blanket bogs appears to be due to the high atmospheric SO_2 pollution of the last 150 years. Interestingly, *Sphagnum imbricatum* was an important component of these bogs until about 1000 AD, as it was in very many lowland raised bogs and upland blanket bogs in the British Isles (see Dickson, 1973).

Although many explanations have been proposed for the demise of *S. imbricatum*, there is little doubt that its disappearance is frequently, but not exclusively (see Aaby and Jacobsen, 1979), associated with a drying-out of the bog surface (Green, 1968). Such desiccation may result from human disturbance (Pearsall, 1956; Pigott and Pigott, 1963), from climatic change (Godwin and Conway, 1939), or from small-scale autogenic bog growth (Morrison, 1959).

Stratigraphic analyses of bryophyte remains in autochthonous peats have shown that rich-fen bryophytes such as *Meesia longiseta* Hedw., *M. triquetra, Calliergon trifarium, C. richardsonii* (Mitt.) Kindb., and *Scorpidium turgescens* were once more widespread in Britain, particularly between 12000 and 7000 BP (see Dickson, 1973; Hulme, 1979). In addition, a very characteristic assemblage of intermediate-fen bryophytes (alliance Sphagno-Tomenthypnion; see Birks, 1973), including *Helodium blandowii, Paludella squarrosa,* and *Homalothecium (Tomenthypnum) nitens*, was once locally common on British mires between 6000 and 10000 BP (see Dickson, 1973; Pigott and Pigott, 1963; Bellamy *et al.*, 1966). As a result of the natural processes of soil acidification, paludification, and the spread of *Sphagnum*-dominated bog, these assemblages have become very rare or, in the case of *M. longiseta,* extinct (Dickson and Brown, 1966; Hall, 1979) in Britain. In the last 150 years, drainage of lowland mires has resulted in the extinction of *Helodium blandowii* and *Paludella squarrosa* and, more recently, of *Meesia triquetra* in its sole Irish locality in West Mayo. The assemblage of intermediate-fen bryophytes occupies a very narrow and precise position in the major ecological gradient of mire-water chemistry (pH 4.7–5.5), and occurs between rich-fens (pH 5.2–7.2) with, for example, *Calliergon giganteum, Scorpidium scorpioides,* and *Drepanocladus revolvens* and poor-fens (pH 4.1–5.6) with *D. fluitans, Sphagnum palustre, S. recurvum,* and *S. girgensohnii.* Intermediate fens are thus inevitably rare and restricted today because of widespread paludification. Every effort should be made to conserve the few remaining examples in Britain.

Kuc (1973) and Kuc and Hills (1971) have described rich autochthonous bryophyte assemblages of Late Miocene or Early Pliocene (Tertiary) age from the Western Canadian arctic. The assemblages contain 45 bryophyte taxa, all but one (*Calliergon aftonianum* Steere) of which are extant. They suggest a regional vegetation of boreal coniferous forests with local mires supporting, *i.a., Meesia triquetra, M. uliginosa, Sphagnum teres, Calliergon giganteum, C. richardsonii, Scorpidium scorpioides, Homalothecium nitens, Pseudobryum cinclidioides, Cinclidium latifolium* Lindb., and *Scorpidium scorpioides.* Open rocks may have supported *Timmia norvegica, T. austriaca, Racomitrium microcarpon, Schistidium apocarpum,* and *Myurella tenerrima.* Recent work by Janssens *et al.* (1979)

on Eocene bryophytes in British Columbia has shown that *Aulacomnium heterostichoides* sp. nov., an extinct taxon closely related to *A. heterostichum* (Hedw.) BSG, formerly grew in eastern North America. Phytogeographically, this is an important discovery, because *A. heterostichum* today grows in deciduous forests of eastern Asia and eastern North America, a disjunct distribution pattern shown by many plants (Gray, 1846; Li, 1952; Iwatsuki and Sharp, 1967, 1968). This fossil discovery provides the first direct bryological evidence that a moss with this modern disjunct distribution was growing in early Tertiary times in western North America, suggesting that it had once had a more widespread, circumboreal distribution which was reduced by climatic changes in the late Tertiary and in the Quaternary.

13.4.2 Allochthonous assemblages

Allochthonous bryophyte assemblages, particularly from lake sediments of late-glacial age, are often very rich in species, which represent a wide range of habitats within the lake's catchment. Recent stratigraphic studies on late-glacial allochthonous assemblages have been made by Burrows (1974) in Wales, by Birks and Mathewes (1978) and H.J.B. Birks (1970) in Scotland, by Birks (1976) in Minnesota, by Miller (1973) in New York State, and by Övstedal and Aarseth (1975) in Norway. Fredskild *et al.* (1975) have described diverse bryophyte assemblages from post-glacial lake sediments in Greenland.

The studies by Birks and Mathewes (1978) at Abernethy Forest provide a detailed picture of vegetational diversity near the site during the late-glacial. Prior to about 11500 BP pioneer communities with *Ceratodon purpureus, Barbula* spp., *Distichium capillaceum,* and *Encalypta rhapdocarpa,* snowbeds with *Polytrichum norvegicum,* and gravel-flushes with *Saxifraga aizoides, Campylium stellatum,* and *Drepanocladus revolvens* were widespread. At 11500 BP shrub-tundra of *Betula nana* and *Empetrum* developed, with *Polytrichum alpinum, Climacium dendroides, Dicranum scoparium, Hylocomium splendens, Pleurozium schreberi,* and *Racomitrium* spp. Open, base-rich areas persisted with *Encalypta rhaptocarpa, Distichium capillaceum, Schistidium apocarpum, Orthothecium rufescens, Campylium stellatum,* and *Scorpidium scorpioides*. Between 11000 and 10000 BP a major climatic deterioration occurred (the so-called Younger *Dryas* or Loch Lomond stadial; see Sissons, 1979), which led to the formation of open, herb-rich vegetation with *Minuartia rubella, Cerastium alpinum, Silene acaulis, Sagina* spp., *Artemisia* spp., *Armeria maritima, Saxifraga cespitosa, S. oppositifolia, Salix herbacea,* and *Luzula spicata*. It included bryophytes such as *Distichium capillaceum, Ceratodon purpureus, Hypnum hamulosum* and *Trichostomum brachydontium*. Snowbeds were also widespread, with abundant *Polytrichum norvegicum.* Moist grassland and flushes were frequent with *Thalictrum alpinum, Saxifraga aizoides, Selaginella selaginoides, Campylium stellatum, Rhizomnium pseudopunctatum, Drepanocladus revolvens,* and *Scorpidium scorpioides.* The largely calcicolous nature of the late-glacial flora and vegetation contrasts with the markedly acid calcifuge character of the heaths, bogs, and pine woods growing on podsols and acid peats overlying the Cairngorm granite in the Abernethy Forest area today. In the early post-glacial, *Betula pubescens* and *Populus tremula* were the first trees to colonize the area. Interestingly, at the time of aspen abundance, there are remains of the epiphyte *Orthotrichum obtusifolium,* a species commonly associated with aspen today.

The bryophyte assemblage recorded by Burrows (1974) from late-glacial lake deposits at Nant Ffrancon in north Wales similarly suggests open, base-rich conditions, with, *i.a., Antitrichia curtipendula, Campylium stellatum, Cratoneuron commutatum, Homalothecium sericeum, Scorpidium turgescens,* and *Tortula ruralis.* Several montane species were also present including *Philonotis seriata, Oligotrichum hercynicum, Polytrichum alpinum,* and *Racomitrium* spp. Aquatic development appears to have been affected by climatic change, with *Fontinalis squamosa* abundant in cold phases, and *Chara* and aquatic macrophytes in warm phases of the late-glacial.

Birks (1976) recorded a diverse assemblage of bryophytes in late-glacial sediments at Wolf Creek, Minnesota, USA from 10000 to 20500 years ago. The assemblage suggests the presence of open, moist, base-rich soils with, for example, *Calliergon trifarium, Campylium polygamum, Bryum pseudotriquetrum, B. neodamense, Meesia triquetra, Scorpidium scorpioides, S. turgescens,* and *Cratoneuron commutatum.* A similar picture of open, base-rich soils in the late-glacial emerges from Miller's (1973, 1976a) work in New York State and Wisconsin, with *Thuidium abietinum, Ditrichum flexicaule, Tortella tortuosa, Meesia triquetra, M. uliginosa, Drepanocladus vernicosus, Scorpidium turgescens, Barbula recurvirostra, Campylium chrysophyllum, C. polygamum, C. stellatum, Distichum capillaceum, Aulacomnium turgidum, A. acuminatum* (Lindb. and Arn.) Kindb., *Myurella julacea,* and *Hypnum bambergeri* (see also Miller, 1976b, Miller and Benninghoff, 1969).

Inwashing of bryophytes into lakes from the surrounding catchment was common in late-glacial times. H.J.B. Birks (1970) analysed in detail a thin (2 cm) but well-marked layer of bryophyte remains interstratified in late-glacial silts at Loch Fada, Isle of Skye. The bryophyte assemblages included *Calliergon cuspidatum, C. sarmentosum, Dichodontium pellucidum, Pohlia wahlenbergii, Rhytidiadelphus squarrosus,* and *Drepanocladus exannulatus.* The bryophyte-rich layer contained a distinctive pollen assemblage, with *Epilobium, Caltha palustris, Stellaria, Saxifraga stellaris,* and *Ranunculus acris,* and an unusual assemblage of testaceous rhizopods, ascomycete fungal hyphae, and fungal fruit bodies. The pollen assemblage differed from those in the overlying and underlying silts in being predominantly corroded, a feature characteristic of pollen preserved in soils rather than lake sediments. H.J.B. Birks (1970) suggested that the bryophyte layer originated from a bryophyte-dominated spring community on the slopes above Loch Fada (cf. Philonoto–Saxifragetum stellaris; Birks, 1973) and that, during a flood or spring melt, the bryophyte carpet was washed into the lake bringing with it not only the bryophytes but a corroded pollen-assemblage of plants that grew locally in the spring, and the microflora and microfauna of the spring. This bryophyte layer clearly provides unique information about the palaeo-ecology of the site at about 12800 BP.

Allochthonous assemblages of bryophytes are often found in archaeological excavations. Dickson (1973) gives a review of bryophytes associated with archaeological deposits and speculates about the use of bryophytes by prehistoric man. *Neckera complanata, Brachythecium rutabulum, Hylocomium splendens, Rhytidiadelphus triquetrus, Thuidium tamariscinum, Lunularia cruciata, Plagiochila asplenioides, Lophocolea cuspidata,* and *Lepidozia reptans* were used as caulking for Bronze Age boats and dug-out canoes; *Polytrichum commune* was used as rope; *Homalothecium sericeum, Isothecium myosuroides, Hylocomium splendens,* and *Thuidium tamariscinum* were used for plugging in wooden dwellings; *Sphagnum* was used as wound-dressings; and *Hylocomium brevirostre* for padding around Mesolithic flint flakes. Seaward and Williams (1976) suggest that mosses such as *Hylocomium splendens, Rhytidiadelphus squarrosus, Pleurozium schreberi,* and *Pseudoscleropodium purum* were deliberately harvested in large quantities for several purposes, including bedding, packing, insulation, and absorption. Tallantire (1979) has suggested that mosses may have been used as packing in soft leather slippers in Viking and early Medieval Trondheim and Dickson (1979) has even proposed that Romans at Bearsden near Glasgow used mosses as toilet paper. Clearly, bryophytes may have played an important role in the ethnology of man.

13.5 CONCLUSIONS

The study of bryophyte remains in Quaternary sediments can provide much valuable information for the palaeo-ecologist in his task of reconstructing past plant communities and past environments. At the same time, detailed stratigraphic studies, particularly of well-dated autochthonous assemblages, can provide information relevant to the testing of ecological theories such as the hydrosere concept, the regeneration-complex of raised bogs, and the continuity of plant communities in time. An enormous amount of work remains to be done in Quaternary bryology, particularly on stratigraphic changes of bryophytes in inter-glaical deposits in north-west Europe (cf. Hall, 1980) and on the composition of bryophyte assemblages in sites south of the limit of glaciation in eastern North America. Recent studies on pollen stratigraphy and plant macrofossils in sites in Georgia, Carolinas, and Florida by Watts (1970, 1980) have shown major changes in the flora and vegetation, with northern boreal vascular plants growing over 1000 km further south than they do today. Bryophyte remains from such deposits, if analysed in detail and measured biometrically, could provide a direct test of Horton and Vitt's (1976) hypothesis for the speciation of *Climacium americanum* Brid. This is endemic to the eastern United States south of the last glacial limit, and may have arisen in isolation from the widespread, circumboreal *C. dendroides* after the last glaciation. Quaternary bryology can clearly contribute not only to bryophyte ecology and Quaternary palaeo-ecology but in this case could also contribute to our understanding of bryophyte evolution. Palaeo-ecologists should never ignore fossil bryophytes in sediments. They are invaluable palaeo-ecological indicators of past floras, vegetation, and environment.

ACKNOWLEDGMENTS

I am indebted to Dr Hilary H. Birks for her critical reading of the manuscript and to Dr Allan Hall, Dr Brian Huntley, and Dr Norton G. Miller for helpful discussions and comments about Quaternary bryophytes.

REFERENCES

Aaby, B. and Jacobsen, J. (1979), *Danm. geol. Unders., Årbog*, **1978**, 5–43.
Aaby, B. and Tauber, H. (1975), *Boreas*, **4**, 1–17.
Bell, F.G. (1970), *Phil. Trans. R. Soc.*, **258**, 347–78.
Bellamy, D.J., Bradshaw, M.E., Millington, G.R. and Simmons, I.G. (1966), *New Phytol.*, **65**, 429–42.
Birks, H.H. (1970), *J. Ecol.*, **58**, 827–46.
Birks, H.H. (1980), *Ergebnisse der Limnologie*, **15**, 60 pp.
Birks, H.H. and Mathewes, R.W. (1978), *New Phytol.*, **80**, 455–84.
Birks, H.J.B. (1970), *New Phytol.*, **69**, 807–20.
Birks, H.J.B. (1973), *Past and Present vegetation of the Isle of Skye. A palaeoecological study*, Cambridge University Press, London.
Birks, H.J.B. (1976), *Ecol. Monogr.*, **46**, 395–429.
Birks, H.J.B. (1981), *Quat. Res.*, **16**, 322–355.

Birks, H.J.B. and Birks, H.H. (1980), *Quaternary Palaeoecology,* Edward Arnold, London.
Bonde, E.K. (1969), *Arc. Alp. Res.,* **1,** 135–40.
Bonny, A.P. (1976), *J. Ecol.,* **64,** 859–87.
Bonny, A.P. (1978), *J. Ecol.,* **66,** 385–418.
Boros, A. and Jarai-Komlodi, M. (1975), *An Atlas of Recent European Moss Spores,* Akademiai Kiado, Budapest.
Brassard, G.R. and Blake, W. (1978), *Can. J. Bot.,* **56,** 1852–9.
Burrows, C.J. (1974), *New Phytol.,* **73,** 1003–33.
Casparie, W.A. (1969), *Vegetatio,* **19,** 146–80.
Clapham, A.R., Tutin, T.G. and Warburg, E.F. (1962), *Flora of the British Isles,* 2nd edn. Cambridge University Press, London.
Clymo, R.S. (1965), *J. Ecol.,* **53,** 747–58.
Conolly, A.P. and Dickson, J.H. (1969), *New Phytol.,* **68,** 197–9.
de Vries, B. and Bird, C.D. (1965), *Can. J. Bot.,* **43,** 947–53.
Dickinson, C.H. and Maggs, G.H. (1974), *New Phytol.,* **73,** 1249–57.
Dickson, J.H. (1967), *Rev. Palaeobot. Palynol.,* **2,** 245–53.
Dickson, J.H. (1973), *Bryophytes of the Pleistocene. The British record and its chorological and ecological implications,* Cambridge University Press, London.
Dickson, J.H. (1979), *Glasgow Naturalist,* **19,** 437–42.
Dickson, J.H. and Brown, P.D. (1966), *Trans. Br. Bryol. Soc.,* **5,** 100–2.
Dixon, H.N. (1927), *Fossilum Catalogus II: Plantae, Part 13: Muscineae,* W. Junk, Berlin.
Fredskild, B., Jacobsen, N. and Røen, U. (1975), *Meddl. om Grønland,* **198(5),** 1–44.
Gams, H. (1932), Quaternary Distribution. In: *Manual of Bryology* (Verdoorn, F., ed.), pp. 297–322. Nijhoff, The Hague.
Glaser, P.H. (1979), *Recent plant-macrofossils from the Alaska interior and their relation to late-glacial landscapes in Minnesota.* Ph.D. thesis, University of Minnesota.
Gleason, H.A. (1939), *Am. Midl. Nat.,* **21,** 92–108.
Godwin, H. and Conway, V.M. (1939), *J. Ecol.,* **27,** 313–59.
Gould, S.J. (1965), *Am. J. Sci.,* **263,** 223–8.
Gray, A. (1846), *Am. J. Sci. Arts,* **II, 2,** 135–6.
Green, B.H. (1968), *J. Ecol.,* **56,** 47–58.
Green, B.H. and Pearson, M.C. (1977), *J. Ecol.,* **65,** 793–814.
Grosse-Brauckmann, G. (1974), *Flora,* **163,** 179–229.
Grosse-Brauckmann, G. (1976), *Flora,* **165,** 415–55.
Hall, A.R. (1979), *J. Bryol.,* **10,** 511–5.
Hall, A.R. (1980), *Phil. Trans. R. Soc.,* B, **289,** 135–64.
Hays, J.D., Saito, T. and Opdyke, H.D. (1969), *Geol. Soc. Am. Bull.,* **80,** 1481–514.
Horton, D.G. and Vitt, D.H. (1976), *Can. J. Bot.,* **54,** 1872–83.
Hulme, P.D. (1979), *J. Bryol.,* **10,** 281–5.
Iversen, J. (1954), *Danm. geol. Unders.,* Series II, **80,** 87–119.
Iwatsuki, Z. and Sharp, A.J. (1967), *J. Hattori bot. Lab.,* **30,** 152–70.
Iwatsuki, Z. and Sharp, A.J. (1968), *J. Hattori bot. Lab.,* **31,** 55–8.

Jankovska, V. (1971), *Folia geobot. phytotax.*, **6,** 281–302.
Janssen, C.R. (1970), *Vegetatio,* **20,** 187–98.
Janssen, C.R., Cup-Uiterwijk, M.J.J., Edelman, H.J., Mekel-Te Riele, J. and Pals, J.P. (1975), *Vegetatio,* **30,** 165–78.
Janssens, J.A.P., Horton, D.G. and Basinger, J.E. (1979), *Can. J. Bot.*, **57,** 2150–61.
Jessen, K. (1949), *Proc. R. Ir. Acad.*, **52B,** 85–290.
Jessen, K. and Milthers, V. (1928), *Danm. geol. Unders.*, **48,** 1–379.
Kuc, M. (1973), *Canadian-Polish Res. Inst. Biol. and Earth Sci.*, Ser. A, **1,** 1–44.
Kuc, M. and Hills, L.V. (1971), *Can. J. Bot.*, **49,** 1089–94.
Kukla, C.J. (1977), *Earth-Science Rev.*, **13,** 307–74.
Lawrence, D.R. (1968), *Geol. Soc. Am. Bull.*, **79,** 1315–40.
Li, H. (1952), *Trans. Am. Philos. Soc.*, **42,** 371–429.
Light, J.J. (1975), *J. Ecol.*, **63,** 937–43.
Light, J.J. and Heywood, R.B. (1973), *Nature,* **242,** 535–6.
Light, J.J. and Heywood, R.B. (1975), *Nature,* **256,** 199–200.
Light, J.J. and Lewis Smith, R.I. (1976), *J. Bryol.*, **9,** 55–62.
Mahony, J.A. (1869), *Geol. Mag.*, **6,** 390–3.
Miller, N.G. (1973), *J. Arnold Arbor.*, **54,** 123–59.
Miller, N.G. (1976a), *Occ. Papers Farlow Herbarium of Harvard University,* **9,** 21–42.
Miller, N.G. (1976b), *J. Hattori bot. Lab.*, **41,** 73–85.
Miller, N.G. (1980), *Bull. Torrey Bot. Club,* **107,** 373–91.
Miller, N.G. and Ambrose, L.J.H. (1976), *Bryologist,* **79,** 55–63.
Miller, N.G. and Benninghoff, W.S. (1969), *Geol. Soc. Amer. Special Paper,* **123,** 225–48.
Morrison, M.E.S. (1959), *Proc. R. Ir. Acad.*, Ser. B, **60,** 291–308.
Osvald, H. (1923), *Svenska Växtsoc. Sallsk. Handl.*, **1,** 1–266.
Övstedal, D.O. and Aarseth, I. (1975), *Lindbergia,* **3,** 61–8.
Paton, J.A. (1965), *Census Catalogue of British Hepatics,* British Bryological Society, Ipswich.
Pearsall, W.H. (1956), *J. Ecol.*, **44,** 493–516.
Persson, H. (1942), *Bot. Notiser,* **1942,** 308–24.
Persson, H. (1944), *Revue bryol. lichenol.*, N.S., **14,** 84–8.
Pigott, C.D. and Pigott, M.E. (1963), *New Phytol.*, **62,** 317–34.
Priddle, J. (1979), *J. Bryol.*, **10,** 517–29.
Priddle, J. and Dartnall, H.J.G. (1978), *Freshwater Biol.*, **8,** 469–80.
Reader, R.J.D. and Stewart, J.M. (1972), *Ecology,* **53,** 1024–37.
Reid, C. (1899), *The Origin of the British Flora,* Dulau & Co., London.
Rigler, F.H. (1972), The Char Lake Project. A study of energy flow in a high Arctic lake. In: *Productivity Problems of Freshwaters* (Kajak, Z. and Hillbricht-Ilkowska, A., eds), pp. 239–69. Polish Scientific Publishers, Warsaw.
Robertson, D. (1881), *Trans. geol. Soc. Glasg.*, **7,** 1–37.
Rybniček, K. (1973), A comparison of the present and past mire communities of central Europe. In: *Quaternary Plant Ecology* (Birks, H.J.B. and West, R.G., eds), pp. 237–61. Blackwell, Oxford.
Rybniček, K. and Rybničková, E. (1968), *Folia geobot. phytotax.*, **3,** 117–42.

Rybniček, K. and Rybničková, E. (1971), *Folia geobot. phytotax.*, **9**, 45–70.
Rybniček, K. and Rybničková, E. (1977), *Folia geobot. phytotax.*, **12**, 245–91.
Rymer, L. (1978), The use of uniformitarianism and analogy in palaeoecology, particularly pollen analysis. In: *Biology and Quaternary Environments* (Walker, D. and Guppy, J.C., eds), pp. 245–57. Australian Academy of Sciences, Canberra.
Seaward, M.R.D. and Williams, D. (1976), *J. Arch. Sci.*, **3**, 173–7.
Shackleton, N.J. and Opdyke, N.D. (1973), *Quat. Res.*, **3**, 39–55.
Shackleton, N.J. and Opdyke, N.D. (1976), *Geol. Soc. Am. Memoir*, **145**, 449–64.
Sissons, J.B. (1979), *Nature*, **280**, 199–203.
Smith, A.J.E. (1978), *The Moss Flora of Britain and Ireland*, Cambridge University Press, Cambridge.
Tallantire, P.A. (1979), *Archaeo-Physika*, **8**, 295–301.
Tallis, J.H. (1964), *J. Ecol.*, **52**, 345–53.
Tallis, J.H. (1973), *J. Ecol.*, **61**, 537–67.
Tauber, H. (1967), *Rev. Palaeobot. Palynol.*, **3**, 277–86.
Tauber, H. (1977), *Dansk Botanisk Arkiv*, **32**, 1–121.
Tralau, H. (1963), *Arkiv Botanik* Ser. 2, **5(3)**, 533–82.
von Post, L. (1916), *Pollen et Spores*, **9**, 375–401.
Walker, D.A. (1970), Direction and rate in some British post-glacial hydroseres. In: *Studies in the Vegetational History of the British Isles* (Walker, D. and West, R.G., eds), pp. 117–39. Cambridge University Press, London.
Walker, D.A. and Walker, P.M. (1961), *J. Ecol.*, **49**, 169–85.
Warren Wilson, J. (1958), *J. Ecol.*, **46**, 191–8.
Wasylikowa, K. (1979), Plant macrofossil analysis. In: *Palaeohydrological Changes in the Temperate Zone in the last 15 000 years* (Berglund, B.E., ed.), Vol. 2, pp. 291–310. Lund.
Watts, W.A. (1970), *Ecology*, **51**, 17–33.
Watts, W.A. (1978), Plant macrofossils and Quaternary palaeoecology. In: *Biology and Quaternary Environments* (Walker, D. and Guppy, J.C., eds), pp. 53–67. Australian Academy of Sciences, Canberra.
Watts, W.A. (1979), *Ecol. Monogr.*, **49**, 427–69.
Watts, W.A. (1980), *Ann. Rev. Ecol. Systematics*, **11**, 387–409.
Watts, W.A. and Bright, R.C. (1968), *Geol. Soc. Am. Bull.*, **79**, 855–76.
Watts, W.A. and Winter, T.C. (1966), *Geol. Soc. Am. Bull.*, **77**, 1339–60.
West, R.G. (1957), *Phil. Trans. R. Soc.*, Ser. B, **241**, 1–31.
West, R.G. (1964), *J. Ecol.*, **52**, (Suppl.), 47–57.
West, R.G. (1973), Introduction. In: *Quaternary Plant Ecology* (Birks, H.J.B. and West, R.G., eds), pp. 1–3. Blackwell, Oxford.
West, R.G. (1977), *Phil. Trans. R. Soc.*, Ser. B, **280**, 229–46.
Wright, H.E. and Watts, W.A. (1969), *Minnesota Geol. Surv. Bull.* S.P., **11**, 1–59.

Author Index

The page numbers in italics refer to the reference lists

Subject Index